Working Analysis

Working Analysis

Jeffery Cooper

Department of Mathematics
University of Maryland at College Park

ELSEVIER
ACADEMIC
PRESS

AMSTERDAM • BOSTON • HEIDELBERG • LONDON
NEW YORK • OXFORD • PARIS • SAN DIEGO
SAN FRANCISCO • SINGAPORE • SYDNEY • TOKYO

Acquisition Editor	Barbara Holland
Project Manager	Christine Brandt
Associate Editor	Tom Singer
Marketing Manager	Linda Beattie
Cover Design	Martha Oatway/Dick Hannas
Composition	Kolam, Inc.
Cover Printer	Phoenix Color
Interior Printer	Maple-Vail Book Manufacturing Group

Elsevier Academic Press
200 Wheeler Road, Burlington, MA 01803, USA
525 B Street, Suite 1900, San Diego, California 92101-4495, USA
84 Theobald's Road, London WC1X 8RR, UK

This book is printed on acid-free paper. ∞

Library of Congress Cataloging-in-Publication Data
Application submitted.

British Library Cataloguing in Publication Data
A catalogue record for this book is available from the British Library

ISBN: 0-12-187604-7

For all information on all Elsevier Academic Press Publications
visit our Web site at www.books.elsevier.com

Printed and bound in Great Britain by
CPI Antony Rowe, Chippenham and Eastbourne

Transferred to Digital Printing, 2010

Contents

Part II

Preface

Why Another Advanced Calculus Book?

Over the years, I have become more and more dissatisfied with our advanced calculus course. In most books used for this type of course, theorems are proved to prove more theorems. An "application" of a theorem is either a trivial calculation or a piece of the proof for another theorem. There are no examples or exercises that use the methods of analysis to solve a real problem. The traditional advanced calculus course has little or no contact with the world outside of mathematics.

A second shortcoming of the traditional advanced calculus course is that, in spite of the abundance of high-quality mathematical software, this kind of course does not use the computer. Although the most important function of an advanced calculus course is to teach students rigorous methods of proof, the computer can be used to illustrate many important concepts. Rates and orders of convergence of sequences, error in various approximations, and numerical differentiation and integration are but a few. Furthermore, many practical problems cannot be considered solved until some numbers are produced. The work of the applied analyst does not end with an existence theorem.

My goals in writing this book are to teach the techniques and results of analysis and to show how they can be applied to solve problems. I have gone outside of the usual range of applications in physics to include examples from biology, sociology, chemistry, and economics.

Prerequisites

I assume that the students using this book have had three semesters of calculus (including multivariable) and one semester of linear algebra. It would be helpful if the students have some experience with a software package. The course will be enhanced if students can graph functions of one and two variables and make simple computations. However, this is not a course in programming. I usually provide the students with MATLAB codes for functional iteration and a Newton solver in one and several variables. A (rather crude) code for the method of steepest descent is also provided. These codes can be found on my web page, `www.math.umd.edu/~jec`.

The Text

The text contains more than enough material for a two semester course. Part I is an introduction to analysis in one dimension. Part II, for the second semester, deals with functions of several variables. Some books begin such a course immediately with several variables, but I feel this approach is too difficult for most students. As a result, there is some repetition of material in Parts I and II. Nevertheless, seeing some ideas (like that of a contraction mapping) in one, and then in several variables, helps in their assimilation by students. Examples and problems that treat applications are sprinkled throughout the text, but the more detailed discussions are found in Part II, which deals with functions of several variables. Exercises of a computational nature are found in almost every chapter, but the more challenging ones are found in the later chapters.

Summary of Part I

Chapter 1 is a brief treatment of the foundations of analysis. One of the reasons analysis is difficult for students is that it involves manipulation of inequalities. I have set aside a section, 1.3, that deals with this issue. Chapter 2 is devoted to sequences, and particular attention is paid to the fact that some sequences converge faster than others. Numerical examples are given and estimates are made to illustrate the difference between linear and quadratic convergence. The examples foreshadow the discussion in Chapter 6 of functional iteration and Newton's method.

Chapter 3 deals with limits of functions and continuity. After establishing that a continuous function on a closed bounded interval must attain its extreme values, I introduce the method of golden-section search to locate the maximizer and minimizer. It is important for the student to realize that there are many optimization problems where the function may not be differentiable or where the derivative is very difficult to calculate. The method of bisection is used to prove the intermediate value theorem.

Linear approximation is the theme of Chapter 4. The O and o notations are introduced to describe the error in linear approximation. In computational exercises, students calculate the errors to see how rapidly they decrease. The mean value theorem is discussed and used to prove a one-dimensional version of the inverse function theorem. The latter is used to justify a change of variable in a differential equations example. l'Hôpital's rule and the second derivative test are followed by an example from economics to close the chapter.

In Chapter 5, higher derivatives and Taylor polynomials are discussed. In several examples, the Lagrange form of the remainder is used to derive useful approximations and to estimate their error. Another use of the Taylor approximation is to derive expressions for numerical differentiation. Centered difference quotients are then used to discretize a nonlinear differential equation. The resulting nonlinear system of equations will be solved later in Part II using Newton's method. Finally, polynomial interpolation with an expression for the error is discussed. These results are used later in Chapter 7 to derive quadrature rules. More immediately, they are used in an introduction to convex functions.

Chapter 6, *Solving Equations in One Dimension*, is devoted to the contraction mapping theorem and to Newton's method. Functional iteration is easy to discuss in one dimension, and it provides a simple computational tool that does not use the derivative. It also provides the central idea in the proof of the inverse function theorem in several dimensions, proved in Part II. The proof of Newton's method shows how quadratic convergence arises, and prepares the student for the treatment in higher dimensions. These topics are also natural situations for computational exercises.

The treatment of the Riemann integral in Chapter 7 is fairly standard. To keep matters simple and practical, I have not stated the most general forms of the fundamental theorem of calculus. The simple quadrature rules are derived using polynomial interpolation. In particular, I think I have given a more conceptual explanation of why Simpson's rule works better than expected; it does, in fact, arise from interpolating by a cubic polynomial. I have also given examples using the change of variable formula for integrals that are of use in numerical calculation. In the exercises for Section 7.4, the student can do certain numerical experiments and then use analysis to explain the observed behavior.

Many ideas from earlier chapters come together in Chapter 8. The all-important concept of uniform convergence is used to justify term-by-term integration and differentiation. To show a use of the latter, I obtain a power series expansion for solutions of the Airy equation. The chapter ends with the example $f(x) = \exp(-1/x^2)$, which is C^∞ but not analytic at $x = 0$. In the exercises, the student is asked to graph the derivatives of f to see if he or she can explain how this function can have a convergent power series expansion for $x \neq 0$ but not at $x = 0$. I think this is an excellent example of the use of the computer in this type of course.

The elementary functions $\sin x$, $\cos x$, e^x, and $\log x$ are used freely in examples. These functions are defined, and their basic properties are developed in the Appendix to Part I.

Summary of Part II

Because I am eager to get to applications, the discussion of norms, topology, and continuity in Chapter 9 is rather brief. Rather than introduce the idea of a metric space, I prefer to deal with three different, and very useful, norms on $\mathbb{R}^n$.

In Chapter 10, the idea of linear approximation is again emphasized, this time for functions of several variables and for vector-valued functions. It is one of the important themes of the course. I do not always state and prove theorems in the most general form. I find that sometimes stronger hypotheses make for a simpler and more transparent proof.

Chapter 11 picks up the material of Chapter 6 and extends it to systems of equations. After proving the contraction mapping theorem in several variables, I use the idea behind Newton's method to prove a simplified form of the Kantorovich existence theorem. Students should learn not only existence theorems, but also constructive algorithms that can be used for numerical calculation. The interplay between the inverse function theorem, or the implicit function theorem, and the calculation of a family of solutions that depend on a parameter is very important. Several examples are explored to show that the "big" theorems can be employed to further the analysis. The examples and the exercises sometimes require the use of mathematical software.

Chapter 12 deals with problems in unconstrained optimization. In addition to the usual second derivative test, we include a short section on convex functions in several variables. This is followed by some algorithms for finding a minimum that do not use the second derivatives. The method of steepest descent has a simple proof of convergence and is easy to visualize. Unfortunately, it is not very efficient, and it is wrong to leave the student with the impression that this method is used in practice. For this reason, I have added a section on conjugate gradient methods. Instructors who feel that this topic takes the course too far in the direction of numerical analysis can state the main idea and then move on. The problems that arise in fitting data with curves that depend in a nonlinear fashion on the parameters are very interesting, and they are important in applications. The Gauss-Newton algorithm is appealing in its simplicity, and it is widely used.

Constrained optimization is the subject of Chapter 13. The Lagrange multiplier method is a good example in which an abstract theorem (the implicit function theorem) is used to derive a system of equations that one can try to solve by the methods of Chapter 11. In applications, there are often one or more parameters present in the problem, and it is important to see how the solutions of the Lagrange equations depend on these parameters. The example of the three bar linkage illustrates how the number of constraints depends on the choice of variables used to describe the problem and how the solutions depend on the parameters present. We also derive the analogue of the second derivative test for constrained maxima and minima. The Karush-Kuhn-Tucker conditions for constraints with inequalities are needed for many problems in economics, in particular the Averch-Johnson effect. The analysis of this problem uses the ideas of Chapter 11 as well.

In Chapter 14, the elements of integration in several variables are treated rather briefly, relying on the treatment in one variable, which the student has already seen. Numerical methods for integrating functions of two and three variables are a very appropriate use of Riemann sums and are easily derived from the standard methods in one variable. The main emphasis in Chapter 14 is on the change of variable theorem. It is an important tool in the numerical evaluation of integrals. It is also useful in finding the probability density function of combinations of random variables.

Chapter 15 explores ways in which the integration theory of Chapter 14 can be applied to partial differential equations. After giving a criterion that justifies differentiation under the integral, we investigate the convolution. The solutions of the diffusion equation can be represented in terms of a convolution, and we explore their properties. We also use convolution to prove the Weier-strass polynomial approximation theorem. Finally, we use the change of variable theorem in a key way to derive the Euler equations of fluid flow.

Acknowledgments

I wish to extend my warm thanks to my colleagues at the University of Maryland for their thoughtful comments and suggestions over the years when I was writing this text: Bruce Kellogg, Robert Pego, Matei Machedon, Tobias von Petersdorff, and Stuart Antman. I am especially grateful for many conversations with John Osborn about the place of computation in a course of analysis. I also wish to thank my editors, Barbara Holland and Tom Singer, for their help and encouragement throughout the development of this project.

Finally I wish to express my appreciation for the work of many reviewers who made important contributions and suggested corrections. They include James Sochacki, James Madison University; Donald Passman, University of Wisconsin Madison; Doug Hundley, Whitman College; Carl Fitzgerald, University of California San Diego; Daniel Flath, University of South Alabama; Alexander J. Smith, University of Wisconsin Eau Claire; Sergei Suslov, Arizona State University; and Seth Oppenheimer, Mississippi State University.

Of course, final responsibility for correctness of the text is mine. Readers who find errors or unduly confusing passages are encouraged to notify me at my e-mail address. I will post errata and clarifications on my webpage.

Jeffery Cooper
Department of Mathematics
University of Maryland
College Park, MD 20742
jec@math.umd.edu

Part I

Part I

Chapter 1

Foundations

Chapter Overview

In this chapter, we discuss the foundations for analysis. The rational numbers are shown to be inadequate for analysis because we cannot solve a simple equation like $x^2 = 2$ in the rational numbers. The real numbers constitute a larger number system that is characterized by the least upper bound property. This latter property ensures that the real numbers do not suffer from the "gaps" present in the rational numbers. The rationals are shown to be dense in the real numbers.

The manipulation of inequalities is central to analysis and we discuss some techniques of estimation. The chapter closes with a review of the concepts of sets and functions as well as a discussion of the properties of the supremum and infimum of a function.

1.1 The Rational Numbers and Ordered Fields

Much of the material in the following chapters revolves around four important problems faced by scientists:

    (1) solving equations;
    (2) finding the maximum and minimum of a function;
    (3) determining integrals; and
    (4) approximating functions that are complicated, or that are not express-ible in closed form, by functions that we can compute easily.

We usually take for granted the number system in which we work to discuss these problems. However, it is essential that the study of analysis have a firm logical foundation; specifically, we must have a careful axiomatic construction of the number system. In any axiomatic construction one starts with certain basic undefined objects to which certain properties are attributed. We are used to computing with the *natural numbers* $\mathbb{N} = \{1, 2, 3, \dots\}$, the *integers* $\mathbb{Z} = \{0, \pm 1, \pm 2, \dots\}$ and the *rational numbers* $\mathbb{Q} = \{m/n : n, m \in \mathbb{Z}, n \neq 0\}$. We shall assume these number structures are well understood. In particular, the natural numbers enjoy the property of being *well ordered*, which means that every nonempty set $S \subset \mathbb{N}$ has a least element. This property is equivalent to the principle of mathematical induction. We discuss the principle of mathematical induction more fully in Section 1.4.

In calculus, when we develop methods to solve problems (1)–(4), we often speak of the *real numbers*, which we identify with points on the "number line." For example, we define functions "for all x in some interval." We also want to solve equations $f(x) = 0$ and we accept answers like $x = \sqrt{2}$ and $x = \pi$. As we shall see shortly, $\sqrt{2}$ is not a rational number, even though it corresponds to the length of a simply defined geometric object, the diagonal of a square of side 1. The Greeks became aware of this fact and found it quite disturbing. It was not until the middle of the nineteenth century that the connection between rational numbers and geometric quantities, such as the diagonal of a square and the area of a circle, was clarified. We shall see that we can find rational numbers that approximate these quantities with arbitrarily small error. However, to speak with confidence about the validity of these approximations, and to carry out calculations with these geometrically defined quantities, we must have the larger real number system.

Before we discuss the real numbers, we show that, indeed, we need to have a number system that is larger than the rational numbers.

Theorem 1.1 There does not exist a rational number x such that $x^2 = 2$.

Proof: We begin by making a simple observation about squares. If p is an integer and p^2 is even, then p is even as well. In fact, if p were odd, $p = 2k + 1$, then $p^2 = 4k^2 + 4k + 1$ is also odd.

Now, to prove the assertion of the theorem, we shall assume the contrary, which is to say that there are integers p and q such that $(p/q)^2 = 2$. We will see that this assumption leads to a contradiction, which means that it is not possible to write $\sqrt{2}$ as a rational number.

After reducing the fraction p/q to lowest terms we can assume that p and q have no factors in common. Now, if $(p/q)^2 = 2$, then we have

$$p^2 = 2q^2 \qquad (1.1)$$

so that p^2 is even. As we have seen, this implies that p is also even. Thus for some integer m, $p = 2m$. We substitute $p = 2m$ in (1.1) and find that $2m^2 = q^2$. However, this implies that q is even as well, and that contradicts the assumption that p and q had no factors in common. The proof is complete.

Theorem 1.1 shows clearly that the rational numbers are not adequate for calculus. A larger number system is needed that has the algebraic properties of the rationals, and that has a unique element that corresponds to each point on the line. Of course, we could just assume that such a number system exists. However, we might go too far and ascribe properties to it that are inconsistent, leading to contradictory results. Fortunately for the beginning analyst, this difficulty has been resolved. In the nineteenth century, mathematicians found rigorous, logically consistent ways to construct a larger number system that includes the rational numbers and that does not have the "gaps" found in the rational numbers. For a discussion of these constructions, the reader may wish to consult the book of S. Krantz [Kr], where the method of Dedekind cuts is described. The method of completion by Cauchy sequences is described in the book of R. Strichartz [St].

These various constructions yield essentially the same mathematical stucture which we call the *real numbers*, denoted $\mathbb{R}$. It is defined axiomatically as a *complete, ordered field*. We now explain what each of these terms means.

A *field* is a set of mathematical objects on which two operations, addition and multiplication, are defined. The axioms of a field stipulate that these operations are associative and commutative and that all of the usual rules of arithmetic are valid; one can add, subtract, multiply and divide (by a nonzero number). We list the axioms for a field here for reference.

Field Axioms

A set F is a *field* if there are two binary operations defined on F, denoted + (addition) and $\cdot$ (multiplication) that satisfy the following axioms:
Additive

    (i) $a + b = b + c$ for all $a, b \in F$;
    (ii) $(a + b) + c = a + (b + c)$ for all $a, b, c \in F$;
    (iii) there exists a unique element $\theta \in F$ such that $\theta + a = a + \theta = a$ for all $a \in F$;

(iv) for each element $a \in F$, there is an element $\bar{a} \in F$ such that $a + \bar{a} = \bar{a} + a = 0$.

Multiplicative

(v) $a \cdot b = b \cdot a$ for all $a, b \in F$;

(vi) $(a \cdot b) \cdot c = a \cdot (b \cdot c)$ for all $a, b, c \in F$;

(vii) there exists a unique element $e \in F$, $e \neq 0$, such that $e \cdot a = a \cdot e = a$ for all $a \in F$;

(viii) for each $a \in F$, $a \neq 0$, there is an element $a' \in F$ such that $a \cdot a' = a' \cdot a = e$; and

Distributive

(ix) $a \cdot (b + c) = a \cdot b + \cdot c$ for all $a, b, c \in F$.

The rational numbers are an example of a field. The integers $\mathbb{Z}$ are not a field, because the quotient of two integers may not be an integer. In a course in abstract algebra one learns that there are many other mathematical structures that satisfy the axioms of a field. These algebraic concepts are discussed in many texts, among them the book of I. N. Herstein [He] and the book of G. Birkhoff and S. MacLane [Bk].

The order on the real numbers is the one inherited from the rational numbers and corresponds to the natural one on the number line. The order relation is defined axiomatically by first describing the properties of the positive elements.

Definition 1.1 The *positive* elements of the real numbers satisfy the following axioms:

(P1) If $a, b \in \mathbb{R}$ and $a, b \in P$, then $a + b \in P$.

(P2) If $a, b \in \mathbb{R}$ and $a, b \in P$, then $a \cdot b \in P$.

(P3) If $a \in \mathbb{R}$, then exactly one of the following is true

$$a \in P, \quad \text{or} \quad a = 0, \quad \text{or} \quad -a \in P.$$

The axiom P3 is referred to as the *trichotomy axiom*.

The order relation is then defined in terms of P.

Definition 1.2

(i) If $a, b \in \mathbb{R}$ and $a - b \in P$, we write $a > b$.

(ii) If $a, b \in \mathbb{R}$ and $a - b \in P \cup \{0\}$, we write $a \geq b$.

All of the usual rules for inequalities can be derived from this definition and from the axioms P1, P2, and P3. We summarize the most important ones here.

Theorem 1.2 Let $a, b, c,$ and d be real numbers.

(i) If $a > b$ and $b > c$, then $a > c$.
(ii) If $a > b$ and $c \geq d$, then $a + c > b + d$.
(iii) If $a > b$ and $c > 0$, then $a \cdot c > b \cdot c$.
(iv) If $a > b$ and $c < 0$, then $a \cdot c < b \cdot c$.
(v) If $a > 0$, then $1/a > 0$.
(vi) If $a > b > 0$, then $(1/a) < (1/b)$.

Proof:

(i) $a > b$ and $b > c$ mean that $b - a \in P$ and $b - c \in P$. Hence $a - c = (a - b) + (b - c) \in P$, so that $a > c$.
(ii) $a > b$ and $c \geq d$ imply that $b - a \in P$ and $c - d \in P$ or equals zero. Hence $(a + c) - (b + d) = a - b + (c - d) \in P$, so that $a + c > b + d$.
(iii) $a > b$ and $c > 0$ imply that $a \cdot c - b \cdot c = (b - a) \cdot c \in P$, or that $a \cdot c > b \cdot c$.
(iv) $a > b$ and $c < 0$ mean that $a - b \in P$ and $-c \in P$. Hence, $b \cdot c - a \cdot c = (a - b) \cdot (-c) \in P$ which proves (iv).

We leave the proof of (v) and (vi) to the exercises.

Manipulation of inequalities is very important in analysis. We shall return to this subject to learn some techniques of estimation in Section 1.3.

Exercises for 1.1

For exercises 1 through 6, use the properties of the order relation on the real numbers as described by Definition 1.2 and Theorem 1.2.

1. Prove part (v) of Theorem 1.2.
2. Prove part (vi) of Theorem 1.2.
3. Show that if a is a nonzero real number, then $a^2 > 0$.
4. Using the fact that $1/2 > 0$, show that if $a, b \in \mathbb{R}$ with $a < b$, then $a < (a + b)/2 < b$.
5. a) For real numbers a and b, show that
$$ab \leq \frac{1}{2}(a^2 + b^2).$$
   b) For real numbers $a, b,$ and c show that
$$ab + bc + ca \leq a^2 + b^2 + c^2.$$

6. For real numbers a and b we have the factorization

$$a^n - b^n = (a - b)(a^{n-1} + a^{n-2}b + \cdots + ab^{n-2} + b^{n-1}).$$

Use this result to show that, if $a \geq b \geq 0$, then $a^n - b^n \geq (a - b)nb^{n-1}$.

7. Following the reasoning of Theorem 1.1, show that $\sqrt{3}$ is not a rational number. You may use the fact that, if 3 divides a product of integers pq, then it must divide either p or q (or perhaps both).

8. Now generalize the result of exercise 7. If p is a prime, then $\sqrt{p}$ is not a rational number.

1.2 Completeness

To define what it means for the real numbers to be complete, we must introduce some additional terms.

Definition 1.3 Let S be a subset of $\mathbb{R}$. A real number b is said to be an *upper bound* for S if $x \leq b$ for all $x \in S$.

The real number b is said to be a *least upper bound* for S if

   (i) b is an upper bound for S, and
   (ii) if b' is any upper bound for S, then $b \leq b'$.

We make some comments about the definitions. A set S may not have any upper bound, but if it has one upper bound then it has many. In fact, if b is an upper bound for S, then any $c \geq b$ is also an upper bound for S. However, the least upper bound, if it exists, is unique. In fact, if b and b' are both supposed to be a least upper bound of S, then both are upper bounds for S. It follows that $b \leq b'$ and that $b' \leq b$; this implies that $b = b'$.

In the special case when S is the empty set, every real number is an upper bound for S.

Definition 1.4 Let S be a set of real numbers. A real number a is said to be a *lower bound* for S if $a \leq x$ for all $x \in S$.

The real number a is said to be a *greatest lower bound* for S if

   (i) a is a lower bound for S, and
   (ii) if a' is any lower bound for S, then $a' \leq a$.

The real numbers are said to be *complete* because they have the following crucial property.

Least Upper Bound Property. If S is a nonempty set of real numbers, bounded above, then S has a least upper bound in the real numbers.

It is not hard to show that the least upper bound property holds if and only if every nonempty subset S of the real numbers that is bounded below has a greatest lower bound in the real numbers.

The least upper bound property does not hold in the rational numbers. For example, the set $S = \{x \in \mathbb{Q} : x^2 < 2\}$ is bounded above by 2, but, as we shall see in Theorem 1.3, the least upper bound of S is $\sqrt{2}$, which we know is not a rational number.

Because the least upper bound property is the property that characterizes the real numbers, we shall look more carefully at the definitions and introduce some new terminology.

If $S \subset \mathbb{R}$ has an upper bound, we shall say that S is *bounded above*. If S is nonempty and bounded above, and b is the least upper bound of S, we say that b is the *supremum* of S and write

$$b = \sup S.$$

Similarly, if S has a lower bound, we say that S is *bounded below*. If S is nonempty and bounded below, and a is the greatest lower bound of S, we say that a is the *infimum* of S and write

$$a = \inf S.$$

If we study the definition of the least upper bound (or supremum) of a nonempty set S, we can see that $b = \sup S$ if and only if

    (i) b is an upper bound for S, and
    (ii) for each $\varepsilon > 0$, there is an element $x \in S$ such that $x > b - \varepsilon$.

The second condition says that, if we depress b the least bit to $b - \varepsilon$, it is no longer an upper bound. The proof of this equivalance follows from a careful reading of the definition. Suppose that $b = \sup S$. Then b is an upper bound for S so that (i) is satisfied. Now for any $\varepsilon > 0$, $b - \varepsilon$ cannot be an upper bound for S, otherwise b would not be the least upper bound. However, if $b - \varepsilon$ is not an upper bound for S, then there must be some element $x \in S$ with $x > b - \varepsilon$.

Now we prove the converse implication. Suppose that a number b satisfies (i) and (ii) for a set S. This means that b is an upper bound for S. If $c < b$, we set $\varepsilon = b - c > 0$. By (ii), there is an $x \in S$ such that $x > b - \varepsilon = c$. Hence c is not an upper bound for S. We see that, if c is an upper bound for S, we must have $c \geq b$, which says that b is the least upper bound for S.

It is important to understand that the supremum of a set may or may not belong to the set. A notion related to the supremum is that of the maximum.

Definition 1.5 Let S be a set bounded above with $b = \sup S$. If $b \in S$, we say that b is the *maximum* of S. If S is bounded below, with $a = \inf S$ and $a \in S$, we say that a is the *minimum* of S.

> **EXAMPLE 1.1:** Let $S = \{x \in \mathbb{R} : 0 \leq x < 1\}$ and $T = S \cup \{1\}$. The number $b = 1$ is obviously an upper bound for both S and T. Furthermore, for each $\varepsilon > 0$, there is an element $x = 1 - \varepsilon/2 \in S \subset T$ with $x > b - \varepsilon = 1 - \varepsilon$. Hence, $b = 1$ is the supremum of both S and T, but $b \in T$ while $b \notin S$. The maximum of T is $b = 1$ and $a = 0$ is the minimum of both S and T.

In Section 1.1, we showed that the rational numbers were inadequate for analysis because $\sqrt{2}$ is not a rational number. We must at least show that the complete ordered field that is the real numbers contains $\sqrt{2}$.

Theorem 1.6 There exists a positive real number r such $r^2 = 2$.

Proof: We let $S = \{x \in \mathbb{R} : x^2 < 2\}$. The set S is nonempty because $1 \in S$. Next, note that 2 is an upper bound for S. In fact, if $x \in S$ and $1 < x$, then $x < x^2 < 2$. If $x \in S$ and $x \leq 1$, then, again, $x < 2$. Therefore, by the least upper bound property of $\mathbb{R}$, we know that S has a least upper bound r. We can see that $r > 1$ because $4/3 \in S$. We will show that $r^2 = 2$ by showing that neither $r^2 < 2$ nor $r^2 > 2$ is possible. It then follows by the trichotomy axiom that $r^2 = 2$.

First suppose that $r^2 < 2$. Let $\varepsilon = (2 - r^2)/(2r + 1)$ and note that $0 < \varepsilon < 2/(2r + 1) < 1/r < 1$ because $r > 1$. Hence,

$$
\begin{aligned}
(r + \varepsilon)^2 &= r^2 + 2r\varepsilon + \varepsilon^2 \\
&< r^2 + \varepsilon(2r + 1) \\
&< r^2 + 2 - r^2 = 2.
\end{aligned}
$$

Thus, $r + \varepsilon \in S$, which contradicts the fact that r is supposed to be an upper bound for S.

Next, suppose that $r^2 > 2$. Choose $\varepsilon = (r^2 - 2)/2r$. We will show that $r - \varepsilon$ is an upper bound for S. We see that

$$
\begin{aligned}
(r - \varepsilon)^2 &= r^2 - 2r\varepsilon + \varepsilon^2 \\
&> r^2 - 2r\varepsilon \\
&> r^2 - (r^2 - 2) = 2.
\end{aligned}
$$

Hence if $0 < x \in S$, then $x^2 < 2 < (r - \varepsilon)^2$. It follows that $x < r - \varepsilon$. If $r - \varepsilon$ is an upper bound for S, then r cannot be the least upper bound for S, which is a contradiction. The proof is complete.

The rational numbers $\mathbb{Q}$ are a proper subset of the real numbers. Real numbers that are not rational are called *irrational*. The numbers π and e have been shown to be irrational. In fact, if we compare orders of infinity, there are far more irrational numbers than rational ones. For a treatment of the intriguing question of comparing orders of infinity, see S. Krantz [Kr] or P. Halmos [Hal].

Consequences of the Least Upper Bound Property. The least upper bound property is a very strong axiom. In particular, it can be used to relate the natural numbers to the other elements of the real numbers. The following result may seem obvious, but it is useful to see how it can be derived from the least upper bound property.

Theorem 1.4 (Archimedean property) For any $a \in \mathbb{R}, a > 0$, there is an $n \in \mathbb{N}$ with $n > a$.

Proof: We suppose the contrary and deduce a contradiction. If the assertion of the theorem is not true, $\mathbb{N}$ is bounded above (and clearly nonempty). Hence, by the least upper bound property, $\mathbb{N}$ has a least upper bound, call it c. From the definition of the least upper bound, we know that, for each $\varepsilon > 0$, there is an $n \in \mathbb{N}$ with $n > c - \varepsilon$. We make a specific choice of ε, taking $\varepsilon = 1/2$. Thus there is an $n \in \mathbb{N}$ such that $n > c - 1/2$. But then, $n + 1 > c + 1/2$, which means that c is not an upper bound for $\mathbb{N}$.

Corollary For each $c > 0$, there is an $n \in \mathbb{N}$ such that $1/n < c$.

Proof: By Theorem 1.4, there is an $n \in \mathbb{N}$ such that $n > 1/c$. Then by (vi) of Theorem 1.2, $1/n < c$.

This corollary leads to an important conclusion about the relationship of the rationals to the real numbers. The rational numbers are extended to the real numbers to be able to solve certain equations. However, we do all computations with rational numbers. For this extension of the rationals to be useful, we must be able to approximate every real number arbitrarily well by a rational. This issue is addressed in the next definition and theorem.

Definition 1.6 A set $S \subset \mathbb{R}$ is said to be *dense* in $\mathbb{R}$ if, for every pair of real numbers $a < b$, there is an element $r \in S$ with $a < r < b$.

Theorem 1.5 The rational numbers $\mathbb{Q}$ are dense in $\mathbb{R}$.

Proof: An appeal to symmetry allows us to reduce the question to the case $0 \leq a < b$. By the corollary to Theorem 1.4, there is an $n \in \mathbb{N}$ such that $1/n < b - a$. Let $S = \{k \in \mathbb{N} : k > na\}$. S is nonempty by Theorem 1.4. By the well-ordering principle of the natural numbers (see Section 1.4), there is a smallest number $m \in S$. Thus, $m/n > a$, but $(m - 1)/n \leq a$. Then we have

$$a < m/n = (m - 1)/n + 1/n \leq a + 1/n < a + b - a = b.$$

For example, for any power $p \in \mathbb{N}$, there is a rational number r (depending on p) such that $\sqrt{2} < r < \sqrt{2} + 10^{-p}$. One of our tasks in later chapters is to find an efficient way to generate these rational approximations.

The irrational numbers are also dense in $\mathbb{R}$.

Corollary Between any two real numbers $a < b$, there is an irrational number.

Proof: We use the fact that, if q is rational, then $q\sqrt{2}$ must be irrational. Now by Theorem 1.5, there must be a rational number q such that

$$a/\sqrt{2} < q < b/\sqrt{2},$$

whence $a < q\sqrt{2} < b$.

Exercises for 1.2

1. Suppose that A is a bounded set of real numbers and that A contains one of its upper bounds, for instance, b. Show that $b = \sup A$.
2. Let S be a finite set of real numbers.

   a) Show that S is bounded above.

b) Show that sup S belongs to S. This is clearly true for a set S that consists of just one element. Then use induction (see Section 1.4) on the number of elements in S.

Hence, we can always speak of the maximum of a finite set of real numbers.

3. Show that the least upper bound property holds if and only if every nonempty set S of real numbers that is bounded below has a greatest lower bound in the real numbers.

4. Show that, if $a, b \in \mathbb{R}$ with $|a - b| \leq 1/n$ for all $n \in \mathbb{N}$, then $a = b$.

5. Find the sup and inf of the following sets when they exist as finite numbers. If either a sup or an inf is not finite, explain why.

   a) $A = \{1 + (-1)^n(n-1)/n : n \in \mathbb{N}\}$.
   b) $B = \{x \in \mathbb{Q} : x^3 < 5x\}$.

6. If A and B are bounded nonempty sets of real numbers, show that $A \cup B$ is bounded with

$$\sup(A \cup B) = \max\{\sup A, \sup B\}.$$

7. Let A be a bounded set of real numbers. If $B \subset A$, show that B is bounded with

$$\inf A \leq \inf B \leq \sup B \leq \sup A.$$

8. Let A be a bounded set of real numbers and define $-A = \{-x : x \in A\}$. Show that $\sup(-A) = -\inf A$.

1.3 Using Inequalities

Inequalities are very important in analysis because we often need to estimate the magnitude of some quantity. A key idea is that of the absolute value of a real number.

Definition 1.7 For $a \in \mathbb{R}$, the *absolute value* of a, denoted $|a|$, is defined to be

$$|a| = a \text{ if } a \geq 0 \text{ and } |a| = -a \text{ if } a < 0.$$

The quantity $|a|$ can be thought of as the geometric distance from the point a to zero. Thus, we always have $-|a| \leq a \leq |a|$. To make an estimate of the form $|a| \leq c$, for example, it suffices to show that $-c \leq a \leq c$.

The important properties of the absolute value are summarized next.

Theorem 1.6 For all $a, b \in \mathbb{R}$,

     (i) $|ab| = |a||b|$,
    (ii) $|a + b| \leq |a| + |b|$,
   (iii) $|a - b| \geq |\,|a| - |b|\,|$.

Inequality (ii) is known as the *triangle inequality*, and (iii) is sometimes called the *reverse triangle inequality*.

Proof: (i) is easily proved by going through the cases: a, b both positive, a, b both negative, and so on. We prove (ii). We have

$$-|a| \leq a \leq |a| \quad \text{and} \quad -|b| \leq b \leq |b|.$$

Adding these two sets of inequalities together yields

$$-(|a| + |b|) \leq a + b \leq |a| + |b|,$$

which is to say $|a + b| \leq |a| + |b|$. Now (iii) follows from two applications of (ii). Indeed,

$$|a| = |a - b + b| \leq |a - b| + |b|$$

so that

$$|a| - |b| \leq |a - b|, \tag{1.2}$$

and

$$|b| = |b - a + a| \leq |b - a| + |a| = |a - b| + |a|$$

so that we have

$$|b| - |a| \leq |a - b|. \tag{1.3}$$

We combine (1.2) with (1.3) to obtain

$$-|a - b| \leq |a| - |b| \leq |a - b|,$$

which is exactly (iii).

It can easily be shown by an induction argument (see Section 1.4) that the triangle inequality is valid for any number of summands:

Corollary Let $a_j \in \mathbb{R}$, $j = 1, \ldots, n$. Then

$$|a_1 + a_2 + \cdots + a_n| \leq |a_1| + |a_2| + \cdots + |a_n|.$$

Here are some examples of how we use the triangle inequality to make estimates.

EXAMPLE 1.2:   To estimate the difference between the products ab and $a'b'$ in terms of the differences $a - a'$ and $b - b'$, the standard device is to subtract and then add a term ab':

$$ab - a'b' = ab - ab' + ab' - a'b' = a(b - b') + (a - a')b'. \qquad (1.4)$$

Then use the triangle inequality:

$$|ab - a'b'| \leq |a||b - b'| + |a - a'||b'|. \qquad (1.5)$$

When a, b, a' and b' are all positive, ab is the area of a rectangle R, and $a'b'$ is the area of another rectangle R'. The left side of (1.4) can be interpreted as the difference in the areas of the two rectangles, and the right side of (1.4) is the sum of the areas of the two shaded regions in Figure 1.1. If, for example, we want to estimate the difference $x \log x - 2 \log 2$, we subtract and add the term $2 \log x$, and we find

$$\begin{aligned}
|x \log x - 2 \log 2| &\leq |x - 2||\log x| + 2|\log x - \log 2| \\
&\leq |x - 2||\log x| + 2|\log(x/2)|.
\end{aligned}$$

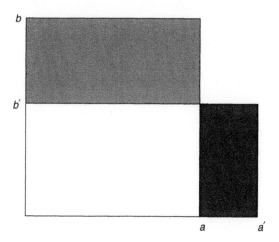

Figure 1.1 Rectangles with side a, b and a', b'

EXAMPLE 1.3: To find an upper bound for a fraction $|a/b|$, we usu-
ally find an upper bound for $|a|$ and a lower bound for $|b|$. We illus-
trate by finding an upper bound for $|f(x)|$ where

$$f(x) = \frac{3x \cos x}{x^2 + \sqrt{1 + x^4}}.$$

The numerator is bounded by $|3x \cos x| \leq 3|x|$ for all $x \in \mathbb{R}$. One
lower bound for the denominator is

$$|x^2 + \sqrt{1 + x^4}| = x^2 + \sqrt{1 + x^4} \geq x^2 + \sqrt{x^4} = 2x^2$$

for all $x \in \mathbb{R}$. Combining these two bounds, we have the estimate

$$|f(x)| \leq \frac{3|x|}{2x^2} = \frac{3}{2|x|} \quad \text{for} \quad x \neq 0. \tag{1.6}$$

While this estimate gives some information for large $|x|$, it is useless
for x near 0. We get a better estimate for $|f(x)|$ near $x = 0$ by using a
different lower bound for the denominator:

$$x^2 + \sqrt{1 + x^4} \geq 1 \quad \text{for} \quad x \in \mathbb{R}.$$

This yields

$$|f(x)| \leq 3|x| \quad \text{for } x \in \mathbb{R}. \tag{1.7}$$

Let $c > 0$. We shall use the estimate (1.6) for $|x| \geq c$ and (1.7) for $|x| \leq c$. To get the best estimate, we choose the value of c so that the two bounds agree at $|x| = c$. This will be the case if $3c = 3/2c$, or $c = 1/\sqrt{2}$. Thus

$$|f(x)| \leq g(x) = \begin{cases} 3|x|, & |x| \leq 1/\sqrt{2} \\ 3/(2|x|), & |x| > 1/\sqrt{2} \end{cases}.$$

The graphs of $|f(x)|$ and $g(x)$ are shown in Figure 1.2.

Exercises for 1.3

1. Let $f(x) = (x + \sqrt{x})/(x^2 + 1)$.

   a) Show that

   $$f(x) \leq \begin{cases} 2\sqrt{x}, & 0 \leq x \leq 1 \\ 2/x, & x \geq 1 \end{cases}.$$

   b) Make similar lower estimates for f on $x \geq 1$ and on $0 \leq x \leq 1$.

2. Find constants A and B such that for $|x| \leq 5$,

   $$|x^2 \cos x - 4\cos 2| \leq A|\cos x - \cos 2| + B|x - 2|.$$

3. Find a constant C such that for all $|x|, |y| \leq 2$,

   $$\left| \frac{1}{x^2 + 1} - \frac{1}{y^2 + 1} \right| \leq C|x - y|.$$

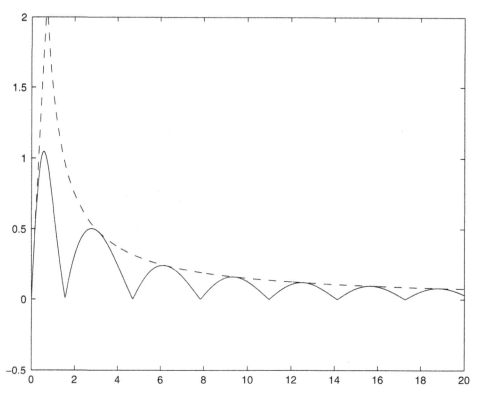

Figure 1.2 Graph of $|f(x)| = 3|x||\cos(x)|/(x^2 + \sqrt{1 + x^4})$ in solid line, with graph of estimating function $g(x)$ in dashed line

4. Find constants A and B such that

$$\frac{A}{x^2} \leq \frac{1}{\sqrt{x^4 + 4}} \leq \frac{B}{x^2} \quad \text{for } x \geq 1.$$

5. Find a constant $M > 0$ such that

$$\left| \frac{x^3 + 2x - 1}{x - 1} \right| \leq M \quad \text{for } 2 \leq |x| \leq 3.$$

6. Let $p(x) = a_n x^n + a_{n-1} x^{n-1} + \cdots + a_0$. Suppose that $a_n > 0$. Use the reverse triangle inequality to show the following: for each $M > 0$, there is an $R > 0$ such that $p_n(x) \geq M$ for $|x| \geq R$.

7. Let f be defined on $[-1, 1]$ satisfying

$$x \leq f(x) \leq x + 2x^2 \quad \text{for } 0 \leq x \leq 1$$

and

$$-2x^2 + x \leq f(x) \leq x \quad \text{for } -1 \leq x \leq 0.$$

Show that

$$\left| \frac{f(x)}{x} - 1 \right| \leq 2|x| \quad \text{for } |x| \leq 1.$$

1.4 Induction

The principle of mathematical induction and the well-ordering principle are often used in mathematical reasoning. We state them here.

Well-ordering Principle. A nonempty set $S \subset \mathbb{N}$ has a least element.

Principle of Mathematical Induction. Let $S \subset \mathbb{N}$ have the two properties:

(i) $1 \in S$.
(ii) If $k \in S$, then $k + 1 \in S$.

Then $S = \mathbb{N}$.

Theorem 1.7 The well-ordering principle and the principle of mathematical induction are equivalent properties of $\mathbb{N}$.

Proof: Suppose that the well-ordering principle holds for $\mathbb{N}$. Let S be a subset of $\mathbb{N}$ such that (i) $1 \in S$, and (ii) if $k \in S$, then $k + 1 \in S$. Let $T = \{n \in \mathbb{N} : n \notin S\}$, and suppose that T is nonempty. In view of the well-ordering principle, T has a least element, say k_0. However, then $k_0 - 1 \in S$ and according to (ii), $k_0 = (k_0 - 1) + 1 \in S$, which is a contradiction.

Now suppose that the principle of mathematical induction holds and that T is a subset of $\mathbb{N}$ that does not have a least element. Let $S = \{n \in \mathbb{N} : n \notin T\}$. Clearly, $1 \notin T$ because $1 \leq k$ for all $k \in \mathbb{N}$. Thus $1 \in S$. Similarly, we can show that $2 \in S$. Suppose that we have shown that $\{1, \ldots, n\} \subset S$. But then

$n+1 \in S$, for otherwise $n+1$ would be the least element of T. Consequently, by the principle of mathematical induction, $S = \mathbb{N}$ so that T is empty. The proof is complete.

We have already used the well-ordering principle of $\mathbb{N}$ in the proof of Theorem 1.7. There we let m be the least natural number such that $m/n > a$.

Now we give some examples of induction arguments.

EXAMPLE 1.4: We find a formula for the sum $\sum_{k=1}^{n} k^2$. We already know the formula

$$\sum_{k=1}^{n} k = \frac{n(n+1)}{2} = (1/2)n^2 + (1/2)n.$$

This formula suggests that we try to find a formula for the sum $\sum_{k=1}^{n} k^2$ in the form

$$S(n) = an^3 + bn^2 + cn.$$

We evaluate the sum for $n = 1, 2, 3$ to get the three equations

$$a + b + c = 1$$
$$8a + 4b + 2c = 5$$
$$27a + 9b + 3c = 14.$$

The solution of these three equations is $a = 1/3, b = 1/2$ and $c = 1/6$. Our tentative formula for $\sum_{k=1}^{n} k^2$ is $S(n) = (2n^3 + 3n^2 + n)/6$. Now we want to prove that, for all $n \in N$,

$$\sum_{k=1}^{n} k^2 = S(n). \tag{1.8}$$

We have chosen the coefficients so that equation (1.8) is satisfied for $n = 1, 2, 3$. Assume now that (1.8) is true for some $n \geq 3$. We want to verify that it is true for $n + 1$. A little manipulation yields

$$\sum_{k=1}^{n+1} k^2 = \sum_{k=1}^{n} k^2 + (n+1)^2$$

$$= S(n) + (n+1)^2$$

$$= \frac{2n^3 + 3n + n + 6(n+1)^2}{6}$$

$$= \frac{2(n+1)^2 + 3(n+1)^2 + n + 1}{6}$$

$$= S(n+1).$$

Thus by the principle of mathematical induction, the formula (1.8) is true for all n.

EXAMPLE 1.5:   For a second example, we shall prove the *Bernoulli inequality*. For all $n \in \mathbb{N}$ and $x \geq -1$,

$$(1+x)^n \geq 1 + nx \tag{1.9}$$

The inequality is actually an equality for $n = 1$. Now assume it is true for some n. Since we assume $x \geq -1$, we have $x + 1 \geq 0$. Thus, when we multiply both sides of (1.9) by $(x + 1)$, the sense of the inequality will be preserved. Hence, we have

$$(1+x)^{n+1} \geq (1+x)(1+nx) = 1 + (n+1)x + nx^2 \geq 1 + (n+1)x,$$

which is just the Bernoulli inequality for $n + 1$.

Exercises for 1.4

1. a) Recall that, for $0 \leq k \leq n$, the binomial coefficient is given by the formula

$$\binom{n}{k} = \frac{n!}{k!(n-k)!}.$$

Do the calculation to verify that, for $1 \le k \le n$, the binomial coefficients satisfy

$$\binom{n}{k-1} + \binom{n}{k} = \binom{n+1}{k}.$$

b) Use part a) and induction to prove the binomial formula

$$(a+b)^n = \sum_{k=0}^{n} \binom{n}{k} a^{n-k} b^k.$$

2. Use induction to show that, for all $n \ge 1$,

$$\sum_{k=1}^{n} k^3 = \left(\sum_{k=1}^{n} k\right)^2.$$

3. Use induction to show that the sum of the first m positive integers is $m(m+1)$.

4. Use induction to show that

   a) $2n+1 \le 2^n$ for $n \ge 3$.
   b) $n^2 \le 2^n + 1$ for $n \ge 1$.

1.5 Sets and Functions

In this section we recall some of the basic notions about sets and functions. If A and B are sets then

  (i) The *intersection* of A and B, denoted $A \cap B$, is the set of all elements $x \in A$ *and* $x \in B$.
 (ii) The *union* of A and B, denoted $A \cup B$, is the set of all elements $x \in A$ *and/or* $x \in B$.
(iii) The empty set, which has no elements, is denoted ϕ.

(iv) If $A \cap B = \phi$, we say that A and B are *disjoint*.

The properties of the set operations are summarized in the following theorem.

Theorem 1.8 Let A, B, and C be sets.
  a) $A \cap B = B \cap A$ and $A \cup B = B \cup A$. commutativity.
  b) $A \cap (B \cap C) = (A \cap B) \cap C$ and $A \cup (B \cup C) = (A \cup B) \cup C$. associativity.
  c) $A \cap (B \cup C) = (A \cap B) \cup (A \cap C)$ and $A \cup (B \cap C) = (A \cup B) \cap (A \cup C)$. distributive property.

A collection of sets can have a quite arbitrary index set. For example, we might consider the collection of sets $A_\gamma = \{x \in \mathbb{R} : x < \gamma\}$ for each irrational number γ. In this case, the index set Γ is the set of irrational numbers.

The associative property allows us to define unambiguously the intersection and union of any collection of sets. If there are sets A_γ with γ in some index set Γ, the set

$$\cup_{\gamma \in \Gamma} A_\gamma$$

consists of the elements x that belong to at least one of the A_γ. The set

$$\cap_{\gamma \in \Gamma} A_\gamma$$

consists of the elements x that belong to each of the A_γ.

If A and B are sets, the *complement of B relative to A* is the set of elements $x \in A$ such that $x \notin B$. This relative complement is denoted $A \backslash B$. The De Morgan laws for three sets are given in the following theorem.

Theorem 1.9 Let A, B and C be sets. Then

$$A \backslash (B \cup C) = (A \backslash B) \cap (A \backslash C)$$

and

$$A \backslash (B \cap C) = (A \backslash B) \cup (A \backslash C).$$

If there is a fixed set U with $A, B, \ldots \subset U$, and all complements are taken with respect to U, the complement $U \backslash A$ may be denoted A'. In this case the De Morgan laws are stated

$$(A \cap B)' = A' \cup B' \quad \text{and} \quad (A \cup B)' = A' \cap B'.$$

If A and B are sets, the *Cartesian product* $A \times B$ is defined to be the set of ordered pairs (x, y) with $x \in A$ and $y \in B$. The Cartesian product $B \times A$ is not the same as $A \times B$, because it consists of ordered pairs (y, x) in which $y \in B$ and $x \in A$. One can also define the Cartesian product of a collection of sets, A_j, $j = 1, \ldots, n$ as the set of ntuples $(x_1, x_2, \ldots, x_n)$ where $x_j \in A_j$.

We shall use extensively the class of sets called *intervals*. They are denoted as follows:

  (i) The *closed*, bounded interval $[a, b]$ is the set of real numbers x such that $a \le x \le b$.

  (ii) The *open*, bounded interval (a, b) is the set of real numbers x such that $a < x < b$.

  (iii) A half-open, (or half-closed) interval $(a, b]$ is the set of real numbers x such that $a < x \le b$.

  (iv) An unbounded open interval is denoted (a, ∞) or $(-\infty, b)$. An unbounded closed interval is denoted $[a, \infty)$ or $(-\infty, b]$.

Functions

The concept of a function, or mapping, is fundamental to mathematics. In loose terms, a function is a rule that to each element x in a set A assigns a unique element y in a set B. We can speak of the existence of a function, even if we cannot write down a formula for it. A more precise definition of a function is

Definition 1.8 Let A and B be sets. A function f is a subset $G(f)$ of ordered pairs $(x, y) \in A \times B$ with the property that, if (x, y) and (x, z) belong to $G(f)$, then $y = z$. The set $G(f)$ is called the graph of f.

We shall assume that for each $x \in A$, there is some $y \in B$ such that $(x, y) \in G(f)$. The set A is called the *domain* of f, sometimes denoted $D(f)$. The set of elements $y \in B$ such that $(x, y) \in G(f)$ for some $x \in A$ is called the *range* of f, denoted $R(f)$. If C is a subset of A, the set $\{f(x) : x \in C\}$ is called the *image of* C under f.

When $(x, y) \in G(f)$, we usually write $y = f(x)$, and we refer to y as the value of f at x. The symbol f, by itself, refers to the whole machine, which consists of $A = D(f)$, $R(f) \subset B$ and $G(f) \subset A \times B$, and we shall often write $f : A \to B$. We shall also use the notation $x \to f(x)$ to refer to the function f.

It is often the case that we wish to restrict the arguments of a function to lie in a subset $E \subset D(f)$. This defines a new function called the *restriction* of f to E; it consists of the pairs $(x, y) \in G(f)$ such that $x \in E$.

It may also be possible to extend the domain of f to a larger subset $F \subset A$ such that $D(f) \subset F$. The new function is called an *extension* of f.

> **EXAMPLE 1.6:** Let A be the square $\{(x, y) : 0 \leq x \leq 1,\ 0 \leq y \leq 1\}$ and $f(x, y) = x^2 + 3y$. We take A as the domain of f. The restriction of f to the diagonal of the square is $\tilde{f}(x) = x^2 + 3x$, $0 \leq x \leq 1$.

Composition of Functions and Inverse Functions

Suppose that A, B, and C are sets and that f and g are functions, $f : A \to B$ and $g : B \to C$. The *composition* of f and g is the function $g \circ f$ defined on A by

$$(g \circ f)(x) = g(f(x)).$$

For any set A, the function $f(x) = x$ is called the *identity* function (or mapping), denoted id_A.

If $f : A \to B$ and if $f(x) = f(x')$ implies $x = x'$, we say that f is *one-to-one*, or sometimes we use the classier term *injective*. If $f : A \to B$ and if, for each $y \in B$, there is an $x \in A$ such that $f(x) = y$, we say that f is *onto* or, sometimes, *surjective*.

In the case in which $f : A \to B$ is one-to-one, it is possible to reverse the action of f on the elements of $R(f)$.

Theorem 1.10 Let $f : A \to B$ be one-to-one. Then there exists a one-to-one function,

$$g : R(f) \to A,$$

such that the composite function $g \circ f = \mathrm{id}_A$ and $f \circ g = id_{R(f)}$. The graph of g,

$G(g)$, is just the set of ordered pairs $(y, x) \in B \times A$ such that $(x, y) \in G(f)$.

Proof: Let $H \subset B \times A$ be defined by

$$H = \{(y, x) : (x, y) \in G(f)\}.$$

Now, if (y, x) and $(y, x') \in H$, then $(x, y), (x', y) \in G(f)$. However, the hypothesis that f is one-to-one means that $x = x'$. Hence, H is the graph of a function that we shall call g. Then $D(g) = R(f)$, and $R(g) = A$.

Furthermore, g is one-to-one. In fact, if $g(y) = g(y') = x$, then the pairs $(x, y), (x, y') \in G(f)$. Since f is a function, we must have $y = y'$.

Finally we note that, if $x \in A$ and $y = f(x)$, then $(x, y) \in G(f)$ and so $(y, x) \in H = G(g)$. Thus, $x = g(y)$, which is to say that $x = g(f(x))$ or, $g \circ f = id_A$. That $f \circ g = id_{R(f)}$ is proved in the same way. The proof is complete.

Definition 1.9 Let $f : D(f) \to R(f)$ be one-to-one. The function $g : R(f) \to D(f)$ such that $g \circ f = id_{D(f)}$ and $f \circ g = id_{R(f)}$ is called the inverse function to f and is denoted f^{-1}.

EXAMPLE 1.7:

a) The function $f(x) = x^3$ is one-to-one on $\mathbb{R}$ and has the inverse function $g(x) = x^{1/3}$, which is defined again on $\mathbb{R}$.

b) The function $f(x) = x^2$, with its domain taken to be $\mathbb{R}$, is not one-to-one because, for any x, $f(x) = f(-x)$. However, we may define a restriction $\tilde{f}$ of f to the set $A = \{x \geq 0\}$. Now, $\tilde{f}$ is one-to-one on A with range equal to A. The inverse function is, of course, $g(y) = \sqrt{y}$, the positive square root function.

c) The function $\sin x$, defined for all real numbers, is not one-to-one because $f(x + 2\pi) = f(x)$ for all x. If we restrict f to the set $-\pi/2 \leq x \leq \pi/2$, we obtain a one-to-one function with the inverse function $g(y) = \arcsin y$ defined on the set $\{-1 \leq y \leq 1\}$.

Properties of Sup and Inf

We will often need to estimate the supremum and/or the infimum of a function. Here are some rules that will be useful.

Definition 1.10 Let $f : A \to \mathbb{R}$. We say that f is *bounded above* if the set of real numbers $\{f(x) : x \in A\}$ is bounded above; this means that there is a number M such that $f(x) \leq M$ for all $x \in A$. In this case, the image $f(A)$ has a least upper bound, and we write

$$\sup_A f = \sup_{x \in A} f(x).$$

Bounded below means that there is an $m \in R$ such that $f(x) \geq m$ for all $x \in A$ and we write

$$\inf_A f = \inf_{x \in A} f(x).$$

If f is bounded above and below on A, we say that f is *bounded* on A.

Theorem 1.11 Let $f : A \rightarrow \mathbb{R}$ be bounded. Then $|f(x)|$ is bounded on A and

$$\sup_A |f| = \max\{|\inf_A f|, |\sup_A f|\}.$$

Proof: (i) If $f(x) \geq 0$ for all $x \in A$, then $0 \leq \inf_A f \leq \sup_A f$. Since $|f(x)| = f(x)$, the result is true. (ii) If $f(x) \leq 0$ for all $x \in A$, $0 \leq -\sup_A f \leq -f(x) \leq -\inf_A f$. In this case, $|f(x)| = -f(x)$ so that $0 \leq |\sup_A f| \leq |f(x)| \leq |\inf_A f|$, and the result is again true. (iii) If f takes on both positive and negative values, $\inf_A f \leq 0$ and $\sup_A f \geq 0$. Hence

$$|f(x)| = \begin{cases} f(x) \leq \sup_A f & \text{when } f(x) \geq 0 \\ -f(x) \leq -\inf_A f = |\inf_A f| & \text{when } f(x) < 0 \end{cases}.$$

Theorem 1.12 Let $f, g : A \rightarrow \mathbb{R}$ be bounded functions. Then $f + g$ is bounded. Furthermore,

$$\sup_A (f + g) \leq \sup_A f + \sup_A g \tag{1.10}$$

and

$$\inf_A (f + g) \geq \inf_A f + \inf_A g. \tag{1.11}$$

Proof: For each $x \in A$, $f(x) \leq \sup_A f$ and $g(x) \leq \sup_A g$. Hence, $f(x) + g(x) \leq \sup_A f + \sup_A g$. Thus, $\sup_A f + \sup_A g$ is an upper bound for the set of values $\{f(x) + g(x) : x \in A\}$; this implies (1.10). The inequality (1.11) is proved in a similar fashion.

EXAMPLE 1.8: Let $f(x) = 2x$ and $g(x) = 1 - 2x^2$, and let $A = \{x : 0 \leq x < 1\}$. We see that

$$\sup_A f = 2, \quad \inf_A f = 0, \quad \sup_A g = 1, \quad \inf_A g = -1.$$

However, $(f + g)(x) = 1 - 2x^2 + 2x = -2(x - 1/2)^2 + 3/2$ so that

$$\sup_A f + g = 3/2 < 2 + 1 = \sup_A f + \sup_A g$$

and

$$\inf_A f + g = 1 > 0 - 1 = \inf_A f + \inf_A g.$$

Exercises for 1.5

1. Show that a set $J \subset \mathbb{R}$ is an interval (of any kind) if and only if, for each pair $x_1, x_2 \in J$ with $x_1 \leq x_2$, the interval $\{x_1 \leq x \leq x_2\} \subset J$.
2. Let $f : A \to \mathbb{R}$ be bounded. Show that for each $c \in \mathbb{R}$, $\sup_A(c + f) = c + \sup_A f$ and $\inf_A(c + f) = c + \inf_A f$.
3. Let $f : A \to \mathbb{R}$ be bounded. Show that for $c > 0$, $\sup_A(cf) = c\sup_A f$ and that $\inf_A(cf) = c\inf_A f$.
4. Let $f : A \to \mathbb{R}$ be a bounded function. The analogue to Theorem 1.11 would appear to be

$$\inf_A |f| = \min\{|\inf_A f|, |\sup_A f|\}.$$

   Is it true? If not, what is the correct result?
5. Let X and Y be sets of real numbers, and let $h(x, y) : X \times Y \to \mathbb{R}$. Let

$$f(x) = \sup_{y \in Y} h(x, y) \quad \text{and} \quad g(y) = \inf_{x \in X} h(x, y).$$

   What is the relationship between $\inf_{x \in X} f(x)$ and $\sup_{y \in Y} g(y)$?

   For some guidance, you may wish to first consider the case of $X = Y = \mathbb{R}$ and the function

$$h(x, y) = \begin{cases} 1, & x + y > 0 \\ 0, & x + y \leq 0 \end{cases}.$$

6. Let $f : A \to \mathbb{R}$ be a bounded function.

   a) Show that

$$\sup_{x \in A} f(x) - \inf_{x \in A} f(x) = \sup_{x, y \in A} [f(x) - f(y)].$$

   b) Use part a) to show that

$$\sup_{x \in A} |f(x)| - \inf_{x \in A} |f(x)| \leq \sup_{x \in A} f(x) - \inf_{x \in A} f(x).$$

Chapter 2

Sequences of Real Numbers

Chapter Overview

The concept of the limit of a sequence is introduced and several techniques for proving convergence of a sequence are discussed. The convergence of bounded, monotone sequences is closely linked to the least upper bound property of the real numbers. An important example for computation is the recursive sequence used to generate rational approximations to the square root of a positive number. This sequence is shown to converge quadratically. Subsequences and the Bolzano-Weierstrass theorem follow. Finally, Cauchy sequences are defined, leading to another characterization of the real number system.

The theorems and examples of this chapter deal with the problem of finding a candidate for the limit of a given sequence, and then proving that the sequence converges. However, it is important to point out that, in applications, we often are given a number, and then want to construct a sequence of approximations that converge to it. We shall see a number of examples of this kind of problem in Chapter 6.

In this chapter, and the remaining chapters of Part I, we use the functions $\cos x$, $\sin x$, e^x and $\log x$ for examples. We assume the standard properties of these functions that are known from calculus. In the appendix to Part I, we provide the definitions of these functions and a derivation of their properties.

2.1 Limits of Sequences

The concept of a limit is fundamental to both analysis and computation. It has taken several centuries to arrive at the definition of limit we use today.

Newton and Leibnitz had a rather vague idea of limit, often relying on physical reasoning to justify a result. It was not until 1820 that Cauchy gave a precise definition.

In this chapter, we will discuss a discrete version of limit that applies to sequences. In later chapters, we will need to deal with limits of continuously defined objects.

Definition 2.1 A *sequence* is a function $f : \mathbb{N} \to \mathbb{R}$. The terms of the sequence are $f(1), f(2), \ldots, f(n), \ldots$. However, the standard method of notating a sequence is to use the index n as a subscript. Thus a sequence is usually written $a_1, a_2, \ldots, a_n \ldots$, and denoted more compactly as $\{a_n\}$.

Sequences are commonly given in two ways.

1) The sequence $a_n = 2n + \cos(n)$ is given by a closed formula. We can calculate $a_{915} = 1830 + \cos(915)$ without having to calculate any other terms of the sequence.
2) A sequence can be given recursively. In this method a formula is given that defines the nth term a_n in terms of terms with lower indices. Here are two examples.

   a) The Fibonacci sequence is given by $a_{n+2} = a_{n+1} + a_n$. In this case, the first elements a_1 and a_2 must be specified and, to calculate a_{915}, we must calculate all a_n with $n < 915$.
   b) The well-known sequence $a_{n+1} = (1/2)(a_n + A/a_n)$ is used to generate approximations to the square root of a positive number A. To calculate a_5, we must specify A and a first term a_1 and then calculate a_2, a_3, and a_4.

Recursive sequences often arise when we are computing approximate solutions of an equation.

What does it mean for a sequence $\{a_n\}$ to converge to a number L? How do we say that a_n approximates L to any desired accuracy? Should this be just for some values of n, or for all n beyond a certain point? The following definition answers these questions.

Definition 2.2 The sequence $\{a_n\}$ *converges* if there is a number L such that: For each $\varepsilon > 0$, there is a natural number $N > 0$ (that may depend on ε) such that $|a_n - L| < \varepsilon$ for all $n \geq N$. We say that L is a *limit* of the sequence $\{a_n\}$, and we write

$$\lim_{n \to \infty} a_n = L.$$

If a sequence does not converge, we say that it *diverges*.

Before giving examples of sequences that converge (and one that does not), we derive a simple consequence of the definition.

Uniqueness of limits. A sequence can have at most one limit.

Proof: Suppose that the sequence $\{a_n\}$ has two limits, L_1 and L_2. We shall show that the definition of a limit forces $L_1 = L_2$. First we use the triangle inequality to estimate the difference

$$|L_1 - L_2| = |L_1 - a_n + a_n - L_2| \leq |L - a_n| + |a_n - L_2|.$$

Because the sequence $\{a_n\}$ converges to L_1, for each $\varepsilon > 0$, there is an $N_1 > 0$ such that $|a_n - L_1| < \varepsilon/2$ for $n \geq N_1$. Similarly, there is an N_2 such that $|a_n - L_2| < \varepsilon/2$ for $n \geq N_2$. Let $N = \max\{N_1, N_2\}$. Now $n \geq N$ implies that $n \geq N_1$ *and* $n \geq N_2$. Hence $n \geq N$ implies that

$$|L_1 - L_2| < \varepsilon/2 + \varepsilon/2 = \varepsilon.$$

Since the choice of ε was arbitrary, we must conclude that $L_1 = L_2$ (see exercise 3 of Section 1.2).

EXAMPLE 2.1:

a) The sequence $\{a_n\} = \sin(n)/n$ converges to $L = 0$. We have $|a_n - L| = |a_n| \leq 1/n$. Let $\varepsilon > 0$ be given. Then, by the corollary to the Archimedean property of the natural numbers, there is a natural number $N > 0$ such that $1/N < \varepsilon$. Finally, $n \geq N$ implies that $1/n \leq 1/N$ so $n \geq N$ implies that $|a_n| < \varepsilon$.

b) Consider the sequence $\{a_n\} = (n+1)/(3n+2)$. We must first come up with a candidate for a limit. If we divide both numerator and denominator by n, we see that

$$a_n = \frac{n+1}{3n+2} = \frac{1+1/n}{3+2/n}.$$

Since $1/n$ and $2/n$ will become very small as n gets large, it seems reasonable to try $L = 1/3$ as a limit for this sequence. Now we calculate

$$a_n - L = \frac{n+1}{3n+2} - \frac{1}{3} = \frac{1}{9n+6}.$$

whence

$$\left| a_n - \frac{1}{3} \right| = \frac{1}{9n+6} \le \frac{1}{9n}.$$

For a given $\varepsilon > 0$, we choose $N > 0$ so large that $N > 1/(9\varepsilon)$. Then for $n \ge N$,

$$\left| a_n - \frac{1}{3} \right| \le \frac{1}{9n} \le \frac{1}{9N} < \varepsilon.$$

c) Let $0 < r < 1$, and let $a_n = r^n$. We claim that $\lim_{n \to \infty} a_n = 0$. Let $c = 1/r - 1$. Because $0 < r < 1$, $c > 0$ and $r = 1/(1+c)$. It follows from the Bernoulli inequality (1.9) that $(1+c)^n \ge 1 + nc$. Hence

$$r^n = \frac{1}{(1+c)^n} \le \frac{1}{1+nc} \le \frac{1}{nc}.$$

For a given $\varepsilon > 0$, we choose $N > 0$ so that $N > 1/(\varepsilon c)$. Then $n \ge N$ implies $nc > 1/\varepsilon$, whence

$$|a_n| = r^n \le \frac{1}{nc} < \varepsilon.$$

d) Let $a_n = (-1)^n$. We will show that this sequence does not converge. First, we must understand how to negate the definition of convergence. The sequence $\{a_n\}$ *does not converge*, if:

For each choice of L, there is an $\varepsilon_0 > 0$ such that, for each $N > 0$, there is an $n \ge N$ with $|a_n - L| \ge \varepsilon_0$.

This means that, no matter how far out we go in the sequence, there is always some index n such that a_n lies outside the interval $(L - \varepsilon_0, L + \varepsilon_0)$. Note that ε_0 may depend on L.

Let L be given, and let $\varepsilon_0 = \max\{|L-1|, |L+1|\}$. It is important that $\varepsilon_0 > 0$. Now for n even, $|a_n - L| = |1 - L|$, and for n odd, $|a_n - L| = |-1 - L| = |1 + L|$. Thus for any n, $\max\{|a_n - L|, |a_{n+1} - L|\} = \varepsilon_0$. No matter how large we choose n, either $|a_n - L| \ge \varepsilon_0$ or $|a_{n+1} - L| \ge \varepsilon_0$.

Empirical Convergence

The type of convergence that we have been discussing might be called *analytical convergence*. A broader meaning of the term convergence is often used with respect to numerical calculations. A scientist or engineer would probably say that the following numerical results are evidence that the sequence of numbers x_n in the middle column is converging to $L = 1$.

n	$x_n = n^2(e^{n^{-2}} - 1)$	$x_n - 1$
10	1.00501670841679	.00501670841679
20	1.00125104231807	.00125104231807
100	1.00005000166714	.00005000166714
500	1.00000199998451	.00000199998451
2000	1.00000012537294	.00000012537294

This might be termed *empirical* convergence. It means that, for reasonably small $\varepsilon > 0$, if we calculate far enough in the sequence to some N, we will have $|x_N - L| < \varepsilon$. Notice how much weaker this is than analytical convergence. For the sequence x_n to converge analytically to L, we require that, for *each* $\varepsilon > 0$, there is an $N > 0$ such that $|x_n - L| < \varepsilon$ for *all* $n \geq N$.

Because of the limitations of machine precision (round-off error), it is possible that a sequence can be shown to converge analytically, but when it is calculated on a machine, it does not converge. In fact, the sequence $x_n = n^2(e^{n^{-2}} - 1)$ does converge analytically to $L = 1$ but, for larger values of N, the calculated values (in double precision) do not get closer to 1.

n	x_n	$x_n - 1$
10000	0.99999999392253	-6.077470970922150e-009
50000	1.00000008274037	8.274037099909037e-008

We must also be on guard that a sequence that appears to be converging empirically may not be converging to the limit we seek. We shall see this later in an example with the method of bisection.

All of these examples underline the fact that we need a good theory of analytic convergence to understand what computations can accomplish.

Rules for Combining Sequences

We want to show that we can add, multiply, and divide convergent sequences to produce convergent sequences. First we must derive another consequence of the definition.

Lemma 2.1 A convergent sequence is bounded.

Proof: By a *bounded* sequence, we mean a sequence whose values lie in a bounded set of $\mathbb{R}$. Suppose that the sequence $\{a_n\}$ converges to the number L. We do not need to use the full strength of the definition. Let $\varepsilon = 1$. Then there is an $N > 0$ such that, for $n \geq N$, $|a_n| = |a_n - L + L| \leq |a_n - L| + |L| \leq 1 + |L|$. Finally, let $M = \max\{|a_1|, \ldots, |a_{N-1}|, |L| + 1\}$. We see that $|a_n| \leq M$ for all n, which means that the sequence $\{a_n\}$ is bounded.

We also need the following technical result.

Lemma 2.2 Let the sequence $\{a_n\}$ converge to $L \neq 0$. Then there is an $N > 0$ such that $|a_n| \geq |L|/2$ for $n \geq N$.

Proof: Again, to prove the assertion, we do not need to use the full strength of the definition. First let us suppose that $L > 0$. We make the particular choice of $\varepsilon_0 = L/2 > 0$. Then, because the sequence converges to L, we know that there is an N_0 such that $|a_n - L| < \varepsilon_0$ for $n \geq N_0$. This means that, for $n \geq N_0$,

$$-L/2 < a_n - L < L/2.$$

Then adding L to each term in this inequality, we deduce that

$$L/2 < a_n < (3/2)L \quad \text{for } n \geq N_0.$$

We can make a similar argument in the case $L < 0$, taking $\varepsilon_0 = -L/2$. However, there is a way to do both cases with one argument. We start with the choice $\varepsilon_0 = |L|/2$. Then there is an N_0 such that $|a_n - L| < \varepsilon_0 = |L|/2$ for $n \geq N_0$. Hence, by the reverse triangle inequality,

$$\big|\,|a_n| - |L|\,\big| \leq |a_n - L| < |L|/2 \quad \text{for } n \geq N_0.$$

Therefore,

$$-|L|/2 < |a_n| - |L| < |L|/2 \quad \text{for } n \geq N_0.$$

We add $|L|$ to each term in this inequality to deduce

$$|L|/2 < |a_n| < (3/2)|L| \quad \text{for } n \geq N_0.$$

The proof is complete.

Theorem 2.3 Let the sequence $\{a_n\}$ converge to L_1 and the sequence $\{b_n\}$ converge to L_2.

(i) For any $c \in R$, the sequence $\{ca_n\}$ converges to cL_1.
(ii) The sum sequence $\{a_n + b_n\}$ converges to the sum of the limits $L_1 + L_2$.
(iii) The product sequence $\{a_n b_n\}$ converges to the product of the limits $L_1 L_2$.
(iv) Assume that $b_n \neq 0$ and that $L_2 \neq 0$. Then the quotient sequence $\{a_n / b_n\}$ converges to the quotient of the limits L_1 / L_2.

Proof: The arguments given here will be used again in several different contexts. The first assertion, (i), is just a special case of (iii).

We start with (ii). First we use the triangle inequality to make the estimate

$$|a_n + b_n - (L_1 + L_2)| \leq |a_n - L_1| + |b_n - L_2|.$$

Let $\varepsilon > 0$ be given. Because $\{a_n\}$ converges to L_1, there is an $N_1 > 0$ such that $|a_n - L_1| < \varepsilon/2$ for $n \geq N_1$. Similarly, there is an N_2 such that $|b_n - L_2| < \varepsilon/2$ for $n \geq N_2$. We take $N = \max\{N_1, N_2\}$. Now for $n \geq N$, *both* inequalities hold so that

$$|a_n + b_n - (L_1 + L_2)| \leq \varepsilon/2 + \varepsilon/2 = \varepsilon.$$

Next we prove (iii) using the device that follows the corollary to Theorem 1.8. We wish to estimate the difference in the products $|a_n b_n - L_1 L_2|$ in terms of the differences $|a_n - L_1|$ and $|b_n - L_2|$. To do this, we subtract and then add the term $a_n L_2$:

$$\begin{aligned} |a_n b_n - L_1 L_2| &\leq |a_n b_n - a_n L_2| + |a_n L_2 - L_1 L_2| \\ &\leq |a_n||b_n - L_2| + |a_n - L_1||L_2|. \end{aligned}$$

We saw in Lemma 2.1 that a convergent sequence is bounded; there is an $M > 0$ such that $|a_n| \leq M$ for all n. Hence, for all n, we have

$$|a_n b_n - L_1 L_2| \leq M|b_n - L_2| + |a_n - L_1||L_2|. \tag{2.1}$$

Now we are ready for the ε-N drill. Given $\varepsilon > 0$, we may choose $N_1 > 0$ so that $|a_n - L_1| < \varepsilon/(2|L_2| + 1)$ for $n \geq N_1$ and $N_2 > 0$ so that $|b_n - L_2| \leq \varepsilon/(2M + 1)$ for $n \geq N_2$. We put the ones in the denominators so that the choices make sense even when $L_2 = 0$ and $M = 0$. Then using (2.1), we see that, for $n \geq N = \max\{N_1, N_2\}$,

$$|a_n b_n - L_1 L_2| < \varepsilon/2 + \varepsilon/2 = \varepsilon.$$

To prove (iv), it will suffice to prove that the sequence $\{1/b_n\}$ converges to $1/L_2$: then part (iii) can be used to finish off the proof. In this argument, we will have to make a preliminary estimate before we can choose the N in terms of the ε. Using Lemma 2.2, we see that there is an N_1 such that, for $n \geq N_1$, $|b_n| \geq |L_2|/2$. It follows that, for $n \geq N_1$,

$$\left| \frac{1}{b_n} - \frac{1}{L_2} \right| = \left| \frac{b_n - L_2}{b_n L_2} \right| \leq \left(\frac{2}{|L_2|^2} \right) |b_n - L_2|.$$

Now, given $\varepsilon > 0$, because b_n converges to L_2, we may choose $N_2 > 0$ so that $|b_n - L_2| < \varepsilon |L_2|^2/2$ for $n \geq N_2$. Finally, to ensure that both inequalities hold, we choose $N = \max\{N_1, N_2\}$. Then for $n \geq N$, we have

$$\left| \frac{1}{b_n} - \frac{1}{L_2} \right| < \left(\frac{2}{|L_2|^2} \right) \varepsilon \frac{|L_2|^2}{2} = \varepsilon.$$

The proof is complete.

EXAMPLE 2.2:   Consider the sequence

$$a_n = (3 + 2^{-n}) \left(\frac{2n + 1}{n} \right).$$

We shall determine the limit of this sequence by applying the parts of Theorem 2.2. The factor $(3 + 2^{-n})$ can be thought of as the sum of two sequences $b_n + c_n$ where $b_n = 3$ for all n and $c_n = 2^{-n}$. Obviously, $\lim_{n \to \infty} b_n = 3$, and $\lim_{n \to \infty} c_n = \lim_{n \to \infty} 2^{-n} = 0$ by part b) of Example 2.1. Hence by part (ii) of Theorem 2.2, $\lim_{n \to \infty} (3 + 2^{-n}) = 3$. Now $(2n + 1)/n = 2 + 1/n$. Consequently, $\lim_{n \to \infty} (2n + 1)/n = 2$. Finally, by part (iii) of Theorem 2.2, $\lim_{n \to \infty} a_n = 3 \cdot 2 = 6$.

Lemma 2.4 Suppose that the sequence a_n converges to L and that, for all n, $a_n \geq 0$. Then $L \geq 0$.

Proof: For a given $\varepsilon > 0$, there is an $N > 0$ such that $|L - a_n| < \varepsilon$ for all $n \geq N$. This means that $L - a_n > -\varepsilon$ for all $n \geq N$. However, $a_n \geq 0$, so $L \geq L - a_n > -\varepsilon$. Since $\varepsilon > 0$ is arbitrary, we must have $L \geq 0$.

Theorem 2.5

    a) Let the sequence $\{a_n\}$ converge to the limit L. Then $\lim_{n \to \infty} |a_n| = |L|$.
    b) Let $\{a_n\}$ be a nonnegative sequence that converges to a. Then $\sqrt{a_n}$ converges to $\sqrt{a}$.

Proof: The proof of a) uses the reverse triangle inequality and we leave it to the reader to do in exercise 4 of this section.

    As for b), we know by Lemma 2.4 that $a \geq 0$. Thus the square root is defined. The proof of b) uses a standard device for making estimates of square roots. We want to estimate the difference $\sqrt{a_n} - \sqrt{a}$ in terms of the difference $a_n - a$. Assuming $a > 0$, we write

$$\sqrt{a_n} - \sqrt{a} = (\sqrt{a_n} - \sqrt{a})\frac{\sqrt{a_n} + \sqrt{a}}{\sqrt{a_n} + \sqrt{a}} = \frac{a_n - a}{\sqrt{a_n} + \sqrt{a}}. \tag{2.2}$$

Hence

$$|\sqrt{a_n} - \sqrt{a}| \leq \frac{|a_n - a|}{\sqrt{a}}.$$

Let $\varepsilon > 0$ be given. Because $\{a_n\}$ converges to a, we know that there is an $N > 0$ such that $|a_n - a| < \varepsilon\sqrt{a}$ for $n \geq N$.

    We use a different argument when $a = 0$. In this case, for any given $\varepsilon > 0$, it suffices to choose $N > 0$ so large that $|a_n| < \varepsilon^2$ for $n \geq N$.

Now we extend Lemma 2.4 to show that the limiting process preserves inequalities.

Theorem 2.6 Let the sequence $\{a_n\}$ converge to L_1, the sequence $\{b_n\}$ converge to L_2, and the sequence $\{c_n\}$ converge to L. If $a_n \leq c_n \leq b_n$ for all n, then $L_1 \leq L \leq L_2$.

Proof: The sequence $\{c_n - a_n\}$ has nonnegative terms and converges to $L - L_1$ by part (ii) of Theorem 2.3. Hence by Lemma 2.4, $L - L_1 \geq 0$. A similar argument shows that $L_2 \geq L$.

    We note that the hypothesis $a_n \geq 0$ for all n of Lemma 2.4 can be weakened. We only need to assume that there is an $N_0 > 0$ with $a_n \geq 0$ for $n \geq N_0$. The same is true for the hypotheses of Theorem 2.6.

    Theorem 2.6 allows us to estimate the limit of the sequence c_n if we know that it converges. The next theorem pursues the same idea to give a proof of convergence of the sequence c_n. However, we must make the stronger hypothesis that $L_1 = L_2$.

Theorem 2.7 Suppose sequences $\{a_n\}$, $\{b_n\}$, and $\{c_n\}$ with $a_n \leq c_n \leq b_n$ and $\lim_{n \to \infty} a_n = \lim_{n \to \infty} b_n = L$. Then the sequence $\{c_n\}$ converges to L.

Proof: Let $\varepsilon > 0$ be given. Because the sequence $\{a_n\}$ converges to L, there is an $N_1 > 0$ such that $|a_n - L| < \varepsilon$ for $n \geq N$. Using only one side of this inequality, we have $a_n - L > -\varepsilon$ for $n \geq N_1$. Similarly, there is an $N_2 > 0$ such that $b_n - L < \varepsilon$ for $n \geq N_2$. Since $a_n \leq c_n \leq b_n$, we have

$$-\varepsilon < a_n - L \leq c_n - L \leq b_n - L < \varepsilon$$

for $n \geq N = \max\{N_1, N_2\}$. Of course, this means that $|c_n - L| < \varepsilon$ for $n \geq N$.

> **EXAMPLE 2.3:** Consider the sequence $\{a_n\} = \{nr^n\}$ where $0 < r < 1$. We will prove that $\lim_{n \to \infty} a_n = 0$ using Theorem 2.7. Because $a_n \geq 0$, it will suffice to find a sequence $\{b_n\}$ such that $a_n \leq b_n$ and $\lim_{n \to \infty} b_n = 0$. We factor the terms a_n as $a_n = n(r^{1/2})^n(r^{1/2})^n$. Now $r^{1/2} < 1$, so we can write $r^{1/2} = 1/(1 + p)$ where $p > 0$. Using Bernoulli's inequality again, we have
>
> $$n(r^{1/2})^n = \frac{n}{(1 + p)^n} \leq \frac{n}{1 + np} \leq \frac{1}{p}.$$
>
> Hence,
>
> $$a_n \leq b_n = \left(\frac{1}{p}\right)(r^{1/2})^n .$$
>
> From Example 2.1, part c), we know that $\{(r^{1/2})^n\}$ converges to zero, and the same is true of the sequence $\{b_n\}$ by Theorem 2.3, part (i). Hence by Theorem 2.7, the sequence $\{a_n\}$ also converges to zero.

Finally, we investigate the convergence of a sequence given recursively.

> **EXAMPLE 2.4:** Let the sequence $\{a_n\}$ be given by
>
> $$a_{n+1} = \frac{3}{2}a_n(1 - a_n) \quad \text{for } n \geq 0. \tag{2.3}$$
>
> We shall start the index at $n = 0$; then a_0 must be specified. It is not immediately obvious that $\{a_n\}$ converges, or what its limit could

be if it did converge. First we need a candidate for a limit. Let us suppose that $\lim_{n\to\infty} a_n = L$ and find out what L must be. Using (2.3) and the rule for limits of products, we deduce

$$L = \lim_{n\to\infty} a_{n+1} = (3/2) \lim_{n\to\infty} a_n(1 - a_n)$$

$$= (3/2)L(1 - L).$$

Thus L must satisfy the quadratic equation,

$$(3/2)L^2 = (1/2)L,$$

which has the two solutions $L = 0$ and $L = 1/3$.

Clearly, if $a_0 = 0$, then the sequence converges to $L = 0$. Let us see if there is a choice of a_0 such that the sequence (2.3) converges to $L = 1/3$. We subtract $L = 1/3$ from both sides of (2.3) and factor the right-hand side:

$$a_{n+1} - 1/3 = -(3/2)(a_n - 2/3)(a_n - 1/3). \tag{2.4}$$

Since we are going to try to prove that $\{a_n\}$ converges to $1/3$, let us suppose that $|a_0 - 1/3| \le c/3$, where $c < 1$, and let us try to prove that a_1 is closer to $1/3$ than is a_0. In fact, if $|a_0 - 1/3| \le c/3$, then

$$|a_0 - 2/3| \le |a_0 - 1/3| + 1/3 \le (c + 1)/3.$$

It follows from (2.4), with $n = 0$, that

$$|a_1 - 1/3| \le (3/2)|a_0 - 2/3||a_0 - 1/3|$$

$$\le \left(\frac{c + 1}{2}\right)|a_0 - 1/3| \le \left(\frac{c + 1}{2}\right)\left(\frac{c}{3}\right).$$

Since $c < 1$, this implies that $|a_1 - 1/3| < |a_0 - 1/3|$.

Now, to get the convergence result, we will use induction to prove that the following inequality holds for all n. Assuming that $|a_0 - 1/3| \le c/3$, with $c < 1$, we have

$$|a_n - 1/3| \le \left(\frac{c + 1}{2}\right)^n \left(\frac{c}{3}\right). \tag{2.5}$$

We have seen that (2.5) holds for $n = 1$. Assume that (2.5) holds for some $n > 1$. First we use (2.5) to make the estimate

$$|a_n - 2/3| \le |a_n - 1/3| + 1/3$$

$$\le \left(\frac{c+1}{2}\right)^n \left(\frac{c}{3}\right) + \frac{1}{3}$$

$$\le \frac{c}{3} + \frac{1}{3} = \frac{c+1}{3}. \tag{2.6}$$

Now using (2.4) and (2.6), we find that

$$|a_{n+1} - 1/3| = (3/2)|a_n - 2/3|\,|a_n - 1/3|$$

$$\le \left(\frac{3}{2}\right)\left(\frac{c+1}{3}\right)|a_n - 1/3|$$

$$\le \left(\frac{c+1}{2}\right)\left(\frac{c+1}{2}\right)^n \left(\frac{c}{3}\right) = \left(\frac{c+1}{2}\right)^{n+1}\left(\frac{c}{3}\right).$$

Thus (2.5) is established for $n+1$. The induction argument is completed, and we see that (2.5) implies that the sequence $\{a_n\}$ converges to $1/3$.

Exercises for 2.1

1. Determine the limits of the following sequences and give an ε-N proof of convergence.

   a) $a_n = 1/(n^2 + 1)$.
   b) $a_n = (2n + 1)/(3n - 2)$.
   c) $a_n = 1/\sqrt{2n + 5}$.
   d) $a_n = \sqrt{n^2 + 1} - n$.

2. Let the sequence $\{a_n\}$ be given by

$$a_n = \sin\left(\frac{2n\pi}{3}\right)\frac{n}{n+1}.$$

Does the sequence $\{a_n\}$ converge?

3. Let $c > 1$. Show that the sequence $\{c^n\}$ does not converge.

4. Suppose the sequence $\{a_n\}$ converges to L. Use the reverse triangle inequality to show that the sequence $\{|a_n|\}$ converges to $|L|$.

5. Suppose that $\{a_n\}$ is a bounded sequence and that the sequence $\{b_n\}$ converges to zero. Show that the sequence $\{a_n b_n\}$ converges to zero.

6. Suppose that $\{a_n\}$ is a sequence with $a_n > 0$, and that the sequence of ratios $\{a_{n+1}/a_n\}$ converges to a number $r < 1$. Show that the sequence $\{a_n\}$ must converge to zero.

7. Let the sequence $\{a_n\}$ converge to A, and let the sequence $\{b_n\}$ converge to B. Let $c_n = \max\{a_n, b_n\}$. Show that the sequence $\{c_n\}$ converges to $\max\{A, B\}$.

8. Let a be an irrational real number. Use the density result Theorem 1.5 to show that there is a sequence $\{x_n\}$ of rational numbers that converges to a.

9. Let the sequence $\{a_n\}$ converge to L. Show that the sequence of averages

$$\sigma_n = \frac{a_1 + a_2 + \cdots + a_n}{n}$$

converges to L as well.

10. Let the sequence $\{a_n\}$ converge to a where $|a| < 1$. Show that the sequence $\{a_n^n\}$ converges to zero.

11. a) Show that, for any $c > 0$, the sequence $\{c^n/n!\}$ converges to zero.
    b) Use part a) to show that $\lim_{n\to\infty}(n!)^{-1/n} = 0$.

12. In this exercise, we show that, for any $p > 0$, $\lim_{n\to\infty} p^{1/n} = 1$. First assume $p > 1$ and write $x_n = p^{1/n} - 1 \geq 0$. Then use Bernoulli's inequality on $p = (1 + x_n)^n$ to show that x_n tends to zero. Finally, treat the cases $0 < p \leq 1$.

13. Let the sequence $\{a_n\}$ be defined recursively by

$$a_{n+1} = a_n(3/2 - a_n), \quad n \geq 0.$$

Follow the steps of Example 2.4 to determine the limits of $\{a_n\}$ and show that $\{a_n\}$ converges.

2.2 Criteria for Convergence

In Section 2.1, we discussed the convergence of sequences without mention of the least upper bound property of the real numbers. In this section, when we look for tests for convergence of a sequence, we shall see a deep connection between these two concepts.

Many sequences have a special property that allows us to conclude that they converge, without knowing the limit.

Definition 2.3 A sequence $\{a_n\}$ is monotone increasing if $a_{n+1} \geq a_n$ for all n. A sequence is *monotone* decreasing if $a_{n+1} \leq a_n$ for all n. We say that a sequence is monotone if it is either monotone increasing, or if it is monotone decreasing.

Theorem 2.8 A bounded monotone sequence converges.

Proof: Let the sequence $\{a_n\}$ be monotone increasing and bounded. By the least upper bound property, the set of values $S = \{a_n : n \in \mathbb{N}\}$ has a least upper bound $L \in R$. We claim that $\{a_n\}$ converges to L. Let $\varepsilon > 0$ be given. From our discussion of the least upper bound, we know that there must be an element, $a_N \in S$ with $L - \varepsilon < a_N \leq L$. By the monotone increasing assumption, we have

$$L - \varepsilon < a_N \leq a_n \leq L$$

for all $n \geq N$. The proof is complete. The argument to show that a bounded, monotone decreasing, sequence must converge is similar.

We make two comments about the theorem. First, we can weaken the hypothesis that $a_{n+1} \geq a_n$ for all n. It is enough to assume that there is some N_0 such that $a_{n+1} \geq a_n$ for all $n \geq N_0$. Second, it may be difficult to determine the least upper bound L, but once we know that the limit of the sequence exists, we can sometimes determine it by other means.

EXAMPLE 2.5:

a) Let $0 < r < 1$ and define

$$a_n = \sum_{k=1}^{n} k r^k.$$

Since $a_{n+1} = a_n + (n+1)r^{n+1}$, we see that the sequence is monotone increasing. To show that it converges, we need to show that it is bounded. In Example 2.3, we showed that, if $0 < r < 1$, then $k r^k \leq (1/p)(r^{1/2})^k$ where $p = -1 + 1/r$. Applying this estimate to each term in the sum, we deduce that

$$a_n \leq \frac{1}{p} \sum_{k=1}^{n} (r^{1/2})^k.$$

The sum is now just a geometric series that can be summed with the formula

$$\sum_{k=1}^{n} c^k = \frac{1 - c^{n+1}}{1 - c} \leq \frac{1}{1 - c}.$$

Hence, with $c = r^{1/2}$, the sequence $\{a_n\}$ can be bounded

$$a_n \leq \frac{(1/p)}{1 - r^{1/2}}.$$

By Theorem 2.8, the sequence converges, although we do not know the value of the limit.

In Chapter 8, we will see that the sequence $a_n = \sum_{k=1}^{n} k^p r^k$ is bounded for any $p > 0$ and any $0 < r < 1$.

b) Theorem 2.8 is useful in proving the convergence of certain sequences that are given recursively. Let $A > 0$. The sequence defined by

$$a_{n+1} = (1/2)(a_n + A/a_n) \tag{2.7}$$

is used to calculate approximations to $\sqrt{A}$. We shall show that, for $n \geq 2$, $\{a_n\}$ is monotone decreasing for any choice of $a_1 > 0$.

Recall the inequality from Section 1.1 (exercise 5) that related the geometric mean and the arithmetic mean of two numbers $a, b > 0$:

$$\sqrt{ab} \leq \frac{(a + b)}{2}.$$

In this inequality, put $a = a_n$ and $b = A/a_n$ to deduce

$$\sqrt{A} \leq (1/2)\left(a_n + \frac{A}{a_n}\right) = a_{n+1}.$$

Thus for $n \geq 2$, $a_n \geq \sqrt{A}$. This implies that a_n is monotone decreasing for $n \geq 2$. In fact, for $n \geq 2$, $A \leq a_n^2$, so that

$$a_{n+1} = (1/2)\left(a_n + \frac{A}{a_n}\right) \leq \frac{a_n + a_n}{2} = a_n.$$

Hence by Theorem 2.8, the sequence a_n converges to some number $L \geq \sqrt{A}$. Taking the limit on both sides of (2.7), we deduce that L satisfies $L = (1/2)(L + A/L)$ and so $L = \sqrt{A}$.

It is of considerable practical importance to have sequences of approximations that converge very rapidly. The sequence (2.7) has this property. To experience this rapid convergence, let us use the sequence to approximate $5 = \sqrt{25}$. We start with $a_1 = 6$, and then find that

$$\begin{aligned}
a_1 &= 6 \\
a_2 &= 5.08333333333333 \\
a_3 &= 5.00068306010929 \\
a_4 &= 5.00000004665074 \\
a_5 &= 5.00000000000000
\end{aligned}$$

These calculations are done in double precision (15 digits). Thus in only four iterations we have obtained full machine precision.

Can we predict this behavior? Knowing that the limit of the sequence (2.7) is $\sqrt{A}$, a short calculation shows that, for $n \geq 2$,

$$\begin{aligned}
a_{n+1} - \sqrt{A} &= \frac{1}{2a_n}(a_n^2 - 2a_n\sqrt{A} + A) \\[2mm]
&= \frac{(a_n - \sqrt{A})^2}{2a_n} \\[2mm]
&\leq \frac{(a_n - \sqrt{A})^2}{2\sqrt{A}}.
\end{aligned}$$

Thus, if the difference $a_n - \sqrt{A} = 10^{-p}$, the difference $a_{n+1} - \sqrt{A} \leq 10^{-2p}/(2\sqrt{A})$. The number of correct digits in the approximation

a_n roughly doubles with each iteration. This kind of convergence is called *quadratic convergence*. We shall discuss it more fully when we study Newton's method.

c) At one time, there were some computing machines that did not have division. However, it suffices to be able to approximate the reciprocal of a number $A > 0$, using only multiplication and subtraction. Division by A is then accomplished by multiplying by the approximation to $1/A$. The approximation to $1/A$ is provided by the following sequence.

Let $A > 0$ and define the sequence

$$a_{n+1} = 2a_n - Aa_n^2.$$

In the exercises you will see that, if $0 < a_1 < 2/A$, then a_n is monotone increasing for $n \geq 2$ and converges quadratically to $1/A$.

We have seen (Lemma 2.1) that every convergent sequence is bounded. However, the example $a_n = (-1)^n$ shows that the converse is not true. Nevertheless, the even terms a_{2n} converge to 1, while the odd terms a_{2n+1} converge to -1. This suggests that there may be a partial converse to Lemma 2.1.

Definition 2.4 Let $\{a_n\}$ be a sequence of real numbers, and let $n_1 < n_2 < n_3 < \dots$ be a strictly increasing sequence of positive integers. Then $\{a_{n_k}\}$ is said to be a *subsequence* of the sequence $\{a_n\}$.

EXAMPLE 2.6:

a) If $\{a_n\}$ is a sequence, then the terms with odd indices form a subsequence, as do the terms with even indices.

b) The interval between n_k and n_{k+1} does not need to be uniform. Indeed, we could take $n_k = k^2$ so that $n_1 = 1$, $n_2 = 4$, $n_3 = 9$, and so on.

Theorem 2.9 Let $\{a_n\}$ be a sequence that converges to L. Then every subsequence a_{n_k} also converges to L.

Proof: Let $\varepsilon > 0$ be given. Because $\{a_n\}$ converges to L, we know that there is an $N > 0$ such that $|a_n - L| < \varepsilon$ for $n \geq N$. Because $\{a_{n_k}\}$ is a subsequence, there is a $K > 0$ such that $n_k \geq N$ for $k \geq K$. Consequently, $|a_{n_k} - L| < \varepsilon$ for $k \geq K$, which means that the subsequence $\{a_{n_k}\}$ converges to L as well.

The convergence of monotone sequences is a consequence of the least upper bound property, which in turn is an expression of the completeness of the real numbers. Another, very intuitive, expression of the completeness property is given in the next theorem. It shows that the real numbers do not have any gaps.

Theorem 2.10 (nested intervals) Let $I_n = [a_n, b_n]$ be a family of closed intervals, $n \in \mathbb{N}$. We assume that the intervals are *nested*, that is, $I_{n+1} \subset I_n$ for all n. Then $\cap_{n=1}^{\infty} I_n$ is nonempty.

Proof: The intervals I_n are all nonempty because they at least contain their endpoints. Next note that $a_n \leq b_m$ for all m and n. In fact, if for some pair of indices n and m we have $a_n > b_m$, then $I_n \cap I_m = \phi$, which is contrary to our assumption that the intervals are nested and nonempty. Thus each b_m is an upper bound for the sequence $\{a_n\}$. Since $I_{n+1} \subset I_n$, we have $a_n \leq a_{n+1}$ for all n. In other words, $\{a_n\}$ is a bounded, increasing monotone sequence. It follows that $a = \sup a_n$ exists and that $a \leq b_m$ for all m. We also have $b_{n+1} \leq b_n$ for all n. Thus $\{b_n\}$ is a monotone decreasing sequence that is bounded below by a. Hence $b = \inf b_n$ exists, and $b \geq a$. If $a \leq x \leq b$, then $x \in I_n$ for all n, and hence $[a, b] \subset \cap_{n=1}^{\infty} I_n$. The proof is complete.

Remarks Note that the result is not true if we relax the hypothesis that each interval I_n is closed. For example, the intersection of the nested half-open intervals $I_n = (0, 1/n]$ is empty.

It is also important to note that Theorem 2.10 is not true in the rational numbers. Let p_n be a monotone increasing sequence of rational numbers with $\sup p_n = \sqrt{2}$, and let q_n be a monotone decreasing of rational numbers with $\inf q_n = \sqrt{2}$. Let $I_n = \{x \in \mathbb{Q} : p_n \leq x \leq q_n\}$. Then each $I_n \neq \phi$ and for all n, $I_{n+1} \subset I_n$, so the intervals I_n are nested and nonempty. However, $\cap_{n=1}^{\infty} I_n = \{\sqrt{2}\}$, which is not a rational number.

The Bolzano-Weierstrass theorem is an important tool for deriving properties of continuous functions and of compact sets.

Theorem 2.11 (Bolzano-Weierstrass) Let $\{a_n\}$ be a bounded sequence. Then $\{a_n\}$ has a convergent subsequence.

Proof: Let $\{a_n\}$ be a bounded sequence. First, suppose that the sequence only takes on a finite number of distinct values. In this case, at least one of the values must be taken on for an infinite sequence of indices $\{n_k\}$. The subsequence $\{a_{n_k}\}$ is trivially convergent.

Now, suppose that the sequence takes on an infinite number of distinct values. Since the sequence is bounded, there is an $M > 0$ such that all of the values of the sequence are contained in the closed interval $[-M, M]$. Let $I_0 = [-M, M]$. We shall use a bisection procedure to select a convergent subsequence. Cut the interval I_0 into closed subintervals $[-M, 0]$ and $[0, M]$. At least one of these two intervals must contain an infinite number of the a_n. Choose one and call it I_1. Repeat the bisection and choice on I_1 and call the chosen interval I_2. In this way we construct a sequence of closed, nested intervals: $\ldots I_{n+1} \subset I_n \subset \ldots \subset I_0$. By Theorem 2.10, $\cap_{n=1}^{\infty} I_n \neq \phi$. In fact, this intersection must consist of a single point because the length of I_n is $2^{-(n+1)}M$. Let the point of intersection be L. We claim that a subsequence of $\{a_n\}$ converges to L. We may choose an index n_1 such that $a_{n_1} \in I_1$ because I_1 contains an infinite number of the a_n. Now suppose that we have chosen indices $n_1 < n_2 < \ldots < n_k$ such that $a_{n_k} \in I_k$. Because there are an infinite number of $a_n \in I_{k+1}$, we may choose an index $n_{k+1} > n_k$ such that $a_{n_{k+1}} \in I_{k+1}$. Because $L \in I_{n_k}$, we have $|L - a_{n_k}| \leq 2^{-n_k}$. Thus for a given $\varepsilon > 0$, there is a $K > 0$ such that $|L - a_{n_k}| < \varepsilon$ for $k \geq K$. The proof is complete.

Corollary Let $I = [a, b]$ be a closed bounded interval. If $a_n \in I$ is a sequence, then there is subsequence a_{n_k} that converges to a point $L \in I$.

Proof: Theorem 2.11 already tells us that a_n has a subsequence a_{n_k} that converges to some real number L. We only need to check that $L \in I$. However, because the points a_{n_k} satisfy $a \leq a_{n_k} \leq b$, it follows by Theorem 2.6 that the limit L satisfies the same inequality.

EXAMPLE 2.7:

a) The sequence $a_n = (-1)^n(1 - 1/n)$ has two convergent subsequences. The terms with n even converge to 1 while the terms with n odd converge to -1.

b) Let the sequence $\{b_n\}$ be given by

$$b_n = \begin{cases} 1/n, & \text{if } n \text{ is not a square} \\ 1 + k^{-2}, & \text{if } n = k^2. \end{cases}$$

This sequence has two subsequences, one converging to zero, and the other converging to 1.

EXAMPLE 2.8:   This example is more substantial.

The sequence $a_n = \sin(n)$ is bounded, and hence by Theorem 2.11, has at least one convergent subsequence. We claim that, for each number $z \in [-1, 1]$, there is a subsequence n_k such that $\lim_{k \to \infty} \sin(n_k) = z$.

In proving this assertion, we shall use the fact that, for all x, y, $|\sin x - \sin y| \leq |x - y|$. This follows from the mean value theorem, which is proved in Chapter 4.

The first step is to show that each $\theta \in [0, 2\pi]$ may be approximated to any accuracy by a number of the form $n - 2\pi q$ where n is a positive integer, and q is an integer. Let $\theta \in [0, 2\pi]$ and $\varepsilon > 0$ be given. Choose a positive integer N such that $1/N < \varepsilon$. Divide the interval $[0, 2\pi]$ into N equal subintervals, $I_j = [2\pi(j - 1)/N, 2\pi j/N]$, $j = 1, \ldots, N$. The number θ lies in one of them, say I_m.

For $j = 0, 1, \ldots, N$, let q_j be largest positive integer such that $2\pi q_j < j$, and let $r_j = j - 2\pi q_j$. (Note that we can never have $2\pi q_j = j$ because $2\pi q_j$ is irrational.) The $N + 1$ numbers r_j are all distinct (why?) with $0 < r_j < 2\pi$. Hence at least two of them must land in one of the subintervals I_j, say r_k and r_l. We assume that $l > k$.

Now we must consider two cases. First assume that $r_l > r_k$ so that $0 < r_l - r_k < \varepsilon$. Then there is a positive integer t such that $t(r_l - r_k) \in I_m$ whence $|t(r_l - r_k) - \theta| \leq \varepsilon$. Let $n = t(l - k)$, and set $q = t(q_l - q_k)$. Then $n - 2\pi q = t(r_l - r_k)$, which implies that

$$|\sin n - \sin \theta| \;=\; |\sin(n - 2\pi q) - \sin \theta| \tag{2.8}$$

$$=\; |\sin(t(r_l - r_k)) - \sin \theta|$$

$$\leq\; |t(r_l - r_k) - \theta| \leq \varepsilon. \tag{2.9}$$

Next assume $r_l < r_k$ so that $-\varepsilon \leq r_l - r_k < 0$. Then there is a positive integer s such that $s(r_l - r_k) + 2\pi \in I_m$. Consequently, $|s(r_l - r_k) + 2\pi - \theta| \leq \varepsilon$. Let $n = s(l - k)$ (again a positive integer) and let $q = s(q_l - q_k)$ so that $n - 2\pi q = s(r_l - r_k)$. This implies that

$$
\begin{aligned}
|\sin n - \sin \theta| &= |\sin(s(r_l - r_k)) - \sin \theta| \\
&= |\sin(s(r_l - r_k) + 2\pi) - \sin \theta| \\
&\leq |s(r_l - r_k) + 2\pi - \theta| \leq \varepsilon
\end{aligned}
\tag{2.10}
$$

Finally we construct the subsequences. Let $z \in [-1, 1]$ be given, and let $\theta \in [0, 2\pi]$ with $\sin \theta = z$. Then, for each $\varepsilon > 0$, there is a positive integer n such that

$$
|\sin(n) - z| \leq \varepsilon.
\tag{2.11}
$$

In fact, there must be an infinite sequence of positive integers such that (2.10) is satisfied. If there were only a finite number, $n_1, \ldots, n_K$, then, for $\varepsilon = \varepsilon_0 < \min_{j=1,K} |\sin(n_j) - z|$, it would not be possible to find a positive integer to satisfy (2.10). Now we take $\varepsilon = 1/k$, $k = 1, 2, \ldots$. Suppose that we have been able to choose $n_1 < n_2 < \cdots < n_k$ so that $|\sin(n_j) - z| < 1/j$ for $j = 1, \ldots, k$. Because there are an infinite number of $n \in \mathbb{N}$ with $|\sin(n) - z| < 1/(k+1)$, we may choose $n_{k+1} > n_k$ with $|\sin(n_{k+1}) - z| < 1/(k + 1)$. This finishes the proof of the assertion.

Exercises for 2.2

1. Let $\{a_n\}$ be a monotone sequence. Show that $\{a_n\}$ converges if and only if the sequence $\{a_n^2\}$ converges. Is the result true if we omit the hypothesis that $\{a_n\}$ is monotone?
2. Let the sequence $\{a_n\}$ be given recursively by the formula

$$
a_1 = 2, \quad a_{n+1} = \frac{2a_n + 4}{3}.
$$

Show that $\{a_n\}$ is monotone increasing and bounded above. Find the limit of $\{a_n\}$.

3. Let the sequence $s_n = a_1 + a_2 + \cdots + a_n$ where $a_k = 1/(k2^k)$. Show that s_n converges.

4. Let the sequence $\{a_n\}$ be given recursively by

$$a_{n+1} = \frac{1}{4} + a_n^2, \quad a_1 = 0.$$

Show that the sequence $\{a_n\}$ converges and find the limit.

5. Discuss the convergence of the sequence $\{y_n\}$ where $y_1 = 1$ and, for $n \geq 1$, $y_{n+1} = \sqrt{y_n + 2}$.

6. Let $a_1 = 1$, $a_2 = 3$, and for $n \geq 3$, set

$$a_n = \frac{a_{n-1} + a_{n-2}}{2}.$$

a) Use induction to show that $2 \leq a_n \leq 3$ for all $n \geq 2$.

b) Show that each of the subsequences $\{a_{2k}\}$ and $\{a_{2k+1}\}$ is monotone and converges to the same limit L.

c) Find a formula for a_n, and use it to determine the limit of the sequence.

7. Let $A > 0$. The following recursive sequence is used to approximate $1/A$:

$$x_{n+1} = x_n(2 - Ax_n).$$

a) Show that if $0 < x_1 < 2/A$, then x_n is monotone increasing for $n \geq 2$ with $0 < x_n \leq 1/A$. Show that the sequence converges to $1/A$.

b) Take $A = 5$ and $x_1 = 0.3$. Compute the iterates x_n for $n = 2, 3, 4, 5$. What is the error $|x_5 - 1/A|$?

c) Show that, as in Example 2.5 b), the convergence is quadratic.

8. Consider the sequence a_n where

$$a_n = \left(\frac{n}{n+1}\right) \sin\left(\frac{2n\pi}{3}\right).$$

What are the convergent subsequences, and what are their limits?

Limit Superior and Limit Inferior

Let $\{a_n\}$ be a bounded sequence. By the Bolzano-Weierstrass theorem, $\{a_n\}$ has at least one convergent subsequence. Let E be the set of all limits of subsequences of $\{a_n\}$. Then

$$A^* = \sup E$$

is called the *limit superior* of the sequence $\{a_n\}$, and

$$A_* = \inf E.$$

is called the *limit inferior* of the sequence $\{a_n\}$. A common notation is

$$A^* = \lim_{n\to\infty} \sup\, a_n, \quad A_* = \lim_{n\to\infty} \inf\, a_n.$$

9. Let $\{a_n\}$ be a bounded sequence.
   a) Show that A^* and A_* belong to E.
   b) Show that, for each $\varepsilon > 0$, there is an $N > 0$ such that $a_n \le A^* + \varepsilon$ for $n \ge N$.
   c) Show that, for each $\varepsilon > 0$, there is a subsequence a_{n_k} with $a_{n_k} > A^* - \varepsilon$.

10. Let $\{a_n\}$ be a bounded sequence. Show that the sequence $\{a_n\}$ converges if and only if $\lim\sup_{n\to\infty} a_n = \lim\inf_{n\to\infty} a_n$.

11. Let $\{a_n\}$ and $\{b_n\}$ be sequences of positive numbers with

$$\lim_{n\to\infty} \sup \frac{a_n}{b_n} = M.$$

Show that there is a constant $\tilde{M}$ such that $a_n \le \tilde{M} b_n$ for $n \ge 1$.

2.3 Cauchy Sequences

The Cauchy criterion for convergence is very important in analysis. It provides us with a test for convergence that does not require a knowledge of the limit, and it does not require that the sequence be monotone. It is also used

in one of the methods of construction of the real numbers from the rational
numbers.

Definition 2.5 A sequence $\{a_n\}$ is said to be a *Cauchy* sequence if, for each $\varepsilon > 0$,
there is an $N > 0$ such that, for all indices $n, m \geq N$, $|a_n - a_m| < \varepsilon$.

The crucial element of this definition is that the indices m and n are independent. They can be close together, or more importantly, far apart. Sometimes
the Cauchy property of a sequence is expressed by the statement

$$\lim_{m,n\to\infty} |a_m - a_n| = 0.$$

The connection between a sequence being Cauchy and being convergent is
not hard to see in one direction. If a sequence converges, then the terms of the
sequence are tending to a limit L, and so it seems reasonable that they should
be getting close to each other as well. To deal with the opposite implication,
we need to make an observation about Cauchy sequences.

Lemma 2.12 A Cauchy sequence is bounded.

Proof: Let $\{a_n\}$ be a Cauchy sequence. To prove that $\{a_n\}$ is bounded, we shall
not need the full strength of the definition. Let $\varepsilon = 1$. Then there is an $N > 0$
such that $|a_n - a_m| \leq 1$ for all $n, m \geq N$. Thus for all $n \geq N$,

$$|a_n| = |a_n - a_N + a_N| \leq |a_n - a_N| + |a_N| \leq 1 + |a_N|.$$

Let $M = \max\{|a_1|, \ldots, |a_{N-1}|, 1 + |a_N|\}$. It follows that $|a_n| \leq M$ for all n.

Now we are ready to make the key link between the notion of a convergent
sequence and that of a Cauchy sequence.

Theorem 2.13 A sequence $\{a_n\}$ is a Cauchy sequence if and only if it converges.

Proof: Suppose first that $\{a_n\}$ is a convergent sequence with limit L. Let $\varepsilon > 0$
be given. There is an $N > 0$ such that $|a_n - L| < \varepsilon/2$ for $n \geq N$. Then for
$m, n \geq N$, we have

$$|a_n - a_m| \leq |a_n - L| + |a_m - L| < \varepsilon/2 + \varepsilon/2 = \varepsilon.$$

The more interesting and deeper implication goes the other way. Now suppose that $\{a_n\}$ is a Cauchy sequence. By Lemma 2.12, the sequence $\{a_n\}$ is

bounded. The Bolzano-Weierstrass Theorem (Theorem 2.11) implies that $\{a_n\}$ has a subsequence, $\{a_{n_k}\}$, that converges to some real number L. We shall use the Cauchy condition to show that the whole sequence $\{a_n\}$ converges to L. We can estimate the difference $|a_n - L|$ by the inequality

$$|a_n - L| = |a_n - a_{n_k} + a_{n_k} - L| \leq |a_n - a_{n_k}| + |a_{n_k} - L|. \tag{2.12}$$

Let $\varepsilon > 0$ be given. Using the full strength of the Cauchy condition, we know that there is an $N > 0$ such that $|a_n - a_m| < \varepsilon/2$ for all $n, m \geq N$. Because the subsequence $\{a_{n_k}\}$ converges to L, there is a number $K > 0$ such that $|a_{n_k} - L| < \varepsilon/2$ for $k \geq K$. By choosing K even larger if necessary, we can be sure that $n_K \geq N$. We fix $k = K$ in (2.12) and deduce that

$$|a_n - L| \leq |a_n - a_{n_K}| + |a_{n_K} - L| < \varepsilon/2 + \varepsilon/2 = \varepsilon.$$

The proof is complete.

Theorem 2.13 does not hold in the system of rational numbers. Indeed, a sequence of rational numbers $\{a_n\}$ that converges to $\sqrt{2}$ satisfies the Cauchy criterion, but it does not converge to a rational number. One method of construction of the real numbers from the rationals does so by "adding" the limits of Cauchy sequences of rational numbers. This method of construction is used in a more abstract context to define the "completion" of a metric space.

EXAMPLE 2.9:

a) We again emphasize that in the Cauchy condition, the indices m and n are independent. For example, it is not sufficient that the successive differences $|a_n - a_{n+1}|$ tend to zero as n tends to infinity. The sequence $a_n = \sqrt{n}$ is unbounded, and clearly does not converge. However, we see that

$$|a_n - a_{n+1}| = |\sqrt{n} - \sqrt{n+1}|$$

$$= \frac{1}{\sqrt{n} + \sqrt{n+1}}$$

$$\leq \frac{1}{\sqrt{n}},$$

which tends to zero as n tends to infinity.

b) A typical way in which a sequence is shown to be Cauchy is to establish an estimate of $|a_n - a_{n+k}|$ that goes to zero as n tends to infinity, independent of k. Consider the sequence a_n given recursively, by

$$a_{n+1} - a_n = (-1)^n |a_n - a_{n-1}|^2. \qquad (2.13)$$

For convenience, we start with $n = 0$. We must specify a_0 and a_1, and we assume that $|a_1 - a_0| < 1$. We observe that

$$a_2 - a_1 = -|a_1 - a_0|^2$$
$$a_3 - a_2 = |a_2 - a_1|^2 = |a_1 - a_0|^4.$$

Using an induction argument, it is not difficult to show that, for all $n \geq 1$,

$$|a_{n+1} - a_n| = |a_1 - a_0|^{2^n}.$$

However, $2^n \geq 2n$ for $n \geq 2$ and $|a_1 - a_0| < 1$. Hence we see that for $n \geq 2$,

$$|a_{n+1} - a_n| \leq |a_1 - a_0|^{2n}.$$

Now to estimate the difference $a_{n+k} - a_n$, we express it in terms of successive differences and use the triangle inequality:

$$|a_{n+k} - a_n| \leq |a_{n+k} - a_{n+k-1}| + |a_{n+k-1} - a_{n+k-2}| + \cdots + |a_{n+1} - a_n|$$
$$\leq |a_1 - a_0|^{2(n+k-1)} + \cdots + |a_1 - a_0|^{2n}.$$

We let $b = |a_1 - a_0|^2 < 1$. Then we have

$$|a_{n+k} - a_n| \leq b^n (b^{k-1} + \cdots + b + 1)$$

$$\leq b^n \frac{1 - b^k}{1 - b} \leq \frac{b^n}{1 - b}.$$

Because $b < 1$, we see that $|a_{n+k} - a_n|$ tends to zero as n tends to infinity, and the estimate is independent of k. Thus when $b = |a_0 - a_1|^2 < 1$, the sequence (2.8) is a Cauchy sequence, and therefore by Theorem 2.13, it must converge.

Exercises for 2.3

1. Let $\{a_n\}$ and $\{b_n\}$ be Cauchy sequences. Using the definition of a Cauchy sequence, show that

   a) the sum sequence $\{a_n + b_n\}$ is a Cauchy sequence; and

   b) the product sequence $\{a_n b_n\}$ is a Cauchy sequence.

2. Let $\{a_n\}$ be a sequence such that for some constant c, $0 < c < 1$,

$$|a_{n+1} - a_n| \leq \frac{c^n}{n}|a_1 - a_0|.$$

   Show that $\{a_n\}$ is a Cauchy sequence, and hence converges.

3. Let $x_n > 0$ with $\lim_{n \to \infty} x_n = L \neq 0$. Suppose that for some $p > 0$, $f : \mathbb{R} \to \mathbb{R}$ satisfies

$$|f(x) - f(y)| \leq \frac{|x - y|^p}{|x| + |y|}$$

   Show that the sequence $f(x_n)$ converges.

4. Let a_n be a monotone increasing sequence, that is bounded above. Show directly, without using Theorems 2.8 and 2.13, that a_n is a Cauchy sequence.

Chapter 3

Continuity

Chapter Overview

Limits of functions are first stated in terms of sequences and then in the (ε, δ) formulation. Continuous functions are defined, and it is shown that sums, products, quotients, and compositions of continuous functions are again continuous. The Bolzano-Weierstrass theorem is used to show that continuous functions on a closed bounded interval attain a maximum and a minimum, and are uniformly continuous. The important computational question of how to find the minimum of a function efficiently, without using derivatives, is addressed with the method of golden-section search. A pseudo-code that can be implemented on a computer is given. Finally, the intermediate value theorem is proved using the method of bisection.

3.1 Limits of Functions

We first define the limit of a function in terms of sequences.

Definition 3.1 We say that a is a *limit point* of a set $A \subset \mathbb{R}$ if there is a sequence $x_n \in A \backslash \{a\}$ such that $\lim_{n \to \infty} x_n = a$.

For example, if the set $A = \{0 < x < 1\} \cup \{2\}$, the point 2 is not a limit point of A, even though it belongs to A. On the other hand, every point in the interval $0 \le x \le 1$ is a limit point of A.

Definition 3.2 Let $f : D(f) \to \mathbb{R}$ and let a be a limit point of $D(f)$. We say that $\lim_{x \to a} f(x) = L$ if there is a number L such that for *every* sequence $x_n \in D(f) \setminus \{a\}$ that converges to a, we have

$$\lim_{n \to \infty} f(x_n) = L.$$

It is important to note that the function f need not be defined at the point a. Indeed, in many important cases, f is not defined at a.

EXAMPLE 3.1:

a) Let $g(x) = \sqrt{1 - x^2}$. $D(g) = \{-1 \leq x \leq 1\}$. We show that, for each $a \in [-1, 1]$, $\lim_{x \to a} g(x) = g(a)$. If $x_n \in [-1, 1]$ is a sequence that converges to $a \in [-1, 1]$, we know from Theorem 2.3 that $1 - x_n^2$ converges to $1 - a^2$. Then by Theorem 2.5, it follows that $\sqrt{1 - x_n^2}$ converges to $\sqrt{1 - a^2}$. Hence $\lim_{x \to a} \sqrt{1 - x^2} = \sqrt{1 - a^2}$.

b) Let $f(x) = (x^2 + 2x)/(x^3 + x^2 + 3x)$ for $x \neq 0$. We show that $\lim_{x \to 0} f(x) = 2/3$. The function is not defined at $x = 0$ and both numerator and denominator tend to zero as x tends to zero. We cannot use the rules of Theorem 2.3 immediately. However, for $x \neq 0$, we may cancel out a factor of x and see that

$$f(x) = \frac{x^2 + 2x}{x^3 + x^2 + 3x} = \frac{x + 2}{x^2 + x + 3}.$$

Now let x_n be any sequence, $x_n \neq 0$, such that $\lim_{n \to \infty} x_n = 0$. Then

$$f(x_n) = \frac{x_n + 2}{x_n^2 + x_n + 3}. \tag{3.1}$$

Since the sequence x_n approaches zero, and the denominator of the fraction on the right side of (3.1) tends to 3, we can use Theorem 2.3 to deduce that

$$\lim_{n \to \infty} f(x_n) = \lim_{n \to \infty} \frac{x_n + 2}{x_n^2 + x_n + 3} = \frac{2}{3}.$$

Hence

$$\lim_{x \to 0} f(x) = \lim_{x \to 0} \frac{x^2 + 2x}{x^3 + x^2 + 3x} = \frac{2}{3}.$$

EXAMPLE 3.2:

a) We show that

$$\lim_{x \to 0} \sin x = 0.$$

In Example 5.2, we show that

$$0 \le \sin \le x \quad \text{for } x \ge 0$$

and

$$x \le \sin x \le 0 \quad \text{for } x \le 0.$$

Consequently $|\sin x| \le |x|$ for all x. It follows from Theorem 2.7 that

$$\lim_{x \to 0} \sin x = 0.$$

b) We show that

$$\lim_{x \to 0} \frac{\sin x}{x} = 1.$$

In Example 5.2, the more detailed estimates are made:

$$x - \frac{x^3}{6} \le \sin x \le x \quad \text{for } x \ge 0$$

and

$$x \le \sin x \le x - \frac{x^3}{6} \quad \text{for } x \le 0.$$

Taking care with the signs, we deduce that

$$1 - \frac{x^2}{6} \le \frac{\sin x}{x} \le 1 \quad \text{for } x \ne 0.$$

It follows, again by Theorem 2.7, that b) is true.

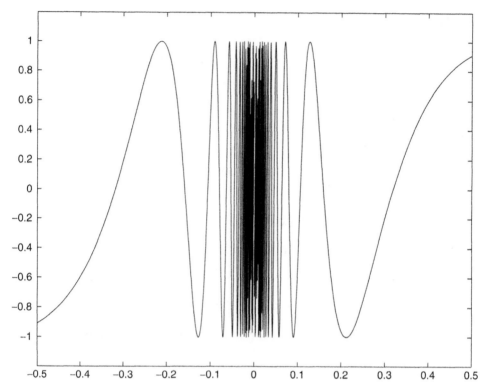

Figure 3.1 Graph of $\sin(1/x)$ showing ever more rapid oscillations near $x = 0$

EXAMPLE 3.3: The $\lim_{x \to 0} \sin(1/x)$ does not exist. In Figure 3.1 we see that the peaks and valleys of the graph come more and more rapidly as x approaches 0.

A simple choice of x_n, which hits the peaks and valleys, is $x_n = ((2n + 1)\pi/2)^{-1}$. We then have

$$\sin(1/x_n) = \sin((2n + 1)\pi/2) = \begin{cases} 1 & \text{for } n \text{ even} \\ -1 & \text{for } n \text{ odd} \end{cases}.$$

Thus x_n converges to 0 as n increases, but clearly $\sin(1/x_n)$ does not converge.

On the other hand, consider the function $f(x) = x \sin(1/x)$. Since $|f(x)| \le |x|$, it is obvious that $\lim_{x \to 0} f(x) = 0$.

An alternate, equivalent, formulation of the limit of a function is stated in terms of the dread ε's and δ's.

Theorem 3.1 Let $f : D(f) \to \mathbb{R}$ and let a be a limit point of $D(f)$. The following are equivalent:

(i) $\lim_{x \to a} f(x) = L$.
(ii) There is a number L such that, for each $\varepsilon > 0$, there is a $\delta > 0$ such that $|f(x) - L| < \varepsilon$ for all $x \in D(f)$ with $0 < |x - a| < \delta$.

Remark It is important to note in (ii) that we require $|x - a| > 0$. The function need not be defined for $x = a$.

Proof: First we show that (i) implies (ii) by showing that, if (ii) is not true, then (i) is not true. The (ε, δ) condition is negated as follows: No matter what number we choose for L, there is some $\varepsilon_0 > 0$ such that, for each $\delta > 0$, there is some $x \in D(f)$ with $0 < |x - a| < \delta$ and $|f(x) - L| \geq \varepsilon_0$. Now let L be any number. We choose a sequence of δs, $\delta_n = 1/n$, $n \in N$. For each n we are supposing that there is a point $x_n \in D(f)$ with $0 < |x_n - a| < \delta_n = 1/n$ and $|f(x_n) - L| \geq \varepsilon_0$. Now it is clear that the sequence x_n converges to a, but the sequence $f(x_n)$ does not converge to L. Thus (i) does not hold.

Now suppose that the (ε, δ) condition holds. Thus, for each $\varepsilon > 0$, there is $\delta > 0$ such that $|f(x) - L| < \varepsilon$ if $x \in D(f)$ and $0 < |x - a| < \delta$. Let $\{x_n\}$ be a sequence of points in $D(f)$ such that x_n converges to a. There must be an $N \in N$ such that, for $n \geq N$, we have $|x_n - a| < \delta$. However, this means that for $n \geq N$, $|f(x_n) - L| < \varepsilon$, or in other words, the sequence $\{f(x_n)\}$ converges to L. Thus (i) holds.

The expected rules for combining limits algebraically are stated in the next theorem.

Theorem 3.2 Suppose $f : A \to \mathbb{R}$ and $g : A \to \mathbb{R}$ and that a is a limit point of A. If $\lim_{x \to a} f(x) = L_1$ and $\lim_{x \to a} g(x) = L_2$, then

(i) For any $c \in \mathbb{R}$, $\lim_{x \to a} cf(x) = cL_1$.
(ii) $\lim_{x \to a}(f + g)(x) = L_1 + L_2$.
(iii) $\lim_{x \to a}(fg)(x) = L_1 L_2$.
(iv) If $g(x) \neq 0$ for $x \in A$, then $\lim_{x \to a}(f/g)(x) = L_1/L_2$.

Proof: Using the sequential Definition 3.2 of limit, each of these results follows immediately from Theorem 2.3 for sequences. To give a feeling for ε, δ proofs, we prove (ii) using the (ε, δ) condition. Let $\varepsilon > 0$ be given. Now $\lim_{x \to a} f(x) = L_1$ means that there is a $\delta_1 > 0$ such that $|f(x) - L_1| < \varepsilon/2$ for all $x \in A$ with $0 < |x - a| < \delta_1$. Similarly, there is a $\delta_2 > 0$ such that $|g(x) - L_2| < \varepsilon/2$ for all $x \in A$ with $0 < |x - a| < \delta_2$. At this point in the proof of Theorem 2.3 (ii), we let $N = \max\{N_1, N_2\}$. In an analogous manner, we let $\delta = \min\{\delta_1, \delta_2\}$. Then for $0 < |x - a| < \delta$, both limitations on x hold so that both $|f(x) - L_1| < \varepsilon/2$ and $|g(x) - L_2| < \varepsilon/2$. Consequently, for $0 < |x - a| < \delta$, we have

$$|f(x) + g(x) - (L_1 + L_2)| \leq |f(x) - L_1| + |g(x) - L_2| < \varepsilon/2 + \varepsilon/2 = \varepsilon.$$

We have proved (ii).

There are many situations in which we want to consider one-sided limits. For example, we may want to study the behavior of a function at the endpoint of an interval.

Definition 3.3 Let $f : D(f) \to \mathbb{R}$, and suppose that an interval $(a, b) \subset D(f)$. We say that the *limit from the left* exists at b, and write

$$\lim_{x \to b^-} f(x) = L \quad \text{or} \quad \lim_{x \uparrow b} f(x) = L,$$

if for, each sequence $x_n \in (a, b)$ that converges to b, the sequence $f(x_n)$ converges to L.

If, for every sequence $x_n \in (a, b)$ that converges to a, the sequence $f(x_n)$ converges to L, we say that the *limit from the right* exists at a, and we write

$$\lim_{x \to a^+} f(x) = L \quad \text{or} \quad \lim_{x \downarrow a} f(x) = L.$$

It is clear that $\lim_{x \to a} f(x) = L$ if and only if the limits from the left and right both exist and are the same.

There are the obvious ε, δ equivalent formulations of one-sided limits.

Definition 3.4 A function f defined on an interval I is said to be *monotone increasing* on I if $f(x) \leq f(y)$ whenever $x \leq y$. The function f is *monotone decreasing* on I if $f(x) \geq f(y)$ whenever $x \leq y$. If f is either monotone increasing or monotone decreasing, we say that f is *monotone* on I.

An analogue to the monotone convergence theorem for sequences holds for monotone functions.

Theorem 3.3 Let f be monotone on the interval (a, b) and bounded. Then

$$\lim_{x \downarrow a} f(x) \quad \text{and} \quad \lim_{x \uparrow b} f(x)$$

both exist.

The proof is similar to that for bounded monotone sequences, and will be seen in an exercise.

Finally, we wish to define what it means for a limit to exist as x becomes arbitrarily large.

Definition 3.5 Let f be defined on some unbounded interval $[a, \infty)$. We say that

$$\lim_{x \to \infty} f(x) = L$$

if, for each $\varepsilon > 0$, there is an R, that may depend on ε, such that $|f(x) - L| < \varepsilon$ for all $x > R$. A similar definition is made for $\lim_{x \to -\infty} f(x) = L$.

EXAMPLE 3.4:

a) Let $f(x) = x/\sqrt{x^2 + 1}$. We claim that $\lim_{x \to \infty} f(x) = 1$. In fact, $f(x) \le 1$ for all $x \ge 0$. Then we note that $\sqrt{x^2 + 1} \le \sqrt{x^2 + 2x + 1} = (x + 1)$ for $x \ge 0$. Hence

$$\frac{x}{x + 1} \le f(x) \le 1$$

for $x \ge 0$. It follows that

$$|f(x) - 1| \le 1 - \frac{x}{1 + x} = \frac{1}{x + 1}.$$

Now for $\varepsilon > 0$ given, we choose $R = 1/\varepsilon$.

b) The limit $\lim_{x \to \infty} f(x)$ does not exist when $f(x) = \sin(\sqrt{x})$. It suffices to find a sequence $\{x_n\}$ that tends to infinity such that the sequence of function values $\{f(x_n)\}$ does not converge. For $f(x) = \sin(\sqrt{x})$, we take $x_n = ((2n + 1)\pi/2)^2$. Then $f(x_n) = (-1)^n$, which does not converge.

Exercises for 3.1

In these exercises, use either formulation of the limit, except where specified.

1. Let a be a limit point of $D(f)$ and suppose that $\lim_{x \to a} f(x) = L$. Show that $\lim_{x \to a} |f(x)| = |L|$.

2. Using the sequential definition of the limit, show that the following limits exist:

a) $\lim\limits_{x\to 0} 2x^2 + x + 5$ b) $\lim\limits_{x\to 0} \dfrac{x^2}{|x|}$ c) $\lim\limits_{x\to 2} \dfrac{x^3 + 3x^2 - 4}{x - 1}$ d) $\lim\limits_{x\to -2} \dfrac{x}{(x + 1)^2}$.

3. Use the (ε, δ) formulation of the limit (Theorem 3.1) to show that the following limits exist:

a) $\lim\limits_{x\to 1} \sqrt{x + 3}$ b) $\lim\limits_{x\to 2} \dfrac{x + 1}{x}$.

4. Suppose that f is defined on $(0, 2)$ and that $\lim_{x\to 1} f(x) = L$. Show that $\lim_{x\to 0} f((1 + x)^2) = L$.

5. Show that the following limits do not exist:

a) $\lim\limits_{x\to 0} \dfrac{x}{|x|}$ b) $\lim\limits_{x\to 0} |x|^{-1/2}$.

6. Show that if f is bounded and monotone on (a, b), then

$$\lim\limits_{x\downarrow a} f(x) \quad \text{and} \quad \lim\limits_{x\uparrow b} f(x)$$

both exist.

7. If f is bounded and monotone on $[0, \infty)$, show that $\lim_{x\to\infty} f(x)$ exists.

8. Suppose that f and g are bounded functions defined on $(-1, 1)$ and suppose that $\lim_{x\to 0} f(x) = 0$. Show that $\lim_{x\to 0} f(x)g(x) = 0$.

9. Give an example where $\lim_{x\to a} f(x)$ exists and $\lim_{x\to a} f(x)g(x)$ exists, but $\lim_{x\to a} g(x)$ does not exist.

10. Suppose that $\lim_{x\to\infty} x^2 f(x) = L$. Show that $\lim_{x\to\infty} f(x) = 0$.

11. Let

$$f(x) = \begin{cases} x & \text{for } x \text{ rational} \\ 0 & \text{for } x \text{ irrational} \end{cases}.$$

a) Show that $\lim_{x\to 0} f(x) = 0$.

b) Show that if $a \neq 0$, then $\lim_{x \to a} f(x)$ does not exist.

12. Suppose that $\lim_{x \to a} f(x) = L \neq 0$, and $\lim_{x \to a} g(x) = 0$. Show that $\lim_{x \to a}(f/g)$ does not exist.

3.2 Continuous Functions

The concept of continuity is important for physical understanding and for mathematical analysis. In a deterministic system, if we have modeled the system properly, we expect that a small change in the data will result in a small change in the result of an experiment. In the mathematical context, we need the idea of continuity to deal with the question of determining where the graph of a function crosses a horizontal axis.

It is difficult to put the notion of continuity into words precisely. In calculus we attempt to give an intuitive idea of continuity by saying that the graph of a continuous function can be drawn without lifting the pencil from the paper. Obviously this definition of continuity is not adequate for careful analysis. It is necessary to find a way to make the transition from the discrete to the continuous. This is accomplished with the use of the quantified phrases "for all sequences" and "for all $\varepsilon > 0$."

Definition 3.6 Let $f : D(f) \to \mathbb{R}$ and let $a \in D(f)$. Then f is *continuous* at $x = a$ if, for all sequences $x_n \in D(f)$ that converge to a, the sequence of function values $f(x_n)$ converges to $f(a)$. When a is a limit point of $D(f)$, this is the same thing as saying

$$\lim_{x \to a} f(x) = f(a).$$

Theorem 3.1, which showed the equivalence of the sequential and (ε, δ) formulations of the limit, immediately implies an alternate, equivalent, (ε, δ) formulation of continuity. This latter way of expressing continuity is easier to visualize (see figure 3.1 where $f(x) = x^2$ and $a = 1$).

Alternate (ε, δ) Formulation of Continuity

Let $f : D(f) \to \mathbb{R}$ and let $a \in D(f)$. Then f is continuous at $x = a$ if, for each $\varepsilon > 0$, there is a $\delta > 0$ such that $|f(x) - f(a)| < \varepsilon$ for all $x \in D(f)$ with $|x - a| < \delta$.

Notice that, in this definition, the condition on x is $|x - a| < \delta$, which allows the possibility of $x = a$, whereas, in the (ε, δ) definition of a limit, we required $0 < |x - a| < \delta$.

Definition 3.7 We will say that f is *continuous on a set A* if f is continuous at each point of A.

> **EXAMPLE 3.5:** We give an example of the (ε, δ) technique by showing that $f(x) = x^2$ is continuous at each point $a \in \mathbb{R}$. First we factor the difference:
>
> $$|f(x) - f(a)| = |x^2 - a^2| = |x + a|\,|x - a| \le (|x| + |a|)|x - a|.$$
>
> We shall bound the product on the right by a quantity of the form $C|x - a|$ where C is a constant that does not depend on x for x close to a. This can be accomplished by making a preliminary restriction on x; we assume that $|x - a| \le 1$. Then, with this restriction, $|x| \le |x - a| + |a| \le 1 + |a|$, so that we have
>
> $$|f(x) - f(a)| \le (1 + 2|a|)|x - a|.$$
>
> Now let $\varepsilon > 0$ be given. We choose $\delta = \min\{1, \varepsilon/(1 + 2|a|)\}$. Then for $|x - a| < \delta$, we have $|f(x) - f(a)| < \varepsilon$. Notice here that the choice of δ depends on a and that the larger the value of a, the smaller the value of δ.

A function f can fail to be continuous at a point $a \in D(f)$ for several reasons:

(i) The limit of the sequence $f(x_n)$ fails to exist for some sequence x_n that converges to a.

(ii) The limits of the sequences $f(x_n)$ are not the same for all sequences x_n that converge to a.

(iii) The limit of the sequence $f(x_n)$ exists and equals L for all sequences x_n that converge to a, but $f(a) \ne L$.

EXAMPLE 3.6:

a) An example of a type (i) discontinuity at $x = 0$ is given by

$$f(x) = \begin{cases} 0, & x = 0 \\ 1/x, & x \ne 0 \end{cases}.$$

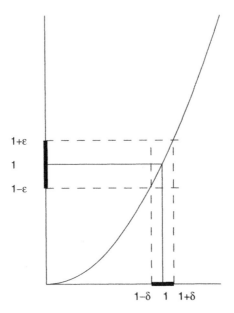

Figure 3.2 Graph of $y = x^2$ together with intervals $(1 - \delta, 1 + \delta)$ on the x axis and $(1 - \varepsilon, 1 + \varepsilon)$ on the y axis

Since the function f is unbounded as x approaches 0, a sequence $f(x_n)$ cannot converge if x_n approaches 0.

b) Another example of a type (i) discontinuity at $x = 0$ is given by $f(x) = \sin(1/x)$, $x \neq 0$ (see Figure 3.2). We have seen that if $x_n = 1/n$, the sequence of function values $f(x_n) = \sin(n)$ does not converge and in fact is dense in the interval $[-1, 1]$.

c) For a type (ii) discontinuity, consider the function

$$f(x) = \begin{cases} 0, & x < 0 \\ 1 + x, & x \geq 0 \end{cases}.$$

In this example, $\lim_{x \to 0^-} f(x) = 0$, while, $\lim_{x \to 0^+} f(x) = 1$.

When the limits from the left and right exist but are different, we say that f has a *jump discontinuity*.

d) Next let

$$f(x) = \begin{cases} \sin x / x, & x \neq 0 \\ 0, & x = 0 \end{cases}.$$

In Example 3.2 we saw that

$$\lim_{x \to 0} \frac{\sin x}{x} = 1.$$

Hence $f(x)$ as defined is not continuous at $x = 0$. However, if we assign the value $f(0) = 1$, f becomes a continuous function. We say that a type (iii) discontinuity is a *removable* discontinuity.

e) Finally, here is an example of a function that is discontinuous at every point. The Dirichlet function is defined to be

$$f(x) = \begin{cases} 1 & \text{for } x \text{ rational} \\ 0 & \text{for } x \text{ irrational} \end{cases}.$$

Theorem 1.7 and its corollary say that every open interval contains both rational and irrational points. Thus for each $a \in \mathbb{R}$, and any choice of $\delta > 0$, there are points $x, y \in (a - \delta, a + \delta)$ with $f(x) = 1$ and $f(y) = 0$.

The following result about combining continuous functions is an immediate consequence of Theorem 3.2.

Theorem 3.4 Let $f, g : A \to \mathbb{R}$ be continuous at $a \in A$. Then

    (i) cf is continuous at a for each $c \in \mathbb{R}$.
    (ii) $f + g$ is continuous at a.
    (iii) fg is continuous at a.
    (iv) If $g \neq 0$ on A, f/g is continuous at a.

Clearly $f(x) = x$ is continuous on $\mathbb{R}$. In view of Theorem 3.4, we can conclude that polynomials are continuous on $\mathbb{R}$ and that a quotient of two polynomials p/q is continuous at all points where $q(x) \neq 0$.

Another important way of combining continuous functions is to use composition.

Theorem 3.5 Let $f : A \to \mathbb{R}$ and $g : B \to \mathbb{R}$ with $R(f) \subset B$. If f is continuous at $a \in A$ and g is continuous at $b = f(a) \in B$, then the composite function $g \circ f : A \to \mathbb{R}$ is continuous at a.

Proof: We shall use the sequential definition of continuity, but the reader should also prove this theorem using the (ε, δ) formulation. Let $x_n \in A$ be a sequence that converges to a. The continuity of f at $x = a$ implies that the sequence $y_n = f(x_n)$ converges to $b = f(a)$. From the continuity of g at b, we conclude that $g(y_n)$ converges to $g(b)$. However, this is the same as saying that $g \circ f(x_n) = g(y_n)$ converges to $g(b) = g \circ f(a)$. The proof is complete.

EXAMPLE 3.7: Theorem 2.4 implies that the function $f(x) = \sqrt{x}$ is continuous on $\{x \geq 0\}$. Appealing to Theorems 3.4 and 3.5, we see that

$$g(x) = \sqrt{\frac{1}{1 + x^2}}$$

is continuous on $\mathbb{R}$.

Exercises for 3.2

1. Show that the following functions are continuous on the indicated domains. State which theorems are used in each case. You may assume that the functions $\sin x$ and $\cos x$ are continuous on $\mathbb{R}$.

   a) $f(x) = \dfrac{x^2 - x + 2}{x^2 - 1}, \quad x \neq \pm 1.$

   b) $f(x) = \sqrt{x^2 + 1} + \cos x, \quad x \in \mathbb{R}.$

   c) $f(x) = \sin(\sqrt{x^2 - 1}), \quad |x| \geq 1.$

2. Use the (ε, δ) formulation of continuity to show that $f(x) = 1/x$ is continuous at each $a \neq 0$. Note in particular how the choice of δ depends on a.

3. The function

$$f(x) = \frac{x^3 - x^2 - 3x + 2}{x - 2}$$

   is defined and continuous for $x \neq 2$. Can you define f at $x = 2$ so that f will also be continuous at $x = 2$?

4. Let $f(x)$ be continuous on $\mathbb{R}$ and suppose that $f(x) \leq 1$ for all x rational. Show that $f(x) \leq 1$ for all $x \in \mathbb{R}$.

5. Let f be continuous on the open interval I, and suppose that $f(x_0) > 0$ for some $x_0 \in I$. Show that there is a $\delta > 0$ such that the interval

$J = (x_0 - \delta, x_0 + \delta) \subset I$ and such that $f(x) > 0$ on J. The argument needed here parallels that of Lemma 2.2.

6. Suppose that f and g are continuous functions on $[a, b]$. Show that the function $h(x) \equiv \max\{f(x), g(x)\}$ is continuous on $[a, b]$.

7. a) Let $f : \mathbb{R} \to \mathbb{R}$ be monotone and let J be an interval. Show that $I = f^{-1}(J)$ is also an interval. You will need the characterization of an interval given in exercise 1 of Section 1.5.

   b) Now suppose that J is an open interval, and suppose that f is both continuous and monotone. Show that $I = f^{-1}(J)$ is an open interval.

   c) Show by example that, if we assume only that f is continuous, and J is an open interval, I may not be an open interval, but rather a union of disjoint open intervals.

8. The solution of the initial value problem

$$y' = y^2, \quad y(0) = a > 0,$$

is the function

$$y(x) = \frac{a}{1 - ax} \quad \text{for } x < \frac{1}{a}.$$

   a) Verify that for fixed x, $y(x)$ is a continuous function of the initial value a for $ax < 1$.

   b) Let y_1 be the solution with $y_1(0) = a$ and y_2 be the solution with $y_2(0) = b$. Show that it is possible to have $|a - b| < 10^{-6}$ but $|y_1(1) - y_2(1)| > 10^6$. Conclude that, even though the solution depends continuously on the initial value a, a small change in a can make a big change in the solution.

3.3 Further Properties of Continuous Functions

Definition 3.8 Let A be a set and $f : A \to \mathbb{R}$. Suppose that the set of real numbers $f(A)$ has a maximum. A point $x^* \in A$ such that $f(x^*) = \max f(A)$ is a *maximizer* for f on A. We say that f attains its maximum value at x^*, and we write

$$f(x^*) = \max_A f.$$

If $f(A)$ has a minimum, a point $x_* \in A$ such that $f(x_*) = \min f(A)$ is a *minimizer* for f on A. We say that f attains its minimum value at x_*, and we write

$$f(x_*) = \min_A f.$$

EXAMPLE 3.8: A function may attain its maximum at more than one point, or it may not have any maximizers. The same is true for minimizers.

a) Let $f(x) = 1/(1 + x^2)$ on $A = \mathbb{R}$. $x^* = 0$ is the maximizer for f on $\mathbb{R}$ with

$$\max_{\mathbb{R}} f = f(0) = 1.$$

We have $f(x) > 0$ for all $x \in \mathbb{R}$ so that f is bounded below with

$$\inf_{\mathbb{R}} f = 0.$$

The function f does not have a minimizer on $\mathbb{R}$. Now restrict f to the smaller set $\tilde{A} = \{|x| \leq 2\}$. We see that $x^* = 0$ is still the only maximizer, but the restriction of f has the two minimizers $x_* = \pm 2$ on $\tilde{A}$.

b) The function $g(x) = (x - 1)^2(x + 1)^2$ on $A = \mathbb{R}$ has two minimizers: $x_* = \pm 1$. It has no maximizer, because g is not bounded above on $\mathbb{R}$.

The Bolzano-Weierstrass theorem is an important tool that we use to derive further properties of continuous functions.

Theorem 3.6 Let $I = [a, b]$ be a closed bounded interval, and let $f : I \to \mathbb{R}$ be continuous. Then f has both a maximizer $x^* \in I$ and a minimizer $x_* \in I$:

$$f(x_*) \leq f(x) \leq f(x^*) \quad \text{for all } x \in I.$$

Proof: First we show that $R(f) = \{f(x) : x \in I\}$ is a bounded set, arguing by contradiction. If $R(f)$ is not bounded above, then there is a sequence of points $x_n \in I$ such that $f(x_n) \geq n$. By the corollary to the Bolzano-Weierstrass theorem, x_n must have a subsequence x_{n_k} that converges to a point z. We have $a \leq x_{n_k} \leq b$ for all $k \geq 1$. Hence by Theorem 2.6, we conclude that $a \leq z \leq b$. Now we invoke the assumption that f is continuous on I. Because $f(x)$ is continuous on I, $f(x_{n_k})$ must converge to $f(z)$. This produces a contradiction because the sequence $f(x_n)$ was assumed to be unbounded. A similar argument shows that $R(f)$ is also bounded below.

Let $M = \sup_I f$ and $m = \inf_I f$. Because M is the least upper bound of $R(f)$, for each n, there is a point, call it y_n, with $y_n \in R(f)$ and $M - 1/n < y_n \leq M$. The sequence y_n converges to M. For each n there is a point $x_n \in I$ such that $f(x_n) = y_n$. Now we again appeal to the Bolzano-Weierstrass theorem to extract a subsequence x_{n_k} such that x_{n_k} converges to a point $x^* \in I$. By the continuity of f, we know that $y_{n_k} = f(x_{n_k})$ converges to $f(x^*)$. Because y_n converges to M, every subsequence of y_n converges to M as well. Hence

$$f(x^*) = \lim_{k \to \infty} f(x_{n_k}) = \lim_{k \to \infty} y_{n_k} = M.$$

Therefore $f(x^*) \geq f(x)$ for all $x \in I$.

The same argument with the inequalities reversed produces the point x_*.

In the (ε, δ) formulation of continuity, we noticed that the choice of δ could depend on both ε and a. That was the case with $f(x) = x^2$ where our choice of δ was $\delta = \min\{1, \varepsilon/(1 + |a|)\}$. The larger $|a|$ is, the smaller δ must be. It is useful in some situations to be able to choose the δ independently of the location of a.

Definition 3.9 Let $f : A \to \mathbb{R}$. We say that f is *uniformly continuous* on A if, for each $\varepsilon > 0$, there is a $\delta > 0$ such that $|f(x) - f(y)| < \varepsilon$ for all $x, y \in A$.

The concept of uniform continuity will be crucial to our development of the Riemann integral in Chapter 7.

EXAMPLE 3.9:

a) We know already that $f(x) = \sqrt{x}$ is continuous on $\{x \geq 0\}$. Since the graph of $\sqrt{x}$ has a vertical tangent at $x = 0$, it would appear that we would have to choose a smaller and smaller δ as a came closer to zero. However, appearances can be deceiving.

We divide the analysis into two cases: $|x - y| < y$ and $|x - y| \geq y$.

(i) Suppose that $|x - y| < y$ (which implies that $y > 0$). Then $\sqrt{|x - y|} < \sqrt{y}$. Using the usual device,

$$|f(x) - f(y)| = |\sqrt{x} - \sqrt{y}| = \frac{|x - y|}{\sqrt{x} + \sqrt{y}}.$$

Then we have the estimate

$$|f(x) - f(y)| \leq \frac{\sqrt{|x - y|}}{\sqrt{y}} \sqrt{|x - y|} \leq \sqrt{|x - y|} . \qquad (3.2)$$

(ii) Suppose that $|x - y| \geq y$. Then $\sqrt{y} \leq \sqrt{|x - y|}$. In addition,

$x \leq y + |x - y| \leq 2|x - y|$ whence $\sqrt{x} \leq \sqrt{2|x - y|}$. It follows that

$$|f(x) - f(y)| \leq \sqrt{x} + \sqrt{y}$$

$$\leq \sqrt{2|x - y|} + \sqrt{|x - y|}$$

$$\leq 3\sqrt{|x - y|}. \qquad (3.3)$$

Now let ε be given. The estimates (3.2) and (3.3) show that if we choose δ so that $3\sqrt{\delta} < \varepsilon$, we will have

$$|f(x) - f(y)| < \varepsilon$$

for any $x, y \geq 0$ with $|x - y| \leq \delta$. Thus $f(x) = \sqrt{x}$ is uniformly continuous on $\{x \geq 0\}$.

b) The function $g(x) = 1/x$, defined on $D(g) = \{0 < x \leq 1\}$, is continuous, but not uniformly continuous. Again the problem appears to be that the graph of g is tending toward the vertical as x approaches zero. To show that g is not uniformly continuous we must find an $\varepsilon_0 > 0$ and a pair of sequences $x_n, y_n \in D(g)$ such that $|x_n - y_n|$ tends to zero as n tends to infinity and yet $|g(x_n) - g(y_n)| \geq \varepsilon_0$ for all n. This is easy to do. Choose $\varepsilon_0 = 1/2$ and let $x_n = 1/n$ and $y_n = 1/n + 1/n^2$. Then $|x_n - y_n| = 1/n^2$ tends to zero, but $|g(x_n) - g(y_n)| = n/(n+1) \geq 1/2$ for all n.

These examples make it seem difficult to tell if a continuous function is uniformly continuous. Fortunately, the Bolzano-Weierstrass theorem will help us sort out this muddle by giving us a criterion for uniform continuity.

Theorem 3.7 Let $I = [a, b]$ be a closed bounded interval, and let $f : I \to \mathbb{R}$ be continuous on I. Then f is uniformly continuous on I.

Proof: We shall have to argue by contradiction. Suppose that f is continuous on I, but not uniformly continuous. Then there must be an $\varepsilon_0 > 0$ such that, no matter how small we choose $\delta > 0$, there is a pair of points $x, y \in I$ such that $|x - y| < \delta$ but $|f(x) - f(y)| \geq \varepsilon_0$. In particular, for each $n \in \mathbb{N}$, there must be points $x_n, y_n \in I$ such that $|x_n - y_n| < 1/n$ and $|f(x_n) - f(y_n)| \geq \varepsilon_0$. Now, by the corollary to the Bolzano-Weierstrass theorem, there must be a subsequence x_{n_k} of x_n that converges to a point $x_0 \in I$. We claim that the corresponding subsequence y_{n_k} must also converge to x_0. Let $\varepsilon > 0$ be given (independent of ε_0). Because x_{n_k} converges to x_0, there is a $K > 0$ such that for $k \geq K$, $|x_{n_k} - x_0| < \varepsilon/2$. By making K larger if necessary, we can also be sure that $1/n_k < \varepsilon/2$ for $k \geq K$. Now using the triangle inequality, we see that, for $k \geq K$,

$$|y_{n_k} - x_0| \leq |y_{n_k} - x_{n_k}| + |x_{n_k} - x_0|$$
$$\leq \varepsilon/2 + \varepsilon/2 = \varepsilon.$$

The contradiction has been a long time in coming, but here it is. Since x_{n_k} and y_{n_k} both converge to x_0 and f is continuous on I, we see that $f(x_{n_k})$ and $f(y_{n_k})$ both converge to $f(x_0)$. This means that, for k sufficiently large, we have

$$|f(x_{n_k}) - f(y_{n_k})| \leq |f(x_{n_k}) - f(x_0)| + |f(x_0) - f(y_{n_k})| < \varepsilon_0/2,$$

which is the contradiction. The proof is complete.

We return to Example 3.9. We see that Theorem 3.7 tells us that $f(x) = \sqrt{x}$ is uniformly continuous on each closed bounded interval $[0, b]$, but it does not allow us to conclude that f is uniformly continuous on all of $\mathbb{R}$. To see that, we had to make the estimates (3.2) and (3.3). With regard to the function $g(x) = 1/x$, Theorem 3.7 tells us that g is uniformly continuous on each closed bounded interval $[\delta, b]$ with $0 < \delta < b$.

Exercises for 3.3

1. Let f be continuous on $I = [a, b]$ with $f(x) > 0$ for all $x \in I$. Show that there is a constant $c > 0$ such that $f(x) \geq c > 0$ for all $x \in I$. Is the result still true if we allow I to be half-open, $I = [a, b)$?

2. Let f be continuous on $\mathbb{R}$ and suppose that $\lim_{x \to \pm\infty} f(x) = 0$. Show that f will attain either a maximum on $\mathbb{R}$ or a minimum on $\mathbb{R}$ or both. Give an example in which f attains a maximum on $\mathbb{R}$ but not a minimum.

3. Show that $f(x) = x^2$ is not uniformly continuous on $\mathbb{R}$.

4. By making direct estimates, show that $f(x) = x/(1 + x^2)$ is uniformly continuous on $\mathbb{R}$.

5. If f is uniformly continuous on $\mathbb{R}$, is f^2 uniformly continuous on $\mathbb{R}$?

6. Let f be uniformly continuous on the half open interval $I = (a, b]$. Let $x_n \in I$ be a sequence that converges to a. Show that $f(x_n)$ is a Cauchy sequence and therefore converges.

7. Let $I = (a, b]$ and let f be uniformly continuous on I. Let $x_n = a + 1/n$. In exercise 6, we saw that $\lim_{n \to \infty} f(x_n) = L$ exists. Show that if $y_n \in I$ is any other sequence converging to a, then $\lim_{n \to \infty} f(y_n) = L$. Finally conclude that if we define $f(a) = L$, f so extended is now continuous on the closed interval $[a, b]$.

8. Let $I = [a, b]$ and let f be continous on I. Divide I into subintervals with mesh points $a = x_0 < x_1 < \cdots < x_n = b$. Assume $x_j - x_{j-1} = h = (b - a)/n$. Construct a step function g that is constant on each subinterval:

$$g_h(x) = f(x_j) \quad \text{for } x_{j-1} < x \leq x_j.$$

Show that for each $\varepsilon > 0$ there is an $h > 0$ such that

$$|f(x) - g_h(x)| \leq \varepsilon \quad \text{for all } x \in I.$$

3.4 Golden-Section Search

Theorem 3.6 guarantees the existence of a maximum and of a minimum of a continuous function on a closed bounded interval, but does not tell us how to find the maximizer or the minimizer. Later when we have introduced the

derivative, we shall seek a minimizer or maximizer at points where the derivative is zero. Now, however, we shall discuss a method to locate an extreme
point approximately that does not even use continuity of the function. This is
the golden-section search method.

We must put some assumptions on the function. We suppose f is defined
on the bounded interval $[a, b]$ and we suppose that there is a unique minimizer
x_* and that $x_* \in (a, b)$. Furthermore, assume that f is strictly decreasing for
$x < x_*$ and strictly increasing for $x > x_*$. Such a function is said to be *unimodal*.

The cost of a search method is measured in the number of times the function
must be evaluated. We shall give an example later where evaluating the function requires a substantial calculation. Thus an efficient search method must
minimize the number of function evaluations.

We shall construct a search method that, at each step, reduces the length of
the interval in which the minimizer lies by a factor of $r < 1$. At each step,
it does this using only one additional functional evaluation. We assume that
$1/2 < r < 1$ and we introduce the comparison points

$$s = a + (1 - r)(b - a) \quad \text{and} \quad t = b - (1 - r)(b - a).$$

Because $1/2 < r < 1$, we have $a < s < t < b$ (see Figure 3.3).

First assume that $f(s) \geq f(t)$. Then we know the minimizer x_* must lie in
the interval $[s, b]$. Indeed, if $x_* \in [a, s)$, then we would have to have $f(s) < f(t)$
because f is strictly increasing to the right of x_*. Let $a_1 = s$ and $b_1 = b$. Note that
$b_1 - a_1 = b - s = r(b - a)$ and that $a_1 < t < b_1$. We shall repeat the comparison procedure on the shorter interval $[a_1, b_1]$. Again we introduce comparison points

$$s_1 = a_1 + (1 - r)(b_1 - a_1) \quad \text{and} \quad t_1 = b_1 - (1 - r)(b_1 - a_1)$$

and note that $a_1 < s_1 < t_1 < b_1$. However, we do not want to have to compute
both values $f(s_1)$ and $f(t_1)$. We shall choose r so that $s_1 = t$. This condition
yields the equation:

$$b - (1 - r)(b - a) = t = s_1 = a_1 + (1 - r)(b_1 - a_1).$$

When we substitute in $b_1 - a_1 = r(b - a)$, we see that r must satisfy

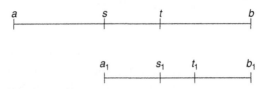

Figure 3.3 Division of interval $[a, b]$ using the golden-section method

$$(r^2 + r - 1)(b - a) = 0.$$

Since $b - a \neq 0$, we see that $s_1 = t$ if and only if r is the positive root of the equation $r^2 + r - 1 = 0$, which is

$$r_* = \frac{\sqrt{5} - 1}{2} \approx 0.6180.$$

On the other hand if, $f(s) < f(t)$, then we know that the minimizer $x_* \in [a, t]$. We set $a_1 = a$ and $b_1 = t$ and again have $b_1 - a_1 = r(b - a)$. We introduce comparison points $a_1 < s_1 < t_1 < b_1$ and require that $t_1 = s$. This requirement leads to the same quadratic equation for r and we again choose the positive root r_*.

Using this value of $r = r_* \approx 0.6180$, we need make only one new function evaluation for each iteration, and we reduce the length of the interval in which x_* must lie by a factor of $r = r_*$ each time. We repeat this procedure until $r_*^n(b-a)$ is small enough to meet our accuracy requirements. Note that the golden-section algorithm generates a nested family of intervals I_n with $I_0 = [a, b]$ and length I_n equal to $r_*^n(b - a)$. Hence by the nested interval lemma, $\cap_{n=0}^{\infty} I_n$ is a single point, the minimizer x_*.

Here is a pseudocode for N iterations of the algorithm:

```
s = a + (1-r)(b-a)
b = b - (1-r)(b-a)
f1 = f(s)
f2 = f(t)
for n = 1:N
     is f1 < f2 ?
     if yes
          b=t
          t=s
          s=a+(1-r)(b-a)
          f2=f1
          f1=f(s)
     if no
          a=s
          s=t
          t=b-(1-r)(b-a)
          f1=f2
          f2=f(t)
     end if
end loop
```

The number $1/r^* = (1 + \sqrt{5})/2 \approx 1.6180$ is known as the *golden ratio*. The Greeks found that a rectangle with the ratio of the long side to the short side given by the golden ratio had the most pleasing proportions. This ratio also appears in many natural objects such as sea shells and pine cones. In the mathematical domain, it is associated with Fibonacci sequences and the order of convergence of the secant method for finding the zeros of a function.

EXAMPLE 3.10:   Consider the function

$$f(x) = \begin{cases} 1.25 - 2x, & x < 0.5 \\ (x - .5)^2, & x \geq 0.5 \end{cases} .$$

The graph of f is shown in Figure 3.4.

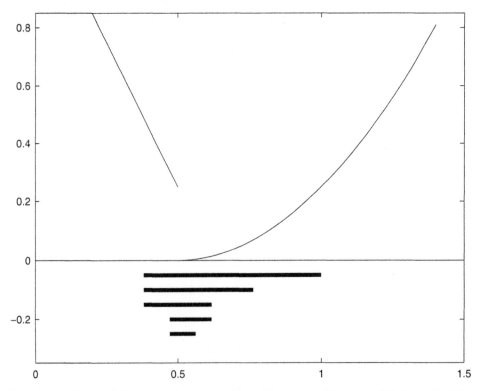

Figure 3.4 Graph of the discontinuous, unimodal function of Example 3.9. Several of the intervals in the golden-section search are shown

We start on the interval $[0, 1]$ and iterate 10 times with the following results:

iterate	a	b	b-a
1	0.3820	1.0000	0.6180
2	0.3820	0.7639	0.3820
3	0.3820	0.6180	0.2361
4	0.4721	0.6180	0.1459
5	0.4721	0.5623	0.0902
6	0.4721	0.5279	0.0557
7	0.4934	0.5279	0.0344
8	0.4934	0.5147	0.0213
9	0.4934	0.5066	0.0132
10	0.4984	0.5066	0.0081

Notice that the last interval has length $r^{10} = 0.0081$. We have used 11 function evaluations. If we evaluate f at the 11 equally spaced points, $0, 0.1, 0.2, \ldots, 0.9, 1$, we can at best locate the minimizer within an interval of length 0.1.

EXAMPLE 3.11: In this example, we encounter a situation where finding the function values is not just a matter of a simple calculation. Let $u(t)$ denote the x coordinate of a projectile and $v(t)$ the y coordinate. The projectile, of mass m, is launched at time $t = 0$ from the point $(0, 0)$. The initial speed is V and the launch angle with respect to the x axis is θ. Thus the initial velocity is $(u'(0), v'(0)) = V(\cos \theta, \sin \theta)$. We assume that the air resistance in the horizontal and vertical components is

$$R_{horizontal} = -ku'\sqrt{(u')^2 + (v')^2}$$

and

$$R_{vertical} = -kv'\sqrt{(u')^2 + (v')^2}$$

where $k > 0$ is a constant. The coordinates u and v satisfy the system of nonlinear differential equations

$$mu'' = -ku'\sqrt{(u')^2 + (v')^2} \tag{3.4}$$
$$mv'' = -kv'\sqrt{(u')^2 + (v')^2} - mg \tag{3.5}$$

with initial conditions
$$u(0) = v(0) = 0, \quad u'(0) = V \cos \theta, \quad v'(0) = V \sin \theta.$$

The function we consider is the range function $f(\theta)$, which is the horizontal distance the projectile travels before it hits the ground. When there is no air resistance (i.e., $k = 0$) it can be calculated that $f(\theta) = (V/g)\sin(2\theta)$, which has a single maximum on the interval $[0, \pi/2]$. Clearly the maximizer is $\theta^* = \pi/4$. However, when there is air resistance, the system (3.4), (3.5) cannot be solved in closed form. The solution must be approximated numerically, and we must then find numerically the value of u when $v = 0$ for the first time. Thus to calculate $f(\theta)$ is much more than simply evaluating a formula. Given a desired tolerance δ, we would like to estimate the maximizer θ^* of f on the interval $[0, \pi/2]$ to within δ, using as few evaluations of f as possible. The method of golden-section search allows us to accomplish this with $n + 1$ evaluations of f where n is the smallest positive integer such that $r_*^n(\pi/2) < \delta$.

Search methods that do not use derivatives are an important topic of research for functions of several variables. A brief introduction and references are given in the book of S. Nash and A. Sofer [N].

3.5 The Intermediate Value Theorem

When does there exist a solution to the equation $f(x) = z$? Our first result is an existence theorem which is intimately related to the least upper bound property.

Theorem 3.8 (Intermediate value theorem) Let $I = [a, b]$ be a closed bounded interval and let $f : I \to \mathbb{R}$ be continuous on I. If $f(a) < z < f(b)$, or $f(b) < z < f(a)$, there is a number $c \in I$ such that $f(c) = z$.

Proof: We consider the case $f(a) < z < f(b)$. Let $S = \{x \in I : f(x) \le z\}$. S is nonempty because $a \in S$, and S is bounded above by b. Let c be the least upper bound of S.

(i) $f(c) \le z$. For each n there is an $x_n \in S$ such that $x_n > c - 1/n$. Hence x_n converges to c, and $f(x_n) \le z$ for all n. Because f is continuous,

$$f(c) = \lim_{n \to \infty} f(x_n) \le z.$$

(ii) $f(c) = z$. Suppose to the contrary that $f(c) < z$. Let $\varepsilon = (z - f(c))/2$. Because f is continuous, there is a $\delta > 0$ such that $|f(x) - f(c)| < \varepsilon$ for $|x - c| < \delta$. Consequently, there is an x_0 such that $c < x_0 < c + \delta$ and such that

$$f(x_0) = f(x_0) - f(c) + f(c)$$
$$< \varepsilon + f(c)$$

$$< \frac{z - f(c)}{2} + f(c) = \frac{z + f(c)}{2}$$

$$< z.$$

However, $f(x_0) < z$ contradicts the fact that c is an upper bound of S. The proof of the case in which $f(b) < z < f(a)$ is similar.

The proof of Theorem 3.8 does not provide any procedure for computing the value of c. We now give an alternate proof using the *method of bisection* which can be used to calculate approximations to c.

Alternate Proof: We replace the function f with $g(x) = f(x) - z$, which is also continuous, and we assume that $g(a) < 0 < g(b)$. Our goal is to show that there is a number $c \in I$ with $g(c) = 0$. We shall construct a family of nested intervals $I_n = [a_n, b_n]$ such that a_n and b_n converge to a common value c with $g(c) = 0$. Let $m = (a+b)/2$ be the midpoint of I. If $g(m) = 0$, we are done. If $g(m) > 0$, set $a_1 = a$ and $b_1 = m$. If $g(m) < 0$, we set $a_1 = a$ and $b_1 = b$. In either case, we have $g(a_1) < 0 < g(b_1)$. Now we proceed inductively. Assume that we have found a_n and b_n with $a_{n-1} \le a_n < b_n \le b_{n-1}$ and $g(a_n) < 0$ and $g(b_n) > 0$. Define $m_n = (a_n + b_n)/2$. If $g(m_n) = 0$ we are done. If $g(m_n) > 0$, set $a_{n+1} = a_n$ and $b_{n+1} = m_n$. If $g(m_n) < 0$, set $a_{n+1} = m_n$ and $b_{n+1} = b_n$. Clearly, $a_n \le a_{n+1} \le b_{n+1}$; the intervals I_n are nested.

Suppose now that the procedure does not stop at any n. The length of I_n is $2^{-n}(b - a)$ so by the nested intervals lemma, $\cap I_n = \{c\}$ for some single point c. The sequence a_n converges upward to c with $g(a_n) < 0$ and the sequence b_n converges downward to c with $g(b_n) > 0$. Finally, using the continuity of g, we deduce that

$$g(c) = \lim_{n \to \infty} g(a_n) \le 0$$

and

$$g(c) = \lim_{n \to \infty} g(b_n) \ge 0.$$

Hence $g(c) = 0$. The proof for the case $g(a) > 0 > g(b)$ is similar.

The method of bisection always converges to a zero of a continuous function that changes sign on an interval. Of course, if there are several zeros in the interval, it will only converge to one of them. The price we pay for such a reliable method is its slow speed of convergence. To find the square root of a number $A > 0$ is the same as solving the equation

$$g(x) \equiv x^2 - A = 0.$$

For $A = 25$, we start with the interval $[4, 6]$. If we require an accuracy on the order of 10^{-15}, the bisection method will require n iterations where $2^{-n+1} = 10^{-15}$. Thus $n = 50$. Compare this with the only four iterations needed with the approximation scheme (2.7).

Finally, we remark that if used carelessly, the method of bisection can give erroneous results. Suppose that a zero of a function occurs near a singularity z. If, by mistake, our first interval brackets z, the method of bisection may converge to z. For example, if we seek the roots of $f(x) = \tan x - x$, which lie close to the singularities of $\tan x$, a carelessly chosen initial bracket may produce a sequence of intervals that converges to an odd multiple of $\pi/2$.

The following result will prove useful in Chapter 4.

Theorem 3.9 Let I be an open interval, and let $f : I \to \mathbb{R}$ be continuous and strictly increasing on I. Then the image $J = f(I)$ is an open interval.

Proof: If $y_1, y_2 \in J$ with $y_1 < y_2$, there are points $x_1, x_2 \in I$ such that $f(x_1) = y_1$ and $f(x_2) = y_2$. If y lies between y_1 and y_2, the intermediate value theorem guarantees that there is a point $x \in I$ such that $f(x) = y$. Thus J must contain the whole interval $[y_1, y_2]$. Because y_1 and y_2 are arbitrary points in J, it follows that J is an interval (see exercise 1 of Section 1.5).

It remains for us to show that J is an open interval, and to do that, we need to exclude both of the following possibilities:

    (i) $d = \sup J$ is finite and belongs to J.
    (ii) $c = \inf J$ is finite and belongs to J.

Suppose that case (i) is true. Then there is an $x_0 \in I$ such that $f(x_0) = d$. Since I is an open interval, there is a $\delta > 0$ such that $x_0 + \delta \in I$ as well. But then $f(x_0 + \delta) \in J$. The function f is strictly increasing so that $f(x_0 + \delta) > f(x_0) = d$ which is impossible. A similar argument excludes case (ii) as well.

Exercises for 3.4 and 3.5

1. Let

$$f(x) = \begin{cases} x & \text{for } 0 \le x \le 1/4 \\ 1/4 - (1/10)(x - 1/4) & \text{for } 1/4 \le x \le 1 \end{cases}.$$

   The minimum of f over $[0, 1]$ is clearly $f(0) = 0$. To what value does the golden-section search converge? What is amiss in this example?

2. Use a computer or a calculator and the method of golden-section search to find the minimizer of the function

$$f(x) = \frac{1}{x} + e^{\sqrt{x}}.$$

   a) Start with the interval $[0.5, 1]$ and make five iterations. What is the interval $[a_5, b_5]$?
   b) How many iterations are needed to shrink the interval that contains the minimizer to less than .005? What is the interval?

3. Consider the initial value problem

$$y' = x^2 - y^2, \quad y(0) = 1.$$

   The solution of this problem attains a minimum near $x = 0.7$. You will need to use a numerical code solver to get the values of the solution $y(x)$. Write a code that combines the numerical code solver with the method of golden-section search, starting with the interval $[0, 1]$. If the code solver has local error E, it is a waste of time to continue iterating the golden-section search when $r_*^n < E$. Why?

4. Let p be a polynomial of odd degree. Show that p must have at least one real root.

5. Let the function

$$f(x) = \frac{1}{\sqrt{x^3 + 2x}} + x^2 - 2x, \quad x > 0.$$

   Show that there are at least two solutions of $f(x) = 0$ in $\{x > 0\}$.

6. Let f be continuous on $\mathbb{R}$ and suppose that f is bounded. Show that there is at least one solution of $f(x) = x$.

7. Let $I = [a, b]$ and suppose that f is continuous on I. Let points $x_j \in I$, $j = 1, \ldots, n$. Show that there must be a point $c \in I$ such that

$$f(c) = \frac{f(x_1) + \cdots + f(x_n)}{n}.$$

8. Let $f(x) = -x^3 + 2x^2 + 1$. Use a calculator or computer and the method of bisection to find the root of $f(x) = 0$ to within 10^{-3}.

Chapter 4

The Derivative

Chapter Overview

The derivative is defined as the slope of the linear approximation. The notations $O(h)$ and $o(h)$ are used to describe the error. We prove the standard rules for differentiation, including the chain rule, emphasizing the role of linear approximation. The mean value theorem is proved, and used in turn to prove a one-dimensional inverse function theorem. This latter theorem is used in an example to justify a change of variable in a differential equation. The Cauchy mean value theorem and l'Hôpital's rule are useful for evaluating limits of quotients. Finally, the second derivative test is derived, and it is applied to a problem in economics.

4.1 The Derivative and Approximation

In calculus the derivative is treated as a rate of change, or as the slope of the tangent line. While keeping this interpretation of the derivative, we shall emphasize another aspect of the derivative, namely that of linear approximation. First we give the usual definition.

Definition 4.1 Let f be defined on an open interval that contains the point a. We say that f is *differentiable* at a if

$$\lim_{h \to 0} \frac{f(a + h) - f(a)}{h}$$

exists. In this case we denote the limit $f'(a)$.

In this definition of the derivative, we interpret the quotients as the slopes of secant lines through the points $(a, f(a))$ and $(a + h, f(a + h))$. We think of these secant lines as tending toward a line tangent to the graph of f at the point $(a, f(a))$ as h tends to zero, and we interpret $f'(a)$ as the slope of the tangent line.

Looking at a plot of the graph of f and drawing a tangent line, it appears that the tangent line approximates the graph of f near the point $(a, f(a))$. How do we make this idea precise? Let $l(x) = f(a) + f'(a)(x - a)$. The graph of l is the tangent line. Thus we are asking how well $l(x)$ approximates $f(x)$ near the point $x = a$. In terms of $h = x - a$, the error is

$$R(h) \equiv f(a + h) - l(a + h).$$

The error $R(h)$ gets smaller as h gets smaller, and indeed $R(0) = 0$. The important issue here is how fast $R(h)$ tends to zero as h tends to zero. If $R(h)$ is proportional to h (cutting h in half cuts $R(h)$ in half), this approximation is not very useful. For the linear approximation to be of some value, we need to know that $R(h)$ tends to zero faster than h does, or, in other words, that $R(h)/h$ tends to zero as h tends to zero. The following notation is useful in describing this situation.

Definition 4.2 Let g be a function defined on $0 < |h| < a$ for some $a > 0$.

   (i) We write

$$g = O(|h|^p) \quad \text{if} \quad \frac{|g(h)|}{|h|^p} \quad \text{is bounded as } h \text{ tends to zero.}$$

     This means that there is a $\delta > 0$ and a constant $M \geq 0$ such that $|g(h)| \leq M|h|^p$ for $0 < |h| < \delta$.

  (ii) We write

$$g = o(|h|^p) \quad \text{if} \quad \frac{|g(h)|}{|h|^p} \quad \text{tends to zero as } h \text{ tends to zero.}$$

     This means that, for each $\varepsilon > 0$, there is a $\delta > 0$ such that $|g(h)| \leq \varepsilon|h|^p$ for $|h| < \delta$.

The expression $O(|h|^p)$ is read "big O of h^p", and $o(|h|^p)$ is read "little o of h^p."

From the definition it is clear that $g = o(|h|^p)$ implies that $g = O(|h|^p)$, and if $g = O(|h|^p)$, then $g = o(|h|^q)$ for $q < p$. For example, let $g(h) = |h|^{3/2} \exp(-h)$. Then $g(h) = o(|h|)$ whereas $g(h) = O(|h|^{3/2})$.

To return to our discussion of the derivative, if we assume that f is differentiable at $x = a$, then for $h \neq 0$,

$$R(h) = f(a + h) - l(a + h) = f(a + h) - (f(a) + f'(a)h)$$

$$= h\left[\frac{f(a + h) - f(a)}{h} - f'(a)\right]. \tag{4.1}$$

Now the expression in the square brackets tends to zero as h tends to zero. Because of the factor h outside the square brackets, we can say that $R(h) = o(|h|)$ as h tends to zero. In fact, the rate at which $R(h)$ tends to zero characterizes the linear approximation.

A note about terminology. A function $l(x) = ax + b$ is strictly speaking an *affine* function, and it is a linear function only when $b = 0$. Nevertheless, in standard usage we refer to $l(x) = f(a) + f'(a)(x - x)$ as a linear approximation to f at $x = a$.

Definition 4.3 Let $f : I \rightarrow \mathbb{R}$ where I is an open interval, and let $a \in I$. Then f *has a linear approximation* at a if there is a slope m such that the error $R(h) \equiv f(a + h) - f(a) - mh$ is $o(|h|)$ as h tends to zero. That is,

$$\lim_{h \to 0} \frac{R(h)}{h} = \lim_{h \to 0} \frac{f(a + h) - f(a) - mh}{h} = 0.$$

In this case the linear approximation is $l(x) = f(a) + m(x - a)$.

Theorem 4.1 Let $f(x)$ be defined on an open interval that contains a. Then f is differentiable at $x = a$ if and only if f has a linear approximation $l(x) = f(a) + m(x - a)$ at a. In this case $m = f'(a)$ so there can be only one linear function with this property.

Proof: We have already seen from (4.1) that if f is differentiable at $x = a$, and we take $m = f'(a)$, then $R(h) = o(|h|)$. On the other hand, suppose that f has a linear approximation $l(x) = f(a) + m(x - a)$ for some m, such that

$$R_1(h) \equiv f(a + h) - l(a + h) = o(|h|),$$

as h tends to zero. Since $l(a + h) = f(a) + mh$,

$$\frac{f(a + h) - f(a)}{h} - m = \frac{R_1(h)}{h}.$$

It follows that

$$\lim_{h \to 0} \frac{f(a + h) - f(a)}{h} = m + \lim_{h \to 0} \frac{R_1(h)}{h} = m. \tag{4.2}$$

From (4.2) we see that we must have $m = f'(a)$. The proof of the theorem is complete.

In general, if f has only one derivative at $x = a$, we cannot say how fast $R(h)/h$ tends to zero; the convergence may be very slow.

EXAMPLE 4.1:

a) Let $f(x) = x^2$. Then $f(a + h) = a^2 + 2ah + h^2$. We read off easily that $l(a + h) = a^2 + 2ah$ and $R(h) = h^2 = o(|h|)$. In this case, if h is cut in half, the error $R(h)$ is reduced by a factor of $1/4$.

b) Let $f(x) = |x|^{1/2} \sin(2x)$. We take $a = 0$. First we verify that f is differentiable at $x = 0$. Because $f(0) = 0$, we have

$$f'(0) = \lim_{h \to 0} \frac{f(h)}{h} = \lim_{h \to 0} |h|^{1/2} \lim_{h \to 0} \frac{\sin(2h)}{h}.$$

Since $\lim_{h \to 0} \sin(2h)/h = 2$, we conclude that $f'(0) = 0$. In this case the linear approximation $l(x) \equiv 0$, whence $R(h) = f(h) = |h|^{1/2} \sin(2h)$. We already know that $R(h) = o(|h|)$, but more precisely, $R(h) = O(|h|^{3/2})$ because $|\sin(2h)| \le 2|h|$. If we cut

h in half, the error $R(h)$ is reduced roughly by a factor of $2^{-3/2}$. Graphs of both examples are shown in Figure 4.1. Notice how the graph in the upper figure, in which the error is $O(|h|^2)$, appears to approach the dashed line faster than it does in the the lower figure where the error is $O(|h|^{3/2})$.

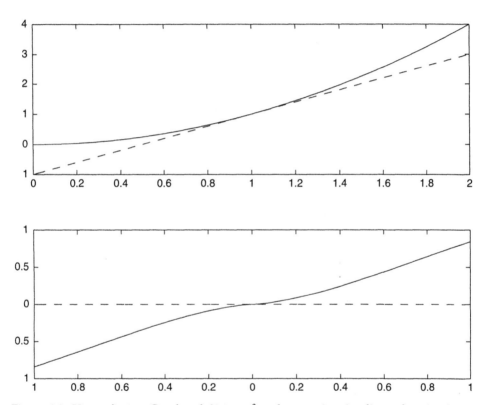

Figure 4.1 Upper figure: Graphs of $f(x) = x^2$ and approximating linear function $l(x) = a^2 + 2(x - a)$, with $a = 1$. Lower figure: graphs of $f(x) = |x|^{1/2} \sin x$ and approximating linear function $l(x) \equiv 0$ at $x = 0$

An immediate consequence of differentiability is

Theorem 4.2 Let f be defined on an open interval that contains a. If f is differentiable at $x = a$, then f is continuous at $x = a$.

Proof: If f is differentiable at $x = a$, then $R(h) = o(|h|)$ and, in particular, $\lim_{h \to 0} R(h) = 0$. Hence

$$\lim_{x \to a} f(x) = \lim_{h \to 0} f(a + h)$$

$$= \lim_{h \to 0} (f(a) + f'(a)h + R(h))$$

$$= f(a) + \lim_{h \to 0} R(h) = f(a).$$

The converse of this theorem is not true. The usual example of a function that is continuous at a point a, but not differentiable there, is $f(x) = |x|$ when $a = 0$. A much more shocking counterexample was exhibited by Weierstrass. His function is continuous everywhere, but it is not differentiable anywhere. Obviously this function is not given by a simple formula. It is constructed by summing an infinite series of sawtooth functions.

Theorem 4.3 Let I be an open interval that contains a and suppose that $f : I \to \mathbb{R}$ and $g : I \to \mathbb{R}$ are both differentiable at $x = a$. Then

(i) For any $c \in R$, the function cf is differentiable at $x = a$ and

$$(cf)'(a) = cf'(a).$$

(ii) The function $f + g$ is differentiable at $x = a$ and

$$(f + g)'(a) = f'(a) + g'(a).$$

(iii) The product function fg is differentiable at $x = a$ and

$$(fg)'(a) = f'(a)g(a) + f(a)g'(a).$$

(iv) If $g(a) \neq 0$, then the quotient function f/g is differentiable at $x = a$ and

$$(\frac{f}{g})'(a) = \frac{f'(a)g(a) - f(a)g'(a)}{g^2(a)}.$$

Proof: The proof of each part follows easily from Theorem 3.2. All we need to do is set up the difference quotients and do a little algebra. We leave (i) and

(ii) to the reader. To prove (iii) we use the device of estimating the difference of products:

$$\frac{f(a+h)g(a+h) - f(a)g(a)}{h} = \frac{f(a+h)g(a+h) - f(a)g(a+h)}{h}$$

$$+ \frac{f(a)g(a+h) - f(a)g(a)}{h}$$

$$= \left[\frac{f(a+h) - f(a)}{h}\right]g(a+h)$$

$$+ f(a)\left[\frac{g(a+h) - g(a)}{h}\right].$$

In the square brackets we see the difference quotients for f and g which we know converge to $f'(a)$ and $g'(a)$. The difference quotient for g is multiplied by the constant $f(a)$ so that this term will converge to $f(a)g'(a)$. The difference quotient for f is multiplied by $g(a + h)$. However, g differentiable at $x = a$ implies that g is continuous at $x = a$. Therefore $g(a + h)$ will converge to $g(a)$ as h tends to zero. Consequently

$$(fg)'(a) = \lim_{h\to 0} \frac{f(a+h)g(a+h) - f(a)g(a)}{h}$$

$$= \lim_{h\to 0} \frac{f(a+h) - f(a)}{h} \lim_{h\to 0} g(a+h)$$

$$+ f(a)\lim_{h\to 0} \frac{g(a+h) - g(a)}{h}$$

$$= f'(a)g(a) + f(a)g'(a).$$

The proof of (iv) proceeds in the same manner. We do the simpler case with $f(x) \equiv 1$ and then combine with part (iii) to get (iv). We have

$$\left(\frac{1}{h}\right)\left[\frac{1}{g(a+h)} - \frac{1}{g(a)}\right] = \left[\frac{g(a) - g(a+h)}{h}\right]\frac{1}{g(a)g(a+h)}.$$

However, these expressions are not defined if $g = 0$. We are assuming that $g(a) \neq 0$, and we use the continuity of g at $x = a$ to deduce that there is a $\delta > 0$ such that $|g(a + h)| \geq |g(a)|/2 > 0$ for $|h| < \delta$. With this preliminary restriction on h, we can take the limit, again using the continuity of g at $x = a$, to deduce

$$\left(\frac{1}{g}\right)'(a) = \lim_{h \to 0} \left(\frac{1}{h}\right)\left[\frac{1}{g(a+h)} - \frac{1}{g(a)}\right]$$

$$= \lim_{h \to 0}\left[\frac{g(a) - g(a+h)}{h}\right]\lim_{h \to 0}\frac{1}{g(a)g(a+h)}$$

$$= \frac{-g'(a)}{g^2(a)}.$$

As a consequence of Theorem 4.3, and the fact that $f(x) = x$ is differentiable for all x, we see that a quotient of polynomials p/q is differentiable at all points a such that $q(a) \neq 0$.

We did not use the idea of a linear approximation in the proof of Theorem 4.3 although it is lurking in the background. To bring out this aspect of the concept of differentiability, and as a warm up for the proof of the chain rule, let us prove the product rule again, this time using the linear approximations. Since f and g are both differentiable at $x = a$, we have

$$f(a + h) = f(a) + f'(a)h + R_1(h)$$

and

$$g(a + h) = g(a) + g'(a)h + R_2(h)$$

where as h tends to zero,

$$R_1(h) = o(h) \quad \text{and} \quad R_2(h) = o(h).$$

If we multiply together the two expressions for $f(a+h)$ and $g(a+h)$, we obtain

$$f(a + h)g(a + h) = f(a)g(a) + [f'(a)g(a) + f(a)g'(a)]h + R(h) \qquad (4.3)$$

where

$$R(h) = [f(a) + f'(a)h]R_2(h) + [g(a) + g'(a)h]R_1(h) + f'(a)g'(a)h^2 + R_1(h)R_2(h).$$

Because of our assumptions about R_1 and R_2, it is clear that $R(h)/h$ tends to zero as h tends to zero. Thus we can see from (4.3) that the product fg

has a linear approximation $l(a + h) = f(a)g(a) + mh$ with $m = f'(a)g(a) + f(a)g'(a)$. Since the linear approximation is unique when it exists, this means that fg is differentiable at $x = a$, and that $(fg)'(a)$ is given by the formula (iii) of Theorem 4.3.

The chain rule is an extremely useful rule of differentiation. It expresses the familiar idea that the rates of change of two processes performed in succession multiply to give the rate of change of the compound process. Let us try to derive the chain rule using difference quotients. We write

$$\frac{g(f(a+h)) - g(f(a))}{h} = \left[\frac{g(f(a+h)) - g(f(a))}{f(a+h) - f(a)}\right] \frac{f(a+h) - f(a)}{h}.$$

The fly in the ointment here is that $f(a + h) - f(a)$ may be zero so that the difference quotient in square brackets may not be defined. A better way, which avoids this problem, uses the idea of linear approximation.

Theorem 4.4 (chain rule). Let I be an open interval and suppose that $f : I \to \mathbb{R}$ is differentiable at $a \in I$. Let J be an open interval with $f(I) \subset J$ and suppose that $g : J \to \mathbb{R}$ is differentiable at $b = f(a)$. Then the composite function $g \circ f$ is differentiable at a and

$$(g \circ f)'(a) = g'(f(a))f'(a).$$

Proof: We will show that the composite function $g \circ f$ has a linear approximation at $x = a$. Thus we must show that there is a number m such that

$$g(f(a + h)) = g(f(a)) + mh + R(h)$$

and $R(h) = o(h)$ as h tends to zero.
  The assumption that f is differentiable at a means that

$$f(a + h) = f(a) + f'(a)h + R_1(h) \tag{4.4}$$

where $R_1(h) = o(h)$ as h tends to zero. The assumption that g is differentiable at $b = f(a)$ means that

$$g(f(a) + k) = g(f(a)) + g'(f(a))k + R_2(k) \tag{4.5}$$

where $R_2(k) = o(k)$ as k tends to zero. If we set $k = f(a+h) - f(a) = hf'(a) + R_1(h)$ and substitute in (4.5) we obtain

$$g(f(a + h)) = g(f(a)) + g'(f(a))f'(a)h + R(h)$$

where

$$R(h) = g'(f(a))R_1(h) + R_2(f'(a)h + R_1(h)).$$

We see that our candidate for the slope m is the number $g'(f(a))f'(a)$.

The theorem will be proved if we can show that $R(h) = o(h)$ as h tends to zero. We have assumed f differentiable at a so that we know

$$\lim_{h \to 0} g'(f(a)) \frac{R_1(h)}{h} = 0.$$

It remains to look at the second term in $R(h)$. (We leave the best part for last.) Because $R_1(h) = o(h)$ as h tends to zero, there is a $\delta_1 > 0$ such that $|R_1(h)| \leq |h|$ for $|h| < \delta_1$. This means that

$$|k| = |f'(a)h + R_1(h)| \leq (|f'(a)| + 1)|h|$$

for $|h| < \delta_1$. Thus as h tends to zero, so will k. We set $M = |f'(a)| + 1$. Now let $\varepsilon > 0$ be given. Because $R_2(k) = o(k)$ as k tends to zero, there is a δ_2 such that

$$|R_2(k)| \leq \frac{\varepsilon|k|}{M} \quad \text{for } |k| < \delta_2. \tag{4.6}$$

Finally we choose $\delta = \min\{\delta_1, \delta_2/M\}$. Now $|h| < \delta$ implies that $|h| < \delta_1$, which in turn implies that $|k| \leq M|h|$. However, $|h| < \delta$ also implies that $|h| < \delta_2/M$, whence $|k| < \delta_2$. We can use the inequality (4.6) to deduce that

$$|R_2(f'(a) + R_1(h))| = |R_2(k)| < \frac{\varepsilon|k|}{M} < \varepsilon \frac{M|h|}{M} = \varepsilon|h|.$$

Because the slope $m = g'(f(a))f'(a)$ yields a linear approximation to $g \circ f$ at $x = a$, we deduce that $g \circ f$ is differentiable at $x = a$ and that

$$(g \circ f)'(a) = m = g'(f(a))f'(a).$$

The theorem is proved.

Exercises for 4.1

1. As preparation for this exercise and the next one, verify that, if $g(x) = C|x|^p$ for $p > 0$, then the ratios $g(x/2)/g(x) = 2^{-p}$.
   Now let $f(x) = \sin(x + \pi/4)$.

   a) Find the linear approximation $l(x)$ to f at $x = 0$. Graph f and l together on the interval $[-0.5, 0.5]$.

   b) Let $R(x) = f(x) - l(x)$. Evaluate $R(x)$ at the points 2^{-k}, $k = 1, 2, \ldots, 5$. By what factor is $R(x)$ reduced when x is cut in half? Does this suggest that $R(x) = o(x)$ as x approaches 0?

   c) Graph $R(x)/x$ on the interval $[-0.5, 0.5]$ to confirm your answer for part b).

2. Let $g(x) = (x^{1/3} + 1) \tan x$.

   a) What is the linear approximation $l(x)$ to g at $x = 0$? You must calculate the derivative of g at $x = 0$ from the definition. Graph g and l together on $[-0.5, 0.5]$ Compare this graph with that of part a) of exercise 1.

   b) Let $R(x) = g(x) - l(x)$. Evaluate $R(x)$ at the points $x_k = 2^{-k}$ for $k = 1, 2, \ldots, 5$. Is there a power $p > 1$ such that $R(x) = O(|x|^p)$? Is $R(x) = o(|x|)$?

   c) Plot $R(x)/x$ on $[-0.5, 0.5]$ to confirm your answer in part b).

3. Let $f(x) = 0$ for $x \le 0$, and $f(x) = 1 - \cos x$ for $x \ge 0$. Is f differentiable at $x = 0$?

4. Suppose that f is differentiable on an open interval I containing x_0, and that $f'(x_0) \ne 0$. Let $y_0 = f(x_0)$. Use the linear approximation to f at x_0 to find an approximate solution to the equation $f(x) = y$ for y near y_0.

5. Let $f(x) = A + Bx + x|x|^{p-1}g(x)$ where g is a continuous function with $g(0) \ne 0$. For what values of $p > 0$ is f differentiable at $x = 0$? If f is differentiable at $x = 0$, what is $f'(0)$? The answer depends on the value of p.

6. If f is differentiable on $\mathbb{R}$, for what values of x is $|f|$ differentiable? Is it true that

$$\frac{d}{dx}|f(x)| = |f'(x)| \, ?$$

7. Let f be differentiable at $x = a$. Find

$$\lim_{h \to 0} \frac{f(a+h) - f(a-h)}{2h}.$$

How do you interpret these quotients in terms of slopes of secant lines to the graph of f?

8. Find constants A and B such that the function

$$f(t) \equiv \sin(A + B(t - 1/2))$$

has the linear approximation $l(t) = t$ at $t = 1/2$.

9. Let $f : I \to \mathbb{R}$ be continuous where I is an open interval containing 0 and suppose that $f(0) = 0$. Suppose that f is differentiable at $x = 0$ with $f'(0) \neq 0$. Show that there is an $\varepsilon > 0$ such that the equation $f(x) = y$ has a solution for $|y| \leq \varepsilon$. Show by an example that the solution need not be unique.

10. Let f be differentiable on $\mathbb{R}$. Define $f_1(x) = f(x)$ and for $n \geq 2$,

$$f_n(x) = f(f_{n-1}(x)).$$

Thus $f_4(x)$ is the composite function $f(f(f(f(x))))$.

a) Calculate $(d/dx)f_n(x)$.

b) If x_0 has the property that $f(x_0) = x_0$ and $|f'(x_0)| < 1$, show that

$$\left. \frac{d}{dx} f_n(x) \right|_{x=x_0}$$

tends to zero as n increases.

4.2 The Mean Value Theorem

The derivative is a quantity that is determined by the local behavior of the function. The derivative $f'(a)$ depends only on values of the function that are very close to a. Nevertheless, the mean value theorem allows us to make estimates of the global change of a function in terms of the derivative.

Definition 4.4 Let I be an open interval. $f : I \to \mathbb{R}$ is *differentiable on I* if f is differentiable at each point $a \in I$.

The set of functions $f : I \to \mathbb{R}$ such that f is differentiable on I and f' is continuous on I is denoted $C^1(I)$. Such functions are said to be *continuously differentiable*, but we shall say that $f \in C^1(I)$, or more simply, f is C^1. When the interval I is not open, for example $I = [a, b)$, $f \in C^1(I)$ will mean that f' is continuous on (a, b) and $\lim_{x \downarrow a} f'(x)$ exists. Thus f' can be extended as a continuous function to $I = [a, b)$. Consistent with this notation, the set of continuous functions on any interval I is denoted $C(I)$ or sometimes $C^0(I)$.

In most circumstances, the functions we deal with are C^1, and some theorems are easier to prove when we use this stronger hypothesis. In Theorem 4.4 (the chain rule), if we assume that f and g are both C^1, the argument to show that $R(h) = o(h)$ is simpler. We shall continue to use the weaker hypothesis of differentiability for a while longer because it is traditional, but in the later chapters we shall opt for efficiency, and use the C^1 hypothesis.

Finding the maxima and minima of a function and estimating the magnitude of change of a function are important topics in analysis.

Definition 4.5 Let I be an interval, and let $f : I \rightarrow \mathbb{R}$. A point $a \in I$ is a *local maximizer* for f if there is a $\delta > 0$ such that $f(x) \le f(a)$ for $x \in I$ and $|x - a| < \delta$. The point $a \in I$ is a *local minimizer* for f if there is a $\delta > 0$ such that $f(a) \le f(x)$ for $x \in I$ and $|x - a| < \delta$.

Theorem 4.5 Let I be an open interval and let $f : I \rightarrow \mathbb{R}$. Suppose that $a \in I$ is a local maximizer or a local minimizer for f and that f is differentiable at a. Then $f'(a) = 0$.

Proof: Suppose that a is a local maximizer for f. Since I is an open interval, there is a $\delta > 0$ such that $f(x) \le f(a)$ for $\{a - \delta < x < a + \delta\} \subset I$. For $h < 0$, the difference quotients

$$\frac{f(a + h) - f(a)}{h} \ge 0.$$

Hence

$$f'(a) = \lim_{h \uparrow 0} \frac{f(a + h) - f(a)}{h} \ge 0.$$

On the other hand, for $h > 0$, the difference quotients are less than or equal to zero and

$$f'(a) = \lim_{h \downarrow 0} \frac{f(a + h) - f(a)}{h} \le 0.$$

Because f is differentiable at a, the limits from the left and the right must agree. Thus $f'(a) = 0$.

A similar argument applies for the case of a local minimizer.

Note that we assumed in Theorem 4.5 that I is an open interval. If I contained an endpoint, and the local maximum occurred there, the (one-sided) derivative need not be zero. An example is given by $f(x) = x$ on $[0, 1]$.

The next step along the way to the mean value theorem is known as Rolle's theorem. It combines a previous result about continuous functions with Theorem 4.5.

Theorem 4.6 (Rolle's theorem) Let $I = [a, b]$ be a closed bounded interval and suppose that $f : I \to \mathbb{R}$ satisfies:

  (i) f is continuous on I;
  (ii) f is differentiable on the open interval $a < x < b$; and
  (iii) $f(a) = f(b) = 0$.

Then there is a number θ, with $a < \theta < b$, where $f'(\theta) = 0$.

Proof: If $f(x) \equiv 0$, the theorem is trivially true. Hence we can assume that $f(x_0) \neq 0$ for some $x_0 \in (a, b)$. First assume that $f(x_0) > 0$. Since f is continuous on I, Theorem 3.5 implies that there is a point $x^* \in I$ such that $f(x^*) \geq f(x)$ for all $x \in I$. In particular, $f(x^*) \geq f(x_0) > 0$, which means that $x^* \in (a, b)$. Now we can apply Theorem 4.5 to conclude that $f'(x^*) = 0$. If $f(x_0) < 0$, we can argue that $\min_I f = f(x_*)$ where $x_* \in (a, b)$ and deduce that $f'(x_*) = 0$.

Theorem 4.7 (Mean value theorem) Let $I = [a, b]$ be a closed bounded interval and suppose that $f : I \to R$ satisfies:

  (i) f is continuous on I;
  (ii) f is differentiable on (a, b).

Then there is a number $\theta \in (a, b)$ such that

$$f(b) - f(a) = f'(\theta)(b - a). \tag{4.7}$$

Proof: We use a simple change of the dependent variable to reduce this case to that of Rolle's theorem. Let $l(x)$ be the linear function

$$l(x) = f(a) + \frac{f(b) - f(a)}{b - a}(x - a).$$

Note that $l(a) = f(a)$ and $l(b) = f(b)$. Then $g(x) = f(x) - l(x)$ satisfies the hypotheses of Rolle's theorem. Hence there is a $\theta \in (a, b)$ such that

$$0 = g'(\theta) = f'(\theta) - l'(\theta).$$

Since $l'(\theta) = (f(b) - f(a))/(b - a)$, the theorem is proved.

The mean value theorem may be expressed in very intuitive geometric terms. It states that for a differentiable function f, there is a number θ in the interval (a, b) where the tangent line to the curve at $(\theta, f(\theta))$ is parallel to the chord through the points $(a, f(a))$ and $(b, f(b))$ (see Figure 4.2).

Several useful corollaries follow easily from the mean value theorem.

Corollary Let f be differentiable on an open interval I and suppose that $f'(x) = 0$ for all $x \in I$. Then f is constant on I.

Proof: Fix $a \in I$, and use Theorem 4.7. For any $x \in I$ we have $f(x) - f(a) = f'(\theta)(b - a)$ for some θ between x and a. However, $f' = 0$ for all $x \in I$ implies $f(x) = f(a)$.

Corollary Let f be differentiable on an open interval I. If $f'(x) > 0$ for all $x \in I$, then f is strictly increasing. If $f'(x) < 0$ for all $x \in I$, then f is strictly decreasing. In either case, f is one-to-one on I.

Proof: We assume that $f'(x) > 0$ on I. If $x, y \in I$ with $x < y$, then there is θ between x and y such that

$$f(y) - f(x) = f'(\theta)(y - x) > 0.$$

This inequality shows that f is one-to-one and strictly increasing on I. The proof when $f'(x) < 0$ is the same.

EXAMPLE 4.2: We often use the mean value theorem together with some inequalities to make estimates of functions.

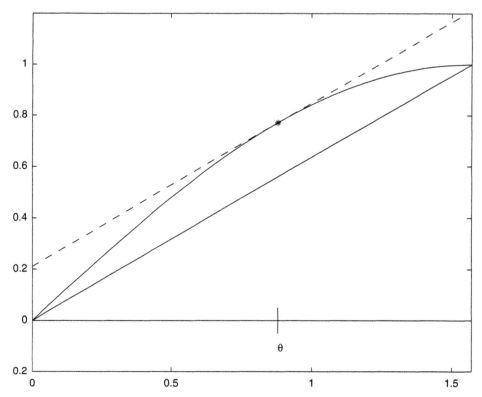

Figure 4.2 Graph of $f = \sin x$, tangent line to graph of f at $(\theta, \sin(\theta))$ where $\theta \approx .8807$, and chord through points $(0,0)$ and $(\pi/2, 1)$

a) We wish to estimate $\log x$ near $x = 1$ in terms of simpler functions, taking advantage of the fact that $\log 1 = 0$ and $(\log x)' = 1/x$. Assuming that $x > 1$, we have

$$\log x = \log x - \log 1 = (x - 1)/\theta$$

where $1 < \theta < x$. This yields the estimates

$$\frac{x - 1}{x} \leq \log x \leq x - 1.$$

Obviously these estimates are not of much use for large x, but they do give us some information for x near 1. For example, we can estimate that

$$0.0476 \leq \frac{0.05}{1.05} \leq \log 1.05 \leq 0.05.$$

b) In Example 1.5, we used induction to prove the Bernoulli inequality

$$(1 + x)^\alpha \geq 1 + \alpha x$$

for $\alpha \in \mathbb{N}$ and $x \geq -1$. The differentiation formula $(x^\alpha)' = \alpha x^{\alpha-1}$ is established for α rational after Theorem 4.8, and for general $\alpha \in \mathbb{R}$ in Appendix I. Assuming this formula, we show that the Bernoulli inequality is valid for all $\alpha \geq 1$ and $x \geq -1$. Suppose $x \geq -1$. We apply the mean value theorem on the interval between x and 0. If $f(x) = (1 + x)^\alpha$, we have $f(0) = 1$ and

$$f(x) - 1 = x f'(\theta)$$

or

$$(1 + x)^\alpha = 1 + \alpha x(1 + \theta)^{\alpha-1}$$

for some θ between 0 and x. Now if $x > 0$, then $\theta > 0$, and because $\alpha \geq 1$, $(1 + \theta)^{\alpha-1} \geq 1$, so that, for $x \geq 0$,

$$(1 + x)^\alpha = 1 + \alpha x(1 + \theta)^{\alpha-1} \geq 1 + \alpha x.$$

When $-1 \leq x < 0$, we have $-1 < \theta < 0$, so that $0 < (1+\theta)^{\alpha-1} \leq 1$. Since $x \leq 0$, we again have

$$(1 + x)^\alpha = 1 + \alpha x(1 + \theta)^{\alpha-1} \geq 1 + \alpha x.$$

Next we use the mean value theorem to answer some questions about inverse functions. Given a function $f : I \to \mathbb{R}$,

(i) When does f have an inverse?
(ii) If f^{-1} exists, will it be differentiable?
(iii) If f^{-1} exists, how do we calculate its derivative in terms of the derivative of f?

These questions can be rephrased as questions about the behavior of the solutions of an equation,

$$f(x) = y.$$

(i) Does this equation have a unique solution for each y in some interval?
(ii) If there is a unique solution $x(y)$ for each y in some interval, does the solution $x(y)$ depend in a differentiable fashion on y? If so, how do we calculate the derivative $dx(y)/dy$?

The next theorem provides some answers to these questions.

Theorem 4.8 Let I be an open interval, and let $f \in C^1(I)$ with $f'(x) \neq 0$ for $x \in I$. Then

(i) The image $J = f(I)$ is an open interval.
(ii) $f : I \to J$ is one-to-one with inverse function $f^{-1} : J \to I$.
(iii) $f^{-1} \in C^1(J)$ and

$$(f^{-1})'(y) = \frac{1}{f'(x)} \quad \text{when } y = f(x).$$

Proof: We assume that $f'(x) \neq 0$ for all $x \in I$. Since f' is continuous on I, the intermediate value theorem tells us that f' cannot change sign on I without passing through zero. Hence either $f'(x) > 0$ for all $x \in I$, or $f'(x) < 0$ for all $x \in I$. We shall assume $f' > 0$ on I.

By the corollary to the mean value theorem, f is strictly increasing (and hence one-to-one). The function f is continuous, so by Theorem 3.8, $J = f(I)$ is an open interval.

Now we use the mean value theorem to show that f^{-1} is continuous. For $x, x_0 \in I$, set $y = f(x)$ and $y_0 = f(x_0)$. Then there is a θ between x and x_0 such that

$$y - y_0 = f(x) - f(x_0) = (x - x_0)f'(\theta).$$

The fact that f' is continuous on I and $f'(x_0) > 0$ implies that there is a $\delta_0 > 0$ such that, for $|x - x_0| < \delta_0$,

$$f'(x) \geq (1/2)f'(x_0).$$

Hence for $|x - x_0| < \delta_0$,

$$|y - y_0| = f'(\theta)|x - x_0| \geq (1/2)f'(x_0)|x - x_0|. \tag{4.8}$$

Let I_0 be the open interval $(x_0 - \delta_0, x_0 + \delta_0)$. Again because f is continuous and strictly increasing, the image of I_0 is another open interval J_0 that contains y_0. Because J_0 is open, there is a $\delta_1 > 0$ such that $\{|y - y_0| < \delta_1\}$ is contained in J_0. This means that when $|y - y_0| < \delta_1$, the inequality (4.8) holds. Finally, to show that f^{-1} is continuous, let $\varepsilon > 0$ be given. We choose $\delta = \min\{\delta_1, \varepsilon f'(x_0)/2\}$. For $|y - y_0| < \delta$, the inequality (4.8) holds whence $|f^{-1}(y) - f^{-1}(y_0)| = |x - x_0| < \varepsilon$.

Showing f^{-1} is differentiable is easy now. Look at the difference quotients

$$\frac{f^{-1}(y) - f^{-1}(y_0)}{y - y_0} = \frac{x - x_0}{f(x) - f(x_0)}. \tag{4.9}$$

We know that if y tends to y_0, then x tends to x_0 and that $f'(x_0) > 0$. Hence we may take the limit in (4.9) to deduce

$$(f^{-1})'(y_0) = \lim_{y \to y_0} \frac{f^{-1}(y) - f^{-1}(y_0)}{y - y_0}$$

$$= \lim_{x \to x_0} \frac{x - x_0}{f(x) - f(x_0)}$$

$$= \frac{1}{f'(x_0)}.$$

The theorem is proved.

In Figure 4.3 we see plotted together the graphs of $f(x)$ and $f^{-1}(x)$. The graph of f^{-1} is found by reflecting the graph of f about the diagonal $y = x$. The slope of the tangent line to the graph of f^{-1} at the point (y_0, x_0) is the reciprocal of the slope of the tangent line at the point (x_0, y_0) on the graph of f.

To extend this result to higher dimensions, we must find the appropriate analogue to the condition $f'(x) > 0$. This will be studied in a later chapter.

Many important functions are in fact defined as inverse functions. The connection between the exponential function and the logarithm function is discussed in Appendix I. Here we observe that, so far, we have defined $f_n(x) = x^n$ for n integer. How should we define x^r for a rational $r = m/n$? It is not hard to verify that, when n is a positive integer, f_n maps $[0, \infty)$ onto itself in a one-to-one fashion. The inverse function, which we denote $f_{1/n}(y) = y^{1/n}$, is the nth

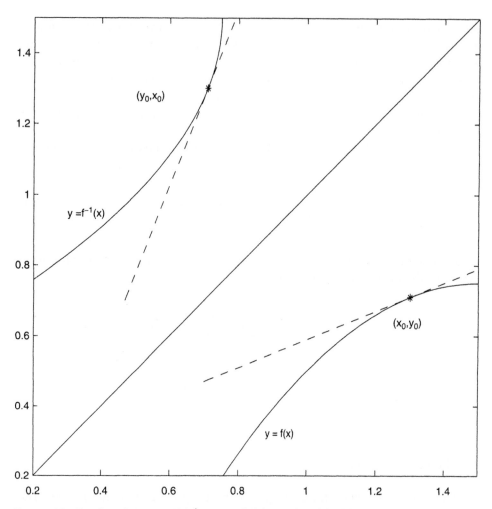

Figure 4.3 Graphs of $f(x)$ and $f^{-1}(x)$ in solid line. Graphs of linear approximations at (x_0, y_0) and at (y_0, x_0) in dashed line

root function. Using Theorem 4.8 we can be sure that $f_{1/n}$ is also C^1 and that

$$f'_{1/n}(y) = \frac{1}{f'_n(f_{1/n}(y))}$$

$$= \frac{1}{n(f_{1/n}(y))^{n-1}}$$

$$= \frac{1}{n} y^{-1+1/n}.$$

For m, n positive integers, $x^{m/n} = f_m(f_{1/n}(x))$ and hence by the chain rule,

$$\frac{d}{dx} x^{m/n} = f'_m(f_{1/n}(x))f'_{1/n}(x)$$

$$= m(x^{1/n})^{m-1}(1/n)x^{-1+1/n}$$

$$= \frac{m}{n} x^{-1+m/n}.$$

We consider several examples that show the necessity of the hypotheses and some of the limitations of the theorem.

EXAMPLE 4.3:

a) Let $f(x) = x^3$. f is one-to-one on all of $\mathbb{R}$ with inverse $f^{-1}(x) = x^{1/3}$. We note that $f'(0) = 0$, so the hypothesis of Theorem 4.8 does not hold for this example on any interval that contains $x = 0$. It is not surprising that the inverse function is not differentiable at $x = 0$, because the graph of the inverse function has a vertical tangent there.

b) If $f : \mathbb{R} \to \mathbb{R}$ satisfies the hypotheses of Theorem 4.8, does f map $\mathbb{R}$ onto $\mathbb{R}$? It may not, as evidenced by the function

$$f(x) = \frac{x|x|}{1 + x^2},$$

whose derivative is

$$f'(x) = \frac{2|x|}{(1 + x^2)^2}.$$

The derivative f' is continuous on $\mathbb{R}$, and $f'(x) > 0$ for all x. However, $|f(x)| < 1$ for all x and, in fact, $f(\mathbb{R})$ is the open interval $(-1, 1)$. If we make the stronger assumption that $f'(x) \geq c > 0$ for all x, then f maps $\mathbb{R}$ onto $\mathbb{R}$ (exercise 6 of this section).

EXAMPLE 4.4: Suppose that we are solving an equation $f(x) = y$ where f satisfies the hypotheses of Theorem 4.8 on an open interval

I. Assume that y depends on a parameter t via a function $y = \phi(t)$ where $\phi(t)$ takes values in $f(I)$ and $t \to \phi(t)$ is C^1. Then the solution x depends on t, $x = x(t)$. By Theorem 4.8 and the chain rule, $t \to x(t)$ will be C^1. Since $x(t) = f^{-1}(\phi(t))$, we have

$$x'(t) = (f^{-1})'(\phi(t))\phi'(t)$$

$$= \frac{\phi'(t)}{f'(x(t))}.$$

EXAMPLE 4.5:   The inverse function theorem can be used to justify a change of variable in a differential equation.

Let R denote the radius of the earth and h the height above the surface of the earth of a rocket shot up vertically at time $t = 0$. From Newton's second law and the inverse square law, we derive the differential equation satisfied by h:

$$h''(t) = -g\left(\frac{R}{h+R}\right)^2 \tag{4.10}$$

with initial conditions

$$h(0) = 0 \quad \text{and} \quad h'(0) = v_0 > 0. \tag{4.11}$$

From the theory of differential equations, we know that there is a unique solution of this initial value problem that has two continuous derivatives. We can transform the equation into a more tractable first order equation in the velocity if we make a change of variable. According to the differential equation, $v(t) = h'(t)$ is decreasing, but $v(0) = v_0 > 0$. Thus there must be some interval $[0, T)$ on which $h'(t) = v(t) > 0$. In view of Theorem 4.8, the inverse function $t = g(h)$ exists on the interval $[0, H)$ where $H = h(T)$. Let $\tilde{v}(h) = v(g(h))$. Then $h'(t) = v(t) = \tilde{v}(h(t))$. We find that

$$h''(t) = \frac{dv(t)}{dt} = \frac{d}{dt}\tilde{v}(h(t))$$

$$= \frac{d\tilde{v}(h)}{dh}\frac{dh}{dt} = \frac{1}{2}\frac{d}{dh}\tilde{v}^2(h).$$

Using h as the dependent variable, the initial value problem (4.10), (4.11) becomes an initial value problem for a first-order differential equation

$$\frac{1}{2}\frac{d}{dh}\tilde{v}^2(h) = -g\left(\frac{R}{R+h}\right)^2, \qquad \tilde{v}(0) = v_0. \tag{4.12}$$

This equation can be integrated immediately. We take the positive root for $\tilde{v}$:

$$\tilde{v}(h) = \sqrt{v_0^2 + \frac{2gR^2}{R+h} - 2gR}. \tag{4.13}$$

The solution (4.13) is valid on the interval $[0, H)$ as long as the following inequality holds:

$$v_0^2 + 2gR^2/(R+h) - 2gR > 0. \tag{4.14}$$

If $v_0^2 < 2gR$, equation (4.14) is satisfied for

$$0 \le h < H = \frac{2gR^2}{2gR - v_0^2} - R.$$

In this case H is the maximum height attained by the rocket. If $v_0^2 \ge 2gR$, (4.14) is satisfied for all $h > 0$. The quantity $v_0 = \sqrt{2gR}$ is the well-known escape velocity.

In the preceding paragraph, we carefully distinguished between the functions $v(t)$, velocity as a function of time, and $\tilde{v}(h)$, velocity as a function of height. In most treatments, however, velocity is a physical quantity that can be thought of as a function of t, or as a function of h, and we write $v(h)$ or $v(t)$. The chain rule is then expressed

$$\frac{dv}{dt} = \frac{dv}{dh}\frac{dh}{dt}.$$

Exercises for 4.2

1. Find the local minimizer(s) and maximizer(s) of the following functions on the indicated domains. Also find the global minimizers and maximizers when they exist.

   a) $f(x) = (x - 1)(x - 2)/(x^2 + 1)$.

   b) $f(x) = |1 + x^2 - (2/3)x^3|$, $-1/2 \le x \le 2$.

2. Let $f(x) = 1/x + 2\sqrt{x + 1}$ for $x > 0$.

   a) Show that $f(x) > f(2)$ for $x \ge 2$ and that $f(x) > f(1)$ for $0 < x \le 1$.

   b) Prove that the global minimum of f occurs in the interval $[1, 2]$.

3. Suppose that f and f' are continuous on $I = [a, b]$. Show that there is a constant $L > 0$ such that $|f(x) - f(y)| \le L|x - y|$ for all $x, y \in I$.

4. Let $f(x)$ be continuous on $\{x \ge 0\}$ and differentiable on $\{x > 0\}$. Suppose that $|f'(x)| \le c < 1$ for all $x > 0$, and that $f(0) > 0$. Show that there is a unique solution $x_* > 0$ to the equation $f(x) = x$.

5. Use the mean value theorem to show that

   a) $\cos x \le \dfrac{1}{\sqrt{2}}(1 + \pi/4 - x)$ for $0 \le x < \pi/2$.

   b) $\cos x \ge \dfrac{1}{\sqrt{2}} + \pi/4 - x$ for $\pi/4 \le x < \pi$.

6. Suppose that f is differentiable on $\mathbb{R}$ and that $f'(x) \ge c > 0$ for all x.

   a) Show that for $x > 0$, $f(x) \ge f(0) + cx$.

   b) Show that for $x < 0$, $f(x) \le f(0) + cx$.

   c) Show that for each $y \in \mathbb{R}$, there is unique x such that $f(x) = y$.

7. Let $f(x) = x \exp(-x^2)$. For which values of y does there exist a solution x of $f(x) = y$? For which values of y is the solution unique?

8. Let $f(x)$ be differentiable on $[0, \infty)$, and suppose that $\lim_{x \to \infty} f'(x) = L$. Show that for each $h > 0$,

$$\lim_{x \to \infty} \frac{f(x + h) - f(x)}{h} = L.$$

9. Use the same hypotheses as in exercise 8. Show that

$$\lim_{x \to \infty} \frac{f(x)}{x} = L.$$

First use the mean value theorem on an interval $[R, x]$ to show that

$$\frac{f(x)}{x} - L = f'(\theta) - L + \frac{f(R) - Rf'(\theta)}{x}$$

for some θ, $R < \theta < x$. Now use the hypothesis that $\lim_{x\to\infty} f'(x) = L$ to finish the proof.

10. Let $f(x) = x - x^2 \sin(1/x)$ for $x \neq 0$, and set $f(0) = 0$.

   a) Show that $f'(0) = 1$.
   b) Show that $\lim_{x\to 0} f'(x)$ does not exist.

11. Let $u(t)$ and $v(t)$ be C^1 functions on $\mathbb{R}$, and suppose that they solve the system of differentiable equations:

$$u' = v$$
$$v' = -\sin u.$$

Show that the function $t \to V(t) = (1/2)v^2(t) + \cos(u(t))$ is constant.

12. Here is an extension of exercise 11. Let $H(u, v)$ be a function of two variables on $\mathbb{R}^2$ with continuous partial derivatives H_u and H_v. Suppose that u and v are C^1 functions on $\mathbb{R}$, and solve the system of differentiable equations:

$$u'(t) = H_v(u(t), v(t))$$
$$v'(t) = -H_u(u(t), v(t)).$$

Show that the function $t \to H(u(t), v(t))$ is constant.

4.3 The Cauchy Mean Value Theorem and l'Hôpital's Rule

The Cauchy mean value theorem is a generalization of the mean value theorem that is especially useful when dealing with quotients.

Theorem 4.9 Suppose that $f : [a, b] \to \mathbb{R}$ and $g : [a, b] \to \mathbb{R}$ are both continuous on $[a, b]$ and differentiable on (a, b). In addition suppose that $g'(x) \neq 0$ for

$x \in (a, b)$. Then there is $\theta \in (a, b)$ such that

$$\frac{f(b) - f(a)}{g(b) - g(a)} = \frac{f'(\theta)}{g'(\theta)}.$$

Proof: The statement of the theorem looks suspicious, because possibly $g(a) = g(b)$. However, the hypothesis that $g'(x) \neq 0$ for $x \in (a, b)$ and the mean value theorem imply that $g(b) \neq g(a)$.

As in the proof of the mean value theorem, we shall reduce the problem to one that can be handled by Rolle's theorem. We look for a function $h(x)$ of the form

$$h(x) = A(g(x) - g(a)) - (f(x) - f(a)).$$

We see that $h(a) = 0$, and we want to choose A so that $h(b) = 0$. This yields the equation

$$f(b) - f(a) = A(g(b) - g(a)). \qquad\qquad (4.15)$$

Since $g(b) \neq g(a)$, we deduce that

$$A = \frac{f(b) - f(a)}{g(b) - g(a)}.$$

It is clear that h is continuous on $[a, b]$ and differentiable on (a, b). Hence by Rolle's theorem, there is a $\theta \in (a, b)$ such that $h'(\theta) = 0$. Now

$$h'(x) = \left(\frac{f(b) - f(a)}{g(b) - g(a)} \right) g'(x) - f'(x).$$

Hence

$$0 = h'(\theta) = \left(\frac{f(b) - f(a)}{g(b) - g(a)} \right) g'(\theta) - f'(\theta),$$

which is the assertion of the theorem.

The Cauchy mean value theorem has a natural geometric interpretation. Let a curve Γ in $\mathbb{R}^2$ be parameterized by $t \to (g(t), f(t))$. $f'(t)/g'(t)$ is the slope of

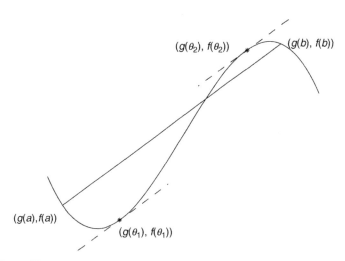

Figure 4.4 Curve Γ parametrized by $t \to (f(t), g(t))$, and two points on the curve where the tangent line has the same slope as that of the chord

the tangent vector to the curve, and $A = (f(b) - f(a))/(g(b) - g(a))$ is the slope of the line through the points $(g(a), f(a))$ and $(g(b), f(b))$. The theorem says that there is a point $(g(\theta), f(\theta))$ on Γ where the slope of the tangent vector equals the slope of the line through the points $(g(a), f(a))$ and $(g(b), f(b))$. In Figure 4.4, we see an example where there are two points on the curve where the tangent line has the same slope as that of the line.

l'Hôpital's rule

With the results at our disposal thus far, we do not know how to determine the limit

$$\lim_{x \to b} \frac{f(x)}{g(x)}$$

when both $\lim_{x \to b} f(x) = 0$ and $\lim_{x \to b} g(x) = 0$. Johann Bernoulli found a way to deal with this problem, and his ideas were published in a book by l'Hôpital; they became known as l'Hôpital's rule. There are several different cases. First we consider the "0/0" case.

Theorem 4.10 (l'Hôpital's rule, 0/0 case) Let $f : (a, b) \to \mathbb{R}$ and $g : (a, b) \to \mathbb{R}$ be differentiable on (a, b) with $g(x) \neq 0$ and $g'(x) \neq 0$ for $x \in (a, b)$. Suppose that the left-hand limits exist,

$$\lim_{x \uparrow b} f(x) = 0 \quad \text{and} \quad \lim_{x \uparrow b} g(x) = 0; \tag{4.16}$$

and that the limit

$$\lim_{x \uparrow b} \frac{f'(x)}{g'(x)} = L.$$

Then

$$\lim_{x \uparrow b} \frac{f(x)}{g(x)} = L.$$

Proof: We extend the definition of f and g to the interval $(a, b]$ by setting $f(b) = g(b) = 0$. Since $\lim_{x \uparrow b} f(x) = \lim_{x \uparrow b} g(x) = 0$, f and g extended this way are continuous on $(a, b]$. For each $x \in (a, b)$, we apply the Cauchy mean value theorem on the interval $[x, b]$: there is a $\theta(x)$, $x < \theta(x) < b$, such that

$$\frac{f'(\theta(x))}{g'(\theta(x))} = \frac{f(b) - f(x)}{g(b) - g(x)} = \frac{f(x)}{g(x)}.$$

Because we have assumed that $\lim_{x \to b}(f'/g')(x) = L$, for each $\varepsilon > 0$, there is a $\delta > 0$ such that

$$\left| \frac{f'(x)}{g'(x)} - L \right| < \varepsilon$$

for $b - \delta < x < b$. If we choose x in this interval, then $\theta(x)$ also lies in this interval. Hence for $b - \delta < x < b$,

$$\left| \frac{f(x)}{g(x)} - L \right| = \left| \frac{f'(\theta(x))}{g'(\theta(x))} - L \right| < \varepsilon.$$

This last inequality shows that $\lim_{x \to b} f(x)/g(x) = L$.

A second version of l'Hôpital's rule deals with the case "∞/∞".

Theorem 4.11 (l'Hôpital's rule, ∞/∞ case). Suppose that f and g are both differentiable on the interval (a, b) and that $g(x) \neq 0$ and $g'(x) \neq 0$ for $x \in (a, b)$. Suppose that the limits from the left are

$$\lim_{x \uparrow b} f(x) = \infty \quad \text{and} \quad \lim_{x \uparrow b} g(x) = \infty; \tag{4.17}$$

and that

$$\lim_{x \uparrow b} \frac{f'(x)}{g'(x)} = L.$$

Then

$$\lim_{x \uparrow b} \frac{f(x)}{g(x)} = L.$$

Proof: We assume the existence of the limit of the quotient f'/g' as x tends to ∞. This means that, for each $\varepsilon > 0$, there is a $\delta_0 > 0$ such that, for $b - \delta_0 < x < b$,

$$\left| \frac{f'(x)}{g'(x)} - L \right| < \varepsilon.$$

Let $x \in (b - \delta_0, b)$. We apply the Cauchy mean value theorem to f/g on the interval $[b - \delta_0, x]$. There is a $\theta \in (b - \delta_0, x)$ such that

$$\frac{f(x) - f(b - \delta_0)}{g(x) - g(b - \delta_0)} = \frac{f'(\theta)}{g'(\theta)}.$$

Hence

$$\left| \frac{f(x) - f(b - \delta_0)}{g(x) - g(b - \delta_0)} - L \right| = \left| \frac{f'(\theta)}{g'(\theta)} - L \right| < \varepsilon.$$

Now we must use the fact that both $f(x)$ and $g(x)$ are getting arbitrarily large as x approaches b. We write the quotient as

$$\frac{f(x) - f(b - \delta_0)}{g(x) - g(b - \delta_0)} - L = \frac{f(x)}{g(x)} h(x) - L$$

where

$$h(x) = \frac{1 - f(b - \delta_0)/f(x)}{1 - g(b - \delta_0)/g(x)}.$$

Assumption (4.17) implies that there is a $\delta_1 > 0$ such that $|h(x) - 1| < \varepsilon$ and $h(x) \geq 1/2$ for $b - \delta_1 < x < b$. Let $\delta = \min\{\delta_0, \delta_1\}$. Then for $b - \delta < x < b$, we have

$$(1/2)\left|\frac{f(x)}{g(x)} - L\right| \leq \left|\left(\frac{f(x)}{g(x)} - L\right)h(x)\right|$$

$$\leq \left|\frac{f(x)}{g(x)}h(x) - L\right| + |L - Lh(x)|$$

$$\leq \varepsilon + \varepsilon|L|.$$

Thus

$$\left|\frac{f(x)}{g(x)} - L\right| \leq 2(1 + |L|)\varepsilon$$

for $b - \delta < x < b$. The theorem is proved.

Theorems 4.10 and 4.11 have been stated and proved for a bounded interval (a, b); the results can also be proved for an unbounded interval with $b = \infty$.

EXAMPLE 4.6:

a) Consider the quotient

$$\frac{f(x)}{g(x)} = \frac{1 - \cos x}{x^2}.$$

We can apply l'Hôpital's rule in the $0/0$ case. We have

$$\lim_{x \to 0} \frac{1 - \cos x}{x^2} = \lim_{x \to 0} \frac{\sin x}{2x} = \frac{1}{2}.$$

b) What is $\lim_{x \downarrow 0} x \log x$? We can apply the ∞/∞ version of l'Hôpital's rule after writing

$$x \log x = \frac{\log x}{1/x}.$$

Then

$$\lim_{x \downarrow 0} x \log x = \lim_{x \downarrow 0} \frac{1/x}{-1/x^2} = 0.$$

4.4 The Second Derivative Test

In Theorem 4.6, we saw that if $f(x)$ is differentiable at $x = a$, and f has a local maximum or local minimum at $x = a$, then $f'(a) = 0$. However, the condition $f'(a) = 0$ is not sufficient to deduce that f has either a local maximum or a local minimum at $x = a$. We need more information about the function.

Definition 4.6 Let f be differentiable at each point of an open interval I. The *second derivative* of f at $a \in I$ is

$$f''(a) = \lim_{h \to 0} \frac{f'(a + h) - f'(a)}{h}$$

when this limit exists.

In the following lemma, we see an alternate way to calculate the second derivative. This will in turn lead us to define a quadratic approximation to f.

Lemma 4.12 Let f be differentiable on an open interval I. Let $a \in I$ and let

$$l(x) = f(a) + f'(a)(x - a)$$

be the linear approximation to f at $x = a$. If $f''(a)$ exists, then

$$\lim_{x \to a} \frac{f(x) - l(x)}{(x - a)^2} = \frac{f''(a)}{2}.$$

Proof: We use l'Hôpital's rule on the fraction $(f(x) - l(x))/(x - a)^2$, since both numerator and denominator tend to zero as x approaches a. Keeping in mind that $l'(x) = f'(a)$, we see that

$$\lim_{x \to a} \frac{f(x) - l(x)}{(x-a)^2} = \lim_{x \to a} \frac{f'(x) - f'(a)}{2(x-a)} = \frac{f''(a)}{2}.$$

This lemma may be restated in a manner that reveals that it deals with a refinement of the linear approximation which we call a quadratic approximation.

Corollary If f is differentiable on an open interval I, $a \in I$, and $f''(a)$ exists, then

$$f(x) = f(a) + f'(a)(x-a) + \frac{f''(a)}{2}(x-a)^2 + R_2(x, a)$$

and $R_2(x, a) = o(|x - a|^2)$ as x approaches a.

An important consequence of Lemma 4.12 is the well-known second derivative test from calculus.

Theorem 4.13 (Second derivative test) Suppose that f is differentiable on an open interval I that contains a. Suppose that $f'(a) = 0$ and that $f''(a)$ exists.

(i) If $f''(a) > 0$, then $x = a$ is a (strict) local minimizer for f.
(ii) If $f''(a) < 0$, then $x = a$ is a (strict) local maximizer for f.
(iii) If $f''(a) = 0$, we cannot draw any conclusion.

Proof: Because f' is differentiable at $x = a$, and $f'(a) = 0$, the linear approximation to f at $x = a$ is $l(x) \equiv f(a)$. Hence Lemma 4.12 says that

$$\lim_{x \to a} \frac{f(x) - f(a)}{(x-a)^2} = \frac{f''(a)}{2}. \qquad (4.18)$$

Now if $f''(a) > 0$, then there is a $\delta > 0$, such that the quotients

$$\frac{f(x) - f(a)}{(x-a)^2} > 0$$

for $0 < |x - a| < \delta$. This implies that $f(x) > f(a)$ for $0 < |x - a| < \delta$. Thus f has a strict minimum at $x = a$. If $f''(a) < 0$, the same argument shows that there is a $\delta > 0$ such that $f(x) < f(a)$ for $0 < |x - a| < \delta$.

The function $f(x) = x^3$, with $f'(0) = f''(0) = 0$ has neither a maximum nor a minimum at $x = 0$, while the function $f(x) = x^4$, also with $f'(0) = f''(0) = 0$, has a minimum at $x = 0$. These examples show that when $f'(a) = f''(a) = 0$, we can not determine if f has a maximum or minimum at $x = a$.

A partial converse is given in the next theorem. We leave the proof as an exercise.

Theorem 4.14 Let I be an open interval and suppose that f is differentiable on I. Suppose that f has a local minimum at $a \in I$. If $f''(a)$ exists, then $f''(a) \geq 0$. If f has a local maximum at a, and $f''(a)$ exists, then $f''(a) \leq 0$.

> **EXAMPLE 4.7:** We take this opportunity to introduce some terminology from economics that we will use here and in later exercises.
> The variable x will represent the output level of some product. $C(x)$ is the cost of producing x units of the product. We shall assume that $C \in C^1([0, \infty))$, that $C(x) \geq 0$ and that the *fixed cost* $C(0) > 0$. The *marginal cost* $MC(x) = C'(x)$. The *average cost* is $AC(x) = C(x)/x$, defined for $x > 0$. The *revenue* function is $R(x)$, and the *profit* function is $\pi(x) = R(x) - C(x)$.
> It is interesting to compare the average cost AC with the marginal cost MC. If $MC > AC$, the average cost should be increasing. If $MC < AC$, the average cost should be decreasing. For example, if a batter gets a hit every time he comes to bat in a game, his batting average will go up, while if he goes hitless, his batting average will go down. This suggests that the average cost should achieve a local minimum when $MC = AC$. In fact, we see that
>
> $$\frac{d}{dx} AC = \frac{d}{dx} \frac{C(x)}{x} = \frac{C'(x)x - C(x)}{x^2}$$
> $$= \frac{MC - AC}{x}.$$
>
> Thus the minima and maxima of AC can only occur at the roots of the equation $MC = AC$.
> To make a statement about a global minimum of AC, we need to make an assumption about the second derivative C''. In addition to the assumption that $C \in C^1([0, \infty))$, we assume that C' is differentiable on $(0, \infty)$ with
>
> (i) $C''(x) > 0$ for all $x > 0$; and that
> (ii) there are constants $\alpha > 0$ and $c > 0$ such that $C''(x) \geq \alpha/x$ for $x \geq c > 0$.

Then AC has a global minimum at x_* where x_* is the unique solution of $MC(x_*) = AC(x_*)$.

First we show that $MC - AC$ has a unique zero. Let $u(x) = x(MC - AC) = xC'(x) - C(x)$. For $x > 0$, we have

$$u'(x) = xC''(x) > 0.$$

For $x \geq c$, we have

$$u'(x) = xC''(x) \geq \alpha.$$

Therefore, u is strictly increasing, and grows without bound as x tends to ∞. However, u is continuous and $u(0) = -C(0) < 0$. By the intermediate value theorem there is an $x_* > 0$ such that $u(x_*) = 0$ and since u is strictly increasing, $u(x) < 0$ for $x < x*$ and $u(x) > 0$ for $x > x_*$. The same conclusion holds for $MC - AC = u(x)/x$. We calculate

$$(AC)'' = \frac{1}{x}C'' - \frac{2}{x^2}(MC(x) - AC(x)).$$

Consequently, $(AC)''(x_*) = C''(x_*)/x_* > 0$, which means that x_* is a local minimum of AC. The global minimum of AC occurs at x_* because $(AC)' = u(x)/x^2$ is negative for $x < x_*$ and positive for $x > x_*$.

As examples, consider cost functions $C(x) = C_0 + x^p$. The assumptions (i) and (ii) are satisfied for $p > 1$. When $p = 1$, the assumptions on C'' are not satisfied, and $AC(x) = 1 + C_0/x$ does not have a global minimum.

Exercises for 4.3 and 4.4

1. Determine if the following limits exist:

a) $\lim\limits_{x \to 0} \dfrac{\log(x + 1)}{\sin x}$

b) $\lim\limits_{x \to 0} \dfrac{\tan x - x}{x^3}$

c) $\lim\limits_{x \uparrow \pi/2} (\pi/2 - x)\tan x$

d) $\lim\limits_{x \to 0} \dfrac{\arctan x}{x}$.

2. We revisit the proof of the Cauchy mean value theorem. Let $f, g :$ $[a, b] \to \mathbb{R}$ be continuous with f and g differentiable on (a, b). We do not assume that $g'(x) \neq 0$. Define the auxilliary function

$$k(x) = (f(b) - f(a))(g(x) - g(a)) - (f(x) - f(a))(g(b) - g(a)).$$

   a) Show that we can apply Rolle's theorem to k.
   b) Show that there is a number $\theta \in (a, b)$ such that

$$(f(b) - f(a))g'(\theta) = (g(b) - g(a))f'(\theta).$$

   c) Suppose that $|f'(x)| \geq |g'(x)|$ for $x \in (a, b)$ and $|f'(x)| > 0$ for $x \in (a, b)$. Show that $|f(x) - f(y)| \geq |g(x) - g(y)|$ for all $x, y \in [a, b]$.

3. In this exercise, we consider an example that shows that the converse to Theorem 4.10 (l'Hôpital's rule) does not hold.
   Let $f(x) = x^2 \sin(1/x)$ for $x \neq 0$ and let $g(x) = \sin x$. Show that $\lim_{x \to 0} f(x) = 0$, that $\lim_{x \to 0} g(x) = 0$, and that $\lim_{x \to 0} (f/g)(x)$ exists. Then show that $\lim_{x \to 0} f'(x)/g'(x)$ does not exist.

4. Determine whether the point $x = 0$ is a maximizer, a minimizer, or neither for the following functions.

   a) $x^2 \tan x$.
   b) $1 - \cos x$.
   c) $(\sin x - x)^2$.

5. Prove Theorem 4.14 using equation (4.18).

6. Let y be twice differentiable on $(0, 1)$, and continuous on the closed interval $[0, 1]$, and suppose that y satisfies the differential equation and boundary conditions:

$$-y'' + y = f, \qquad y(0) = y(1) = 0.$$

   Use Theorem 4.14 to show that if $f(x) > 0$ for $0 < x < 1$, then $y(x) > 0$ for $0 < x < 1$.

7. Let f be differentiable on an open interval I, and let $a \in I$. Assume that $f''(a)$ exists. Use the corollary to Lemma 4.12 to show that

$$\lim_{h \to 0} \frac{f(a + h) - 2f(a) + f(a - h)}{h^2} = f''(a).$$

8. Let f be twice differentiable on an open interval I. Suppose that there are three distinct points $x_1, x_2, x_3 \in I$ with $f(x_1) = f(x_2) = f(x_3) = 0$. Show that there is a point $z \in I$ such that $f''(z) = 0$.

Chapter 5

Higher Derivatives and Polynomial Approximation

Chapter Overview

The subject of this chapter is polynomial approximation. Taylor polynomials and the Lagrange form of the remainder are presented along with several applications. Error bounds are derived for one of the standard approximations to the square root function. An asymptotic formula is derived to locate the eigenvalues of a boundary value problem. Numerical differentiation formulas illustrate another application of Taylor polynomials. The centered difference approximation for the second derivative is used to approximate a nonlinear second order differential equation with a system of nonlinear algebraic equations. Polynomial interpolation is discussed, along with the form of the error. These results will be used later in Chapter 7 to derive formulas for numerical integration. The last section in the chapter is devoted to convex functions. We show that convex functions are continuous, and we use the error results for polynomial interpolation to characterize convex functions with two derivatives.

5.1 Taylor Polynomials

Up to this point, we have used linear approximations, with the exception of Section 4.4, where we used the second derivative to get information about the local maxima and minima of a function. It seems plausible that if we could approximate a function by a polynomial of higher degree, we could get even more information about the function. However, to make higher order approximations, we must make more assumptions about the function.

Definition 5.1 Let I be an open interval and let $a \in I$. We say that f is k *times differentiable* at a if f is $k - 1$ times differentiable on I and if the limit

$$f^{(k)}(a) = \lim_{h \to 0} \frac{f^{(k-1)}(a + h) - f^{(k-1)}(a)}{h}$$

exists.

Definition 5.2 Let I be an open interval. We shall say that $f \in C^k(I)$ if $f, f', \cdots, f^{(k)}$ all exist and are continuous on I. We say that $f \in C^\infty(I)$ if f has derivatives of all orders at each point of I.

Definition 5.3 Let I be an open interval and let $a \in I$. Let $f \in C^n(I)$. The *Taylor polynomial* of f at a of degree $n \geq 0$ is

$$P_n(x, a) = f(a) + f'(a)(x - a) + \frac{1}{2}f''(a)(x - a)^2 + \cdots + \frac{1}{n!}f^{(n)}(a)(x - a)^n.$$

We note that the Taylor polynomial of degree 1 is just the linear approximation that we have used before. The Taylor polynomial of degree 2 is the quadratic approximation that we used in Section 4.4 to investigate the behavior of a function near a critical point. In general, the Taylor polynomial of degree n is constructed so that

$$
\begin{aligned}
P(a) &= f(a) \\
P'(a) &= f'(a) \\
P''(a) &= f''(a) \\
\cdots &= \cdots \\
P^{(n)}(a) &= f^{(n)}(a).
\end{aligned}
$$

In Figure 5.1, the function $f(x) = \sin x + 8(x - \pi/4)^3$ is approximated by Taylor polynomials of degree 1 and 2 at $x = \pi/4$.

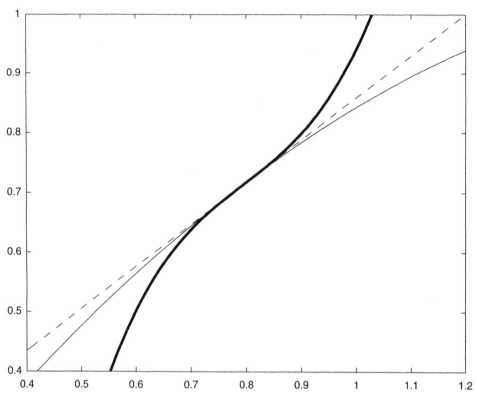

Figure 5.1 Function $f(x) = \sin x + 8(x - \pi/4)^3$, heavy solid line. Taylor polynomials $P_1(x, \pi/4)$, dashed line, and $P_2(x, \pi/4)$, thin line

We will investigate how well a Taylor polynomial approximates a function. Our results so far show that if $f \in C^1(I)$, and $a \in I$, then

$$f(x) = f(a) + f'(a)(x - a) + R_1(x, a) = P_1(x, a) + R_1(x, a)$$

where $R_1(x, a) = o(|x - a|)$. In Section 4.4, we saw that if $f \in C^2(I)$,

$$\begin{aligned} f(x) &= f(a) + f'(a)(x - a) + (1/2)f''(a)(x - a)^2 + R_2(x, a) \\ &= P_2(x, a) + R_2(x, a) \end{aligned}$$

and $R_2(x, a) = o((x - a)^2)$. In general, it can be shown that if $f \in C^n(I)$, then

$$f(x) = P_n(x) + R_n(x, a)$$

and $R_n(x, a) = o(|x - a|^n)$. We will not prove this result. Rather, we will obtain a more explicit form of the remainder $R_n(x, a)$ by requiring that f have one more derivative. As a consequence of the following theorem, we shall see that if $f \in C^{n+1}(I)$, then $R_n(x, a) = O(|x - a|^{n+1})$.

Theorem 5.1 (Lagrange) Let I be an open interval. Let $f \in C^{n+1}(I)$ and $a \in I$. Then for $x \in I$,

$$f(x) = P_n(x, a) + R_n(x, a)$$

where the remainder

$$R_n(x, a) = \frac{1}{(n + 1)!} f^{n+1}(\theta)(x - a)^{n+1}.$$

The number θ lies between x and a. The longer form of this equation is

$$f(x) = f(a) + f'(a)(x - a) + \cdots + \frac{1}{n!} f^{(n)}(a)(x - a)^n$$
$$+ \frac{1}{(n + 1)!} f^{(n+1)}(\theta)(x - a)^{n+1}. \tag{5.1}$$

Proof: Let $M = \max\{f^{(n+1)}(t) : |t - a| \le |x - a|\}$ and $m = \min\{f^{(n+1)}(t) : |t - a| \le |x - a|\}$. Let us first show that for $x > a$,

$$\frac{m}{(n + 1)!}(x - a)^{n+1} \le f(x) - P_n(x, a) \le \frac{M}{(n + 1)!}(x - a)^{n+1}. \tag{5.2}$$

To do this, we consider the auxiliary function

$$g(x) \equiv f(x) - P_n(x, a) - \frac{M}{(n + 1)!}(x - a)^{n+1}.$$

We note that

$$g(a) = g'(a) = \cdots = g^{(n)}(a) = 0$$

and

$$g^{(n+1)}(t) = f^{(n+1)}(t) - M \le 0, \quad a \le t \le x.$$

It follows by the mean value theorem that $g^{(n)}(t) \leq 0$ for $a \leq t \leq x$. This in turn implies that $g^{(n-1)}(t) \leq 0$ for $a \leq t \leq x$. Repeating this argument, we deduce that $g(t) \leq 0$ for $a \leq t \leq x$, and in particular, $g(x) \leq 0$. This means that

$$f(x) - P_n(x) \leq \frac{M}{(n+1)!}(x-a)^{n+1}. \tag{5.3}$$

To get the other inequality, let

$$h(x) = f(x) - P_n(x,a) - \frac{m}{(n+1)!}(x-a)^{n+1}.$$

Arguing as before, $h^{(n+1)}(t) \geq 0$, $a \leq t \leq x$, and we deduce that $h(x) \geq 0$. Hence

$$f(x) - P_n(x,a) \geq \frac{m}{(n+1)!}(x-a)^{n+1}. \tag{5.4}$$

We have proved (5.2). Now multiply (5.2) by $(n+1)!/(x-a)^{n+1}$ to arrive at

$$m \leq \frac{(n+1)!}{(x-a)^{n+1}}[f(x) - P_n(x,a)] \leq M. \tag{5.5}$$

We have assumed that $f^{(n+1)}$ is continuous so that there are values t_* and t^* in the interval $[a,x]$ where $f^{(n+1)}(t_*) = m$ and $f^{(n+1)}(t^*) = M$. Hence by the intermediate value theorem, there is a value $\theta \in [a,x]$, that depends on x, such that (5.1) is true.

The proof for $x < a$ is similar, but some care must be taken with the signs.

The estimates (5.3) and (5.4) are illustrated in the case that $f(x) = x + \sin x$, with $n = 1$ at $x = 0$ (see Figure 5.2). The linear approximation is $P_1(x) = 2x$. Since $f''(x) = -\cos x$, $M = 1$ and $m = -1$. The upper quadratic approximation is $2x + x^2/2$ and the lower quadratic approximation is $2x - x^2/2$.

The Taylor polynomials are useful in dealing with the situation when the second derivative test is inconclusive (i.e., $f''(a) = 0$).

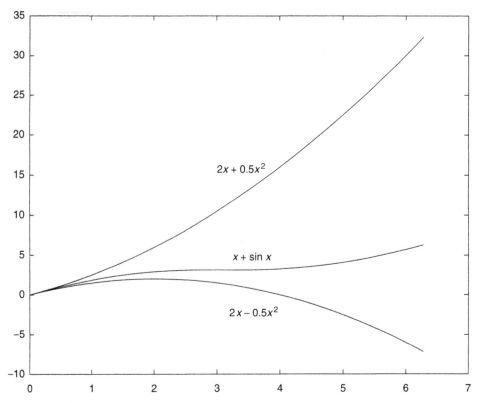

Figure 5.2 Comparison of $f(x) = x + \sin x$ with upper quadratic approximation, $2x + x^2/2$ and lower quadratic approximation $2x - x^2/2$

EXAMPLE 5.1: Suppose that $f \in C^4(I)$ where I is an open interval, $a \in I$. If

$$f'(a) = f''(a) = f'''(a) = 0$$

and $f^{(iv)}(a) > 0$, then f has a local minimum at $x = a$. This result follows easily from Theorem 5.1. In fact, the Taylor polynomial $P_3(x, a) \equiv f(a)$ so that

$$f(x) = f(a) + \frac{1}{4!} f^{(iv)}(\xi)(x - a)^4 \tag{5.6}$$

for some ξ between x and a. Because $f \in C^4(I)$, and $f^{(iv)}(a) > 0$, there is a $\delta > 0$ such that $f^{(iv)}(x) > 0$ for $|x - a| \leq \delta$. It follows from (5.6) that $f(x) > f(a)$ for $|x - a| \leq \delta$.

In the next two examples, we see how Taylor polynomials can be used to generate some well-known approximations, and estimate the error made when using them.

EXAMPLE 5.2:   The function $\sin x$ has Taylor polynomials at $x = 0$,

$$P_1(x) = x, \quad P_3(x) = x - \frac{x^3}{6}, \quad P_5(x) = x - \frac{x^3}{6} + \frac{x^5}{120}.$$

It is clear from the graphs of these functions (shown in Figure 5.3), that the approximation is getting better as n increases. Note that, because the even derivatives of $\sin x$ are zero at $x = 0$, $P_2(x) = P_1(x)$ and $P_4(x) = P_3(x)$.

We often see the approximation $\sin x \approx x$ for small x. What is the error in this approximation? We know that $(\sin x)''' = -\cos x$. Hence using Theorem 5.1 with $n = 2$ and $a = 0$, we have

$$\sin x = x - \frac{\cos \xi}{3!} x^3$$

where ξ lies between 0 and x. Since $|\cos x| \leq 1$, we conclude that

$$|\sin x - x| \leq \frac{|x|^3}{6}.$$

The approximation $\sin x \approx x$ for small x is used in differential equations. The differential equation for the nonlinear pendulum equation is

$$\theta'' + \sin \theta = 0$$

where θ is the angle from the vertical of the pendulum (see Figure 5.4).

It is not possible to get closed-form solutions to this equation. However, for small values of θ, we have just seen that we may approximate $\sin \theta$ by θ and that the error in this approximation is bounded by $|\theta|^3/6$. If we are only interested in small amplitude solutions of the differential equation, near equilibrium at

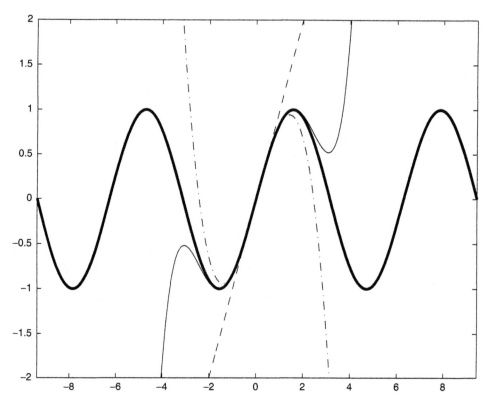

Figure 5.3 The function $\sin x$, heavy line, and Taylor polynomials at $x = 0$. P_1, dashed line, P_3, dash-dot curve, and P_5, solid curve

$\theta = 0$, it seems reasonable to replace the nonlinear differential equation with the linear equation

$$\theta'' + \theta = 0,$$

which does have solutions in closed form. We expect, but it must be proved, that solutions of the linear equations are good approximations to those of the nonlinear differential equation for small amplitudes.

EXAMPLE 5.3: Since a square root can be difficult to work with, the quantity $\sqrt{x^2 + c}$ is often approximated by $x + c/(2x)$. How much error do we make in using this approximation?

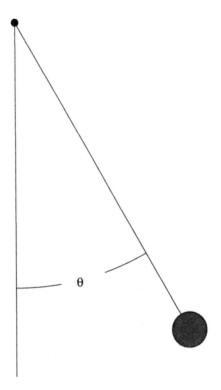

Figure 5.4 Diagram of a pendulum, with the pivot at the top

We assume $x, c > 0$. First we complete the square to find that

$$\sqrt{x^2 + c} \leq \sqrt{x^2 + c + (c/2x)^2} = \sqrt{(x + c/(2x))^2} = x + c/(2x).$$

Next, we need to get a lower estimate on $\sqrt{x^2 + c}$. We consider x as fixed, and we define $f(t) = \sqrt{x^2 + t}$. The Taylor expansion using the Taylor polynomial of degree 1, with error term, is

$$\sqrt{x^2 + c} = f(c) = f(0) + f'(0)c + (1/2)f''(\theta)c^2$$

$$= x + \frac{c}{2x} - \frac{1}{8}\frac{c^2}{(x^2 + \theta)^{3/2}}$$

where $0 < \theta < c$. Since $c^2/(x^2 + \theta)^{3/2} \le c^2/x^3$, we find that

$$x + \frac{c}{2x} - \frac{c^2}{8x^3} < \sqrt{x^2 + c} < x + \frac{c}{2x}. \tag{5.7}$$

Thus the error we commit in using $x + c/2x$ as an approximation to $\sqrt{x^2 + c}$ is not greater than $c^2/(8x^3)$. We can use this approximation for values of x which are much larger than c. Of course, we can get better approximations by using the higher degree Taylor polynomials of $f(t)$.

In the next example, we see how we can use the Taylor polynomials to estimate the location of the eigenvalues of a boundary value problem.

EXAMPLE 5.4:   The eigenvalues λ of the boundary value problem

$$-u'' = \lambda u, \quad 0 < x < 1$$

$$u(0) = 0, \quad u'(1) + u(1) = 0$$

satisfy the equation

$$\sqrt{\lambda} \tan \sqrt{\lambda} = 1.$$

Set $s = \sqrt{\lambda}$ and $f(s) = s \tan s$ so that the equation becomes

$$f(s) = 1.$$

We can roughly locate the roots by looking at the intersections of the curves $\tan s$ and $1/s$ (see Figure 5.5). There is a root s_n in the interval $(n\pi, n\pi + \pi/2)$. Let us find an estimate for the location of the root that gets better as n increases. We make a Taylor expansion for f around $s = n\pi$, and we assume that $n\pi < s < n\pi + \pi/2$:

$$f(s) = f(n\pi) + f'(n\pi)(s - n\pi) + (1/2)f''(\theta)(s - n\pi)^2$$

where θ satisfies $n\pi < \theta < s < n\pi + \pi/2$. Now

$$f'(s) = \tan s + s \sec^2 s$$

and

$$f''(s) = 2 \sec^2 s(1 + s \tan s).$$

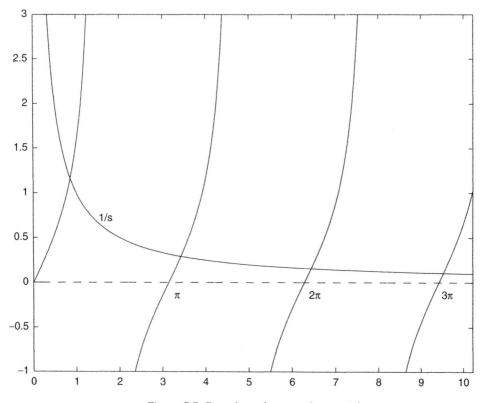

Figure 5.5 Branches of $\tan s$ and curve $1/s$

Thus the expansion for f is

$$f(s) = n\pi(s - n\pi) + \sec^2(\theta)(1 + \theta \tan \theta)(s - n\pi)^2.$$

We see that $f''(s) \geq 2$ for $n\pi \leq s < n\pi + \pi/2$. Hence for s in this range,

$$f(s) \geq g(s) \equiv n\pi(s - n\pi) + (s - n\pi)^2.$$

Since the graph of f lies above the graph of g for s to the right of $n\pi$, the root s_n of $f(s) = 1$ must lie to the left of the solution of $g(s) = 1$. The solution of the latter equation is

$$\frac{n\pi + \sqrt{n^2\pi^2 + 4}}{2}.$$

We have seen how to estimate this square root before. Use the right side of the inequality (5.7) to deduce

$$s_n < \frac{n\pi + \sqrt{n^2\pi^2 + 4}}{2} < n\pi + \frac{1}{n\pi}.$$

Now that we know that the root lies between $n\pi$ and $n\pi + 1/(n\pi)$, we can estimate f'' from above. It can be shown that there is a constant M, independent of n, such that

$$\max\{f''(s) : n\pi \leq s \leq n\pi + 1/(n\pi)\} \leq 2M.$$

Hence on the interval $n\pi \leq s \leq n\pi + 1/(n\pi)$, we have

$$f(s) \leq h(s) \equiv n\pi(s - n\pi) + M(s - n\pi)^2.$$

It follows that the root s_n lies to the right of the root of the equation $h(s) = 1$, which is

$$\frac{(2M - 1)n\pi + \sqrt{n^2\pi^2 + 4M}}{2M}.$$

To get a lower estimate of this quantity, we use the left side of the inequality (5.7). This yields

$$s_n > \frac{(2M - 1)n\pi + \sqrt{n^2\pi^2 + 4M}}{2M} > n\pi + \frac{1}{n\pi} - \frac{M}{(n\pi)^3}.$$

Putting our two estimates together, we have shown

$$n\pi + \frac{1}{n\pi} - \frac{M}{(n\pi)^3} < s_n < n\pi + \frac{1}{n\pi}.$$

Taylor Polynomials and l'Hôpital's Rule

We have seen in these several examples that Taylor polynomials, and the Lagrange form of the remainder, provide a good description of the local behavior of functions, e.g. for x near 0, $\sin x = x + O(x^3)$. Taylor polynomials can also give us a better understanding of l'Hôpital's rule when the functions are C^1 or better.

Suppose that $f, g : I \to \mathbb{R}$ are C^2 functions where I is an open interval that contains the point a. Assume that $f(a) = g(a) = 0$, and that $g(x) \neq 0$ for $x \neq a$. Then for $x \neq a$ the quotient f/g is defined. We use the Taylor polynomial approximations at $x = a$ for both f and g to write

$$\frac{f(x)}{g(x)} = \frac{f'(a)(x-a) + (1/2)f''(\xi_x)(x-a)^2}{g'(a)(x-q) + (1/2)g''(\theta_x)(x-a)^2}$$

for $x \neq a$. The points ξ_x and θ_x both lie between x and a. Now cancel out the factor $(x-a)$ to deduce

$$\frac{f(x)}{g(x)} = \frac{f'(a) + (1/2)f''(\xi_x)(x-a)}{g'(a) + (1/2)g''(\theta_x)(x-a)}.$$

Hence if $g'(a) \neq 0$,

$$\lim_{x \to a} \frac{f(x)}{g(x)} = \frac{f'(a)}{g'(a)} = \lim_{x \to a} \frac{f'(x)}{g'(x)}.$$

If $g'(a) = 0$ and $f'(a) \neq 0$, we can see that the quotient will become unbounded as x approaches a. On the other hand, if both $f'(a) = 0$ and $g'(a) = 0$, but $g''(a) \neq 0$, we can cancel out another factor of $(x-a)$ to deduce that

$$\lim_{x \to a} \frac{f(x)}{g(x)} = \lim_{x \to a} \frac{f''(\xi_x)}{g''(\theta_x)} = \frac{f''(a)}{g''(a)} = \lim_{x \to a} \frac{f''(x)}{g''(x)}$$

because $f, g \in C^2(I)$.

Exercises for 5.1

1. Let
$$f(x) = \begin{cases} 0 & \text{for } x \leq 0 \\ x^2 & \text{for } x \geq 0 \end{cases}.$$
Show that $f \in C^1(\mathbb{R})$ but that $f \notin C^2(\mathbb{R})$.

2. This exercise involves the chain rule and higher derivatives. Suppose that $f, g \in C^3(\mathbb{R})$. Let $h(x) = f(g(x))$. Compute $h''(x)$ and $h'''(x)$.

3. Find the Taylor polynomial P_4 for $f(x) = 5 + 2x - x^3 + x^4$ at each of the three points $a = 0$, $a = -1$, and $a = 2$.

4. Find the Taylor polynomials P_n for $\cos x$ at $x = \pi/4$ for $n = 0, 1, 2, 3$. Plot them together with $\cos x$ on the interval $[-\pi, \pi]$.

5. Write a short program to graph the Taylor polynomial $P_{25}(x)$ for $\sin x$ at $x = 0$. How large must you take x to make the difference $|\sin x - P_{25}(x)| \geq 0.1$?

6. Let $f(x) = 1/\sqrt{1 - x^2}$. Find a polynomial $g(x)$ of degree less than or equal to 2 such that $f(x) = g(x) + O(x^4)$ for x near 0. Plot f and g together on the interval $[-.75, .75]$.

7. a) Use the mean value theorem to show that $e^{-x} \geq 1 - x$ for $x \geq 0$. This estimate is only useful for $0 \leq x \leq 1$.

   b) Use the Taylor polynomial $P_1(x, 0)$ for e^{-x} and Theorem 5.1 to get the sharper estimate $e^{-x} \geq 1 - x + x^2/(2e)$ for $0 \leq x \leq 1$.

8. Use Taylor polynomials to show that, for $x \geq 0$,

$$x^2 - (1/2)x^3 \leq x \log(1 + x) \leq x^2 - (1/2)x^3 + (1/3)x^4.$$

Why is the lower estimate only useful for $0 \leq x \leq 2$?

9. Derive the binomial formula for $(1 + x)^n$, $n \in \mathbb{N}$, by finding the Taylor polynomial P_n at $x = 0$. Then set $x = b/a$ and derive the formula for $(a + b)^n$. Refer to exercise 1 of Section 1.4 for the binomial formula.

10. We needed the following estimate in Example 5.3. Show that there is a constant M, independent of n, such that

$$\sec^2(s)(1 + s \tan s) \leq M \quad \text{for} \quad n\pi \leq s \leq n\pi + 1/(n\pi).$$

11. a) Use the intermediate value theorem to prove that there is a root x_0 of the equation $\sin x = x \cos x$ between π and $3\pi/2$.

    b) Show that $x_0 \leq 3\pi/2 - 2/(3\pi)$.

12. If f does not have the extra derivative needed in Theorem 5.1, we can still get some information about the remainder $f - P_n$. Suppose that $f \in C^n(I)$, where I is an open interval, and $a \in I$. Use l'Hôpital's rule to show that $f_n(x) - P_n(x, a) = o(|x - a|^n)$.

13. Let $C(x) > 0$ be a cost function that is continuously differentiable for $x \geq 0$, with the marginal cost $MC(x) \leq AC(x)$ for $x \geq 1$. Show that $C(x) \leq C(1)x$ for $x \geq 1$.

5.2 Numerical Differentiation

Taylor polynomials and Theorem 5.1 are also used to derive formulas to approximate derivatives. We know already that when f is differentiable, the difference quotient

$$\frac{f(a + h) - f(a)}{h} = f'(a) + o(1).$$

However, if we make a stronger hypothesis on f, we get a better approximation. Indeed, if we assume that f is C^2 in an open interval that contains a, then by Theorem 5.1, we have

$$f(a + h) = f(a) + f'(a)h + \frac{1}{2}f''(\theta)h^2.$$

Hence

$$\frac{f(a + h) - f(a)}{h} = f'(a) + O(h) \tag{5.8}$$

as h tends to zero. Equation (5.8) is known as a *forward difference* approximation to f'. If we replace h with $-h$, we obtain

$$\frac{f(a) - f(a - h)}{h} = f'(a) + O(h). \tag{5.9}$$

This expression is called a *backward difference* approximation.

If we consider a sketch of the graph of f near $x = a$, we see that the forward difference approximation (with $h > 0$), is the slope of the secant line to the right, and the backward difference approximation is the slope of the secant line to the left. It appears plausible that an average of the forward and backward difference approximations would give a more accurate approximation to $f'(a)$. This average is called a *centered difference* approximation, and it is given by the expression

$$\frac{f(a + h) - f(a - h)}{2h}.$$

To estimate the error, we assume that $f \in C^3$ and use the longer Taylor expansion

$$f(a + h) = f(a) + f'(a)h + \frac{1}{2}f''(a)h^2 + \frac{1}{6}f'''(\theta_1)h^3 \qquad (5.10)$$

where θ_1 lies between a and $a + h$. We replace h by $-h$ in (5.10) and obtain

$$f(a - h) = f(a) - f'(a)h + \frac{1}{2}f''(a)h^2 - \frac{1}{6}f'''(\theta_2)h^3 \qquad (5.11)$$

where θ_2 lies between a and $a - h$. We subtract (5.11) from (5.10) and divide by $2h$ to arrive at

$$\frac{f(a + h) - f(a - h)}{2h} = f'(a) + O(h^2). \qquad (5.12)$$

As expected, the centered difference approximation is more accurate than either the forward or the backward difference approximation.

Next we show how to approximate f'' using a centered difference quotient. We saw in exercise 7 of Section 4.4 that, if f is differentiable and $f''(a)$ exists, then

$$\lim_{h \to 0} \frac{f(a + h) - 2f(a) + f(a - h)}{h^2} = f''(a).$$

However, to use these difference quotients as an approximation to $f''(a)$, we need to be able to estimate the error. To do so we assume that $f \in C^3$. We make the Taylor expansion

$$f(a + h) = f(a) + f'(a)h + \frac{1}{2}f''(a)h^2 + \frac{1}{6}f'''(\theta_1)h^3. \qquad (5.13)$$

where θ_1 lies between a and $a + h$. Again we make a similar expansion for $f(a - h)$:

$$f(a - h) = f(a) - f'(a)h + \frac{1}{2}f''(a)h^2 - \frac{1}{6}f'''(\theta_2)h^3. \qquad (5.14)$$

where θ_2 lies between a and $a - h$. This time we add equations (5.13) and (5.14) and divide by h^2 to find

$$\frac{f(a+h) - 2f(a) + f(a-h)}{h^2} = f''(a) + \frac{1}{6}\left[f^{(3)}(\theta_1) - f^{(3)}(\theta_2)\right]h$$

$$= f''(a) + O(h). \tag{5.15}$$

EXAMPLE 5.5:   We show in this example how the difference approximations to a derivative are used to transform a differential equation into a system of algebraic equations.

We consider the nonlinear boundary value problem

$$-u''(x) + u^3(x) = f(x), \quad 0 < x < 1 \tag{5.16}$$

with the boundary conditions

$$u(0) = u(1) = 0. \tag{5.17}$$

Here f is the given data of the problem. This problem cannot be solved using analytic techniques. The first step in the analysis of this problem is to prove the existence of a solution for each f in a reasonable class of functions. Truth be told, most physicists and engineers would skip this part, and relying on physical intuition, assume that there is a solution.

The next step is to attempt to compute numerical approximations to the solution. We introduce a mesh, $x_j = jh$, $j = 0, 1, \ldots, N$ where $h = 1/N$ is the length of the subintervals $[x_{j-1}, x_j]$. We assume that we have a solution $u \in C^4$ of (5.16), (5.17). Using (5.15) we approximate u'' at x_j, keeping in mind that $x_j - h = x_{j-1}$ and $x_j + h = x_{j+1}$. This yields, for each $j = 1, \ldots, N - 1$, the equations

$$-\left[\frac{u(x_{j+1}) - 2u(x_j) + u(x_{j-1})}{h^2}\right] + u^3(x_j) = f(x_j) + O(h^2).$$

Now we drop the error term, and look for a vector of numbers $(u_0, \ldots, u_N)$ that satisfy the system of equations:

$$-u_{j-1} + 2u_j - u_{j+1} + h^2 u_j^3 = h^2 f_j, \quad j = 1, \ldots N - 1$$

where we have set $f_j = f(x_j)$. When h is small, we expect that the numbers u_j will be a good approximation to the exact values $u(x_j)$.

This is a system of $N - 1$ equations in the $N + 1$ unknowns $u_0, \ldots, u_N$. Motivated by our knowledge of linear systems, we try to have systems in which the number of unknowns equals the number of equations. Here is where the boundary conditions come to the rescue. To satisfy the boundary conditions, we shall take $u_0 = u_N = 0$. Now we have a system of $N - 1$ equations in $N - 1$ unknowns:

$$
\begin{aligned}
2u_1 - u_2 + h^2 u_1^3 &= h^2 f_1 \\
-u_1 + 2u_2 - u_3 + h^2 u_2^3 &= h^2 f_2 \\
-u_2 + 2u_3 - u_4 + h^2 u_3^3 &= h^2 f_3. \\
\cdots &= \cdots \\
-u_{N-2} + 2u_{N-1} + h^2 u_{N-1}^3 &= h^2 f_{N-1}.
\end{aligned}
\tag{5.18}
$$

An exact solution of this system provides an approximation to the solution of the boundary value problem (5.16), (5.17). We expect that u_j is a good approximation to $u(x_j)$, and that as h tends to zero, the approximations u_j converge in some sense to the exact solution $u(x)$. It is the task of the numerical analyst to prove this convergence. However, it is not possible to solve the algebraic system (5.18) exactly. In Chapter 11, we will learn about methods to find approximate solutions to the system (5.18). Thus there are two steps of approximation involved: approximation of the differential equation by the system (5.18), and approximation of the solutions of (5.18).

Exercises for 5.2

1. Use a calculator or computer to compare the forward difference approximation and the centered difference approximations to the derivative of $\exp(x)$ at $x = 0$.

   a) Let $E_1(h) = \frac{e^h - 1}{h} - 1$. Compute E_1 for $h = 0.04, 0.02, 0.01$, and 0.005. What is the effect on E_1 of cutting h in half?

   b) Let $E_2(h) = \frac{e^h - e^{-h}}{2h} - 1$. Compute E_2 for $h = 0.04, 0.02, 0.01, 0.005$. Now what is the result of cutting h in half?

2. Consider the ordinary differential equation $y'(x) = f(x, y)$ on $[a, b]$ with initial condition $y(a) = y_0$. Introduce the mesh points $x_n = a + nh$, $n = 0, \ldots, N$ where $h = (b - a)/N$.

   a) Derive a numerical procedure for computing approximations, call them y_n, to the value of the exact solution $y(x_n)$ by replacing y' in the equation with the forward difference approximation. What well-known method have you found?

   b) Show that, if the exact value $y(x_n)$ and the approximate value y_n agree, then $y(x_{n+1}) - y_{n+1} = O(h^2)$. Use the Taylor polynomial P_1 for $y(x)$ at $x = x_n$ and the error expression given in Theorem 5.1.

   c) If we replace y' in the differential equation with its backward difference approximation, what is the resulting numerical method?

3. To prove (5.15), we assumed that $f \in C^3(I)$. Show that if $f \in C^4(I)$, we get the better result:

$$\frac{f(a + h) - 2f(a) + f(a - h)}{h^2} = f''(a) + O(h^2).$$

4. Here is another differencing scheme to approximate f''. Use the Taylor polynomials to show that if $f \in C^3$, then

$$\frac{f(a + 2h) - 2f(a + h) + f(a)}{h^2} = f''(a) + O(h).$$

5.3 Polynomial Interpolation

The Taylor polynomial of a function f at a point a uses information about the function for values of x that are very close to a. It provides an approximation to f that is good near $x = a$, but can be very poor further away from a. We shall discuss a method of approximation that uses the values of f at several points.

Definition 5.4 Let f be defined on an interval I, and let $n + 1$ distinct points, $x_0, x_1, \ldots, x_n$, be given in I. We say that a polynomial $p(x)$ interpolates the function f at the points x_i if $p(x_i) = f(x_i)$, $i = 0, \ldots, n$.

Theorem 5.2 Let I be an interval and $f : I \to \mathbb{R}$. Let $n + 1$ distinct points, $x_0, x_1, \ldots, x_n$, be given in I. Then there is a unique polynomial $Q_n(x)$ of degree $d \leq n$ that interpolates f at the points x_i.

Proof: A polynomial of degree $d \leq n$ has the form

$$Q(x) = a_n x^n + \cdots + a_1 x + a_0.$$

The interpolating conditions $p(x_i) = f(x_i)$ provide $n + 1$ linear equations in the $n + 1$ unknown coefficients a_j,

$$
\begin{aligned}
Q_n(x_0) &= a_0 + a_1 x_0 + a_2 x_0^2 + \cdots + a_n x_0^n = f(x_0) \\
Q_n(x_1) &= a_0 + a_1 x_1 + a_2 x_1^2 + \cdots + a_n x_1^n = f(x_1) \\
\cdots &= \cdots \\
Q_n(x_n) &= a_0 + a_1 x_n + a_2 x_n^2 + \cdots + a_n x_n^n = f(x_n).
\end{aligned}
\tag{5.19}
$$

The matrix for this system is the well-known Vandermonde matrix

$$
V = \begin{bmatrix}
1 & x_0 & x_0^2 & \cdots & x_0^n \\
1 & x_1 & x_1^2 & \cdots & x_1^n \\
. & . & . & \cdots & . \\
1 & x_n & x_n^2 & \cdots & x_n^n
\end{bmatrix}.
$$

This matrix has determinant

$$\det V = \Pi_{i<j}(x_j - x_i) \neq 0$$

because of our assumption that the interpolating points x_i are distinct. Hence the system (5.19) has a unique solution.

The interpolating polynomial can be represented in various ways. For our purposes later in Chapter 7, we shall use the *Lagrange* representation. First we construct the Lagrange basis polynomials. Let $L_i(x)$ denote the polynomial of degree $d = n$ such that

$$L_i(x_i) = 1, \quad L_i(x_j) = 0 \quad \text{for } i \neq j.$$

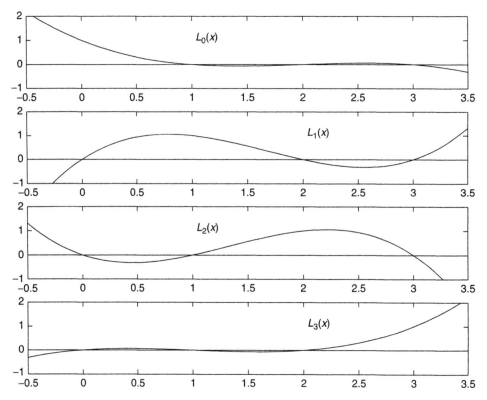

Figure 5.6 Lagrange basis polynomials $L_0, \ldots, L_3(x)$ at interpolation points $x = 0, 1, 2, 3$

Because of Theorem 5.2 we know there is such a polynomial L_i and it is unique. In fact, we can write it down,

$$L_i(x) = \frac{\Pi_{j \neq i}(x - x_j)}{\Pi_{j \neq i}(x_i - x_j)}.$$

The Lagrange basis polynomials (cubics) for the points $0, 1, 2, 3$ are shown in Figure 5.6. Thus the polynomial that interpolates f at x_i, $i = 0, \ldots, n$ is

$$Q_n(x) = \sum_{i=0}^{n} f(x_i) L_i(x). \qquad (5.20)$$

Now we address the question: What is the error $f(x) - Q_n(x)$ when Q_n interpolates f at $n + 1$ points? We shall need the function

$$w_{n+1}(x) \equiv (x - x_0)(x - x_1)\ldots(x - x_n),$$

which is a polynomial of degree $n + 1$, that vanishes at each of the x_i.

Theorem 5.3 Let I be an open interval, and let $f \in C^{(n+1)}(I)$. Let points $x_0, x_1, \ldots, x_n$ belong to I and let $Q_n(x)$ be the polynomial of degree $d \le n$ that interpolates f at the x_i. Then for each $x \in I$, there is a point ξ_x such that

$$f(x) - Q_n(x) = \frac{1}{(n+1)!} f^{(n+1)}(\xi_x) w_{n+1}(x) \tag{5.21}$$

and such that ξ_x lies in the interval $J = [a, b]$ where

$$a = \min\{x_0, x_1, \ldots, x_n, x\} \quad \text{and} \quad b = \max\{x_0, x_1, \ldots, x_n, x\}.$$

Remark The function $x \to \xi_x$ may not be continuous, but the composite function $g(x) \equiv f^{(n+1)}(\xi_x)$ is continuous on I. In fact, from equation (5.21) we see that

$$g(x) = (n + 1)! \frac{f(x) - Q_n(x)}{(x - x_0)\cdots(x - x_n)}.$$

Because f is continuous, g is clearly continuous on I for all $x \ne x_j$. Using l'Hôpital's rule, g can be extended to be continuous at each x_j.

Proof: The presence of the $(n + 1)!$ in the denominator of the error expression suggests that we may use an argument that depends on Rolle's theorem, or the mean value theorem, in some way.

Note that because $w_{n+1}(x_i) = 0$, the error is zero at each of the interpolating points, as it must be. Thus we may assume that x is not one of the points x_i. Let us use Q_n to construct a polynomial $r(t)$ of degree $d \le n + 1$ that interpolates f at the $n + 2$ points $\{x_0, x_1, \ldots, x_n, x\}$. In fact we can take

$$r(t) = Q_n(t) + \frac{f(x) - Q_n(x)}{w_{n+1}(x)} w_{n+1}(t).$$

Since $w_{n+1}(x_i) = 0$ we see that $r(x_i) = Q_n(x_i) = f(x_i)$ for each $i = 0, \ldots, n$. In addition, $r(x) = Q_n(x) + (f(x) - Q_n(x)) = f(x)$. This means that the difference $G(t) = f(t) - r(t)$ vanishes at all of the points $\{x_0, x_1, \ldots, x_n, x\}$. It is also not difficult to verify that $G \in C^{(n+1)}(I)$. Using Rolle's theorem, we deduce that $G'(t) = 0$ in at last one point between each of the interpolating points. Thus $G'(t) = 0$ for at least $n + 1$ points in the interval J. Applying Rolle's theorem again, we see that $G''(t) = 0$ for at least n points in J. Continuing in this fashion we see that there must be at least one point $t = \xi_x \in J$ where $G^{(n+1)}(t) = 0$. What is $G^{(n+1)}$? Since $Q_n(t)$ has degree $d \leq n$, the $(n + 1)$th derivative of $Q_n(t)$ is zero. In addition, the $(n + 1)$th derivative of $w_{n+1}(t)$ is $(n + 1)!$. Hence

$$G^{(n+1)}(t) = f^{(n+1)}(t) - r^{(n+1)}(t)$$

$$= f^{(n+1)}(t) - (n + 1)! \frac{f(x) - Q_n(x)}{w_{n+1}(x)}.$$

Therefore $G^{(n+1)}(\xi_x) = 0$ means

$$f(x) - Q_n(x) = \frac{f^{(n+1)}(\xi_x)}{(n + 1)!} w_{n+1}(x).$$

The theorem is proved.

Now let us compare the Lagrange form of the error for the Taylor polynomial of degree n, $P_n(x, a)$, and the error expression that we have just found for $f(x) - Q_n(x)$. By Theorem 5.6,

$$f(x) - P_n(x) = \frac{f^{(n+1)}(\theta_x)}{(n + 1)!}(x - a)^{(n+1)}$$

and by Theorem 5.3

$$f(x) - Q_n(x) = \frac{f^{(n+1)}(\xi_x)}{(n + 1)!} w_{n+1}(x).$$

We see that both polynomials are based on $n+1$ bits of information about the function f. In the case of $P_n(x, a)$, we use the values $f(a), f'(a), \ldots, f^{(n)}(a)$ to construct $P_n(x, a)$, and the error is in terms of $f^{(n+1)}$, multiplied by a polynomial $(x - a)^{(n+1)}$ of degree $n + 1$. $Q_n(x)$ uses the $n + 1$ pieces of information $f(x_i)$, $i = 0, 1, \ldots n$. The error again involves $f^{(n+1)}$ multiplied by a polynomial $w_{n+1}(x)$ of degree $n + 1$. It can be shown that if the points $x_1, x_2, \ldots, x_n$ all tend to x_0, the interpolating polynomial Q_n tends to the Taylor polynomial $P_n(x, x_0)$.

EXAMPLE 5.6:

a) When there are just two interpolating points, x_0 and x_1, the Lagrange basis functions are

$$L_0(x) = \frac{x - x_1}{x_0 - x_1} \quad \text{and} \quad L_1(x) = \frac{x - x_1}{x_1 - x_0}$$

and $Q_1(x)$ is the linear function

$$Q_1(x) = f(x_0)L_0(x) + f(x_1)L_1(x)$$

$$= \frac{f(x_1)(x - x_0) - f(x_0)(x - x_1)}{x_1 - x_0}.$$

b) For three interpolating points, $Q_2(x)$ is a parabola given by

$$Q_2(x) = \frac{f(x_0)(x - x_1)(x - x_2)}{(x_0 - x_1)(x_0 - x_2)}$$

$$+ \frac{f(x_1)(x - x_0)(x - x_2)}{(x_1 - x_0)(x_1 - x_2)}$$

$$+ \frac{f(x_2)(x - x_0)(x - x_1)}{(x_2 - x_0)(x_2 - x_1)}.$$

From the form of the error $f - Q_n$, it can be seen that the error will generally get larger if x is outside of the interval $[x_0, x_n]$. Even inside the interval, but near the endpoints, the error can develop very large oscillations as shown by the example of Runge. As a rule, high order polynomial interpolation is not a good idea. There are much better ways to approximate by polynomials.

EXAMPLE 5.7: Let $f(x) = 1/(1 + x^2)$, and let us use 11 interpolation points: $0, \pm 1, \pm 2, \pm 3, \pm 4, \pm 5$. The graph of Q_{10}, together with that of f, is shown in Figure 5.7. Although Q_{10} fits f fairly well in the middle of the interval, the fit near the endpoints is terrible. Increasing the number of interpolation points does not necessarily make the fit better. However, the problem can be dealt with by not

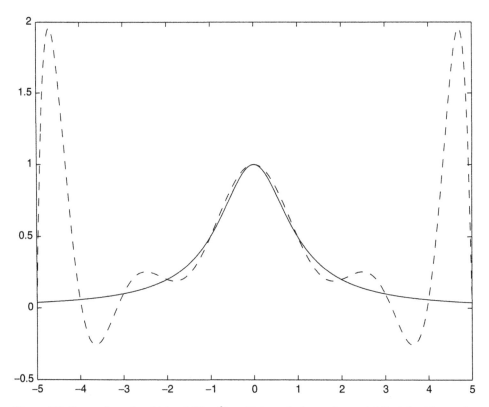

Figure 5.7 Runge function $f(x) = 1/(1+x^2)$ and interpolating polynomial $Q_{10}(x)$ with evenly spaced interpolation points on the interval $[-5, 5]$

making the interpolation points evenly spaced. There is a special choice of points that makes $\max_{[-1,1]} |w_{n+1}(x)|$ as small as possible. This choice gives a much better fit across the whole interval. The book of D. Kincaid and W. Cheney [Ki] addresses these questions.

The Weierstrass Approximation Theorem

Actually, in the case of interpolation, the interpolating polynomials $p_n(x)$ do not necessarily converge to the function as the number of interpolating points $n + 1 \to \infty$. However, an important result in analysis states that a sequence of approximating polynomials can be generated without assuming that the function is even differentiable.

Theorem (Weierstrass approximation theorem) Let $f : [a, b] \rightarrow \mathbb{R}$ be continuous. Then there is a sequence of approximating polynomials $p_n(x)$ with the following property. For each $\varepsilon > 0$, there is an N such that

$$\max_{a \leq x \leq b} |f(x) - p_n(x)| \leq \varepsilon$$

for all $n \geq N$. We say that polynomials p_n approximate f *uniformly* on the interval $[a, b]$.

There are several different ways of proving this theorem. One method uses a special family of polynomials, called the Bernstein polynomials, to construct the approximating sequence. Another method, which we present in Chapter 14, uses integration, specifically convolution, to construct the approximating sequence.

The subject of approximation theory deals with these kinds of questions. Given a continuous function f on $[a, b]$, is there a best approximation to f among all the polynomials of degree less than or equal to n? If there is, is it unique? How can one estimate the error in the best approximation? For an introduction to the subject and references, see the book of K. R. Davidson and A. P. Donsig [Da].

5.4 Convex Functions

The class of convex functions is important in the subject of optimization.

Definition 5.5 Let $I \subset \mathbb{R}$ be an interval and let $f : I \rightarrow \mathbb{R}$. The function f is *convex* on I, if for each pair $x_0, x_1 \in I$,

$$f((1 - t)x_0 + tx_1) \leq (1 - t)f(x_0) + tf(x_1) \quad \text{for } 0 \leq t \leq 1.$$

Note that if $g(t) = (1 - t)f(x_0) + tf(x_1)$, we have $g(0) = f(x_0)$ and $g(1) = f(x_1)$. Furthermore, the graph of $g(t)$ is just the straight line (chord) from the point $(x_0, f(x_0))$ to the point $(x_1, f(x_1))$. Thus f is convex if its graph always lies beneath the straight line through the points $(x_0, f(x_0))$ and $(x_1, f(x_1))$ for any pair $x_0, x_1 \in I$.

We say that a function f is *concave* if $-f$ is convex. When f is concave, we have

$$f((1-t)x_0 + tx_1) \geq (1-t)f(x_0) + tf(x_1), \quad 0 \leq t \leq 1.$$

Convex functions need not be differentiable, as can be seen from the example of $f(x) = |x|$, but they are always continuous.

Theorem 5.4 Let I be an open interval, and let $f : I \to \mathbb{R}$ be convex on I. Then f is continuous on I.

Proof: Let $x_0 \in I$. Because I is open, there is an $r > 0$ such that $[x_0 - r, x_0 + r] \subset I$.

First we derive an estimate for $f(x) - f(x_0)$ when $x_0 < x < x_0 + r$. For x in this interval, we can write

$$x = (1-t)x_0 + t(x_0 + r)$$

with $t = (x - x_0)/r$. Since $0 < t \leq 1$, the convexity implies

$$f(x) \leq (1-t)f(x_0) + tf(x_0 + r).$$

Thus for x between x_0 and $x_0 + r$, we have

$$f(x) - f(x_0) \leq t[f(x_0 + r) - f(x_0)]$$
$$\leq \left(\frac{x - x_0}{r}\right)[M - f(x_0)] \tag{5.22}$$

where $M = \max\{f(x_0 + r), f(x_0 - r)\}$.

To get an inequality in the opposite direction, note that $x_0 - r < x_0 < x$. Hence we may write

$$x_0 = (1-s)(x_0 - r) + sx$$

with $s = r/(x - x_0 + r)$. Again by the convexity of f,

$$f(x_0) \leq (1-s)f(x_0 - r) + sf(x)$$

whence

$$f(x) - f(x_0) \geq \left(\frac{1-s}{s}\right)[f(x_0) - f(x_0 - r)].$$

Since $(1 - s)/s = (x - x_0)/r$, we deduce that

$$f(x) - f(x_0) \geq \left(\frac{x - x_0}{r} \right)[f(x_0) - M]. \qquad (5.23)$$

Now $f(x_0) \leq M$ because of the convexity of f. Thus we may combine inequalities (5.22) and (5.23) to conclude that, for $x_0 < x \leq x_0 + r$,

$$|f(x) - f(x_0)| \leq \left(\frac{|x - x_0|}{r} \right)[M - f(x_0)]. \qquad (5.24)$$

If we take x to lie between $x_0 - r$ and x_0, a similar argument will produce the same estimate. Thus (5.24) holds for all x with $|x - x_0| \leq r$. It follows that f is continuous at x_0.

The next result gives a criterion for convexity when f has two derivatives.

Theorem 5.5 Let $I \subset \mathbb{R}$ be an open interval and let $f \in C^2(I)$. Then f is convex on I if and only if $f''(x) \geq 0$ for $x \in I$.

Proof: First we show that $f''(x) \geq 0$ implies convexity. This assertion may be proved using Taylor polynomials, or even more easily using Theorem 5.3. Let $x_0, x_1 \in I$ with $x_0 < x_1$. Then x satisfies $x_0 \leq x \leq x_1$ if and only if $x = (1 - t)x_0 + tx_1$ with $0 \leq t \leq 1$. In fact, $t = (x - x_0)/(x_1 - x_0)$. To show f is convex, we must show that, for $x_0 \leq x \leq x_1$,

$$f(x) \leq (1 - t)f(x_0) + tf(x_1) = \left(\frac{x_1 - x}{x_1 - x_0} \right)f(x_0) + \left(\frac{x - x_0}{x_1 - x_0} \right)f(x_1).$$

The right-hand side of this equation is just the polynomial Q_1 of degree 1 that interpolates f at x_0 and x_1. According to Theorem 5.3, if $x \in I$,

$$f(x) = Q_1(x) + \frac{f''(\xi)}{2}(x - x_0)(x - x_1)$$

for some point $\xi \in I$. Since $f''(x) \geq 0$ on I, if we restrict x to lie between x_0 and x_1, we see that $f(x) \leq Q_1(x)$. We leave the converse implication to the exercises.

Exercises for 5.3 and 5.4

1. Find the quadratic polynomial $Q_2(x)$ that interpolates $f(x) = \sin x$ at $x = 0, x = \pi/4$ and $x = \pi/2$. Estimate the error $E(x) = \sin x - Q_2(x)$ over the interval $[0, \pi/2]$ by finding the maximum of $w_3(x)$ over this interval and then using (5.21). What happens to the size of E if we go outside of the interval $[0, \pi/2]$, for example, to the interval $[0, \pi]$?

2. Let $f \in C^1$. Let $Q_1(x, h)$ be the linear polynomial such that $Q_1(\pm h, h) = f(\pm h)$, and let $P_1(x) = f(0) + f'(0)x$ be the Taylor polynomial of f. Show that the coefficients of Q_1 converge to the coefficients of $P_1(x)$ as h tends to zero.

3. Let g be a function on $[-1, 1]$. Use Example 5.6 to find the polynomial $Q_2(x)$ that interpolates g at points $x = 0$ and $x = \pm 1$.

4. Let g be a C^1 on $[-1, 1]$. Find the cubic polynomial $\tilde{Q}_3$ such that $\tilde{Q}_3(0) = g(0)$, $\tilde{Q}_3(\pm 1) = g(\pm 1)$, and $\tilde{Q}_3'(0) = g'(0)$. Look for $\tilde{Q}_3$ in the form

$$\tilde{Q}_3(s) = Q_2(x) + Ax(x^2 - 1).$$

Note that for any value of A, $\tilde{Q}_3(x) = Q_2(x)$ for $x = 0, \pm 1$. Now choose A so that $\tilde{Q}_3'(0) = g'(0)$.

5. Next we find an expression for the error $g - \tilde{Q}_3$. Following the proof of Theorem 5.3 as a template, we let $r(t)$ be the quartic polynomial such that

$$r(0) = g(0), \quad r(\pm 1) = g(\pm 1), \quad r'(0) = g'(0)$$

and $r(x) = g(x)$ at some fourth point $x \neq 0, x \neq \pm 1$. r is given by

$$r(t) = \tilde{Q}_3(t) + \frac{g(x) - \tilde{Q}_3(x)}{x^2(x^2 - 1)}t^2(t^2 - 1).$$

Show that $g' - r' = 0$ at four points $t_1, t_2, t_3, 0$. Eventually show that

$$g(x) = \tilde{Q}_3(x) + g^{(4)}(\xi_x)x^2(x^2 - 1)$$

for some point ξ_x.

6. Let f and g be convex functions on I. Show that for $a, b \geq 0$, the function $af(x) + bg(x)$ is convex.

7. Let I be an open interval, and that suppose f and g are convex functions on I. Show that the function $h(x) = \max\{f(x), g(x)\}$ is also convex.

8. Let I be an open interval, and suppose that $f \in C^2(I)$. Use exercise 7 of Section 4.4 to show that if f is convex on I, then $f''(x) \geq 0$ on I.

9. Let I be an open interval and suppose that $f \in C^1(I)$. Show that f is convex if and only if for all $x, y \in I$,

$$f(y) - f(x) \geq (y - x)f'(x).$$

Chapter 6

Solving Equations in One Dimension

Chapter Overview

Solving equations is one of the central problems of analysis. The proof of the intermediate value theorem (Theorem 3.7) using the bisection method provides a procedure for computing the solutions of certain equations. On the other hand, the inverse function theorem (Theorem 4.8) gives conditions that ensure the existence of solutions, but it does not give any indication of how to compute them. In this chapter, we will investigate some simple, but effective, computational methods that use the principles of analysis that we have developed so far. We first study fixed-point problems, the contraction mapping principle, and functional iteration. For computational purposes, we study how an equation $f(x) = 0$ can be put in fixed-point form, in a way that anticipates the inverse function theorem in higher dimensions. We consider Newton's method and we show that it converges quadratically.

6.1 Fixed-Point Problems

Often we need to solve an equation of the form

$$g(x) = x. \tag{6.1}$$

A solution of (6.1) is called a *fixed point* of the function g. Graphically, a fixed point occurs when the graph of g crosses the diagonal $y = x$.

Theorem 6.1 Let g be continuous on the closed bounded interval $I = [a, b]$ and suppose that $R(g) \subset I$ (i.e., $a \le g(x) \le b$ for all $x \in I$). Then there is an $x_* \in I$ such that $g(x_*) = x_*$.

Proof: We let $f(x) = g(x) - x$, which is again a continuous function on I. $f(a) = g(a) - a \ge 0$, and $f(b) = g(b) - b \le 0$. It follows by the intermediate value theorem that there must be an $x_* \in I$ such that $f(x_*) = 0$. The desired fixed point is x_*.

This simple result establishes the existence of a fixed point, but it does not provide any procedure for computation. Let us look at an example to see how we might proceed.

EXAMPLE 6.1: Let $g(x) = 2 \sin x$ and consider the equation

$$x = g(x). \tag{6.2}$$

The function g maps $[0, \pi]$ into itself, so by Theorem 6.1, g has at least one fixed point. Of course $x = 0$ is a fixed point, but we look for a solution $x_* \in (0, \pi]$. If we set $f(x) = x - 2 \sin x$, we see that $f(1.8) < 0$, whereas $f(2) > 0$. With the use of the intermediate value theorem (Theorem 3.9) we conclude that there is at least one solution x_* of (6.2) in the interval $[1.8, 2.0]$.

The graphs of g and of x are displayed in Figure 6.1; they intersect over the interval $[1.8, 2.0]$. We can generate a sequence of approximations to the fixed point x_* as follows. Make a first guess x_0 to the left of, and not too far from, x_*. Let us approximate the function $g(x)$ using the constant function $y \equiv g(x_0)$. The horizontal line $y = g(x_0)$ crosses the line $y = x$ at the point $x_1 = g(x_0)$, and it appears that x_1 lies closer to x_0. Emboldened by this success, let us now approximate $g(x)$ using the constant function $y \equiv g(x_1)$. The horizontal line $y = g(x_1)$ crosses the diagonal $y = x$ at the point $x_2 = g(x_1)$. It appears that we now have a procedure for generating approximations to the fixed point x_*:

$$x_{n+1} = g(x_n).$$

This procedure is called *functional iteration*. Here are the results for 20 iterations, starting with $x_0 = 1.7$:

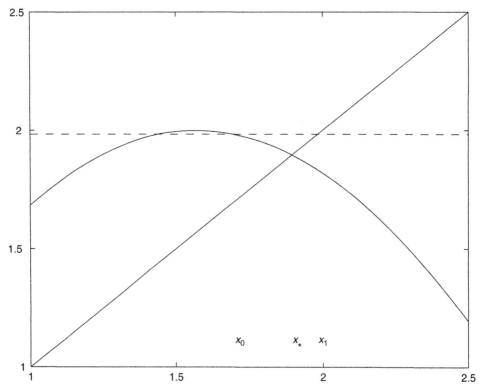

Figure 6.1 Graphs of $g(x) = 2 \sin x$ and x, solid curves, graph of $y \equiv g(x_0)$ in dashed line

x	g(x)
1.70000000000000	1.98332962090494
1.98332962090494	1.83221617277555
1.83221617277555	1.93204797873196
1.93204797873196	1.87091033192695
1.87091033192695	1.91060558419431
1.91060558419431	1.88563651714635
1.88563651714635	1.90169175686794
1.90169175686794	1.89150361285717
1.89150361285717	1.89802538425563
1.89802538425563	1.89387323015816
1.89387323015816	1.89652607068072
1.89652607068072	1.89483492095389
1.89483492095389	1.89591454748403

1.89591454748403	1.89522594177589
1.89522594177589	1.89566540239786
1.89566540239786	1.89538504720604
1.89538504720604	1.89556394289859
1.89556394289859	1.89544980609257
1.89544980609257	1.89552263324400
1.89552263324400	1.89547616735155

The convergence appears to be taking place, but rather slowly. After all, the approximation of $g(x)$ by a constant function is rather crude. Nevertheless, we can prove a very useful result about functional iteration. First we must introduce a class of continuous functions.

Definition 6.1 A function $f : A \to \mathbb{R}$ is said to be *Lipschitz* continuous on A if there is a constant $L \geq 0$ such that

$$|f(x) - f(y)| \leq L|x - y| \text{ for all } x, y \in A.$$

The constant L is called the *Lipschitz constant* for the function f and the domain A.

EXAMPLE 6.2:

a) The function $f(x) = x^2$ is Lipschitz continuous on each bounded interval $|x| \leq b$. In fact,

$$|f(x) - f(y)| = |x + y| \, |x - y| \leq 2b|x - y|$$

for all $|x|, |y| \leq b$. The Lipschitz constant for the interval $[-b, b]$ is $L = 2b$.

b) The function $f(x) = |x|$ is Lipschitz continuous on $\mathbb{R}$ with Lipschitz constant $L = 1$.

c) The function $f(x) = \sqrt{x}$ is *not* Lipschitz continuous on any interval $(0, b]$ or $[0, b]$. Intuitively, we can see that this is true because the slope of the graph becomes steeper and steeper as x tends to 0.

It appears plausible that if we restrict the slope of the graph, the function should be Lipschitz continuous. The next theorem provides a simple criterion for Lipschitz continuity.

Theorem 6.2 Let I be an open interval, and let $f : I \to \mathbb{R}$ be differentiable on I. Suppose that there is a constant $L \geq 0$ such that $|f'(x)| \leq L$ for all $x \in I$. Then f is Lipschitz continuous on I with Lipschitz constant L.

Proof: The mean value theorem yields

$$|f(x) - f(y)| = |f'(\theta)||x - y| \leq L|x - y|.$$

Definition 6.2 A function $g : A \to \mathbb{R}$ is a *contraction* on A if it is Lipschitz continuous on A with Lipschitz constant $L < 1$.

The next theorem is a one-dimensional version of the contraction mapping principle that we will develop in Chapter 11 to solve systems of equations.

Theorem 6.3 Let I be a closed interval, and suppose that $g : I \to \mathbb{R}$ is a contraction on I such that $g(I) \subset I$. Then g has a unique fixed point $x_* \in I$. Furthermore, for any starting point $x_0 \in I$, the sequence of functional iterates,

$$x_{n+1} = g(x_n), \tag{6.3}$$

converges to x_*.

Proof: Let the Lipschitz constant be c, $0 \leq c < 1$, so that

$$|g(x) - g(y)| \leq c|x - y| \quad \text{for all } x, y \in I. \tag{6.4}$$

First we prove that if a fixed point exists, it must be unique. Indeed, suppose that there are two fixed points, x_* and y_*. Then by (6.4), we have

$$|x_* - y_*| = |g(x_*) - g(y_*)| \leq c|x_* - y_*|$$

whence $(1 - c)|x_* - y_*| \leq 0$. Since $c < 1$, we must have $x_* = y_*$.

Now to prove the existence of the fixed point x_*, we shall show that the functional iterates x_n form a Cauchy sequence. The contraction inequality (6.4) implies that the successive differences satisfy

$$|x_{n+1} - x_n| = |g(x_n) - g(x_{n-1})| \leq c|x_n - x_{n-1}|.$$

Thus

$$|x_2 - x_1| \leq c|x_1 - x_0| \quad \text{and}$$
$$|x_3 - x_2| \leq c|x_2 - x_1| \leq c^2|x_1 - x_0|.$$

In general, an induction argument shows that

$$|x_{n+1} - x_n| \leq c^n|x_1 - x_0|. \tag{6.5}$$

Now to estimate the difference $x_{n+k} - x_n$ for an arbitrary $k > 0$, we express it in terms of the successive differences:

$$x_{n+k} - x_n = (x_{n+k} - x_{n+k-1}) + (x_{n+k-1} - x_{n+k-2}) + \cdots + (x_{n+1} - x_n).$$

It follows from the triangle inequality and (6.5) that

$$
\begin{aligned}
|x_{n+k} - x_n| &\leq |x_{n+k} - x_{n+k-1}| + |x_{n+k-1} - x_{n+k-2}| + \cdots + |x_{n+1} - x_n| \\
&\leq (c^{n+k-1} + c^{n+k-2} + \cdots + c^n)|x_1 - x_0| \\
&\leq c^n(c^{k-1} + \cdots + 1)|x_1 - x_0| \\
&\leq c^n \frac{1 - c^k}{1 - c}|x_1 - x_0| \\
&\leq \frac{c^n}{1 - c}|x_1 - x_0|. \tag{6.6}
\end{aligned}
$$

The estimate (6.6) is independent of k so that, for any pair of indices n and $m = n + k$, we have

$$|x_n - x_m| \leq \frac{c^n}{1 - c}|x_1 - x_0|.$$

The crucial hypothesis that $c < 1$ now implies that, for given $\varepsilon > 0$, we can choose $N > 0$ such that $|x_n - x_m| < \varepsilon$ for $m \geq n \geq N$. Therefore x_n is a Cauchy sequence and must converge to a real number x_*. The number $x_* \in I$ because I is a closed interval. Now we take the limit on both sides of (6.3):

$$x_* = \lim_{n \to \infty} x_{n+1} = \lim_{n \to \infty} g(x_n) = g(x_*)$$

because g is continuous. Thus x_* is a fixed point and we know it is unique. The theorem is proved.

Figure 6.2 displays the graph of $g(x) = (1/2)x + 1/(x + 1)$ and the functional iterates starting with $x_0 = 5$.

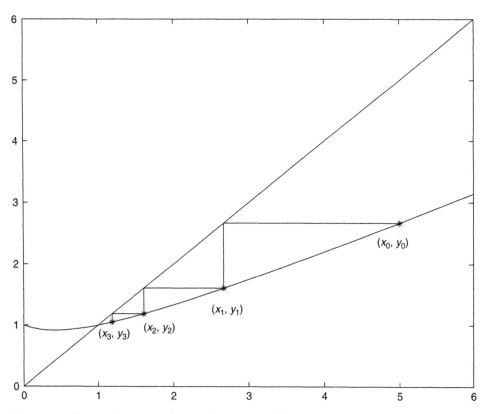

Figure 6.2 Graph of $g(x) = (1/2)x + 1/(1 + x)$ and functional iterates x_0, x_1, x_2, x_3 starting with $x_0 = 5$. The values $y_n = g(x_n)$

How fast do the functional iterates converge to the fixed point? We can make the estimate:

$$|x_n - x_*| = |g(x_{n-1}) - g(x_*)| \leq c|x_{n-1} - x_*|.$$

Thus the distance $|x_n - x_*|$ is reduced by a factor of c with each iteration. This is called *linear* convergence with rate c. The method of functional iteration produces rather slow convergence for computational purposes, but it has the advantage that we do not have to compute the derivative for each iteration.

Corollary Let g be continuous on a closed interval I with $g(I) \subset I$. Suppose in addition that g is differentiable on the interior of I with $|g'(x)| \leq c < 1$. Then g has a unique fixed point $x_* \in I$.

The proof is a simple consequence of Theorem 6.2 and Theorem 6.3.

6.2 Computation with Functional Iteration

A useful variation on Theorem 6.3 and its corollary is possible when the location of the fixed point is known approximately, and we can estimate the value of g' at the fixed point.

Theorem 6.4 Let I be an open interval, and let $g \in C^1(I)$. Suppose that g has a fixed point $x_* \in I$ and that $|g'(x_*)| < 1$. Then there is a $\delta > 0$ such that g is a contraction on the closed interval $I_0 = [x_* - \delta, x_* + \delta]$, and g maps I_0 into itself. If $x_0 \in I_0$, then the functional iterates $x_{n+1} = g(x_n)$ converge to x_*.

Proof: Since $g \in C^1(I)$ and $|g'(x_*)| < 1$, there is a $\delta > 0$ such that $|g'(x)| \leq c < 1$ on $I_0 = [x_* - \delta, x_* + \delta]$. If $x \in I_0$, then

$$\begin{aligned}
|g(x) - x_*| &= |g(x) - g(x_*)| \\
&= |g'(\theta)||x - x_*| \\
&\leq |x - x_*| < \delta.
\end{aligned}$$

Hence $g(I_0) \subset I_0$. The remaining assertion follows by an appeal to the corollary of Theorem 6.3.

We return to Example 6.1 where $g(x) = 2\sin x$. We saw that g has a fixed point in the interval $I = [1.8, 2.0]$. Now $g'(x) = 2\cos x$ is decreasing on I with $g'(1.8) = 2\cos(1.8) = -0.4544$ and $g'(2) = 2\cos(2) = -0.8323$. Hence $|g'(x)| < 0.8323$ on the whole interval I. We can apply Theorem 6.4 and deduce that there is a closed interval $I_0 \subset I$ such that if $x_0 \in I_0$, the functional iterates, starting with x_0, must converge to the fixed point x_*.

In fact, in this example, for any starting point x_0 with $0 < x_0 \leq \pi$, the functional iterates will converge to the fixed point x_*, which lies between 1.8 and 2.0. On the other hand, $x = 0$ is also a fixed point, and $g'(0) = 2 > 1$. No matter how close we start to $x = 0$, the functional iterates will not converge to $x = 0$.

Equations can often be put in fixed-point form, but the result is not always useful.

EXAMPLE 6.3: Let $f(x) = e^x - 2\sqrt{x} - 1$. Now $f(1) = e - 3 < 0$ and $f(2) = e^2 - 2\sqrt{2} - 1 > 0$. Hence there is a root of the equation $f(x) = 0$ in the interval $1 < x < 2$. One way to put the equation

$f(x) = 0$ into fixed-point form is to eliminate the square root. We obtain the fixed-point form:

$$x = g(x) \equiv \frac{(e^x - 1)^2}{4}.$$

We know that g has a fixed point in the interval $(1, 2)$. However,

$$g'(x) = \frac{e^x(e^x - 1)}{2} \geq \frac{e^2 - e}{2} > 1 \quad \text{for } x \geq 1,$$

so we cannot apply Theorem 6.4.

On the other hand, we can also solve for x by using the logarithm to obtain the equivalent equation:

$$x = h(x) \equiv \log(1 + 2\sqrt{x}).$$

The function h is just the inverse function of g. Because

$$h'(x) = \frac{1}{g'(h(x))},$$

we conclude that $|h'(x_*)| < 1$ at the fixed point. In fact, if we calculate directly, we find

$$|h'(x)| = \frac{1}{2x + \sqrt{x}} \leq \frac{1}{3} \quad \text{for } x \geq 1.$$

Hence there can be only one root x_* of $f(x) = 0$ in the interval $1 < x < 2$, and starting with any point x_0 in this interval, the iterates $x_{n+1} = h(x_n)$ will converge to x_*.

It may not always be possible to put the equation in fixed-point form by solving for x. Therefore, we consider a more general approach to converting equations to fixed-point form. This treatment uses an idea that is central to the proof of the inverse function theorem in higher dimensions.

Suppose that $f(x)$ is C^1 on some open interval I and that there is a solution $x_* \in I$ of $f(x) = 0$. Now for any $\alpha \neq 0$, an equivalent equation is

$$x = x - \alpha f(x) \equiv g_\alpha(x).$$

Can we pick α such that $|g'_\alpha(x)| \leq c < 1$ on some interval that contains x_*? We do not know the exact value of x_*, but we may be able to make a reasonable estimate of $f'(x_*)$. Assuming that $f'(x_*) \neq 0$, we choose α to be an estimate of $1/f'(x_*)$. With this choice of α, we have

$$g'_\alpha(x) = 1 - \alpha f'(x) \approx 1 - f'(x)/f'(x_*).$$

Because $f'(x)$ is continuous, it is reasonable to hope that there is some $\delta > 0$ such that $|g'(x)| \leq c < 1$ for $|x - x_*| \leq \delta$. Thus if we start close enough to x_*, the functional iterates of g_α should converge to x_*.

EXAMPLE 6.4:   We return to the function $f(x) = e^x - 2\sqrt{x} - 1$ of Example 6.3. From a graph, we see that the root x_* of $f(x) = 0$ that lies in $(1, 2)$ lies closer to 1. We take

$$\alpha = \frac{1}{f'(1)} = \frac{1}{e-1}$$

whence

$$g_\alpha(x) = x - \frac{f(x)}{e-1}$$

$$= x - \frac{e^x - 2\sqrt{x} - 1}{e - 1},$$

and

$$g'_\alpha(x) = 1 - \frac{e^x - x^{-1/2}}{e-1}.$$

We see that $|g'_\alpha(x)| < 1$ for $1 \leq x \leq 1.4$. Hence we can apply Theorem 6.4 on this interval.

Exercises for 6.1 and 6.2

1. For $c > 0$, let $f(x) = cx(1 - x)$.

   a) For what values of c does f map the interval $[0, 1]$ into itself?

b) What are the fixed points of f, expressed in terms of c?

c) What is the derivative of f at the fixed points?

2. Now compute the functional iterates for the function f of exercise 1 for various values of c. Let N be the number of iterations.

   a) Try $c = 2$ and $x_0 = 0.1$, and then $x_0 = 0.8$ and make $N = 10$ iterations. Describe the behavior of the iterates x_k.

   b) Now try $c = 3.2$, the same starting points, and the same N. How is the behavior different this time? Explain the difference in behavior.

   c) What is the value of c between $c = 2$ and $c = 3.2$ where the behavior of the functional iterates changes?

   d) What kind of behavior do you see for c between 3 and 3.5? Look at the numbers for x and $f(x)$; notice how they alternate.

3. The function $f(x) = 2.5x(1 - x)$ maps $[0, 1]$ into $[0, 1]$ and has fixed points at $x = 0$ and $x = 0.6$. Does this lack of uniqueness contradict Theorem 6.3?

4. Let $g(x) = \cos(\sqrt{x})$.

   a) Show that g has a unique fixed point in the interval $[0, \pi^2/4]$ and that the functional iterates $x_n = g(x_{n-1})$ converge to the fixed point x_* for any starting point $x_0 \in [0, \pi^2/4]$.

   b) Starting with $x_0 = 1$, use a calculator or a computer to compute functional iterates $x_n = g(x_{n-1})$ for $n = 1, \ldots, 10$.

5. Consider the equation $\sin x = e^{-x}$.

   a) Show that the equation $f(x) = \sin(x) - e^{-x} = 0$ has a root x_* in the interval $[0, \pi/2]$. Graph the functions $\sin x$ and e^{-x} together to get a rough idea of where the root lies.

   b) Put the equation in fixed point form $x = g(x)$ where $g(x) = -\log(\sin x)$. Calculate $g'(x)$ and estimate it near the root x_*. Can you use functional iteration with g to calculate the root?

   c) Put the equation in the form $x = h(x)$ where h is the inverse of g. Verify that $|h'(x)| < 1$ near the fixed point x_*. With $x_0 = 0.5$, make 10 functional iterations $x_n = h(x_{n-1})$.

6. Let $f(x) = \sin x - e^{-x}$. From exercise 5 we know that f has a root x_* in the interval $[0, \pi/2]$. Use the graphs of f to estimate the location of f and then estimate $f'(x)$ for x near x_*. Let $\alpha \approx 1/f'(x_*)$ and set

$$g(x) = x - \alpha f(x).$$

Starting with $x_0 = 0.5$, make 10 functional iterations $x_n = g(x_{n-1})$.

7. Let $g(x, \lambda) = \lambda \sin x$. We assume that $\lambda \geq \pi/2$.

a) Apply the intermediate value theorem to $g(x, \lambda) - x$ to deduce that $g(x, \lambda)$ has a fixed point in the interval $[\pi/2, \pi]$. Show that there is only one fixed point of $g(x, \lambda)$ in this interval. We shall label it $x_*(\lambda)$. Note that $x_*(\pi/2) = \pi/2$.

b) Use the fact that $g(x, \lambda) \leq \lambda$ to show that $\pi/2 \leq x_*(\lambda) \leq \lambda$ when $\pi/2 \leq \lambda \leq \pi$.

c) Explain why part b) implies that, for $\pi/2 \leq \lambda \leq \pi$,

$$g'(x_*(\lambda), \lambda) = \lambda \cos x_*(\lambda) \geq \lambda \cos \lambda.$$

d) Use a Taylor polynomial at $\lambda = \pi/2$ to show that $\lambda \cos \lambda \geq \lambda(\pi/2 - \lambda)$ for $\lambda \geq \pi/2$.

e) For what range of λ is it true that $|g'(x_*(\lambda), \lambda)| < 1$? In this case we can apply Theorem 6.4 to deduce that if we start close enough to $x_*(\lambda)$, the functional iterates will converge to $x_*(\lambda)$.

8. Use functional iteration with a computer or a calculator to compute the fixed point $x_*(\lambda)$ of exercise 7 for $\lambda = 1.6, 1.8, 2$. What happens when you try functional iteration to compute the fixed point when $\lambda = 2.5$?

9. Let g be C^1 in an open interval I. Suppose that g has a fixed point $x_* \in I$ and that $x_n \in I$ is a sequence of functional iterates, $x_{n+1} = g(x_n)$, such that $\lim_{n \to \infty} x_n = x_*$. Show that

$$\lim_{n \to \infty} \frac{|x_{n+1} - x_*|}{|x_n - x_*|} = |g'(x_*)|.$$

Note that, if $g'(x_*) \neq 0$, this shows that the convergence is linear, with rate of convergence $|g'(x_*)|$.

6.3 Newton's Method

Let us take another look at Example 6.1, where we sought to compute a root of equation (6.2), $x = g(x) = 2 \sin x$. Perhaps a better approximation would produce a sequence that converges more rapidly. We could instead use the tangent line approximation to $g(x)$, which is

$$l(x) = g(x_0) + g'(x_0)(x - x_0).$$

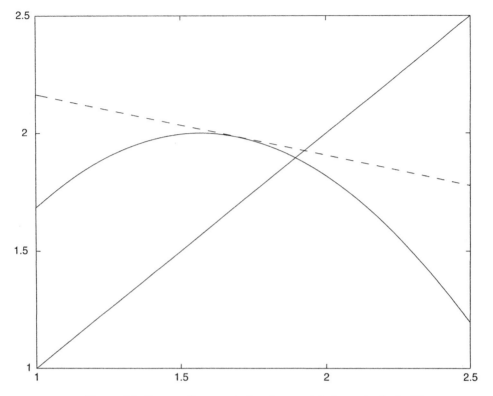

Figure 6.3 Tangent line approximation to $g(x)$ shown in dashed line

The intersection of this line with the diagonal $y = x$ would provide us with the next iterate. In Figure 6.3, we can see that the intersection of the tangent line with the diagonal line is closer to the root than was the horizontal line $y \equiv g(x_0)$.

Newton's method uses this approach. It is concerned with finding the roots of an equation

$$f(x) = 0.$$

After making a first guess, x_0, Newton's method uses the tangent line approximation:

$$l(x) = f(x_0) + f'(x_0)(x - x_0).$$

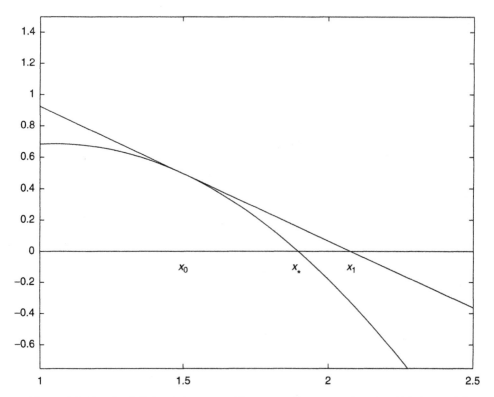

Figure 6.4 Graph of $f(x) = 2 \sin x - x$ with tangent line approximation $l(x)$ at $x_0 = 1.5$

We replace the equation $f(x) = 0$ by the linear equation $l(x) = 0$ (see Figure 6.4). The solution of this linear equation is

$$x_1 = x_0 - \frac{f(x_0)}{f'(x_0)}.$$

The sequence of Newton iterates is given by the formula

$$x_{n+1} = x_n - \frac{f(x_n)}{f'(x_n)}. \tag{6.7}$$

How well does this method work for equation (6.2)? We set $f(x) = g(x) - x = 2 \sin x - x$. The Newton iteration scheme (6.7) becomes

$$x_{n+1} = x_n - \frac{2 \sin x_n - x_n}{2 \cos x_n - 1}.$$

With the same starting point $x_0 = 1.7$ we have the results:

x	$f(x)$
1.70000000000000	0.28332962090494
1.92527796893757	-0.04962489086190
1.89598707057448	-0.00080746455200
1.89549440745065	-0.00000023000884
1.89549426703399	-0.00000000000002
1.89549426703398	0
1.89549426703398	0

We see that after only five iterations, there is no change in the first 16 digits. Newton's method converges much more quickly than functional iteration, but it does require a knowledge of the first derivative of f.

Theorem 6.5 Let I be an open interval, and let $f \in C^1(I)$ and assume further that f' is Lipschitz continuous on I with Lipschitz constant L. Suppose that $x_* \in I$ is a root of $f(x) = 0$ with $f'(x_*) \neq 0$. Then there is a $\delta > 0$ such that, if $|x_0 - x_*| < \delta$, the Newton iterates, as defined by (6.7), converge to x_*. Furthermore, there is a constant K such that the error satisfies

$$|x_{n+1} - x_*| \leq K|x_n - x_*|^2. \tag{6.8}$$

Proof: For the proof in this chapter, we shall make the slightly stronger hypothesis that $f \in C^2(I)$. Since $f'(x_*) \neq 0$, there is a $\delta_0 > 0$ such that the closed interval $I_0 = [x_* - \delta_0, x_* + \delta_0] \subset I$ and such that $|f'(x)| \geq (1/2)|f'(x_0)| > 0$ for $x \in I_0$.

Now suppose $x_n \in I_0$. The next Newton iterate is found by using the linear approximation to f at $x = x_n$:

$$l(x) = f(x_n) + f'(x_n)(x - x_n).$$

The iterate x_{n+1} is the solution of $l(x) = 0$ and therefore satisfies

$$0 = f(x_n) + f'(x_n)(x_{n+1} - x_n). \tag{6.9}$$

The error in using this linear approximation is R_1:

$$f(x) = f(x_n) + f'(x_n)(x - x_n) + R_1(x, x_n).$$

Because $f(x_*) = 0$, we deduce that

$$0 = f'(x_n) + f'(x_n)(x_* - x_n) + R_1(x_*, x_n). \tag{6.10}$$

Now subtract (6.9) from (6.10) to find

$$0 = f'(x_n)(x_* - x_{n+1}) + R_1(x_*, x_n). \tag{6.11}$$

Equation (6.11) expresses the relationship between the sides of the triangle shown with dotted lines in Figure 6.5. Now we bring in our two main assumptions. Because $f \in C^2(I)$, Theorem 5.1 (Lagrange form of the remainder) says that

$$R_1(x_*, x_n) = \frac{f''(\theta)(x_* - x_n)^2}{2}$$

where θ lies between x_n and x_*. The second assumption allows us to divide by $f'(x_n)$ because $x_n \in I_0$. Hence,

$$x_{n+1} - x_* = \frac{f''(\theta)(x_n - x_*)^2}{2f'(x_n)}.$$

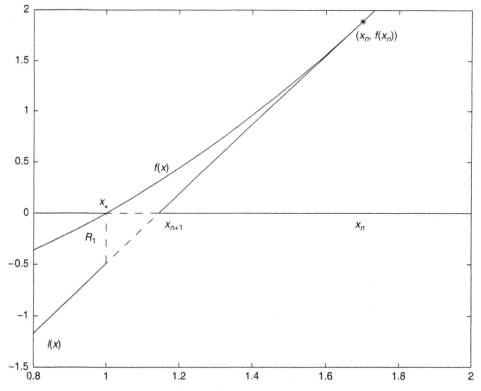

Figure 6.5 The Newton approximation used to generate the $(n + 1)$st iterate

We conclude that

$$|x_{n+1} - x_*| \le K|x_n - x_*|^2$$

where

$$K = \frac{\max_{I_0} |f''(x)|}{2\min_{I_0} |f'(x)|}.$$

The crucial element in the derivation of this inequality is the fact that the error R_1 is $O(|x_n - x_*|^2)$. We see that if $x_n \in I_0$, then (6.8) is satisfied. However, we cannot be sure that $x_{n+1} \in I_0$; we need to know this to be able to repeat the argument. To be able to continue, we shall reduce the interval I_0. Choose $0 < \delta \le \delta_0$ so that $c \equiv \delta K < 1$, and let $J = [x_* - \delta, x_* + \delta] \subset I_0$. Then if $x_n \in J$, we have

$$\begin{aligned}
|x_{n+1} - x_*| &\le K|x_n - x_*|^2 \\
&\le K\delta|x_n - x_*| \\
&\le c|x_n - x_*|.
\end{aligned} \tag{6.12}$$

This last inequality has two consequences. First, it tells us that if $x_n \in J$, then the next Newton iterate x_{n+1}, given by (6.9), also lies in J because $c < 1$. Hence if $x_0 \in J$, the sequence of Newton iterates x_n will all lie in J. Second, we can use the inequality (6.12) and a simple induction argument to show that

$$|x_n - x_*| \le c^n|x_0 - x_*|.$$

Because $c < 1$, $\lim_{n \to \infty} x_n = x_*$, The theorem is proved.

Remark Let $e_n = x_n - x_*$. Then (6.8) says that $|Ke_{n+1}| \le |Ke_n|^2$. Consequently, if for some n, $|Ke_n| \le 10^{-2}$, then $|Ke_{n+1}| \le 10^{-4}$ and $|Ke_{n+2}| \le 10^{-8}$. This means that if our starting point x_0 is close enough to the root x_*, the number of correct digits will approximately double with each iteration. This kind of rapid convergence is called *quadratic* convergence.

EXAMPLE 6.5: Recall Example 2.3 b), where the sequence a_n was defined recursively by

$$a_{n+1} = (1/2)(a_n + A/a_n).$$

Here $A > 0$ and the sequence was shown to converge to $\sqrt{A}$. In fact this sequence is generated by Newton's method when we take $f(x) = x^2 - A$. The derivative $f'(\sqrt{A}) = 2\sqrt{A} > 0$ so Theorem 6.5 implies that the sequence converges quadratically.

The sequence in Example 2.3 c) is generated by Newton's method and the function $f(x) = A - 1/x$. Even though the function f involves division by x, the Newton iteration formula,

$$x_{n+1} = 2x_n - x_n^2 A,$$

uses only multiplication and subtraction. It also converges quadratically when a correct choice of starting point x_0 is made.

EXAMPLE 6.6:   Theorem 6.4 says that, *if the starting point x_0 is close enough to the root x_**, then Newton's method will converge. However, if the starting point is not close enough to the desired root, Newton's method may converge to a different root, or not converge at all. Recall the Example 6.1, with $f(x) = 2\sin x - x$. A local maximum of f is attained at $x = \pi/3$. If the starting point $x_0 < \pi/3$, the Newton iterates may converge to x_*, they may converge to zero, they may converge to $-x_*$, or they may not converge at all. The same may be said for $x_0 > \pi/3$ but very close to $\pi/3$. Figure 6.6 shows two tangent line approximations for x_0 close to $\pi/3$. It can be seen that the next iterate, x_1, will be far from the root x_*.

Some interesting comments about the history of Newton's method can be found in the book of S. Nash and A. Sofer [N].

Exercises for 6.3

1. Let $A > 0$ be a positive number.

   a) Let $f(x) = x^2 - A$. Write out the Newton iteration scheme for the equation $f(x) = 0$. Verify that it agrees with the formula in the first part of Example 6.5.

   b) Let $f(x) = A - 1/x$. Write out the Newton iteration scheme for the equation $f(x) = 0$. Verify that it agrees with the formula in the second part of Example 6.5.

2. Let $f(x) = \sin x - e^{-x}$. We saw in exercise 5 of the previous section that there is a root $x_* \in [0, \pi/2]$ of the equation $f(x) = 0$.

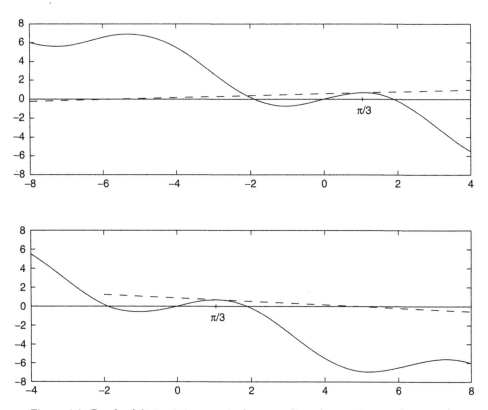

Figure 6.6 Graph of $f(x) = 2 \sin x - x$ and tangent lines from points x_0 close to $\pi/3$

a) Use the starting point $x_0 = 0.5$. Use a calculator or a computer to compute two steps in the Newton iteration scheme for the equation $f(x) = 0$.

b) Write a code that uses Newton's method, starting with $x_0 = 0.5$, to approximate the value of x_* to within 10^{-12}. How many iterations are needed? Compare with the number of iterations needed when using functional iteration.

c) Explain what happens if the starting value $x_0 \geq 1.8$.

3. Let $f(x) = \log x - x + 2$.

a) Show that $f(x) = 0$ has two roots in $x > 0$.

b) Show that for Newton's method to converge to the root closer to zero, we must take $0 < x_0 < e^{-1}$.

c) Using a Newton code, approximate both roots to within 10^{-12}.

4. Let $g(x, \lambda) = \lambda \sin x$. In exercise 7 of Section 6.2, we verified that there is a fixed point $x_*(\lambda)$ of g in $(0, \pi]$ for $\lambda > 1$.

   Use a Newton code on the function $f(x, \lambda) = g(x, \lambda) - x$ to approximate the fixed point $x_*(\lambda)$ for $\lambda = 1.6, 1.8, 2$. Starting at $x_0 = \pi/2$, how many iterations are needed in each case to approximate the fixed point to within 10^{-12}? Compare these results with the functional iteration results of exercise 8 of Section 6.2.

5. The proof of the quadratic convergence of Newton's method requires the hypothesis that $f'(x_*) \neq 0$. If $f'(x_*) = 0$, Newton's method still converges, but much more slowly. Assume that $f \in C^2$ with $f''(x_*) \neq 0$. Start with equation (6.11), and show that

$$\lim_{n \to \infty} \frac{|x_{n+1} - x_*|}{|x_n - x_*|} = \frac{1}{2}.$$

   Thus in this case, Newton's method converges linearly with constant multiple $1/2$.

6. Let $f(x) = \sin^2(\pi x)$. Then $f(1) = f'(1) = 0$, but $f''(1) \neq 0$. Examine the Newton iterates x_n that converge to $x_* = 1$. Verify the rate of convergence predicted by exercise 5.

7. Here is another way to explain quadratic convergence for Newton's method. Let $g(x) \in C^2$, let x_* be a fixed point of g, and suppose that $g'(x_*) = 0$. By Theorem 6.4, we know that if x_0 is close enough to x_*, the functional iterates $x_n = g(x_{n-1})$ will converge to x_*.

   a) Using the fact that $g'(x_*) = 0$, show that there is a constant K such that

$$|x_n - x_*| \leq K|x_{n-1} - x_*|^2.$$

   b) Verify that Newton's method can be viewed as functional iteration of the function $g(x) = x - f(x)/f'(x)$ and that $g'(x_*) = 0$ if $f(x_*) = 0$.

Chapter 7

Integration

Chapter Overview

We shall assume that the reader is familiar with the standard techniques of integration that are seen in a first course in calculus. We begin with a study of the Riemann-Darboux integral and derive the usual properties of the integral. The fundamental theorem of calculus and the change of variable rule are proved in the next section. We use the results of polynomial interpolation to derive several numerical methods for approximating integrals, along with expressions for the error. Finally, we discuss improper integrals.

7.1 The Definition of the Integral

The integral is an extension of our notion of area for rectangular sets, and we shall be careful to frame the definition so that $\int_a^b f$ has the following properties:

(i) $\int_a^b f = c(b-a)$ when f is a constant function, $f(x) \equiv c$.

(ii) $\int_a^b (f+g) = \int_a^b f + \int_a^b g$.

(iii) $\int_a^b f = \int_a^c f + \int_c^b f$ when $a < c < b$.

Notice that we have not indicated a variable of integration because, at this stage in our discussion, it is not necessary. Later on, when it is useful, we shall write $\int_a^b f(x)dx$ or $\int_a^b f(s)ds$.

When we divide a closed bounded interval $[a, b]$ into n subintervals $a = x_0 < x_1 < x_2 < \cdots < x_n = b$, we call the set of points $P = \{x_0, \ldots, x_n\}$ a *partition* of $[a, b]$. Note that we do not assume that the lengths of intervals $x_j - x_{j-1}$ are equal.

Definition 7.1 Let $f : [a, b] \to \mathbb{R}$ be a bounded function. For a given partition P of $[a, b]$, we define

$$M_j = \sup_{[x_{j-1}, x_j]} f(x) \text{ and } m_j = \inf_{[x_{j-1}, x_j]} f(x).$$

Then the *upper Darboux sum* is defined to be

$$U(f, P) = \sum_{j=1}^n M_j(x_j - x_{j-1})$$

and the *lower Darboux sum* is defined to be

$$L(f, P) = \sum_{j=1}^n m_j(x_j - x_{j-1}).$$

Let $M = \sup_{[a,b]} f$ and $m = \inf_{[a,b]} f$. On each subinterval we have,

$$m \le m_j = \inf_{[x_{j-1}, x_j]} f \le \sup_{[x_{j-1}, x_j]} f = M_j \le M.$$

Hence

$$m(b - a) \le L(f, P) \le U(f, P) \le M(b - a). \tag{7.1}$$

However, we wish to establish the stronger result that (7.1) holds when we use one partition P for the lower sum and a different partition Q for the upper sum. We shall need the following notion.

Definition 7.2 A partition Q is a *refinement* of the partition P if every subdivision point of P is also a subdivision point of Q.

Lemma 7.1 Let P and Q be partitions of $[a, b]$, and suppose that Q is a refinement of P. Then for any bounded function $f : [a, b] \to \mathbb{R}$,

$$L(f, P) \le L(f, Q) \quad \text{and} \quad U(f, Q) \le U(f, P).$$

Proof: Let $[x_{k-1}, x_k]$ be one of the subintervals of the partition P. Suppose that the refinement Q adds one point z to this interval, $x_{k-1} < z < x_k$. We set

$$m'_k = \inf_{[x_{k-1}, z]} f \quad \text{and} \quad m''_k = \inf_{[z, x_k]} f.$$

Then $m_k \le m'_k$ and $m_k \le m''_k$. Hence,

$$m_k(x_k - x_{k-1}) \le m'_k(z - x_{k-1}) + m''_k(x_k - z).$$

We conclude that

$$
\begin{aligned}
L(f, P) &= \sum_{j \ne k} m_j(x_j - x_{j-1}) + m_k(x_k - x_{k-1}) \\
&\le \sum_{j \ne k} m_j(x_j - x_{j-1}) + m'_k(z - x_{k-1}) + m''_k(x_k - z) \\
&\le L(f, Q).
\end{aligned}
$$

Now a partition Q that is a refinement of P can be thought of as constructed by adding one additional subdivision point at a time to P. As each point is added, the lower sum will either stay the same, or increase. Hence we can deduce that $L(f, P) \le L(f, Q)$. A similar argument is used to prove the assertion about the upper sums.

Lemma 7.2 Let $f : [a, b] \to \mathbb{R}$ be a bounded function. Then for any two partitions P and Q of $[a, b]$,

$$L(f, P) \le U(f, Q).$$

Proof: Let $R = P \cup Q$. R is the partition of $[a, b]$ that consists of all of the subdivision points of both P and Q. Hence R is a refinement of both P and Q. Then by Lemma 7.1,

$$L(f, P) \le L(f, R) \le U(f, R) \le U(f, Q).$$

A consequence of Lemma 7.2 is that every upper sum $U(f, Q)$ is an upper bound for the set of all lower sums $L(P, f)$. Thus for each partition Q,

$$\sup_P L(f, P) \leq U(f, Q).$$

Now we see that $\sup_P L(f, P)$ is a lower bound for the set of upper sums. We deduce that

$$\sup_P L(f, P) \leq \inf_Q U(f, Q). \tag{7.2}$$

Definition 7.3 For a bounded function $f : [a, b] \to \mathbb{R}$, the *upper Darboux integral* is defined to be

$$\overline{\int_a^b} f = \inf_Q U(f, Q)$$

and the *lower Darboux integral* is defined to be

$$\underline{\int_a^b} f = \sup_P L(f, P).$$

Note that we always have $\underline{\int_a^b} f \leq \overline{\int_a^b} f$.

Definition 7.4 A bounded function $f : [a, b] \to \mathbb{R}$ is *Riemann integrable* on $[a, b]$ if

$$\overline{\int_a^b} f = \underline{\int_a^b} f.$$

and, in this case we set

$$\int_a^b f = \underline{\int_a^b} f = \overline{\int_a^b} f.$$

In this text, we shall only be dealing with the integral defined by Definition 7.4 and, consequently, we shall drop the adjective Riemann.

Of course, we hope that with this definition, a large useful class of functions will turn out to be integrable. Before we can answer this question, we must provide a more convenient test for a function to be integrable.

Theorem 7.3 (Integrability criterion) Let $f : [a, b] \to \mathbb{R}$ be bounded. Then f is integrable on $[a, b]$ if and only if, for each $\varepsilon > 0$, there is a partition P of $[a, b]$ such that

$$U(f, P) - L(f, P) < \varepsilon.$$

Proof: First suppose that f is integrable on $[a, b]$. Let $\varepsilon > 0$ be given. Using our descriptions of the infimum and the supremum, we know that there is a partition P such that

$$L(f, P) \geq \int_a^b f - \varepsilon/2$$

and there is also a partition Q such that

$$U(f, Q) \leq \int_a^b f + \varepsilon/2.$$

Now let $R = P \cup Q$ be the common refinement of P and Q. By Lemma 7.1,

$$L(f, P) \leq L(f, R) \leq U(f, R) \leq U(f, Q).$$

It follows that

$$U(f, R) - L(f, R) \leq \int_a^b f + \varepsilon/2 - \left(\int_a^b f - \varepsilon/2 \right) = \varepsilon.$$

Now we prove the converse. Assume that for each $\varepsilon > 0$, there is a partition P such that

$$U(f, P) - L(f, P) < \varepsilon.$$

We have

$$U(P, f) \geq \overline{\int_a^b} f \quad \text{and} \quad L(P, f) \leq \underline{\int_a^b} f$$

If we subtract these inequalities, the result is

$$\varepsilon > U(P, f) - L(P, f) \geq \overline{\int_a^b} f - \underline{\int_a^b} f \geq 0.$$

Because ε is arbitrary, we conclude that $\overline{\int_a^b} f = \underline{\int_a^b} f$. We have proved that f is integrable.

We will now see two important classes of functions that are integrable.

Recall that f is monotone increasing if $f(x) \leq f(y)$ whenever $x \leq y$ and monotone decreasing if $f(x) \geq f(y)$ whenever $x \leq y$.

Theorem 7.4 Let $f : [a, b] \to \mathbb{R}$ be a monotone function. Then f is integrable.

Proof: Suppose that f is monotone increasing on $[a, b]$. We choose a partition P, with subdivision points $x_j = a + j(b - a)/n$, $j = 0, 1, 2, \ldots n$. Because f is increasing, on each subinterval $[x_{j-1}, x_j]$, we have $M_j = f(x_j)$ and $m_j = f(x_{j-1})$. Consequently, we have

$$U(f, P) - L(f, P) = \sum_{j=1}^{n} (M_j - m_j)(x_j - x_{j-1})$$

$$= \sum_{j=1}^{n} (f(x_j) - f(x_{j-1})) \frac{(b - a)}{n}$$

$$= \frac{(f(b) - f(a))(b - a)}{n}.$$

Now given $\varepsilon > 0$ we choose n so large that $(f(b) - f(a))(b - a)/n < \varepsilon$. The argument for monotone decreasing functions is similar. In both cases, we then appeal to Theorem 7.3 to deduce that f is integrable.

To determine a second class of integrable functions, we start with the continuous functions.

Theorem 7.5 If $f : [a, b] \to \mathbb{R}$ is continuous, then f is integrable.

Proof: Because f is continuous on $[a, b]$ we know by Theorem 3.6 that f is bounded. In addition, we finally get to use the concept of uniform continuity. Let $\varepsilon > 0$ be given. By Theorem 3.7, we know that f is uniformly continuous on $[a, b]$. Hence there is a $\delta > 0$ such that $|f(x) - f(y)| < \varepsilon/(b - a)$ whenever $x, y \in [a, b]$ with $|x - y| < \delta$. Let P be any partition such that $x_j - x_{j-1} < \delta$ for all j. Then $M_j - m_j \leq \varepsilon/(b - a)$. Consequently,

$$U(f, P) - L(f, P) = \sum_{j} (M_j - m_j)(x_j - x_{j-1})$$

$$\leq \frac{\varepsilon}{b - a} \sum_{j} (x_j - x_{j-1}) = \varepsilon.$$

Thus f is integrable by the criterion of Theorem 7.3.

Finally, we expand our second class of integrable functions to include functions that are continuous except at a finite number of points.

Theorem 7.6 Let $f : [a, b] \to \mathbb{R}$ be bounded, and suppose that f is continuous except possibly at a finite number of points in $[a, b]$. Then f is integrable on $[a, b]$.

Proof: To keep the notation simple, we shall consider a special case. Let $a < z < b$, and suppose that f is continuous on the subintervals $[a, z)$ and $(z, b]$. Let $\varepsilon > 0$ be given. Now let $\delta > 0$ be so small that $(z - \delta, z + \delta) \subset [a, b]$ and $2\delta(M - m) < \varepsilon/3$ where $M = \sup_{[a,b]} f$ and $m = \inf_{[a,b]} f$. Since f is continuous on $[a, z - \delta]$, f is uniformly continuous on $[a, z - \delta]$. We may use the procedure of Theorem 7.5 to construct a partition P of $[a, z - \delta]$ such that $U(f, P) - L(f, P) < \varepsilon/3$. Similarly, f is continuous on $[z + \delta, b]$, and so there is a partition Q of $[z + \delta, b]$ such that $U(f, Q) - L(f, Q) < \varepsilon/3$. Finally, we let the partition R of $[a, b]$ be $R = P \cup Q$. Then with $M_\delta = \sup_{[z-\delta, z+\delta]} f \le M$ and $m_\delta = \inf_{[z-\delta, z+\delta]} f \ge m$, we have

$$U(f, R) - L(f, R) = U(f, P) - L(f, P) + 2\delta(M_\delta - m_\delta) + U(f, Q) - L(f, Q).$$

Because of our choice of δ and the partitions P and Q, we conclude that

$$U(f, R) - L(f, R) \le \varepsilon/3 + \varepsilon/3 + \varepsilon/3 = \varepsilon.$$

Therefore f is integrable on $[a, b]$ according to Theorem 7.3. The argument can be extended readily to the general case of a finite number of discontinuities.

EXAMPLE 7.1:

a) Let $f(x) = x$. We compute the integral $\int_0^1 f$ using the upper and lower sums. Take a partition P_n of equally spaced points $x_j = (j/n), j = 0, \ldots, n$. Then $M_j = j/n$ and $m_j = (j - 1)/n$. We compute the upper and lower sums for this partition:

$$L(P_n, f) = \sum_{j=1}^{n} (j - 1)/n^2$$

$$= \frac{1}{n^2} \frac{n(n - 1)}{2},$$

and

$$U(P_n, f) = \sum_{j=1}^{n} j/n^2$$

$$= \frac{1}{n^2} \frac{n(n+1)}{2}.$$

Thus $U(f, P_n) - L(f, P_n) = 1/n$. We see that $L(P_n, f)$ converges upward to $1/2$ and that $U(P_n, f)$ converges downward to $1/2$.

b) It is possible for a function to have an infinite number of discontinuities and still be integrable. Let $f(1/n) = 1$ for $n = 1, 2, \ldots$, and let $f(x) = 0$ otherwise. We claim that f is integrable on $[0, 1]$. Let $\varepsilon > 0$ be given. Observe that f is continuous except at a finite number of points on the interval $[\varepsilon/2, 1]$. Hence by Theorem 7.6, f is integrable on $[\varepsilon/2, 1]$ and by Theorem 7.3, there is a partition Q of $[\varepsilon/2, 1]$ such that $U(f, Q) - L(f, Q) < \varepsilon/2$. Now let P be the partition obtained by adding the point 0 to Q. Then

$$U(f, P) = \varepsilon/2 + U(f, Q) \quad \text{and} \quad L(f, P) = L(f, Q).$$

It follows that

$$\begin{aligned} U(f, P) - L(f, P) &= \varepsilon/2 + U(f, Q) - L(f, Q) \\ &\le \varepsilon/2 + \varepsilon/2 = \varepsilon. \end{aligned}$$

Therefore f is integrable by Theorem 7.3.

c) We have seen that the Dirichlet function, given by

$$f(x) = \begin{cases} 1, & x \text{ rational} \\ 0, & x \text{ irrational} \end{cases},$$

is nowhere continuous. It is also not integrable on any interval $[a, b]$ if $a < b$. Let P be any partition of $[a, b]$. Because every open interval contains an irrational number, $L(f, P) = 0$. On the other hand, every interval also contains a rational number. Thus $M_j = 1$ for all j and therefore, $U(f, P) = \sum M_j(x_j - x_{j-1}) = \sum_j (x_j - x_{j-1}) = b - a$. Hence $\underline{\int_a^b} f = 0$ but $\overline{\int_a^b} f = 1$.

It is a limitation of the Riemann-Darboux theory of integration that it cannot assign a value to the integral of the Dirichlet function. As a result, one cannot always be sure that the limit of a sequence of integrable functions is integrable. To address this problem, a more advanced theory of integration, called Lebesgue integration, was developed. Every function that is integrable in the Riemann-Darboux sense is Lebesgue integrable, but in the Lebesgue theory, the Dirichlet function is integrable and its integral is zero. An introduction to Lebesgue integration can be found in the book of R. Strichartz [Str].

Exercises for 7.1

1. Let $f(x) = x^2$. Use an equally spaced partition P_n of the interval $[0, b]$, with $h = b/n$, $n \in \mathbb{N}$. Compute the upper sum $U(P_n, f)$ and the lower sum $L(P_n, f)$, and show that they both converge to $b^3/3$. You may use the formula $\sum_{k=1}^{n} k^2 = (2n^3 + 3n^2 + 3n)/6$.

2. Let $a = y_0 < y_1 < \cdots < y_m = b$ be a partition of $[a, b]$. Suppose that f is a bounded function on $[a, b]$ such that f has the constant value c_i on the open subinterval (y_{i-1}, x_i). Show that f is integrable and that

$$\int_a^b f = c_1(x_1 - x_0) + \cdots c_n(x_n - x_{n-1}).$$

3. Let f be a bounded integrable function on $[a, b]$. Let c be a constant.
   a) Show that cf is integrable with

$$\int_a^b cf = c \int_a^b f.$$

   b) Show that $f(x) + c$ is integrable with

$$\int_a^b (f + c) = \int_a^b f + c(b - a).$$

4. Let f and g be bounded integrable functions on $[a, b]$ with $f(x) \le g(x)$ for all $x \in [a, b]$. Show that $\int_a^b f \le \int_a^b g$.

5. Let f and g be bounded integrable functions on $[a, b]$, and suppose that

$$|f(x) - g(x)| \le \delta \quad \text{for all } x \in [a, b].$$

Show that

$$\left| \int_a^b f - \int_a^b g \right| \le \delta(b - a).$$

6. Let f be a bounded integrable function on $[a, b]$. Let $h = (b - a)/n$ where $n \in \mathbb{N}$. Let P_n be the partition $x_j = a + jh$, $j = 1, \ldots, n$. Let R_n be the sums

$$R_n = \sum_{j=1}^{n} f(y_j)h$$

where $y_j = (x_j + x_{j-1})/2$ is the midpoint of each subinterval in the partition. Show that

$$\lim_{n \to \infty} R_n = \int_a^b f.$$

7. Let $f : [a, b] \to \mathbb{R}$ be continuous with $f(x) \ge 0$ for all $x \in [a, b]$. Suppose that for some point $c \in [a, b]$, $f(c) > 0$. Show that $\int_a^b f > 0$.

7.2 Properties of the Integral

Although we have established two classes of functions that are integrable in the Riemann-Darboux sense, we have not as yet verified that the integral we have defined has the properties (i), (ii), and (iii) that were listed at the beginning of Section 7.1. We could do this, working only with the definition and Theorem 7.3. However, we choose to show that the integral we have defined may also be considered a limit of the Riemann sums that are familiar from calculus. We then derive properties (i), (ii), and (iii) using Riemann sums.

Definition 7.5 Let $P = \{x_0, \ldots, x_n\}$ be a partition of $[a, b]$. The *norm* of the partition P is

$$\|P\| = \max_j (x_j - x_{j-1}).$$

We shall need a somewhat technical lemma that provides another criterion for integrability.

Lemma 7.7 Let $f : [a,b] \to \mathbb{R}$ be bounded. Then f is integrable if and only if, for each $\varepsilon > 0$, there is a $\delta > 0$ (depending on ε) such that

$$U(f,P) - L(f,P) < \varepsilon$$

for all partitions P with $\|P\| < \delta$. \hfill (7.3)

Proof: It is clear that if the new criterion (7.3) is satisfied, then by Theorem 7.3, f is integrable.

Now suppose that f is integrable. By Theorem 7.3, for each $\varepsilon > 0$, there is a partition $Q = \{x_0, x_1, \ldots, x_n\}$ such that

$$U(f,Q) - L(f,Q) < \varepsilon.$$

Let $P = \{y_0, \ldots, y_m\}$ be another partition of $[a,b]$. We group the subintervals of P into two categories. Let J consist of the indices i such that the subinterval $[y_{i-1}, y_i]$ is contained in some subinterval $[x_{j-1}, x_j]$ of Q. Let K be the set of indices i such that the open interval (y_{i-1}, y_i) contains at least one of the x_j of Q. The set of indices $\{1, \ldots, m\}$ is a disjoint union of J and K. Hence, we have

$$U(f,P) - L(f,P) = \sum_{i=1}^{m} (M_i - m_i)(y_i - y_{i-1})$$

$$= \sum_{i \in J} (M_i - m_i)(y_i - y_{i-1}) + \sum_{i \in K} (M_i - m_i)(y_i - y_{i-1}).$$

The subintervals $[y_{i-1}, y_i]$ with $i \in J$ form a refinement of Q. Hence by Lemma 7.1,

$$\sum_{i \in J} (M_i - m_i)(y_i - y_{i-1}) \leq U(Q,f) - L(Q,f) < \varepsilon. \hfill (7.4)$$

Let the norm of P be $\delta = \max_i(y_i - y_{i-1})$, and let $M = \sup_{[a,b]} f$ and $m = \inf_{[a,b]} f$. Now the key point to observe is that K contains no more than n indices. Hence

$$\sum_{i \in K} (M_i - m_i)(y_i - y_{i-1}) \leq n\delta(M - m). \tag{7.5}$$

We combine (7.4) and (7.5) to arrive at

$$U(f, P) - L(f, P) \leq \varepsilon + n\delta(M - m).$$

It suffices then to choose δ so that $n\delta(M - m) < \varepsilon$. The proof of the lemma is complete.

Definition 7.6 Let $f : [a, b] \to \mathbb{R}$ and let $P = \{x_0, \ldots, x_n\}$ be a partition of $[a, b]$. For each $j = 1, \ldots, n$, let $s_j \in [x_{j-1}, x_j]$. The sum

$$S(f, P, s) = \sum_{j=1}^{n} f(s_j)(x_j - x_{j-1})$$

is called a *Riemann sum* for the function f and the partition P. Usually we will suppress the indication for the choice of points s_j and simply write $S(f, P)$.

We note that, for any partition P and all choices of the points s_j,
$$L(f, P) \leq S(f, P) \leq U(f, P).$$

Theorem 7.8 Let $f : [a, b] \to \mathbb{R}$ be a bounded function. The following are equivalent:

(i) f is integrable on $[a, b]$.
(ii) There is a number $\mathcal{I}$ such that, for each $\varepsilon > 0$, there is a $\delta > 0$ (depending on ε) such that

$$|S(f, P) - \mathcal{I}| < \varepsilon \tag{7.6}$$

for all Riemann sums $S(f, P)$ with $\|P\| < \delta$.

In this case, $\mathcal{I} = \int_a^b f$.

Proof: First suppose that f is integrable. Then for a given $\varepsilon > 0$, Lemma 7.7 tells us that there is $\delta > 0$ such that
$$U(f, P) - L(f, P) < \varepsilon$$

for all partitions P with $\|P\| < \delta$. Thus $\int_a^b f$ lies in the interval $\{L(P, f) \leq y \leq U(f, P)\}$ which has length less than or equal to ε. On the other hand, all Riemann sums $S(f, P)$ for such a partition also lie in this interval. Hence $|S(f, P) - \int_a^b f| \leq \varepsilon$ whenever $\|P\| < \delta$. Thus (i) implies (ii).

Now suppose that (ii) holds. Let $\varepsilon > 0$ be given, let δ be as prescribed in (ii), and let P be a partition with $\|P\| < \delta$. Then $|S(f, P) - \mathcal{I}| < \varepsilon$ for any choice of the points s_j. First choose $s_j \in [x_{j-1}, x_j]$ such that $f(s_j) \geq M_j - \varepsilon/(b - a)$. Next choose t_j in the same interval so that $f(t_j) \leq m_j + \varepsilon/(b - a)$. It follows that

$$
\begin{aligned}
S(f, P, s) &= \sum_{j=1}^{n} f(s_j)(x_j - x_{j-1}) \\
&\geq \sum_{j=1}^{n} (M_j - \varepsilon/(b - a))(x_j - x_{j-1}) \\
&\geq U(f, P) - \varepsilon.
\end{aligned} \tag{7.7}
$$

Similarly,

$$
S(P, f, t) \leq L(f, P) + \varepsilon. \tag{7.8}
$$

Then we use (7.7) and (7.8) to deduce

$$
\begin{aligned}
U(f, P) - L(f, P) &\leq S(f, P, s) - (S(f, P, t) + 2\varepsilon \\
&\leq S(f, P, s) - \mathcal{I} + \mathcal{I} - S(f, P, t) + 2\varepsilon \\
&\leq \varepsilon + \varepsilon + 2\varepsilon = 4\varepsilon.
\end{aligned}
$$

Theorem 7.3 implies that f is integrable. We have already seen that (i) implies (ii), so we conclude that $\mathcal{I} = \int_a^b f$.

When f is integrable on $[a, b]$, we shall write

$$
\lim_{\|P\| \to 0} S(P, f) = \int_a^b f.
$$

However, this is not a limit in the usual sense, because it does not depend on a single parameter. The precise meaning is given by (7.6).

Now we address some basic properties of the integral. First we establish additivity over intervals.

Theorem 7.9 Let $f : [a, b] \rightarrow \mathbb{R}$ be bounded and let $a < c < b$. Then f is integrable on $[a, b]$ if and only if f is integrable on $[a, c]$ and on $[c, b]$ and in this case,

$$\int_a^b f = \int_a^c f + \int_c^b f. \tag{7.9}$$

Proof: Assume first that f is integrable on both $[a, c]$ and on $[c, b]$. Let $\varepsilon > 0$ be given. By Theorem 7.3, there are partitions P' of $[a, c]$ and P'' of $[c, b]$ such that

$$U(f, P') - L(f, P') < \varepsilon$$

and

$$U(f, P'') - L(f, P'') < \varepsilon.$$

Then $P = P' \cup P''$ is a partition of $[a, b]$, and

$$L(f, P) = L(f, P') + L(f, P'')$$

while

$$U(f, P) = U(f, P') + U(f, P'').$$

It follows that $U(f, P) - L(f, P) < 2\varepsilon$. Hence by Theorem 7.3, f is integrable on $[a, b]$. We leave the implication in the other direction as an exercise.

Now having established that f is integrable on $[a, b]$ if and only if f is integrable on both $[a, c]$ and $[c, b]$, we use Riemann sums to show (7.9). In fact, if $P' = \{x_0, \ldots, x_r\}$ (with $x_r = c$) is a partition of $[a, c]$ and $P'' = \{y_0, \ldots, y_p\}$ (with $y_0 = c$) is a partition of $[c, b]$, then $P = P' \cup P''$ is a partition of $[a, b]$. For any choice of $s_j \in [x_{j-1}, x_j]$ and $t_k \in [y_{k-1}, y_k]$, we have

$$S(f, P) = S(f, P', s) + S(f, P'', t)$$

is a Riemann sum for f on $[a, b]$. Consequently,

$$\int_a^b f = \lim_{\|P\| \to 0} S(f, P)$$

$$= \lim_{\|P'\| \to 0} S(f, P', s) + \lim_{\|P''\| \to 0} S(f, P'', t)$$

$$= \int_a^c f + \int_c^b f.$$

Up to this point, we have always assumed $a < b$. However we can exploit the one-dimensional nature of the integral to assign an orientation. When $a > b$, we set

$$\int_a^b f = - \int_b^a f. \tag{7.10}$$

Now the formula

$$\int_a^b f = \int_a^c f + \int_c^b f \tag{7.11}$$

holds even when c does not lie between a and b.

Theorem 7.10 Let f and g be bounded integrable functions on $[a, b]$. Then,

(i) for any $c \in \mathbb{R}$, cf is integrable with $\int_a^b cf = c \int_a^b f$;
(ii) $f + g$ is integrable with $\int_a^b (f + g) = \int_a^b f + \int_a^b g$; and
(iii) if $f(x) \leq g(x)$ on $[a, b]$, then $\int_a^b f \leq \int_a^b g$.

Proof: The assertions are easily proved using Riemann sums. We leave the proofs to the exercises.

EXAMPLE 7.2:   Recall that we already know that if f is continuous, then f is integrable (Theorem 7.5), and if g is monotone, then g is integrable (Theorem 7.4). Hence by Theorem 7.10, $f + g$ is integrable. Now g may have have an infinite number of discontinuous steps. Let g be defined stepwise, for $k = 1, 2, \ldots n$ by

$$g(x) = 1 - 2^{-k} \quad \text{for} \quad 1 - 2^{-k-1} \le x < 1 - 2^{-k}.$$

Then $\sin x + g(x)$ is integrable.

This suggests that perhaps the class of Riemann integrable functions includes any function that is continuous except at an infinite sequence of points $\{x_k\}$. In fact, this is true as shown in the more general Lebesgue theory of integration.

Theorem 7.11 If f is a bounded integrable function on $[a, b]$, then $|f(x)|$ is also bounded and integrable on $[a, b]$, and

$$\left| \int_a^b f \right| \le \int_a^b |f|. \tag{7.12}$$

Proof: Let P be a partition of $[a, b]$, $P = \{a = x_0 < x_1 < \cdots < x_n = b\}$. Let $M_j = \sup_{[x_{j-1}, x_j]} f$ and $m_j = \inf_{[x_{j-1}, x_j]} f$. Let $\hat{M}_j = \sup_{[x_{j-1}, x_j]} |f|$ and $\hat{m}_j = \inf_{[x_{j-1}, x_j]} |f|$. From exercise 6 of Section 1.5, we have

$$\hat{M}_j - \hat{m}_j \le M_j - m_j.$$

Hence for any partititon P, we have

$$U(|f|, P) - L(|f|, P) \le U(f, P) - L(f, P).$$

We assume that f is integrable so that, by Theorem 7.3, for each $\varepsilon > 0$, we may choose the partition P so that $U(f, P) - L(f, P) < \varepsilon$. However, then $U(|f|, P) - L(|f|, P) < \varepsilon$ as well. Making another appeal to Theorem 7.3, we conclude that $|f|$ is integrable.

Finally, for all $x \in [a, b]$, we have $-|f(x)| \le f(x) \le |f(x)|$ so that by Theorem 7.10,

$$- \int_a^b |f| \le \int_a^b f \le \int_a^b |f|.$$

The theorem is proved.

The converse of Theorem 7.11 is not true. If we tinker with the Dirichlet function a bit, we can get an example of a function f such that $|f|$ is integrable but f is not integrable.

Exercises for 7.2

1. Prove Theorem 7.10 using Riemann sums.
2. Using the Dirichlet function, give an example of a function f such that $|f|$ is integrable, but f is not integrable.
3. Let f be continuous on an open interval I.

   a) Suppose that $\int_a^b f \geq 0$ for every closed interval $[a, b] \subset I$. Show that $f(x) \geq 0$ for every $x \in I$.

   b) Suppose that $\int_a^b f = 0$ for every closed interval $[a, b] \subset I$. Show that $f(x) = 0$ for all $x \in I$.

4. Let f be continuous on $[a, b]$, and suppose that $\int_a^b fg = 0$ for all functions g that are continuous on $[a, b]$. Show that $f(x) = 0$ for all $x \in [a, b]$.
   This result is often used in the following way. Suppose that $L(u)$ is a differential operator, for example, $L(u) = (d/dx)(p(x)u'(x)) + q(x)u(x)$, and that we can show that

   $$\int_a^b L(u)(x)\phi(x)dx = 0$$

   for all continuous functions $\phi(x)$. Then we can conclude that u satisfies the differential equation $L(u) = 0$.

5. Let f be bounded and integrable on $[a, b]$. Let P_n be the usual partition with equally spaced points $x_j = a + jh$, $j = 0, 1, \ldots n$ where $h = (b-a)/n$, and choose $s_j \in [x_{j-1}, x_j]$. The average of the function values $f(s_j)$ is given by

   $$\sigma_n(f) = \frac{f(s_1) + \cdots + f(s_n)}{n}.$$

   Show that

   $$\lim_{n \to \infty} \sigma_n(f) = \frac{1}{b-a} \int_a^b f.$$

6. Let $f \in C^1[a, b]$. Let P_n be the partition of equally spaced points of Exercise 5. Let $S(f, P_n, s) = \sum_{j=1}^n f(s_j)h$ be a Riemann sum for f. Show that

   $$\left| \int_a^b f - S(f, P_n, s) \right| \leq M(b-a)h$$

where

$$M = \max_{a \le x \le b} |f'(x)|.$$

7. Let f be a bounded integrable function on $[a, b]$. Set

$$f_+(x) = \frac{f(x) + |f(x)|}{2} \quad \text{and} \quad f_-(x) = \frac{f(x) - |f(x)|}{2}.$$

a) Sketch the graphs of $f_\pm$ for $f(x) = \sin(x)$.
b) Without using Theorem 7.11, show directly that both $f_\pm$ are integrable.

8. Theorem 7.11 may be generalized as follows. Let $f : [a, b] \to \mathbb{R}$ be bounded and integrable. Let g be Lipschitz continuous on the range of f. Then the composite function $x \to g(f(x))$ is integrable on $[a, b]$.

9. Riemann sums are often used to derive integral formulas for physical and geometrical quantities. Let $f \in C^1[a, b]$. We derive the formula for the arc length of the graph of f. Let P_n be the partition of $[a, b]$ with equally spaced points $x_j = a + jh$, $j = 0, 1, \ldots, n$, where $h = (b - a)/n$. Approximate f by the piecewise linear function $l_n(x)$ where

$$l_n(x) = f(x_{j-1}) + \frac{f(x_j) - f(x_{j-1})}{h}(x - x_{j-1})$$

for $x_{j-1} \le x \le x_j$.

a) The graph of l_n is a polygonal approximation to the graph of f. Calculate the length L_n of the graph of l_n.
b) Use the mean value theorem to show that L_n is a Riemann sum for the function $\sqrt{(f'(x))^2 + 1}$.
c) Conclude that the arc length is given by the formula

$$L = \lim_{n \to \infty} L_n = \int_a^b \sqrt{(f'(x))^2 + 1} \, dx.$$

7.3 The Fundamental Theorem of Calculus and Further Properties of the Integral

The intermediate value theorem for integrals expresses the connection between the point values of a function f and its average value. It can be used to prove the fundamental theorem of calculus, although we shall use a slightly different approach.

Theorem 7.12 Let f be continuous on $[a, b]$. Let g be continuous on $[a, b]$ with $g(x) \geq 0$. Then there is a value θ between a and b such that

$$\int_a^b fg = f(\theta) \int_a^b g. \qquad (7.13)$$

In the special case that $g(x) \equiv 1$, (7.13) states that there is a value θ where f attains its average value on $[a, b]$:

$$f(\theta) = \frac{1}{b - a} \int_a^b f.$$

Proof: Because f is continuous on $[a, b]$, f attains a maximum value M and a minimum value m on $[a, b]$. Because $g(x) \geq 0$,

$$mg(x) \leq f(x)g(x) \leq Mg(x)$$

for all $x \in [a, b]$. Then using Theorem 7.10,

$$m \int_a^b g \leq \int_a^b fg \leq M \int_a^b g.$$

Theorem 7.10 also implies that $\int_a^b g dx \geq 0$. If $\int_a^b g = 0$, then $\int_a^b fg = 0$ and the result is true. If $\int_a^b g > 0$, we can divide to deduce

$$m \leq \frac{\int_a^b fg}{\int_a^b g} \leq M.$$

Now the intermediate value theorem for continuous functions (Theorem 3.8) tells us that there is a point θ between a and b such that

$$f(\theta) = \frac{\int_a^b fg}{\int_a^b g}.$$

The theorem is proved.

Integration is a "smoothing process" because it involves averaging the values of a function. Here is a first example.

Let $f : [a, b] \to \mathbb{R}$ be a bounded integrable function. Then for $x \in [a, b]$, we define

$$F_0(x) = \int_a^x f. \tag{7.14}$$

Theorem 7.13 Let $f : [a, b] \to \mathbb{R}$ be bounded and integrable. Then F_0 defined by (7.14) is Lipschitz continuous on $[a, b]$ with Lipschitz constant $L = \sup_{[a,b]} |f(x)|$.

Proof: For $a \leq x \leq y \leq b$, we use the additivity of the integral (Theorem 7.9) to deduce

$$
\begin{aligned}
F_0(y) - F_0(x) &= \int_a^y f - \int_a^x f \\
&= \int_a^x f + \int_x^y f - \int_a^x f \\
&= \int_x^y f.
\end{aligned}
$$

Then by Theorem 7.11, we have

$$|F_0(y) - F_0(x)| = \left| \int_x^y f \right| \leq \int_x^y |f| \leq L(y - x).$$

Although the function F_0 is continuous in a strong sense, it may not be differentiable at every point. For example, let f be the unit step function:

$$f(x) = \begin{cases} 1, & \text{for } x \geq 0 \\ 0, & \text{for } x < 0 \end{cases}.$$

Let

$$F_0(x) = \int_0^x f = \begin{cases} x, & \text{for } x \geq 0 \\ 0, & \text{for } x < 0 \end{cases}.$$

Clearly, F_0 is not differentiable at $x = 0$.

To ensure that F_0 is differentiable, we make a stronger assumption about the integrand f. The result is the crucial link between differential and integral calculus.

Theorem 7.14 (Fundamental theorem of calculus) Let $f : [a, b] \to \mathbb{R}$ be bounded and integrable and let $F_0(x)$ be defined by (7.14). If f is continuous at $x_0 \in [a, b]$, then F_0 is differentiable at x_0, with

$$F_0'(x_0) = f(x_0).$$

If $x_0 = a$ or $x_0 = b$, the appropriate one-sided derivative exists and equals $f(x_0)$.

Proof: Suppose that $x_0 \in (a, b)$ and that h is small enough so that $x_0 + h \in (a, b)$. We consider the difference quotients:

$$\frac{F_0(x_0 + h) - F_0(x_0)}{h} = \frac{1}{h} \left(\int_a^{x_0+h} f - \int_a^{x_0} f \right)$$

$$= \frac{1}{h} \int_{x_0}^{x_0+h} f.$$

Now

$$f(x_0) = \frac{1}{h} \int_{x_0}^{x_0+h} f(x_0)ds.$$

Hence

$$\frac{F_0(x_0 + h) - F_0(x_0)}{h} - f(x_0) = \frac{1}{h}\int_{x_0}^{x_0+h} [f(s) - f(x_0)]ds.$$

Let $\varepsilon > 0$ be given. Because f is continuous at x_0, there is a $\delta > 0$ such that

$$|f(s) - f(x_0)| < \varepsilon \quad \text{for} \quad |s - x_0| < \delta.$$

Then for $|h| < \delta$,

$$\left| \frac{F_0(x_0 + h) - F_0(x_0)}{h} - f(x_0) \right| \leq \frac{1}{|h|}\left| \int_{x_0}^{x_0+h} [f(s) - f(x_0)]ds \right|$$

$$\leq \frac{1}{|h|}\varepsilon|h| = \varepsilon.$$

This means that

$$\lim_{h \to 0} \frac{F_0(x_0 + h) - F_0(x_0)}{h} = f(x_0).$$

Thus F_0 is differentiable at $x = x_0$ and $F_0'(x_0) = f(x_0)$. The theorem is proved.

Corollary If f is continuous on $[a, b]$, then $F_0(x)$ defined by (7.14) is an antiderivative for f on $[a, b]$, and we write

$$\frac{d}{dx}\int_a^x f(s)ds = f(x).$$

Using the sign convention (7.10) we also have

$$\frac{d}{dx}\int_x^b f = -f(x).$$

This form of the fundamental theorem is essentially a statement about averages. It says that if f is continuous at x_0, then its average value over an interval I that contains x_0 will converge to the point value $f(x_0)$ as the length of I shrinks to zero (see Figure 7.1). A slightly different proof of the fundamental theorem can be constructed using the intermediate value theorem for integrals. We leave this matter to the exercises.

The second form of the fundamental theorem, which is presented next, is the well-known tool for evaluating integrals.

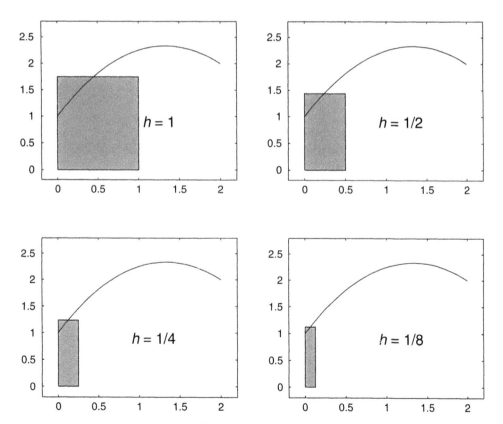

Figure 7.1 Graphs of $f(x) = -0.75x^2 + 2x + 1$ with rectangles with widths h and areas equal to $\int_0^h f(x)dx$, for $h = 1, 1/2, 1/4$ and $h = 1/8$. The heights of the rectangles approach the value $f(0) = 1$ as h gets smaller

Theorem 7.15 (Fundamental theorem of calculus: second form) Let f be continuous on $[a, b]$. Then for any antiderivative F of f,

$$\int_a^b f = F(b) - F(a).$$

Proof: Theorem 7.14 says that $F_0(x)$ is an antiderivative of f. Let $F(x)$ be any other antiderivative of f. Then by the corollary to the mean value theorem, we know that $F(x) = F_0(x) + C$ where C is a constant. Because $F_0(a) = 0$, we have $C = F(a)$ and $F_0(b) = F(b) - F(a)$. Hence

$$\int_a^b f = F_0(b) = F(b) - F(a).$$

The proof is complete.

Another useful tool for the evaluation of integrals is the substitution rule (change of variable formula) for integrals.

Theorem 7.16 Let $f : [c, d] \to \mathbb{R}$ be continuous and let $g : [a, b] \to \mathbb{R}$ be C^1 with $g([a, b]) \subset [c, d]$. Then

$$\int_a^b f(g(y))g'(y)dy = \int_{g(a)}^{g(b)} f(x)dx. \tag{7.15}$$

Proof: Let F be an antiderivative of f. The chain rule says that

$$f(g(y))g'(y) = F'(g(y))g'(y)$$

$$= \frac{d}{dy}(F \circ g)(y).$$

Then apply Theorem 7.15 (second form of the FTC) to this equation. There results

$$\int_a^b f(g(y))g'(y)dy = \int_a^b \frac{d}{dy}(F \circ g)(y)dy$$

$$= (F \circ g)(b) - (F \circ g)(b)$$

$$= \int_{g(a)}^{g(b)} f(x)dx.$$

In a more informal manner, we say that we make a substitution $x = g(y)$ in the integral $\int_{g(a)}^{g(b)} f(x)dx$, setting $dx = g'(y)dy$, and changing the limits of integration.

The following example shows how the change of variable formula is used with formulas of numerical integration.

EXAMPLE 7.3: A *quadrature rule* M is a method of approximating the integral of a function f on an interval $[a, b]$ using the values of f at several points $x_1, \ldots, x_n \in [a, b]$:

$$\int_a^b f \approx M(f) = A_1 f(x_1) + \cdots + A_n f(x_n).$$

The points x_j are called the nodes, and the coefficients A_j are the weights. Since any quadrature rule is exact on the constant function $f(x) \equiv 1$, we always have $\sum_j A_j = b - a$. An example is the single-panel Simpson's rule:

$$\int_a^b f \approx S(f) = \frac{(b-a)}{6} \left[f(a) + 4f(\frac{a+b}{2}) + f(b) \right].$$

Often the weights and nodes for a quadrature rule are given in manuals for the interval $[-1, 1]$. The user must convert the formula for use on a particular interval $[a, b]$. Suppose that M is given for functions $g(t)$ on the interval $[-1, 1]$:

$$M(g) = A_1 g(t_1) + \cdots A_n g(t_n)$$

where $t_j \in [-1, 1]$. We want to use this quadrature rule for a function $f(x)$ defined on $[a, b]$. To do so, we make the linear change of variable

$$x = \varphi(t) = a + \left(\frac{b-a}{2}\right)(t+1)$$

that maps the interval $[-1, 1]$ onto the interval $[a, b]$. Using Theorem 7.16, we have

$$\int_a^b f(x)dx = \left(\frac{b-a}{2}\right) \int_{-1}^1 g(t)dt$$

where $g(t) = f(\varphi(t)) = f(a + (\frac{b-a}{2})(t+1))$. Then we apply the quadrature rule to g to obtain

$$\int_a^b f dx = \left(\frac{b-a}{2}\right) \int_{-1}^1 g dt$$

$$\approx \left(\frac{b-a}{2}\right) M(g) = \left(\frac{b-a}{2}\right) \sum_j A_j g(t_j)$$

$$= \left(\frac{b-a}{2}\right) \sum_j A_j f\left(a + \frac{b-a}{2}(t_j + 1)\right).$$

Exercises for 7.3

1. Let the function f be given as

$$f(x) = \begin{cases} 0, & 0 \le x < 1 \\ 1, & 1 \le x < 2 \ . \\ x - 1, & x \ge 2 \end{cases}$$

Compute $F(x) = \int_0^x f(s)ds$. At which points is F differentiable?

2. Let f be continuous, and let $\varphi(x)$ be continuously differentiable. Use the chain rule to calculate

$$\frac{d}{dx} \int_a^{\varphi(x)} f(s)\, ds.$$

3. Let f be continuous, and let $\alpha(x)$ and $\beta(x)$ be continuously differentiable. Calculate

$$\frac{d}{dx} \int_{\alpha(x)}^{\beta(x)} f(s)\, ds.$$

4. Calculate the derivatives with respect to x of the following functions:

a) $\displaystyle\int_x^{x^2} \frac{1}{1+t^2}\, dt$

b) $\displaystyle\int_{\sqrt{x}}^{\sqrt{x^2+1}} \sin t\, dt.$

5. Here is another proof of Theorem 7.15. Suppose $F \in C^1[a, b]$ and $F' = f$. Let P be a partition of $[a, b]$, $a = x_0 < x_1 < \cdots < x_n = b$. Write $F(b) - F(a) = \sum_{j=0}^{n}(F(x_j) - F(x_{j-1}))$. Use the mean value theorem on each interval $[x_{j-1}, x_j]$ to show that $F(b) - F(a)$ is a Riemann sum, and take the limit.

6. A function $f : \mathbb{R} \to \mathbb{R}$ is said to be *periodic* with period p if $f(x+p) = f(x)$ for all x. Assuming that f is continuous and periodic with period p, show that, for all a,

$$\int_a^{a+p} f(x)\, dx = \int_0^p f(x)\, dx.$$

This is readily seen to be true when you draw the graph of f.

7. Gronwall's inequality. Let u be continuous on $\{t \geq 0\}$. Assume that there are constants A and c with $A, c \geq 0$ such that

$$|u(t)| \leq A + c \int_c^t |u(s)|\, ds \quad \text{for } t \geq 0.$$

Show that

$$|u(t)| \leq Ae^{ct} \quad \text{for } t \geq 0.$$

8. Suppose that $f : [0, \infty) \to \mathbb{R}$ is continuous and satisfies the equation:

$$f^2(t) = 2 \int_0^t f(s) \, ds \quad \text{for } t \geq 0.$$

Show that either $f \equiv 0$, or that there is a $t_0 \geq 0$ such that

$$f(t) = \begin{cases} 0, & 0 \leq t \leq t_0 \\ (t - t_0), & t \geq t_0 \end{cases}.$$

9. Let $f \in C^{n+1}$.

a) Use integration by parts to deduce

$$f(x) = f(0) + x f'(0) - \int_0^x (s - x) f''(s) \, ds.$$

b) Use integration by parts and an induction argument to show that

$$f(x) = f(0) + x f'(0) + \frac{x^2}{2} f''(0) + \cdots$$

$$+ \frac{1}{n!} \int_0^x (x - s)^n f^{(n+1)}(s) \, ds.$$

c) The last term is the integral form of the remainder for the Taylor polynomial P_n at $x = 0$. Deduce the Lagrange form of the remainder from this expression.

10. In this exercise, you will construct a second proof of the fundamental theorem, using a stronger hypothesis. Let $f : [a, b] \to \mathbb{R}$ be continuous. Use Theorem 7.12 to show that

$$\lim_{h \to 0} \frac{F_0(x + h) - F_0(x)}{h} = f(x)$$

for each $x \in [a, b]$.

11. Let f be continuous on $[0, 1]$ and suppose that $f(0) = 0$. Show that

$$\lim_{n \to \infty} \int_0^1 f(x^n)dx = 0.$$

7.4 Numerical Methods of Integration

Although the fundamental theorem of calculus is a great tool for evaluating integrals, we often must consider integrals when it is impossible to find an anti-derivative of the integrand. The error function $\text{erf}(x) = (2/\sqrt{\pi})\int_0^x \exp(-s^2)ds$ is a well-known example.

The *Newton-Cotes* rules are based on polynomial interpolation at equally spaced points. An $n + 1$ point Newton-Cotes rule for the interval $[a, b]$ uses the values of the integrand f at the points $x_j = a + jh$, $j = 0, \ldots, n$ where $h = (b - a)/n$. The single-panel form of the rule is constructed as follows. Let Q_n be the polynomial of degree less than or equal to n that interpolates f at these points. We shall use Q_n as an approximation to f under the integral sign. Recall that Q_n has the representation in terms of the Lagrange basis functions

$$Q_n(x) = f(x_0)L_0(x) + f(x_1)L_1(x) + \cdots + f(x_n)L_n(x).$$

Then the resultant rule is

$$\int_a^b f(x)dx \approx \int_a^b Q_n(x)dx = a_0 f(x_0) + a_1 f(x_1) + \cdots + a_n f(x_n). \qquad (7.16)$$

where

$$a_k = \int_a^b L_k(x)dx.$$

We shall investigate the three most commonly used elementary rules: the midpoint rule, the trapezoid rule, and Simpson's rule. First we derive the single-panel form of the rule, and then we derive the compound rule.

The Midpoint Rule

The simplest way to approximate a function $f(x)$ on an interval $[a, b]$ is to use a constant. We interpolate with a polynomial of degree 0 at the midpoint $c = (a+b)/2$. We do not need to use the Lagrange representation of the interpolating polynomial in this simple case. The midpoint approximation is

$$\int_a^b f(x)dx \approx f(c)(b - a).$$

In the exercises, you will use a Taylor expansion about $x = c$ to find that the error in the midpoint rule, on a single panel, is given by the following expression. There is a number $\eta \in [a, b]$ such that

$$\int_a^b f(x)dx = f(c)(b - a) + \frac{1}{24}f''(\eta)(b - a)^3. \tag{7.17}$$

The second term on the right in equation (7.17) is an expression for the error in a single-panel midpoint rule. To get a *compound* midpoint rule, we partition the interval $[a, b]$ with subdivisions $a = x_0 < x_1 < \cdots < x_n = b$ which we will assume have equal length $x_j - x_{j-1} = h = (b - a)/n$. Let c_j be the midpoint of the interval $[x_{j-1}, x_j]$. Now we apply the single-panel rule on each subinterval $[x_{j-1}, x_j]$, and collect all of the errors into a single term, $E_M(f)$. The result is:

$$\begin{aligned}
\int_a^b f(x)dx &= \sum_{j=1}^n \int_{x_{j-1}}^{x_j} f(x)dx \\
&= \sum_{j=1}^n f(c_j)h + E_M(f).
\end{aligned} \tag{7.18}$$

The following result is also discussed in the exercises.

Theorem 7.17 Let $f \in C^2[a, b]$. There is a point $\eta \in [a, b]$ such that the compound midpoint rule and its error are given by

$$\int_a^b f(x)dx \;=\; \sum_{j=1}^n f(c_j)h + E_M(f),$$

$$\text{with } E_M(f) \;=\; \frac{(b-a)}{24}h^2 f''(\eta). \tag{7.19}$$

We note that $\sum_{j=1}^n f(c_j)h$ is just a Riemann sum, and we know that the Riemann sums converge to the integral as h tends to zero. However, (7.19) provides an estimate of how rapidly the sums converge. In this case

$$|E_M(f)| \;\leq\; \frac{(b-a)}{24}h^2 \max_{[a,b]} |f''|.$$

In particular, we see that, although we interpolated f with a polynomial of degree 0, the midpoint rule is in fact exact on polynomials of degree 1. We get a bit more bang for the buck because we chose c as the midpoint of $[a, b]$.

The Trapezoid Rule

In the single-panel trapezoid rule, we again assume that f is C^2, and we interpolate $f(x)$ on $[a, b]$ with the polynomial

$$Q_1(x) = f(a) + \frac{f(b) - f(a)}{b - a}(x - a).$$

$Q_1(a) = f(a)$ and $Q_1(b) = f(b)$. From Theorem 5.3, we see that

$$f(x) = Q_1(x) + \frac{1}{2}f''(\xi_x)w_2(x).$$

Then

$$\int_a^b f\,dx \;=\; \int_a^b Q_1(x)dx + E_T(f)$$

$$=\; \frac{(f(a) + f(b))}{2}(b - a) + E_T(f)$$

where

$$E_T(f) = \frac{1}{2} \int_a^b f''(\xi_x) w_2(x) dx.$$

We wish to apply the mean value theorem for integrals to this last expression. First we verify that $w_2(x) = (x - a)(x - b) \leq 0$ on $[a, b]$ and

$$\int_a^b w_2(x) dx = -\frac{1}{6}(b - a)^3 < 0.$$

In the remark following Theorem 5.3 we observed that the function $x \to f''(\xi_x)$ is continuous on $[a, b]$. Hence there is a point $\eta \in [a, b]$ such that

$$E_T(f) = \frac{1}{2} f''(\eta) \int_a^b w_2(x) dx$$
$$= -\frac{1}{12}(b - a)^3 f''(\eta). \qquad (7.20)$$

We construct the compound trapezoid rule, with equally spaced points, $x_j = a + (j/n)(b - a)$, $j = 0, \ldots, n$. With $h = (b - a)/n$, we have

$$\int_a^b f(x) dx = \sum_{j=1}^n \int_{x_{j-1}}^{x_j} f(x) dx$$

$$= \frac{h}{2} \sum_{j=1}^n [f(x_{j-1}) + f(x_j)] - \frac{1}{12} \sum_{j=1}^n f''(\eta_j) h^3$$

where the point $\eta_j \in [x_{j-1}, x_j]$. Now we note that if one of the factors of h is written $h = (b - a)/n$, the error can be expressed

$$E_T(f) = -\left(\frac{b - a}{12}\right) h^2 \frac{1}{n} \sum_{j=1}^n f''(\eta_j).$$

The sum, divided by n, is an average of the values of the continuous function $f''(x)$ at the points η_j, and hence, by the intermediate value theorem, there is a number $\eta \in [a, b]$ where

$$f''(\eta) = \frac{1}{n} \sum_{j=1}^{n} f''(\eta_j).$$

We state this result as a theorem.

Theorem 7.18 Let $f \in C^2[a, b]$. Then there is a point $\eta \in [a, b]$ such that the compound trapezoid rule and its error are given by

$$\int_a^b f(x)dx = \frac{h}{2}[f(x_0) + 2f(x_1) + \cdots + 2f(x_{n-1}) + f(x_n)] + E_T(f), \quad (7.21)$$

$$E_T(f) = -\frac{(b-a)}{12}h^2 f''(\eta).$$

If we compare the error terms for the midpoint rule and the trapezoid rule, we see that both methods are exact on linear functions, but not on polynomials of higher degree. If f is convex, $f'' > 0$, the midpoint rule produces a number $M(f) < \int_a^b f$, and the trapezoid rule produces a number $T(f) > \int_a^b f$.

Simpson's Rule

The errors of the midpoint rule and of the trapezoid rule are both $O(h^2)$ as h tends to zero. This means that if we cut h in half, thereby roughly doubling the number of function evaluations, we expect the error to decrease by a factor of $1/4$. Perhaps if we interpolate the function with a higher degree polynomial, we can produce a quadrature rule that is more accurate, and that requires fewer function evaluations. We should try using a parabola.

We derive Simpson's rule on the interval $[-1, 1]$. Let $Q_2(t)$ be the polynomial of degree 2 that interpolates a function $g(t)$ at points $t = -1, 0, 1$. Using the Lagrange representation (5.20), we see that

$$Q_2(g) = \frac{1}{2}g(-1)(t^2 - t) - g(0)(t^2 - 1) + \frac{1}{2}g(1)(t^2 + t). \quad (7.22)$$

By Theorem 5.8, we know that there is some point $\eta \in [-1, 1]$ such that

$$g(t) = Q_2(t) + \frac{1}{6}g^{(3)}(\eta_t)t(t^2 - 1).$$ (7.23)

We integrate (7.23) over $[-1, 1]$ to derive Simpson's rule and an expression for its error,

$$
\begin{aligned}
\int_{-1}^{1} g(t)dt &= \int_{-1}^{1} Q_2(t)dt + \frac{1}{6}\int_{-1}^{1} g^{(3)}(\eta_t)(t^3 - t)dt \\
&= \frac{1}{3}[g(-1) + 4g(0) + g(1)] + \frac{1}{6}\int_{-1}^{1} g^{(3)}(\eta_t)(t^3 - t)dt.
\end{aligned}
$$ (7.24)

However, we cannot proceed as we did in the derivation of the error expression for the trapezoid rule, because the function $w_3(t) = t(t - 1)(t + 1)$ changes sign on $[-1, 1]$. We also make the curious observation that Simpson's rule is exact on cubic polynomials (try it), even though we derived the rule by interpolating with a parabola. This suggests that Simpson's rule may be the result of interpolating with a cubic polynomial. Let $\tilde{Q}_3$ be the cubic polynomial such that

$$\tilde{Q}_3(0) = g(0), \quad \tilde{Q}_3(\pm 1) = g(\pm 1), \quad \tilde{Q}_3'(t) = g'(0).$$

In exercise 4 of Section 5.4, it was found that

$$\tilde{Q}_3(t) = Q_2(t) + Aw_3(t)$$

where

$$A = -g'(0) - (1/2)g(-1) + (1/2)g(1).$$

Because $\int_{-1}^{1} w_3(t)dt = 0$, we have

$$\int_{-1}^{1} \tilde{Q}_3(t)dt = \int_{-1}^{1} Q_2(t)dt = \frac{1}{3}[g(-1) + 4g(0) + g(1)].$$

Thus Simpson's rule also arises when we interpolate with the cubic polynomial $\tilde{Q}_3$. In exercise 5 of Section 5.4, an expression for the error was found:

$$g(t) = \tilde{Q}_3(t) + \frac{1}{4!}g^{(4)}(\eta_t)t^2(t^2 - 1). \tag{7.25}$$

Note that the function $t^2(t^2 - 1) \leq 0$ for $t \in [-1, 1]$. When we integrate (7.25) over $[-1, 1]$, and use the intermediate value theorem for integrals, we see that there is number $\eta \in [-1, 1]$ such that

$$\int_{-1}^{1} g(t)dt = \frac{1}{3}[g(-1) + 4g(0) + g(1)] + \frac{1}{4!}g^{(4)}(\eta) \int_{-1}^{1} t^2(t^2 - 1)dt \tag{7.26}$$

$$= \frac{1}{3}[g(-1) + 4g(0) + g(1)] - \frac{1}{90}g^{(4)}(\eta).$$

To transform this rule and the error expression to the general interval $[a, b]$, we use the linear mapping (Example 7.3):

$$x = \varphi(t) = a + \frac{b - a}{2}(t + 1).$$

For f given on $[a, b]$ we let $g(t) = f(\varphi(t))$. Then $g(-1) = f(a)$, $g(0) = f((a+b)/2)$ and $g(1) = f(b)$. Furthermore, by the chain rule,

$$g^{(4)}(t) = \left(\frac{b - a}{2}\right)^4 f^{(4)}(\varphi(t)).$$

Thus the single-panel Simpson's rule and its error expression on the general interval $[a, b]$ are:

$$\int_{a}^{b} f(x)dx = \left(\frac{b - a}{2}\right) \int_{-1}^{1} g(t)dt$$

$$= \left(\frac{b - a}{6}\right)\left[f(a) + 4f\left(\frac{a + b}{2}\right) + f(b)\right]$$

$$-\frac{1}{90}\left(\frac{b - a}{2}\right)^5 f^{(4)}(\xi) \tag{7.27}$$

where $\xi \in [a, b]$.

To construct the compound Simpson's rul, we divide the interval $[a, b]$ into equal subintervals with length $h = (b - a)/n$. Let $x_j = a + jh$, $j = 0, \ldots, n$ and let $c_j = (x_{j-1} + x_j)/2$. We apply the single-panel Simpson's rule on the intervals $[x_{j-1}, x_j]$ and sum:

$$\int_a^b f(x)dx = \sum_{j=1}^{n} \int_{x_{j-1}}^{x_j} f(x)dx$$

$$= \left(\frac{h}{6}\right) \sum_{j=1}^{n} [f(x_{j-1}) + 4f(c_j) + f(x_j)] + E_S(f)$$

where ξ_j is in the jth panel and

$$E_S(f) = -\frac{1}{90} \left(\frac{h}{2}\right)^5 \sum_{j=1}^{n} f^{(4)}(\xi_j).$$

We use the averaging procedure that we used to get the final form of the error expression for the trapezoid rule and arrive at

Theorem 7.19 Let f be in $C^4[a, b]$. Let $h = (b-a)/n$. Then the n-panel Simpson's rule on $[a, b]$ is:

$$S(f) = \left(\frac{h}{6}\right) \sum_{j=1}^{n} [f(x_{j-1}) + 4f(c_j) + f(x_j)].$$

The error is:

$$E_S(f) = \int_a^b f(x)dx - S(f).$$

There is a number $\xi \in [a, b]$ such that

$$E_S(f) = -\frac{(b - a)}{2880} h^4 f^{(4)}(\xi).$$

The form of the error for Simpson's rule indicates that when we cut h in half, we expect the error to be reduced by a factor of $1/16$.

EXAMPLE 7.4:   We consider the performance of the three methods on the integral

$$\int_0^1 e^{-x^2/2} dx.$$

number of panels	midpoint error	ratio	trapezpoid error	ratio	Simpson's error	ratio
1	.02687251		-.05235906		.00046199	
2	.00641203	0.2386	-.01274328	0.2434	.00002693	0.0583
4	.00158529	0.2472	-.00316562	0.2484	.00000165	0.0614
8	.00039524	0.2493	-.00079017	0.2496	.00000010	0.0622

The ratios of errors are recorded as the number of panels is doubled, and they behave as predicted. Since $\exp(-x^2/2)$ is convex down on $[0, 1]$, the midpoint error is positive and the trapezoid error is negative. Note that Simpson's rule with one panel uses three function evaluations, and yields results with the same order of error as the eight-panel trapezoid rule that uses nine function evaluations.

The three rules we have considered all use an equal spacing between the points where the function is evaluated. More effective quadrature rules can be developed if we allow these points to be unequally spaced. Gaussian quadrature is an example of this kind of rule. Another way in which the basic rules can be improved upon is by the use of adaptive quadrature. Instead of reducing the spacing of the points uniformly across the interval $[a, b]$, for example, by cutting h in half, adaptive quadrature refines the mesh only in those places where the function is changing rapidly. Most numerical integration software routines use adaptive quadrature.

Exercises for 7.4

1. Suppose that $f \in C^2[a, b]$. The Taylor polynomial at $x = c = (a + b)/2$ with the Lagrange form of the remainder is

$$f(x) = f(c) + f'(c)(x - c) + \frac{f''(\eta)}{2}(x - c)^2.$$

a) Using this expansion, find an expression for the error for the single-panel midpoint rule.

b) Find an expression for the error of the compound midpoint rule.

2. a) Derive the single-panel Newton-Cotes rule for $\int_{-1}^{1} g \, dt$ using the points $-1, -1/3, 1/3, 1$. Using the procedure of Example 7.3, transform this rule to the general interval $[a, b]$. Call the resulting rule $M(f)$.

3. It can be shown that the error in the quadrature rule $M(f)$ of exercise 2 has the form

$$\int_a^b f \, dx = M(f) + \frac{K}{4!}(b - a)^5 f^{iv}(\xi)$$

where K is a constant independent of f and of the interval $[a, b]$. What is the constant K?

4. a) Write a short code to implement the compound trapezoid rule on the interval $[a, b]$.

b) Use the trapezoid code to compute approximations to

$$\log 2 = \int_1^2 \frac{dx}{x}$$

for $n = 5, 10, 20, 40$.

c) Compare your approximate values to "exact" values for $\log 2$ found on your calculator or computer. Compute the error for $n = 5, 10, 20, 40$. Does it behave as predicted by Theorem 7.18?

5. Let $f(x) = 30(x^4 - 2x^3 + x^2)$.

a) Use your trapezoid code on the integral $\int_0^{.8} f(x) \, dx$ for $n = 5, 10, 20, 40$. When h is halved (n is doubled), what happens to the error? Are these results in agreement with Theorem 7.18?

b) Use your trapezoid code to approximate the integral $\int_0^1 f(x) \, dx$. Compute the error for the same values of n. Now observe what happens when h is halved. How is this different from what happened in part a)?

6. This exercise is an attempt to explain the difference in results of parts a) and b) of the previous exercise. Recall that the error in the compound trapezoid rule could be written

$$E_T(f) = -\left(\frac{1}{12}\right)h^3 \sum_{j-1}^{n} f''(\eta_j)$$

where $\eta_j \in [x_{j-1}, x_j]$. Use Exercise 6 of Section 7.2 to show that

$$E_T(f) = -\frac{1}{12}[f'(b) - f'(a)] + O(h^3).$$

Conclude that when $f'(a) = f'(b)$, the error in the compound trapezoid rule is $O(h^3)$ instead of the usual $O(h^2)$. Calculate $f'(0)$ and $f'(1)$ for the f of exercise 5.

In fact, using the Euler-Maclaurin expansion for the trapezoid rule (see the book [Ki]) one can show that when $f'(a) = f'(b)$, the error is $O(h^4)$. This is what happens in part b) of exercise 5. The Euler-Maclaurin expansion predicts that if f is C^∞ and has period $b - a$, the error in the compound trapezoid rule is $O(h^n)$ for all $n \in \mathbb{N}$.

7. For $c \geq 0$, define the function

$$f(c) = \int_0^1 \frac{dx}{1 + e^{cx}}.$$

We want to prove the existence of a unique value $c_* > 0$ such that $f(c_*) = 1/4$, and then compute c_*.

a) Prove that $f(c)$ is continuous and monotone decreasing to 0 as $c \to \infty$.

b) Prove that there is unique value c_* such that $f(c_*) = 1/4$ and that $0 < c_* \leq 4$.

c) What combination of algorithms would you use to compute c_*? Write the codes to do so, and compute c_* to within 10^{-4}. Remember that there will be errors in the numbers you calculate for the integral.

7.5 Improper Integrals

Several mathematical objects carry adjectives that imply that they are not re-
spectable. Imaginary numbers, degenerate differential equations, and improper integrals are examples. Nevertheless, being open minded, we shall
enter into our discussion without any prejudice.

Up to now we have only considered integrals of functions that are bounded
on a bounded interval. Improper integrals arise when we relax one or both of
these conditions. For example, we want to know how to define an integral for
$f(x) = x^{-1/2}$ on $(0, 1]$, and for $g(x) = \sin x / x^2$ on $[1, \infty)$.

Definition 7.7 Let f be integrable on $[r, b]$ for each $r > a$. Then we set

$$\int_a^b f = \lim_{r \downarrow a} \int_r^b f$$

when this limit exists.

Definition 7.8 Let f be integrable on $[a, r]$ for each $r > a$. Then we set

$$\int_a^\infty f = \lim_{r \to \infty} \int_a^r f$$

when this limit exists.

In both cases, we say that the improper integral "converges" when the limit
exists, and we say that it "diverges" otherwise.

For the rest of this section, we shall assume that the functions are integrable
on every closed subinterval of $(a, b]$ or (a, ∞).

EXAMPLE 7.5:

a) We consider first the improper integral of x^{-p} on $(0, 1]$. For $p \neq 1$,

$$\int_r^1 x^{-p} dx = \frac{1}{1-p}(1 - r^{1-p})$$

and for $p = 1$,

$$\int_r^1 \frac{dx}{x} = -\log r.$$

We see that the integral converges for $p < 1$ with

$$\int_0^1 x^{-p}dx = \lim_{r \downarrow 0} \int_r^1 x^{-p}dx = \frac{1}{1-p}.$$

For $p \geq 1$ the integral diverges.

b) Next we consider the integral of x^{-p} on $[1, \infty)$. Again, we see that for $p \neq 1$,

$$\int_1^r x^{-p}dx = \frac{1 - r^{1-p}}{p-1}$$

and for $p = 1$,

$$\int_r^1 \frac{dx}{x} = \log r.$$

Hence

$$\lim_{r \to \infty} \int_1^r x^{-p}dx = \frac{1}{p-1}$$

for $p > 1$ and the integral diverges for $p \leq 1$.

c) For $k > 0$,

$$\int_0^\infty e^{-kx}dx = \lim_{r \to \infty} \int_0^r e^{-kx}dx$$

$$= \lim_{r \to \infty} \frac{1}{k}(1 - e^{-kr})$$

$$= \frac{1}{k}.$$

Comparison Tests

To deal with a larger class of integrands, we must develop some criteria for convergence that use comparison with the cases a), b), and c) of Examples 7.5. In all of these examples the integrand is nonnegative and the integral can be interpreted as the area under the curve. That is the motivation for the first comparison test.

Theorem 7.20 Let $0 \leq f(x) \leq g(x)$ on $[a, b)$.

    a) If the improper integral $\int_a^b g(x)dx$ converges, then the integral $\int_a^b f(x)dx$ also converges.

    b) If the improper integral $\int_a^b f(x)dx$ diverges, then the integral $\int_a^b g(x)dx$ also diverges.

    The same results apply to the improper integrals $\int_a^\infty f dx$ and $\int_a^\infty g dx$ when $0 \leq f(x) \leq g(x)$ on $[a, \infty)$.

Proof:

    a) For each r, $a \leq r < b$,

$$\int_a^r f(x)dx \leq \int_a^r g(x)dx \leq \int_a^b g(x)dx.$$

Hence $F(r) = \int_a^r f(x)dx$ is a monotone increasing function, bounded above, for $a \leq r < b$. By Theorem 3.3, $\lim_{r \uparrow b} F(r)$ exists, which is to say that the integral $\int_a^b f(x)dx$ converges.

    b) If the integral $\int_a^b f(x)dx$ does not converge, it is because $F(r)$ is unbounded as $r \uparrow b$. It follows that $G(r) = \int_a^r g(x)dx$ is unbounded as $r \uparrow b$, and therefore, that the integral $\int_a^b g(x)dx$ diverges.

EXAMPLE 7.6: The improper integral

$$\int_0^\infty \frac{x \, dx}{1 + x^3} = \int_0^1 \frac{x \, dx}{1 + x^3} + \int_1^\infty \frac{x \, dx}{1 + x^3}$$

converges because, for $x \geq 1$,

$$\frac{x}{1+x^3} \le \frac{x}{x^3} = \frac{1}{x^2}$$

and we know that $\int_1^\infty x^{-2}dx$ converges.

The convergence or divergence of an improper integral can be strongly influenced by the amount of cancellation between the regions where the integrand is positive and those where it is negative. To remove this aspect of convergence, we consider the absolute value of the integrand.

Definition 7.9 Let f be defined on $(a, b]$. We say that the improper integral $\int_a^b f(x)dx$ converges *absolutely* if

$$\int_a^b |f(x)|dx$$

converges. Similarly the improper integral $\int_a^\infty f(x)dx$ converges absolutely if $\int_a^\infty |f(x)|\,dx$ converges.

If the integral $\int_a^b f(x)dx$ converges but $\int_a^b |f(x)|dx$ diverges, we say that the integral converges *conditionally*.

Theorem 7.21 Let $f(x)$ be defined on $(a, b]$, and suppose that $\int_a^b f(x)dx$ converges absolutely. Then $\int_a^b f(x)dx$ also converges.

Proof: We have $0 \le f(x) + |f(x)| \le 2|f(x)|$. Hence by Theorem 7.20, the improper integral

$$\int_a^b [f(x) + |f(x)|]dx$$

converges. Then

$$\lim_{r\downarrow a} \int_r^b f(x)dx = \lim_{r\downarrow a} \int_r^b [f(x) + |f(x)|]dx - \lim_{r\downarrow a} \int_r^b |f(x)|dx$$

$$= \int_a^b [f(x) + |f(x)|]dx - \int_a^b |f(x)|dx$$

which proves the assertion.

EXAMPLE 7.7:

a) Here is an example of an improper integral that converges absolutely. Let $f(x) = \sin x/x^2$ for $x \geq 1$. Then $|f(x)| \leq x^{-2}$ so that $\int_1^\infty |f(x)|dx$ converges. By Theorem 7.21, $\int_1^\infty \sin x/x^2 \, dx$ converges.

b) The following integral converges conditionally. Let $g(x) = \sin x/x$. We know that $\int_1^\infty dx/x$ does not converge. However, the changing sign of $\sin x$ provides enough cancellation to make the integral $\int_1^\infty g(x)dx$ convergent. In fact, if we integrate by parts we see that

$$\int_1^r \frac{\sin x}{x} dx \;=\; -\frac{\cos x}{x}\Big|_1^r - \int_1^r \frac{\cos x}{x^2} dx$$

$$=\; \cos(1) - \frac{\cos(r)}{r} - \int_1^r \frac{\cos x}{x^2} dx.$$

Because $|\cos x/x^2| \leq x^{-2}$, the integral $\int_1^\infty \cos x/x^2 dx$ converges absolutely. Using the fact that $|\cos r| \leq 1$, we may take the limit to deduce

$$\int_1^\infty \frac{\sin x}{x} dx = \cos(1) - \int_1^\infty \frac{\cos x}{x^2} dx.$$

On the other hand, to show that $\int_1^\infty \sin /x dx$ does not converge absolutely, we must show that

$$\int_1^r \Big|\frac{\sin x}{x}\Big| dx$$

is unbounded as r tends to ∞. We shall let r run through the values $r = m\pi$, $m = 1, 2, \ldots$, because it will be easy to get a lower bound on the integral over the intervals $[(k-1)\pi, k\pi]$. In fact

$$\int_{(k-1)\pi}^{k\pi} \frac{|\sin x|}{x} dx \geq \frac{1}{k\pi} \int_{(k-1)\pi}^{k\pi} |\sin x| dx = \frac{1}{k\pi} \int_0^\pi \sin x dx$$

$$= \frac{2}{k\pi} \geq 2 \int_{k\pi}^{(k+1)\pi} \frac{dx}{x}.$$

Therefore,

$$\int_1^{m\pi} \frac{|\sin x|}{x} dx \geq \int_\pi^{m\pi} \frac{|\sin x|}{x} dx = \sum_{k=2}^m \int_{(k-1)\pi}^{k\pi} \frac{|\sin x|}{x} dx$$

$$\geq 2 \sum_{k=2}^m \int_{k\pi}^{(k+1)\pi} \frac{dx}{x} = 2 \int_{2\pi}^{(m+1)\pi} \frac{dx}{x}$$

$$= 2(\log((m+1)\pi) - \log(2\pi)) = 2\log\left(\frac{m+1}{2}\right).$$

Now it is clear that $\int_1^{m\pi} |\sin x|/x dx$ becomes unbounded as $r = m\pi$ tends to ∞.

The change of variable formula (Theorem 7.16) can be useful for evaluating improper integrals numerically.

EXAMPLE 7.8: The improper integral

$$\int_0^1 \frac{\cos x}{x^p} dx$$

converges (absolutely) for $p < 1$, but because the integrand is not defined at $x = 0$, the numerical procedures disucussed in Section 7.4 cannot be used. We shall seek a change of variable in this integral (using Theorem 7.16) that eliminates the singularity at $x = 0$.

Let $x = t^\alpha$. Then for $r > 0$,

$$\int_r^1 \frac{\cos x}{x^p}\, dx = \alpha \int_{r^{1/\alpha}}^1 \cos(t^\alpha) t^{\alpha(1-p)-1}\, dt.$$

We see that if $\alpha = 1/(1 - p)$, the integrand is continuous on $[0, 1]$. Hence, with $\alpha = 1/(1 - p)$,

$$\int_0^1 \frac{\cos x}{x^p}\, dx = \alpha \lim_{r \downarrow 0} \int_{r^{1/\alpha}}^1 \cos(t^\alpha)\, dt = \alpha \int_0^1 \cos(t^\alpha)\, dt.$$

The transformed integral can now be estimated using numerical methods.

Exercises for 7.5

1. Discuss the convergence of the following improper integrals:

a) $\displaystyle\int_0^1 \frac{\cos x}{\sqrt{x}}\, dx$

b) $\displaystyle\int_0^1 \frac{\sin x}{x^{5/2}}\, dx$

c) $\displaystyle\int_1^\infty \frac{x}{x^2 + x + 1}\, dx$

d) $\displaystyle\int_1^\infty \frac{\sin x}{x\sqrt{x^2 - 1}}\, dx$

e) $\displaystyle\int_0^{\pi/2} \frac{dx}{\sqrt{1 - \sin x}}$

f) $\displaystyle\int_0^1 x \log x\, dx$

2. Show that $\int_0^\infty x^p e^{-kx}\, dx$ converges for $k > 0$ and for all $p > 0$.

3. For what values of α and β do the following integrals converge?

a) $\displaystyle\int_0^1 x^\alpha |x - 1|^\beta\, dx$

b) $\displaystyle\int_0^{\pi/2} x^\alpha \sin^\beta x\, dx$

4. A function $f(t)$ defined for $t \geq 0$ is said to be of exponential type if there are constants A and k such that $|f(t)| \leq Ae^{kt}$.

a) Let $f(t) = te^t$. Show that f is of exponential type for $k > 1$.

b) The Laplace transform of a function of exponential type is defined to be

$$F(s) = \mathcal{L}(f)(s) = \int_0^\infty f(t)e^{-st}\, dt.$$

Show that if f is of exponential type for some k, then $F(s)$ is defined for $s > k$.

c) Use integration by parts to compute $\mathcal{L}(t^n)$ for $n \in \mathbb{N}$.

5. The gamma function $\Gamma(x)$ is defined by the improper integral:

$$\Gamma(x) = \int_0^\infty u^{x-1}e^{-u}\, du.$$

a) Show that the integral, which defines $\Gamma(x)$, converges for all $x > 0$.

b) Integrate by parts to show that, for any $x > 0$,

$$\Gamma(x + 1) = x\Gamma(x).$$

c) Deduce that for $x = n \in \mathbb{N}$, $\Gamma(n + 1) = n!$.

d) Using the fact that $2 \int_0^\infty e^{-u^2}\, du = \sqrt{\pi}$, show that

$$\Gamma(n + 1/2) = \frac{(2n)!\sqrt{\pi}}{4^n n!}.$$

6. To numerically integrate an improper integral $\int_1^\infty f(x)dx$, we must approximate it by the truncated integral $\int_1^T f(x)dx$. Then we can use numerical methods on $\int_1^T f(x)dx$.

a) Consider the integral

$$\int_1^\infty \frac{\cos x}{1 + x^2}\, dx.$$

How large must we choose r so that the error

$$\left| \int_1^\infty \frac{\cos x}{1 + x^2} dx - \int_1^r \frac{\cos x}{1 + x^2} dx \right| \leq 10^{-6}?$$

b) Since the choice of r that satisfies the tolerance of part a) is quite large, we must see if there is another way to deal with the integral before truncation. Make the change of variable $x = t^p$ transforming the integral into a form $\int_1^\infty g(t)dt$. Choose p so that

$$\int_{10}^\infty |g(t)|dt \leq 10^{-6}.$$

This change of variable reduces the numerical problem to the interval $[0, 10]$, but what additional difficulty now presents itself?

7. Make the change of variable $x = t^2 + 1$ in the improper integral:

$$\int_0^1 \frac{e^x \, dx}{\sqrt{|x - 1|}}.$$

Show that the integrand of the transformed integral is now a bounded function on $[0, 1]$.

8. Consider the improper integral

$$\int_0^1 \frac{x^{1/4}(e^x - 1)}{\sin^2(x)} dx.$$

a) Use Taylor polynomials to find the order of the singularity at $x = 0$.
b) Find a power $\alpha > 0$ such that the change of variable $x = t^\alpha$ transforms the integral into one whose integrand is a bounded function on $[0, 1]$.

Chapter 8

Series

Chapter Overview

Infinite series arise often in situations in which a quantity (e.g., the solution of an equation) is represented as the limit of a sequence of sums. The development of infinite series has much in common with that of improper integrals. Convergence results for improper integrals lead to convergence results for series. The crucial concept of uniform convergence is used in the study of sequences and series of functions. An important instance of uniform convergence occurs in power series. A typical use of power series is to represent solutions of differential equations and we give an example with Airy's equation. The chapter closes with the well-known example of a function that is C^∞, but not analytic.

8.1 Infinite Series

An infinite series consists of two sequences:

    (i) a sequence of *terms*, which we shall write a_k; and

    (ii) a sequence of *partial sums* $s_n = \sum_{k=1}^{n} a_k$.

Definition 8.1 We say that an infinite series *converges* or is *summable* if the sequence of partial sums converges. Otherwise, we say that the infinite series diverges.

There is some ambiguity in the notation for a series. The symbol $\sum_{k=1}^{\infty} a_k$ sometimes means the sum of the series when it converges, as in

$$\sum_{k=1}^{\infty} a_k = \lim_{n \to \infty} \sum_{k=1}^{n} a_k,$$

and sometimes it just means the series itself as in "does the series $\sum_{k=1}^{\infty} a_k$ converge?" The meaning is usually clear from context.

The first obvious topic in the study of series is that of criteria that determine when a series converges. The treatment in many ways parallels that given to improper integrals.

A first consequence of the definition is

Theorem 8.1 If the series $\sum_{k=1}^{\infty} a_k$ converges, then $\lim_{k \to \infty} a_k = 0$.

Proof: Let $S = \lim_{n \to \infty} s_n$ where $s_n = \sum_{k=1}^{n} a_k$. Since we also have $\lim_{n \to \infty} s_{n-1} = S$, it follows that

$$\lim_{n \to \infty} a_n = \lim_{n \to \infty} s_n - s_{n-1}$$

$$= \lim_{n \to \infty} s_n - \lim_{n \to \infty} s_{n-1} = S - S = 0.$$

EXAMPLE 8.1:

a) A well-known example is the geometric series:

$$\sum_{k=0}^{\infty} r^k.$$

When $r \neq 1$, the partial sums are

$$\sum_{k=0}^{n} r^k = \frac{1 - r^{n+1}}{1 - r}$$

and it is clear that the geometric series converges to $1/(1 - r)$ if $|r| < 1$. By Theorem 8.1, it must diverge for $|r| \geq 1$ because the terms do not converge to zero.

b) The condition that the terms a_k tend to zero is not sufficient to ensure convergence. We shall see shortly that the harmonic series $\sum_{k=1}^{\infty} 1/k$ does not converge, even though the terms $a_k = 1/k$ do tend to zero.

For the moment, let us concentrate on series with nonnegative terms. In this case, the partial sums s_n form a monotone increasing sequence. The next theorem is an immediate consequence of the monotone convergence theorem for sequences.

Theorem 8.2 Let $a_k \geq 0$. Then the series $\sum_{k=1}^{\infty} a_k$ converges if and only if the partial sums $s_n = \sum_{k=1}^{n} a_k$ are bounded.

Thus a series with nonnegative terms will converge if and only if the terms tend to zero rapidly enough. This is similar to the question of when an improper integral $\int_1^{\infty} f(x)dx$ will converge. If $f \geq 0$, the integral will converge if and only if $f(x)$ tends to zero rapidly enough as x tends to ∞. The integral comparison test makes this connection clear.

Theorem 8.3 Let $f(x)$ be nonnegative on $[1, \infty]$ and monotone decreasing. Then the following are equivalent:

(i) The improper integral $\int_1^{\infty} f(x)dx$ converges.
(ii) The series $\sum_{k=1}^{\infty} a_k$ with $a_k = f(k)$ converges.

Proof: We first make an estimate that relates the partial sums of the series with the integrals $\int_1^n f$. Because f is monotone decreasing,

$$a_{k+1} = f(k+1) \leq \int_k^{k+1} f(x)dx \leq f(k) = a_k.$$

Then summing from $k = 1$ to $n - 1$,

$$s_n - a_1 = \sum_{k=1}^{n-1} a_{k+1} \leq \int_1^n f(x)dx \leq \sum_{k=1}^{n-1} a_k = s_{n-1}. \tag{8.1}$$

Let $F(r) = \int_1^r f(x)dx$. Now if the series converges to a sum S, the partial sums $s_{n-1} \leq S$ for all n. Hence the integrals $F(n) = \int_1^n f\,dx \leq S$ for all n. For any $r > 0$, there is a positive integer n with $r \leq n$. Because the function $F(r) = \int_1^r f(x)dx$ is monotone increasing, we have $F(r) \leq F(n) \leq S$. Thus by a variant of Theorem 3.3 (exercise 7 of Section 3.1), the improper integral

$$\int_1^\infty f(x)dx = \lim_{r \to \infty} F(r)$$

converges.

On the other hand, if the integral converges, then for all n, we have

$$s_n \leq \int_1^\infty f(x)dx + a_1.$$

Since the s_n are bounded, the series must converge by Theorem 8.2.

An immediate consequence of Theorem 8.3 and Example 7.5 is that $\sum_{k=1}^p k^{-p}$ converges if and only if $p > 1$. In particular the *harmonic* series,

$$1 + \frac{1}{2} + \frac{1}{3} + \cdots + \frac{1}{k} + \cdots$$

diverges. However, the partial sums grow rather slowly, on the order of the logarithm function.

To use these examples, we need some comparison tests.

Theorem 8.4 Let $0 \leq a_k \leq b_k$. If the series $\sum_k b_k$ converges, then the series $\sum_k a_k$ also converges. If the series $\sum_k a_k$ diverges, then the series $\sum_k b_k$ diverges.

Proof: If the series $\sum_k b_k$ converges to B, the partial sums $\sum_{k=1}^n a_k \leq \sum_{k=1}^n b_k \leq B$ for all n. By Theorem 8.2, the series $\sum_k a_k$ must converge.

The partial sums in the series $\sum_k a_k$ are monotone increasing. Hence if the series $\sum_k a_k$ diverges, the sequence of partial sums $s_n = \sum_{k=1}^n a_k$ is growing without bound. The same must hold for the partial sums of the series $\sum_k b_k$, so it must also diverge.

We note that the hypothesis of Theorem 8.4 may be relaxed somewhat, but with the same conclusion. It suffices that there is some integer $K > 0$ such that $0 \leq a_k \leq b_k$ for all $k \geq K$.

Theorem 8.5 (Limit comparison test) Suppose $a_k > 0$ and $b_k > 0$ for all k and that

$$\lim_{k \to \infty} \frac{a_k}{b_k} = L > 0.$$

Then the series $\sum_k a_k$ converges if and only if the series $\sum_k b_k$ converges.

Proof: Since the ratios a_k/b_k converge to the positive number L, there is a K such that $L/2 \le a_k/b_k \le 2L$ for $k \ge K$; or what is the same thing, $(L/2)b_k \le a_k \le 2Lb_k$ for $k \ge K$. The result then follows by an appeal to Theorem 8.4.

EXAMPLE 8.2: Let

$$a_k = \frac{1 + e^{-k}}{3k^2 - k + 4}.$$

Since $3k^2 - k + 4 \ge 2k^2$, we have

$$a_k \le \frac{2}{2k^2} = \frac{1}{k^2}.$$

In view of Theorem 8.4, with $b_k = 1/k^2$, we conclude that the series $\sum_k a_k$ converges. We could also draw the same conclusion using the limit comparison test, Theorem 8.5. With $b_k = 1/k^2$,

$$\lim_{k \to \infty} \frac{a_k}{b_k} = \lim_{k \to \infty} \frac{k^2(1 + e^{-k})}{3k^2 - k + 4}$$

$$= \lim_{k \to \infty} \frac{1 + e^{-k}}{3 - k^{-1} + 4k^{-2}} = \frac{1}{3}.$$

The next two tests for convergence involve comparisons with the geometric series.

Theorem 8.6 (Ratio test) Suppose $a_k > 0$ and

$$\lim_{k \to \infty} \frac{a_{k+1}}{a_k} = r.$$

Then $\sum_k a_k$ converges if $r < 1$ and diverges if $r > 1$. If $r = 1$, no conclusion can be made.

Proof: Suppose that $r < 1$. Because $\lim_{k \to \infty} a_{k+1}/a_k = r$, there is a $K > 0$ such that $|a_{k+1}/a_k - r| \leq (1 - r)/2$ for $k \geq K$. Let $c = (r + 1)/2 < 1$. Then

$$\frac{a_{k+1}}{a_k} \leq c = \frac{c^{k+1}}{c^k} \quad \text{for } k \geq K.$$

By Theorem 8.5, the series $\sum_k a_k$ converges because the geometric series $\sum_k c^k$ converges for $c < 1$.

Now suppose that $r > 1$. Then there is a K such that

$$\frac{a_{k+1}}{a_k} \geq 1 \quad \text{for } k \geq K.$$

In this case the a_k are not tending to zero, and hence the series cannot converge. The ambiguity when $r = 1$ is illustrated by the p series, $a_k = k^{-p}$. We have

$$\lim_{k \to \infty} \frac{a_{k+1}}{a_k} = \lim_{k \to \infty} \left(\frac{k}{k + 1}\right)^p = 1.$$

When $p = 1$ the series $\sum_k a_k$ diverges, and when $p > 1$ it converges.

Remark The key element in the proof of Theorem 8.6 in the case when $r < 1$ is the existence of a constant $c < 1$ such that $a_{k+1}/a_k \leq c$ for k is sufficiently large. In the case $r > 1$, $a_{k+1}/a_k \geq 1$ for all k sufficiently large. A more general version of the ratio test, that does not require the existence of the limit of the ratios, can be stated in terms of the lim sup and lim inf.

Ratio Test Suppose that $a_k > 0$.

    a) If $\limsup_{k \to \infty} \frac{a_{k+1}}{a_k} < 1$, the series converges.
    b) If $\liminf_{k \to \infty} \frac{a_{k+1}}{a_k} > 1$, the series diverges.

Theorem 8.7 (Root test) Let $a_k \geq 0$, and suppose that

$$\lim_{k \to \infty} a_k^{1/k} = r.$$

If $r < 1$, the series $\sum_k a_k$ converges. If $r > 1$, the series diverges. If $r = 1$, no conclusion can be made.

Proof: If $r < 1$, there is a $K > 0$ such that

$$(a_k)^{1/k} \leq c < 1 \quad \text{for } k \geq K.$$

Hence $a_k \leq c^k$ for $k \geq K$. The geometric series $\sum c^k$ converges when $c < 1$ and, by Theorem 8.4, so does $\sum_k a_k$.

On the other hand, if $r > 1$, then there is a $K > 0$ such that $a_k \geq 1$ for $k \geq K$. Since the terms of the series do not tend to zero, the series must diverge.

The case $r = 1$ is ambiguous by reason of the same example that was given in Theorem 8.6.

The extended version of the root test is

Root Test Let $a_k \geq 0$.

    a) If $\limsup_{k \to \infty} |a_k|^{1/k} < 1$, the series converges.
    b) If $\liminf_{k \to \infty} |a_k|^{1/k} > 1$, the series diverges.

EXAMPLE 8.3:

    a) If $r < 1$, the series $\sum_k a_k$ with $a_k = k^p r^k$ converges for any $p > 0$. We use the ratio test.

$$\lim_{k \to \infty} \frac{a_{k+1}}{a_k} = \lim_{k \to \infty} r \left(\frac{k+1}{k} \right)^p = r.$$

In particular, we see that when $r < 1$, r^k tends to zero faster than k^{-p} for any $p > 0$. This means that the geometric series converges much more rapidly than does the p series.

b) Consider the series:

$$\frac{1}{2} + 1 + \frac{1}{8} + \frac{1}{4} + \frac{1}{32} + \frac{1}{16} + \frac{1}{128} + \frac{1}{64} + \cdots .$$

We can express the terms as

$$a_k = \begin{cases} 2^{-k} & \text{for } k \text{ odd} \\ 2^{-(k-2)} & \text{for } k \text{ even} \end{cases} .$$

Then the ratios

$$\frac{a_{k+1}}{a_k} = \begin{cases} 2 & \text{for } k \text{ odd} \\ 1/8 & \text{for } k \text{ even} \end{cases} .$$

Thus the ratio test cannot be applied in this case. However, the root test is applicable:

$$(a_k)^{1/k} = \begin{cases} 2^{-1} & \text{for } k \text{ odd} \\ 2^{-1-2/k} & \text{for } k \text{ even} \end{cases} .$$

We have $\lim_{k \to \infty} (a_k)^{1/k} = 1/2$ so the series converges.

In general, the root test has a wider scope, but the ratio test is usually easier to apply.

Definition 8.2 The series $\sum_k a_k$ is said to *converge absolutely* if the series $\sum_k |a_k|$ converges. If the series $\sum_k a_k$ converges but the series $\sum_k |a_k|$ does not converge, we say that the series $\sum_k a_k$ converges *conditionally*.

Theorem 8.8 If a series $\sum_k a_k$ converges absolutely, then it converges.

Proof: Let $s_n = \sum_{k=1}^{n} a_k$ and $t_n = \sum_{k=1}^{n} |a_k|$. Then for $m > n$,

$$|s_m - s_n| = \left| \sum_{k=n+1}^{m} a_k \right| \leq \sum_{k=n+1}^{m} |a_k| = t_m - t_n. \tag{8.2}$$

Because the series converges absolutely, the sequence of partial sums t_n converges, and hence is a Cauchy sequence. The inequality (8.2) shows that the sequence of partial sums s_n is also a Cauchy sequence. Hence the sequence of partial sums s_n converges, which means that the series $\sum_k a_k$ converges.

Theorem 8.8 allows us to apply the comparison theorems and tests to see if a series converges absolutely and to deduce that it converges.

EXAMPLE 8.4: Consider the series

$$\sum_{k=1}^{\infty} \frac{k \sin k}{k^3 + 2}.$$

This series is absolutely convergent because

$$\left| \frac{k \sin k}{k^3 + 2} \right| \leq \frac{k}{k^3 + 2} \leq \frac{1}{k^2}.$$

Absolute convergence does not take into account any cancellation that can occur when the terms have differing signs. When we take this effect into consideration, the convergence questions can be quite subtle. We consider a class of series in which the terms have differing signs, called *alternating series*.

Theorem 8.9 Let $a_k \geq 0$ and suppose that a_k is monotone decreasing to 0. Then the alternating series $\sum_k (-1)^k a_k$ converges. Let S be the sum of the series. The error we make in using the partial sum $\sum_{k=1}^{n} (-1)^k a_k$ as an approximation to S is

$$\left| \sum_{k=1}^{n} (-1)^k a_k - S \right| \leq a_{n+1}.$$

Proof: We look at the partial sums. For n even,

$$s_n = s_{n-2} - a_{n-1} + a_n \leq s_{n-2}$$

because $a_n \leq a_{n-1}$. Furthermore, for n even,

$$s_n = -a_1 + (a_2 - a_3) + (a_4 - a_5) + \cdots + (a_{n-2} - a_{n-1}) + a_n \geq -a_1.$$

On the other hand, for n odd,

$$s_n = s_{n-2} + a_{n-1} - a_n \geq s_{n-2}$$

and

$$s_n = -a_1 + a_2 - (a_3 - a_4) - (a_5 - a_6) - \cdots - (a_n - a_{n-1}) \leq a_2.$$

Thus the even partial sums form a decreasing subsequence that is bounded below by $-a_1$, and the odd partial sums form an increasing subsequence, that is bounded above by a_2. Consequently, both subsequences converge.

They must converge to the same limit S, because $|s_n - s_{n-1}| = a_n$ tends to zero as n tends to infinity. We have shown that the series converges.

To prove the assertion about the error, let n be odd. We know that $s_n \leq S \leq s_{n+1}$. Hence

$$0 \leq S - s_n \leq s_{n+1} - s_n = a_{n+1}.$$

If n is even, we have $s_{n+1} \leq S \leq s_n$ so that

$$-a_{n+1} = s_{n+1} - s_n \leq S - s_n \leq 0.$$

Putting these two inequalities together, we deduce that $|S - s_n| \leq a_{n+1}$ for all n. The theorem is proved.

EXAMPLE 8.5: Theorem 8.9 tells us that the alternating harmonic series

$$\sum_{k=1}^{\infty} \frac{(-1)^k}{k}$$

converges. This series is conditionally convergent because we know that the harmonic series $\sum_k (1/k)$ does not converge.

Exercises for 8.1

1. Suppose two series $\sum_k a_k$ and $\sum_k b_k$ such that $a_k = b_k$ for all but a finite number of indices. Show that $\sum_k a_k$ converges if and only if $\sum_k b_k$ converges.

2. Suppose that the series $\sum_k a_k$ converges absolutely and that b_k is a bounded sequence. Show that the series $\sum_k a_k b_k$ converges absolutely.

3. Suppose that the series $\sum_k a_k$ converges absolutely.

   a) Show that the series $\sum_k a_k^2$ converges.

   b) Give an example of a convergent series $\sum_k a_k$ such that $\sum_k a_k^2$ does not converge.

4. If $a_k \geq 0$ and $\sum_k a_k$ converges, will the series $\sum_k \sqrt{a_k}$ converge?

5. Establish the convergence or divergence of the series $\sum_k a_k$ where a_k is given below:

   a) $\dfrac{1}{k(k+1)}$

   b) $\dfrac{k}{(k+1)(k+2)}$

   c) $k^2 2^{-k}$

   d) $\dfrac{2 + \sin(k\pi)}{2^k}$

   e) $k! e^{-k}$

   f) $k! e^{-k^2}$

6. We have seen that the harmonic series $\sum_k k^{-1}$ diverges. Is there a factor $r(k)$ that grows more slowly than k^ε for any $\varepsilon > 0$ and such that the series

$$\sum_{k=1}^{\infty} \frac{1}{kr(k)}$$

converges?

a) Show that for any, $p > 1$ and $\varepsilon > 0$,

$$\lim_{k \to \infty} \frac{(\log k)^p}{k^\varepsilon} = 0.$$

b) Use the integral comparison test to show that

$$\sum_{k=2}^{\infty} \frac{1}{k(\log k)^p}$$

converges for $p > 1$.

7. Show that, for $n \geq 1$,

$$\log(n + 1) \leq \sum_{k=1}^{n} \frac{1}{k} \leq 1 + \log n.$$

8. Show that the series

$$\sum_{k=1}^{\infty} (-1)^k \frac{k - 5}{1 + k^2}$$

converges. In view of exercise 1, you only need to show that there is a K such that a_k satisfy the hypotheses of Theorem 8.9 for $k \geq K$.

8.2 Sequences and Series of Functions

In many situations, we attempt to construct the solution of a problem by constructing approximate solutions, and then passing to the limit. For example, we may wish to find a solution $y(x)$ to a differential equation. We first find approximate solutions $y_n(x)$. Then we define

$$y(x) = \lim_{n \to \infty} y_n(x).$$

We need to be precise about the nature of the convergence and what properties of the approximations $y_n(x)$ carry over to the limiting function $y(x)$. If each of the $y_n(x)$ is differentiable, will $y(x)$ be continuous, or differentiable? If each of the y_n is integrable, will y be integrable?

We begin by defining two kinds of convergence of functions.

Definition 8.3 Let $\{f_n\}$, $n \in \mathbb{N}$, be a sequence of functions defined on a set A. The sequence converges *pointwise* to a limit function $f(x)$ defined on A if for each $x \in A$,

$$\lim_{n \to \infty} f_n(x) = f(x).$$

Definition 8.3 says that, for given a $\varepsilon > 0$ and each $x \in A$, there is an $N > 0$ such that

$$|f_n(x) - f(x)| < \varepsilon \quad \text{for } n > N. \tag{8.3}$$

The important point to note here is that N can depend on both ε and x. It may be the case that, for some ε, there does not exist a single value of N such that this inequality holds for all $x \in A$.

Definition 8.4 Let $\{f_n\}$ be a sequence of functions defined on a set A. The sequence converges *uniformly* on A to a limit function $f(x)$ defined on A if, for each $\varepsilon > 0$, there is an $N > 0$, such that

$$|f_n(x) - f(x)| < \varepsilon \quad \text{for all } n > N \text{ and all } x \in A .$$

Uniform convergence may be understood visually from Figure 8.1. When we specify an $\varepsilon > 0$, we think of sketching a band of vertical height 2ε on the graph of the limit function $f(x)$. If $\{f_n\}$ converges uniformly to f, it must be possible to choose an N such that, for all $n \geq N$, the graphs of f_n are contained in this ε band.

Remark The concept of uniform convergence can be restated as follows: The sequence of functions $\{f_n\}$ converges uniformly on A to $f : A \to \mathbb{R}$ if and only if

$$\lim_{n \to \infty} \sup_A |f(x) - f_n(x)| = 0.$$

We leave the proof to the exercises.

EXAMPLE 8.6: Let $f_n(x) = x + 1/(1 + nx)$ for $x \in I = \{x \geq 0\}$. We see that we have pointwise convergence on I:

$$\lim_{n \to \infty} f_n(x) = f(x) = \begin{cases} 1, & \text{for } x = 0 \\ x, & \text{for } x > 0 \end{cases}.$$

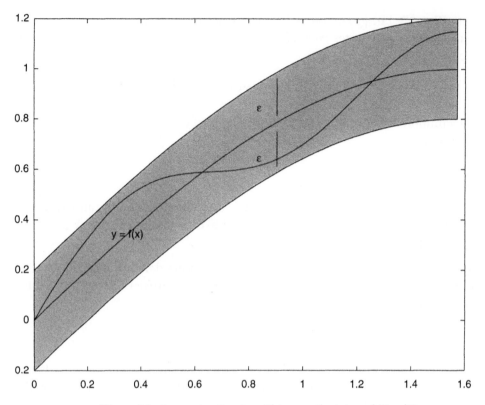

Figure 8.1 Approximation to within ε on the interval $[0, \pi/2]$

The convergence is not uniform on I. To demonstrate this, we must show that, for some $\varepsilon_0 > 0$, there is a sequence $x_n \in I$ such that $|f_n(x_n) - f(x)| \geq \varepsilon_0$ for all n sufficiently large. In this example, we may take $\varepsilon_0 = 1/2$ and note that

$$f_n(x) - f(x) = \frac{1}{1 + nx} = \frac{1}{2}$$

for $x = x_n = 1/n$.

However, the convergence is uniform on any interval $I_a = \{x \geq a\}$ when $a > 0$. We have

$$0 \leq f_n(x) - f(x) = \frac{1}{1 + nx} \leq \frac{1}{1 + na}$$

for $x \geq a$. For a given ε with $0 < \varepsilon < 1$, if we choose

$$n \geq N_\varepsilon = \frac{1 - \varepsilon}{a\varepsilon}$$

then we will have

$$0 \leq f_n(x) - f(x) \leq \varepsilon$$

for all $x \in I_a$.

Next we turn to the question of what properties of the approximating functions f_n are inherited by the limit function f. If each of the f_n are continuous, is the limit function f continuous? Example 8.6 shows that this is not always the case. Each of the functions $f_n(x) = x + 1/(1+nx)$ is continuous on $x \geq 0$, but the limit function f is not continuous at $x = 0$. Pointwise convergence is not enough to ensure that the limit function is continuous. We need the stronger hypothesis of uniform convergence.

Theorem 8.10 Let $f_n : A \to \mathbb{R}$ be continuous, and suppose that $\{f_n\}$ converges uniformly on A to f. Then f is continuous on A.

Proof: Fix $a \in A$, and let $x \in A$. We split the difference $f(x) - f(a)$ into three parts:

$$|f(x) - f(a)| \leq |f(x) - f_n(x)| + |f_n(x) - f_n(a)| + |f_n(a) - f(a)|. \qquad (8.4)$$

Let $\varepsilon > 0$ be given. Because the convergence is uniform, there is an $N > 0$ such that the first and third terms in the right side of (8.4) are less than $\varepsilon/3$ for all $n \geq N$, and all $x \in A$. Pick any $n_0 > N$ and fix it. We have

$$|f(x) - f(a)| \leq 2\varepsilon/3 + |f_{n_0}(x) - f_{n_0}(a)| \qquad (8.5)$$

for all $x \in A$. Now we use the continuity of f_{n_0}. There is a $\delta > 0$ such that $|f_{n_0}(x) - f_{n_0}(a)| < \varepsilon/3$ for $|x - a| \leq \delta$. It follows from (8.5) that $|f(x) - f(a)| < \varepsilon$ for $|x - a| < \delta$.

Returning again to Example 8.6, because the convergence is uniform on $I_a = \{x \geq a\}$, Theorem 8.10 guarantees that f will be continuous on I_a for each $a > 0$. Hence f is continuous on $\cup_a I_a = \{x > 0\}$.

One can also interpret Theorem 8.10 as a result about interchanging limit operations. In fact the theorem says that when $\{f_n\}$ converges uniformly to f, and the f_n are continuous, then

$$\lim_{n\to\infty} \lim_{x\to a} f_n(x) = \lim_{n\to\infty} f_n(a) = f(a)$$

$$= \lim_{x\to a} f(x)$$

$$= \lim_{x\to a} \lim_{n\to\infty} f_n(x).$$

Uniform convergence also preserves the class of integrable functions.

Theorem 8.11 Let $\{f_n\}$ be a sequence of bounded and integrable functions on $[a, b]$ and suppose that $\{f_n\}$ converges uniformly on $[a, b]$ to f. Then f is bounded and integrable and

$$\lim_{n\to\infty} \int_a^b f_n = \int_a^b f.$$

Proof: Because $\{f_n\}$ converges uniformly to f, there is an n_0 such that $|f(x) - f_{n_0}(x)| \leq 1$ for all $x \in [a, b]$. The function f_{n_0} is bounded, and therefore f is bounded.

To show that f is integrable we shall use the integrability criterion of Theorem 7.3. Let $\varepsilon > 0$ be given. We use the uniform convergence again: there is an N such that $|f(x) - f_n(x)| \leq \varepsilon$ for all $n > N$ and all $x \in [a, b]$. Pick one such n and fix it. f_n is integrable so that there is a partition P of $[a, b]$ such that $0 \leq U(f_n, P) - L(f_n, P) < \varepsilon$. Now we estimate the upper and lower sums for f using those of f_n. For each j,

$$M_j(f) \leq M_j(f_n) + \varepsilon$$

and

$$m_j(f) \geq m_j(f_n) - \varepsilon.$$

Hence

$$U(f, P) = \sum_j M_j(f)(x_j - x_{j-1})$$

$$\leq \sum_j (M_j(f_n) + \varepsilon)(x_j - x_{j-1})$$

$$\leq U(f_n, P) + \varepsilon(b - a).$$

Similarly,

$$L(f, P) = \sum_j m_j(f)(x_j - x_{j-1})$$

$$\geq \sum_j (m_j(f_n) - \varepsilon)(x_j - x_{j-1})$$

$$\geq L(f_n, P) - \varepsilon(b - a).$$

Therefore,

$$U(f, P) - L(f, P) \leq U(f_n, P) - L(f_n, P) + 2\varepsilon(b - a) \leq \varepsilon(1 + 2(b - a)).$$

We conclude that f is integrable by an appeal to Theorem 7.3.

To show convergence of the integrals, let $\varepsilon > 0$ be given. Then there is an $N > 0$ such that $|f_n(x) - f(x)| < \varepsilon$ for $n \geq N$ and $a \leq x \leq b$. Hence for $n \geq N$,

$$\left| \int_a^b f_n(x)dx - \int_a^b f(x)dx \right| \leq \int_a^b |f_n(x) - f(x)|dx \leq \varepsilon(b - a).$$

The theorem is proved.

Finally, we consider the question of differentiability. Uniform convergence of the sequence $\{f_n\}$ does not imply convergence of the derivatives.

EXAMPLE 8.7:

a) Let $f_n(x) = \sqrt{x^2 + 1/n}$. These functions converge uniformly on $\mathbb{R}$ to the function $f(x) = |x|$ (exercise 4 of this section). Although

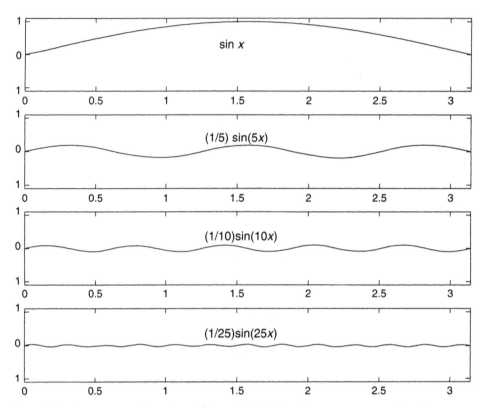

Figure 8.2 A sequence of functions $f_n(x) = (1/n)\sin(nx)$ that converges uniformly to zero, but where the derivatives f_n' do not converge

each $f_n \in C^1(R)$, the limit function $f(x)$ is not differentiable at $x = 0$.

b) Let $f_n(x) = (1/n)\sin nx$. Since $|f_n(x)| \leq 1/n$, these functions converge uniformly on $\mathbb{R}$ to $f(x) \equiv 0$. However, $f_n'(x) = \cos(nx)$ does not converge for each x. In particular, $f_n'(\pi) = (-1)^n$. This is an example in which the amplitude of a function tends to zero but the frequency of oscillation is increasing without bound. See Figure 8.2.

To ensure that the limit function is differentiable, we must make an assumption about the convergence of the derivatives.

Theorem 8.12 Let $\{f_n\}$ be a sequence of functions on $I = [a, b]$ with $f_n \in C^1(I)$. Suppose that the sequence $\{f_n\}$ converges pointwise on $[a, b]$ to a function

$f(x)$, and that the sequence of derivatives $\{f_n'\}$ converges uniformly on $[a, b]$ to a function $g(x)$. Then $f \in C^1([a, b])$, $\{f_n\}$ converges uniformly to f and $\{f_n'\}$ converges uniformly to f'.

Proof: For each n and all $x \in [a, b]$, we use the fundamental theorem of calculus to write

$$f_n(x) = f_n(a) + \int_a^x f_n'(s)ds. \tag{8.6}$$

Because the sequence $\{f_n'\}$ converges uniformly to g on $[a, b]$, Theorem 8.10 implies that g is continuous. Then we use Theorem 8.11 to conclude that, for each $x \in [a, b]$,

$$\lim_{n \to \infty} \int_a^x f_n'(s) = \int_a^x g(s)ds.$$

The hypothesis that $\{f_n\}$ converges pointwise to f and the fact that the integrals converge, together with (8.6), imply that, for each x,

$$f(x) = \lim_{n \to \infty} f_n(x)$$

$$= f(a) + \int_a^x g(s)ds.$$

Next we show that the sequence $\{f_n\}$ converges uniformly to f. Given $\varepsilon > 0$, there is an $N_1 > 0$ such that $|f(a) - f_n(a)| < \varepsilon$ for $n \ge N_1$. There is an $N_2 > 0$ such that for $n \ge N_2$, $|g(s) - f_n'(s)| < \varepsilon/(b - a)$ for $n \ge N_2$ and $a \le s \le b$. Hence for $n \ge \max\{N_1, N_2\}$, and for all x with $a \le x \le b$, we have

$$|f(x) - f_n(x)| \le |f(a) - f_n(a)| + \int_a^x |g(s) - f_n'(s)|ds \le 2\varepsilon.$$

Finally, we see that $f \in C^1([a, b])$, because $f(x) = f(a) + \int_a^x g(s)ds$ and, by the fundamental theorem of calculus, $f'(x) = g(x)$. The theorem is proved.

The following analogue of the Cauchy criterion is useful for discussing series of functions.

Definition 8.5 Let $f_n : A \to \mathbb{R}$. The sequence $\{f_n\}$ is *uniformly Cauchy* on A if, for each $\varepsilon > 0$, there is an $N > 0$ such that $|f_n(x) - f_m(x)| < \varepsilon$ for all $x \in A$ and all $n, m \geq N$.

Theorem 8.13 The sequence $\{f_n\}$ converges uniformly on a set A if and only if it is uniformly Cauchy on A.

Proof: This is largely a matter of manipulating the definitions. Suppose that $\{f_n\}$ converges uniformly on A to f. Let $\varepsilon > 0$ be given. This means there is an $N > 0$ such that, for $n \geq N$ and for all $x \in A$,

$$|f_n(x) - f(x)| \leq \varepsilon/2.$$

Then for $n, m > N$ and for all $x \in A$,

$$|f_n(x) - f_m(x)| \leq |f_n(x) - f(x)| + |f(x) - f_m(x)| < \varepsilon/2 + \varepsilon/2 = \varepsilon.$$

To prove the reverse implication, suppose that the sequence of functions $\{f_n\}$ is uniformly Cauchy on A. Then, in particular, for each $x \in A$, the sequence of real numbers $\{f_n(x)\}$ is a Cauchy sequence and, hence must converge to a real number, which we will call $f(x)$. It remains to be shown that $\{f_n\}$ converges uniformly to f. Let $\varepsilon > 0$ be given. Because the sequence $\{f_n\}$ is uniformly Cauchy, there is an $N > 0$ such that

$$|f_n(x) - f_m(x)| < \varepsilon$$

for all $n, m > N$ and for all $x \in A$. In other words,

$$-\varepsilon < f_n(x) - f_m(x) < \varepsilon.$$

We take the limit as m tends to infinity to deduce that

$$-\varepsilon \leq f_n(x) - f(x) \leq \varepsilon$$

or

$$|f_n(x) - f(x)| \leq \varepsilon,$$

holding for all $x \in A$ and all $n > N$. The theorem is proved.

Series of Functions

Since the sum of a series is defined as the limit of the partial sums, the results on uniform convergence of sequences of functions also apply to series of functions.

Definition 8.6 Let $f_k : A \to \mathbb{R}$ be a sequence of functions. The series $\sum_k f_k$ converges *pointwise* on A if the sequence of partial sums $\sum_{k=1}^{n} f_k$ converges pointwise on A. The series converges *uniformly* on A if the sequence of partial sums converges uniformly on A.

Theorems 8.10, 8.11, and 8.12 can be reformulated for series.

Theorem 8.14 Let $I = [a, b]$, and let $f_k : I \to \mathbb{R}$ be a sequence of functions. Assume that the series $\sum_k f_k$ converges uniformly on I with sum

$$F(x) = \sum_{k=1}^{\infty} f_k(x).$$

a) If each of the f_k is continuous on I, then F is continuous on I.
b) If each of the f_k is bounded and integrable on I, then F is bounded and integrable on I, and

$$\int_a^b F(x)dx = \sum_{k=1}^{\infty} \int_a^b f_k(x)dx.$$

In this case we say that we can integrate the series "term-by-term."

Theorem 8.15 Let $I = [a, b]$ and let $f_k \in C^1(I)$. Assume that the series $\sum_k f_k$ converges pointwise to F on I, and that the series of derivatives $\sum f_k'$ converges uniformly on I. Then $F \in C^1(I)$, the series $\sum_k f_k$ converges uniformly to F, and

$$F'(x) = \frac{d}{dx} \sum_{k=1}^{\infty} f_k(x) = \sum_{k=1}^{\infty} f_k'(x).$$

In other words, to see if we can differentiate a series term-by-term, we do the differentiation. If the resulting series converges uniformly, the term-by-term differentiation is justified.

Finally, we need a convenient criterion to determine whether a series of functions converges uniformly.

Theorem 8.16 (Weierstrass M test) Let $f_k : A \to \mathbb{R}$ and suppose that there is a sequence of constants $M_k \geq 0$ such that

(i) for each k, $\sup_{x \in A} |f_k(x)| \le M_k$; and
(ii) the series $\sum_{k=1}^{\infty} M_k$ converges.

Then the series of functions $\sum_k f_k$ converges uniformly on A.

Proof: Let $F_n(x) = \sum_{k=1}^{n} f_k(x)$ be the sequence of partial sums. We will show that $\{F_n\}$ is uniformly Cauchy on A, and thus by Theorem 8.13, that $\{F_n\}$ converges uniformly on A. We have assumed that the series $\sum_k M_k$ converges and, hence that the partial sums $\sum_{k=1}^{n} M_k$ form a Cauchy sequence. If $\varepsilon > 0$ is given, there is an $N > 0$ such that, for all $m > n > N$,

$$\sum_{k=n+1}^{m} M_k < \varepsilon.$$

Hence for $m > n$, we have

$$|F_m(x) - F_n(x)| = \left| \sum_{n+1}^{m} f_k(x) \right| \le \sum_{k=n+1}^{\infty} |f_k(x)| \le \sum_{k=n+1}^{m} M_k < \varepsilon.$$

Thus the series is uniformly Cauchy on A, and therefore it converges uniformly on A.

EXAMPLE 8.8: Consider the series:

$$\sum_{k=1}^{\infty} \frac{x}{k(k+x)}. \tag{8.7}$$

For each $R > 0$ and $0 \le x \le R$, the terms may be bounded

$$\left| \frac{x}{k(k+x)} \right| \le \frac{R}{k^2}.$$

Since the series $\sum_k R/k^2$ converges, we conclude by the Weierstrass test that the series (8.7) converges uniformly on each interval $[0, R]$. Let

$$F(x) \equiv \sum_{k=1}^{\infty} \frac{x}{k(k+x)}.$$

Theorem 8.14 then tells us that $F(x)$ is continuous on each interval $[0, R]$ and therefore on $\cup_{R>0}[0, R] = [0, \infty]$.

However, we cannot use the Weierstrass test to prove that this series converges uniformly on all of $[0, \infty)$, because for each k,

$$\sup_{[0,\infty)} \frac{x}{k(k+x)} = \frac{1}{k}$$

and we know that the harmonic series does not converge.

In view of Theorem 8.14, we can integrate the series (8.7) term-by-term on any interval $[0, R]$. Let F_n denote the partial sums of the series:

$$F_n(x) = \sum_{k=1}^{n} \frac{x}{k(k+x)}.$$

We obtain

$$\int_0^1 F_n(x)dx = \sum_{k=1}^{n} \int_0^1 \frac{x}{k(k+x)}$$

$$= \sum_{k=1}^{n} \left(\frac{1}{k} - \log(1+k) + \log(k) \right)$$

$$= \sum_{k=1}^{n} \frac{1}{k} - \log(n+1).$$

Therefore

$$\int_0^1 F(x)dx = \lim_{n\to\infty} \int_0^1 F_n(x)dx$$

$$= \lim_{n\to\infty} \left(\sum_{k=1}^{n} \frac{1}{k} - \log(n+1) \right).$$

If we let $\gamma = \int_0^1 F(x)dx$, we see that

$$\gamma = \lim_{n\to\infty} \left[\sum_{k=1}^{n} \frac{1}{k} - \log n + \log(n + 1/n) \right]$$

which is to say that

$$\sum_{k=1}^{n} \frac{1}{k} = \log n + \gamma + \sigma_n \qquad (8.8)$$

where σ_n tends to zero as n tends to ∞. The constant γ is called the Euler constant, and the computed value is $\gamma \approx 0.57721$.

EXAMPLE 8.9: Now consider the series:

$$G(x) = \sum_{k=1}^{\infty} \frac{x}{1 + x^2 k^2}. \qquad (8.9)$$

For $x \geq a > 0$,

$$\frac{x}{1 + x^2 k^2} \leq \frac{1}{ak^2}.$$

Hence, for each $a > 0$, the series (8.9) converges uniformly for $x \geq a$, which implies that $G(x)$ is continuous for all $x > 0$. Is the convergence uniform on the closed interval $[0, \infty)$? If so, then G would be continuous at $x = 0$. We see from (8.9) that $G(0) = 0$. To determine if G is continuous at $x = 0$, we estimate the terms in the series in the manner of the integral comparison test, Theorem 8.3. For each k,

$$\frac{x}{1 + (k + 1)^2 x^2} \leq \int_{k}^{k+1} \frac{x}{1 + t^2 x^2} dt \leq \frac{x}{1 + k^2 x^2}.$$

Now we sum these terms, and deduce that

$$\sum_{k=0}^{\infty} \frac{x}{1 + (k + 1)^2 x^2} \leq \int_0^{\infty} \frac{x}{1 + x^2 t^2} dt \leq \sum_{k=0}^{\infty} \frac{x}{1 + k^2 x^2}$$

or

$$G(x) \leq \int_0^\infty \frac{x}{1 + x^2 t^2} dt \leq x + G(x).$$

The middle integral can be evaluated. Make the change of variable $y = xt$. This yields

$$\int_0^\infty \frac{x}{1 + x^2 t^2} dt = \int_0^\infty \frac{dy}{1 + y^2}$$

$$= \lim_{R \to \infty} \int_0^R \frac{dy}{1 + y^2}$$

$$= \lim_{R \to \infty} [\arctan(R) - \arctan(0)] = \pi/2.$$

Thus

$$G(x) \leq \pi/2 \leq x + G(x).$$

We conclude that

$$\lim_{x \downarrow 0} G(x) = \pi/2 \neq G(0) = 0.$$

Therefore the convergence of the series (8.8) cannot be uniform on $[0, \infty)$.

Exercises for 8.2

1. Prove that a sequence of functions $f_n : A \to \mathbb{R}$ converges uniformly to a function $f : A \to \mathbb{R}$ if and only if

$$\lim_{n \to \infty} \sup_A |f_n(x) - f(x)| = 0.$$

2. Let $f_n(x) = x^n$.

   a) Show that the sequence $\{f_n\}$ converges to zero uniformly on $[0, 1 - \delta]$ for each $\delta > 0$.

b) Show directly, by negating the definition of uniform convergence, that $\{f_n\}$ does not converge uniformly on $[0, 1]$.

3. Let $g_n(x) = (1 - x)x^n$. Show that the sequence $\{g_n\}$ converges uniformly to zero on $[0, 1]$.

4. Let $f_n(x) = \sqrt{x^2 + 1/n}$. Show that the sequence $\{f_n\}$ converges uniformly on all of $\mathbb{R}$ to $f(x) = |x|$.

5. a) Let $f_n(x) = \sin(x/n)$. For each $R > 0$, show that the sequence $\{f_n\}$ converges uniformly on $[-R, R]$ to $f(x) \equiv 0$.

   b) Does $\{f_n\}$ converge to f uniformly on all of $\mathbb{R}$?

6. Let $h_n(x) = n^2 x e^{-nx}$.

   a) Show that $\{h_n(x)\}$ converges to zero for each $x \in [0, 1]$.

   b) Show that $\lim_{n\to\infty} \int_0^\infty h_n(x)dx = 1$.

   c) Find the maximum of h_n on $[0, 1]$. Explain why the result of part b) does not contradict Theorem 8.11.

7. Let $f_n : [a, b] \to \mathbb{R}$ be a sequence of functions that are each Lipschitz continuous on $[a, b]$ with the same Lipschitz constant: $|f_n(x) - f_n(y)| \leq L|x - y|$ for all $x, y \in [a, b]$. Suppose that the sequence $\{f_n\}$ converges pointwise on $[a, b]$ to a function $f(x)$. Show that $\{f_n\}$ converges uniformly on $[a, b]$ to f.

8. Let $f_n(x) = \sin(nx)$.

   a) Show that $\lim_{n\to\infty} \int_0^1 f_n(x)dx = 0$.

   b) Show that $\{f_n\}$ does not converge to zero even pointwise on $[0, 1]$. Draw graphs of f_n for several values of n to explain the result of part a).

9. a) Let $f : [0, 1] \to \mathbb{R}$ and $f_n : [0, 1] \to \mathbb{R}$ be continuous functions such that

   (i) $|f(x)|, |f_n(x)| \leq M$ for all n and $0 \leq x \leq 1$ and

   (ii) for each $\delta > 0$, f_n converges uniformly to f on each closed interval $[\delta, 1]$.

   Show that $\lim_{n\to\infty} \int_0^1 f_n(x)dx = \int_0^1 f dx$.

   b) Apply part a) to the functions $f_n(x) = \cos((1 - x)^n \pi/2)$ to determine $\lim_{n\to\infty} \int_0^1 f_n(x)dx$.

10. a) Suppose that the series $\sum_k a_k$ and $\sum_k b_k$ both converge absolutely. Show that the series

$$\sum_k a_k \cos(kx) + b_k \sin(kx)$$

converges uniformly on $\mathbb{R}$. Let $f(x)$ denote the sum of the series. Is f continuous on $\mathbb{R}$?

b) Make the stronger hypothesis that the series $\sum_k ka_k$ and $\sum_k kb_k$ both converge absolutely. Show that the function $f(x)$ defined in part a) is now in $C^1(\mathbb{R})$ and that

$$f'(x) = \sum_{k=1}^{\infty} -ka_k \sin(kx) + kb_k \cos(kx).$$

11. Consider the series:

$$\sum_{k=0}^{\infty} \frac{x^2}{(1+x^2)^k}.$$

a) Show that the series converges uniformly on $[\delta, \infty]$ for each $\delta > 0$. Let $f(x)$ denote the sum of the series. Show that $f(x) = 1 + x^2$ for $x > 0$.

b) Since $f(0) = 0$, does the series converge uniformly on $[0, \infty)$?

8.3 Power Series and Analytic Functions

Polynomials are easy to differentiate and integrate. We have already seen how it is useful to approximate functions by polynomials. In this section, we shall investigate what kinds of functions can be represented as the sum of a convergent series of the form $\sum a_k(x - x_0)^k$ (i.e., a "polynomial of infinite degree"). Because a series of this form converges rapidly, functions of this kind have very special properties. We begin by studying power series. A general power series has the form:

$$\sum_{k=0}^{\infty} a_k(x - c)^k.$$

Since the convergence results do not depend on the parameter c, it suffices to consider the case $c = 0$.

Theorem 8.17 Suppose that the power series $\sum_k a_k x^k$ converges for some $x_0 \neq 0$. Then for each $0 < r < |x_0|$, it converges absolutely and uniformly for all $|x| \leq r$.

Proof: If $\sum_k a_k x^k$ converges for x_0, then $\lim_{k \to \infty} a_k x_0^k = 0$. In particular, they are bounded: there is a constant M such that $|a_k x_0^k| \leq M$ for all k. Then for $|x| \leq r < |x_0|$,

$$|a_k x^k| = |a_k x_0^k| \left| \frac{x^k}{x_0^k} \right| \leq M(r/|x_0|)^k.$$

Because $r/|x_0| < 1$, the geometric series

$$\sum_{k=0}^{\infty} \left(\frac{r}{|x_0|} \right)^k$$

converges. Hence by the Weierstrass test, the series $\sum_k a_k x^k$ converges absolutely and uniformly for $|x| \leq r$.

Definition 8.7 Let $\sum_k a_k x^k$ be a power series. Let

$$R = \sup\{r : \sum_k a_k r^k \text{ converges}\}$$

when this quantity is finite. Otherwise we set $R = +\infty$. R is the *radius of convergence* of the power series. Note that $R \geq 0$ because the series always converges for $x = 0$. The interval $(-R, R)$ is often called the *interval of convergence*.

The radius of convergence may be determined from the behavior for large k of the coefficients a_k.

Theorem 8.18 Let the power series $\sum_k a_k x^k$ have radius of convergence $R \geq 0$.

    (i) If the sequence $|a_k|^{1/k}$ is unbounded, then $R = 0$.
    (ii) If $\lim_{k \to \infty} |a_k|^{1/k} = L > 0$, then $R = 1/L$.
    (iii) If $\lim_{k \to \infty} |a_k|^{1/k} = 0$, then $R = \infty$.

Proof: Case (i) For any $x \neq 0$, the sequence of kth roots $|a_k x^k|^{1/k} = |a_k|^{1/k}|x|$ is unbounded; in particular $|a_k x^k|^{1/k} \geq 1$ for sufficiently large k. This means that the terms do not tend to zero as $k \to \infty$. The series must diverge for each $x \neq 0$.

Case (ii) Let $L = \lim_{k \to \infty} |a_k|^{1/k}$ and assume that $L > 0$. Then apply Theorem 8.7 (root test). We have

$$\lim_{k \to \infty} |a_k x^k|^{1/k} = |x| \lim_{k \to \infty} |a_k|^{1/k} = |x|L.$$

If $|x| < 1/L$, the series converges. If $|x|L > 1$, then there is a $K > 0$ such that $|a_k x^k|^{1/k} \geq 1$ for $k \geq K$. This means that when $|x| > 1/L$, the terms in the series do not tend to zero; it must diverge.

Case (iii) If $L = 0$, the series will converge for all x so that $R = \infty$.

Remark In the case in which the sequence $|a_k|^{1/k}$ is bounded but $\lim_{k \to \infty} |a_k|^{1/k}$ does not exist, it can be shown that $R = 1/L$ where

$$L = \limsup_{k \to \infty} |a_k|^{1/k}.$$

This expression is often given as the definition of the radius of convergence.

In an exercise, we will see that the radius of convergence is also given by $1/L$ when the following limit exists:

$$L = \lim_{k \to \infty} \left| \frac{a_{k+1}}{a_k} \right|.$$

This latter limit is often easier to determine.

The question of convergence at the endpoints $\pm R$ must be determined on a case by case basis.

Theorem 8.17 says that a power series with radius of convergence $R > 0$ converges uniformly on each closed subinterval $[-r, r] \subset (-R, R)$. Thus by Theorem 8.16 (Weierstrass test), the sum of the series must be a continuous function on $(-R, R)$. The next theorem shows that far more is true.

Theorem 8.19 Let the power series $\sum_k a_k x^k$ have radius of convergence $R > 0$ and let

$$f(x) = \sum_{k=0}^{\infty} a_k x^k$$

be the sum of the series for $|x| < R$.

(i) Then f is infinitely differentiable on $(-R, R)$, and its derivatives can be calculated by differentiating term by term.

(ii) For a, $x \in (-R, R)$, $\int_a^x f(s)ds$ can be calculated by integrating term by term.

The power series that results from differentiating or integrating term by term converges uniformly on each closed interval $[-r, r] \subset (-R, R)$.

Proof: To show that f is differentiable, we shall apply Theorem 8.15. It will suffice to show that the formally differentiated series

$$\sum_{k=1}^{\infty} k a_k x^{k-1} \tag{8.10}$$

converges uniformly on closed subintervals of $(-R, R)$. Let $0 < r_1 < r_0 < R$. Since the original series converges for $x = r_0$, there is an $M > 0$ such that the terms $|a_k r_0^k| \leq M$ for all k. Hence for $|x| \leq r_1$,

$$|k a_k x^{k-1}| \leq k|a_k| \frac{|x|^{k-1}}{r_0^k} r_0^k$$

$$\leq k \left(\frac{r_1}{r_0}\right)^k \frac{M}{r_1}.$$

Therefore

$$\sum_{k=1}^{\infty} k|a_k x^{k-1}| \leq \frac{M}{r_1} \sum_{k=1}^{\infty} k \left(\frac{r_1}{r_0}\right)^k.$$

Now to this last series, we apply the key result (Example 8.2) that, for any $p > 0$ and $0 \leq r < 1$, the series $\sum_k k^p r^k$ converges. Because $(r_1/r_0) < 1$, the Weierstrass test allows us to conclude that the differentiated series (8.10) converges uniformly on each closed interval $[-r, r] \subset (-R, R)$. This implies that f' is given by the series (8.10). Now by induction, it follows that the series can be differentiated term-by-term infinitely often.

That the series may be integrated term-by-term follows from Theorem 8.14.

EXAMPLE 8.10: Our constant companion, the geometric series, has radius of convergence $R = 1$ and, for $|x| < 1$, can be summed to

$$f(x) = \frac{1}{1-x} = \sum_{k=0}^{\infty} x^k.$$

Using Theorem 8.19, we see that for $|x| < 1$ we can differentiate term-by-term to obtain the series for f' and for f'':

$$\frac{1}{(1-x)^2} = f'(x) = \sum_{k=0}^{\infty} \frac{d}{dx} x^k = \sum_{k=1}^{\infty} k x^{k-1},$$

and

$$\frac{2}{(1-x)^3} = f''(x) = \sum_{k=2}^{\infty} k(k-1) x^{k-2}.$$

EXAMPLE 8.11: We use term-by-term integration to obtain a power series for $f(x) = \log(x+1)$. Replace x with $-x$ in the geometric series to obtain

$$\frac{1}{1+x} = \sum_{k=0}^{\infty} (-1)^k x^k.$$

This series has radius of convergence $R = 1$. Now for $|x| < 1$, we integrate term-by-term to obtain

$$\log(1 + x) = \int_0^x \frac{1}{1+s} ds$$

$$= \sum_{k=0}^{\infty} \int_0^x (-1)^k s^k ds$$

$$= \sum_{k=1}^{\infty} (-1)^{k-1} \frac{x^k}{k}.$$

Note that the original series for $1/(1+x)$ does not converge at $x = \pm 1$. The integrated series converges conditionally at $x = 1$, but does not converge at $x = -1$. If we integrate a second time, we obtain the series $\sum_{k=2}^{\infty}(-1)^k x^k/(k(k-1))$, which converges at both endpoints $x = \pm 1$.

EXAMPLE 8.12:   The ability to differentiate a series term-by-term is very important in the construction of solutions of differential equations in the form of power series. Airy's equation arises in the subject of optics. It is

$$y''(x) = xy(x). \tag{8.11}$$

We shall assume a solution in the form of a convergent power series

$$y(x) = \sum_{k=0}^{\infty} a_k x^k.$$

In the interval of convergence, which is unknown at the present, we can differentiate term-by-term. Thus we substitute into (8.11) and find that the series, when it converges, must satisfy

$$\sum_{k=2}^{\infty} k(k-1)a_k x^{k-2} = \sum_{k=0}^{\infty} a_k x^{k+1}.$$

We replace the index k by $k+3$ on the left, and we combine the sums to arrive at

$$2a_2 + \sum_{k=0}^{\infty}[(k+3)(k+2)a_{k+3} - a_k]x^{k+1} = 0.$$

For the power series to be a solution, we must have $a_2 = 0$, and a_k must satisfy a recursion relation:

$$a_{k+3} = \frac{a_k}{(k+3)(k+2)}, \quad k \geq 0. \tag{8.12}$$

A solution of this recursion relation is determined by specifying a_0 and a_1. One solution arises by setting $a_1 = 0$. Then $a_k \neq 0$ only for $k = 0, 3, 6, 9, \ldots$, that is, for $k = 3l$, $l = 0, 1, 2, \ldots$. A second solution arises by setting $a_0 = 0$. In this case $a_k \neq 0$ only for $k = 1, 4, 7, \ldots$, which is to say, for $k = 3l + 1$, $l = 0, 1, 2, 3, \ldots$.

We show first that the series

$$y_1(x) = \sum_{l=0}^{\infty} a_{3l} x^{3l}$$

converges, and hence defines a solution of the differential equation. Because $(k+3)/(k+2) \leq 3/2$ for $k \geq 0$, we have the simple estimate:

$$|a_{k+3}| \leq \left(\frac{3}{2}\right) \frac{|a_k|}{(k+3)^2}. \tag{8.13}$$

If we put $k = 3l - 3$ in (8.13), we arrive at

$$|a_{3l}| \leq \frac{|a_{3(l-1)}|}{6l^2}, \quad l \geq 1.$$

Then we can use induction to show that

$$|a_{3l}| \leq \frac{|a_0|}{6^l (l!)^2}, \quad l \geq 1.$$

Consequently, for $C = |a_0|^{1/3l} 6^{-1/3}$, we have

$$|a_{3l}|^{1/3l} \leq C[(l!)^{-1/l}]^{2/3}.$$

Recall from exercise 11 of Section 2.1 that $\lim_{n \to \infty} (n!)^{-1/n} = 0$. Hence

$$\lim_{l \to \infty} |a_{3l}|^{1/3l} = 0.$$

Thus the series $y_1(x)$, with the coefficients defined by a choice of a_0 and by the recursion relation (8.12), converges for all $x \in \mathbb{R}$, and therefore is a solution of the Airy equation. If we take $a_0 = 1$, we find

$$y_1(x) = 1 + \frac{x^3}{2 \cdot 3} + \frac{x^6}{2 \cdot 3 \cdot 5 \cdot 6} + \frac{x^9}{2 \cdot 3 \cdot 5 \cdot 6 \cdot 8 \cdot 9} + \cdots .$$

Similarly if we take the second series ($k = 3l + 1$) and choose $a_1 = 1$, we obtain

$$y_2(x) = x + \frac{x^4}{3 \cdot 4} + \frac{x^7}{3 \cdot 4 \cdot 6 \cdot 7} + \frac{x^{10}}{3 \cdot 4 \cdot 6 \cdot 7 \cdot 9 \cdot 10} + \cdots ,$$

convergent for all $x \in \mathbb{R}$. The general solution of the Airy equation is a linear combination $y(x) = Ay_1(x) + By_2(x)$.

Analytic Functions

We have seen that a power series defines a C^∞ function in the interval $(-R, R)$ where it converges. Now we wish to change our point of view, and we ask what kind of functions can be represented as the sum of convergent power series.

Definition 8.8 We shall say that a function f is (real) *analytic* on a set A if, at each $x_0 \in A$, f is represented as the sum of a power series

$$f(x) = \sum_{k=0}^{\infty} a_k(x - x_0)^k$$

that converges in some interval $I_\delta = (x_0 - \delta, x_0 + \delta)$, $\delta > 0$.

As a consequence of Theorem 8.19, an analytic function is C^∞. It also follows that the power series expansion at each point is unique. In fact, if we calculate $f^{(n)}(x)$ using the series, and evaluate at $x = x_0$, we find that

$$a_k = \frac{f^{(k)}(x_0)}{k!}. \tag{8.14}$$

We are familiar with the following property of polynomials: if $p(x)$ is a polynomial such that $p(x)$ is constant on some open interval, then $p(x)$ is constant everywhere. The same holds true of analytic functions. If f is analytic on an open interval I and f is constant on some open subinterval $J \subset I$, then it can be shown that f is constant on I. We do not give a proof of this result. However, note that if $x_0 \in J$, then (8.14) implies that f will be constant on the interval of convergence of the power series expansion centered at x_0.

> **EXAMPLE 8.13:** We consider again $f(x) = 1/(1 - x)$ and we take
> the set $A = \{x \in \mathbb{R} : x \neq 1\}$. Given a point $x_0 \in A$, we could
> calculate the coefficients using (8.14). However, an easier method is
> to manipulate the fraction and use the geometric series expansion.
> We rewrite $f(x)$ as
>
> $$f(x) = \frac{1}{1 - x} = \frac{1}{x - x_0 - (x - x_0)}$$
>
> $$= \left(\frac{1}{1 - x_0}\right)\frac{1}{1 - (\frac{x - x_0}{1 - x_0})}.$$
>
> Now we substitute the fraction $(x - x_0)/(1 - x_0)$ into the geometric
> series to obtain
>
> $$f(x) = \frac{1}{1 - x} = \left(\frac{1}{1 - x_0}\right)\sum_{k=0}^{\infty}\left(\frac{x - x_0}{1 - x_0}\right)^k.$$
>
> This series will converge for $|x - x_0| < |1 - x_0|$. Thus in this case, the
> radius of convergence is just the distance from x_0 to the singularity
> at $x = 1$. This will not always be true for real analytic functions. The
> complete picture can be found in the theory of analytic functions of
> a complex variable.

What kinds of C^{∞} functions are analytic? These will be functions with Taylor polynomials P_n that become better and better approximations as n tends to infinity. Recall Taylor's theorem with the Lagrange form of the remainder. If I is an open interval, $f \in C^{n+1}(I)$ and $x_0 \in I$ then

$$f(x) = P_n(x) + \frac{f^{(n+1)}(\xi)}{(n+1)!}(x - x_0)^{n+1}$$

where ξ lies between x_0 and x, and

$$P_n(x) = f(x_0) + f'(x_0)(x - x_0) + \frac{f''(x_0)}{2}(x - x_0)^2 + \cdots + \frac{f^{(n)}(x_0)}{n!}(x - x_0)^n$$

is the Taylor polynomial of degree n at x_0. A function f can be represented as the sum of a series $\sum_{k=0}^{\infty} a_k(x - x_0)^k$ if the remainder term converges to zero as n tends to infinity.

A sufficient condition for f to be analytic is that the derivatives of f do not grow too fast.

Theorem 8.20 Let I be an open interval with $f \in C^\infty(I)$. Suppose that for each $x_0 \in I$, there is an interval $I_\delta = (x_0 - \delta, x_0 + \delta) \subset I$, with the following property: there is a constant $M > 0$ such that for all $n \in \mathbb{N}$,

$$\max_{I_\delta} |f^{(n)}(x)| \le M^n.$$

Then f is analytic on I.

Proof: From the Lagrange form of the remainder we see that for $x \in I_\delta$,

$$|f(x) - P_n(x)| \le \frac{(M\delta)^{n+1}}{(n+1)!}.$$

Now for any $c > 0$,

$$\lim_{n \to \infty} \frac{c^n}{n!} = 0.$$

Consequently, the Taylor polynomials P_n converge to f uniformly on I_δ.

The power series expansion

$$f(x) = \sum_{k=0}^{\infty} \frac{f^{(k)}(x_0)}{k!}(x - x_0)^k$$

is often called a Tayor series. When $x_0 = 0$, it is called a Maclaurin series.

EXAMPLE 8.14:

a) We compute the Taylor series for $f(x) = e^x$ at $x = x_0$. We have $f^{(n)}(x) = e^x$ and for any $r > 0$,

$$\max_{[-r,r]} |f^{(n)}(x)| = \max_{[-r,r]} |e^x| \le e^r.$$

We can apply Theorem 8.20 to deduce that e^x is analytic on $\mathbb{R}$. The Taylor series for e^x at x_0,

$$e^x = e^{x_0}[1 + (x - x_0) + \frac{(x - x_0)^2}{2!} + \cdots + \frac{(x - x_0)^n}{n!} + \cdots],$$

converges for each $x \in \mathbb{R}$. In particular at $x_0 = 0$, we have

$$e^x = 1 + x + \frac{x^2}{2!} + \cdots + \frac{x^n}{n!} + \cdots. \tag{8.15}$$

b) If $f(x) = \sin(x)$,

$$f^{(n)}(x) = \begin{cases} (-1)^{n/2} \sin x & \text{for } n \text{ even} \\ (-1)^{(n-1)/2} \cos x & \text{for } n \text{ odd} \end{cases}.$$

Thus again by Theorem 8.20, $\sin x$ is analytic on $\mathbb{R}$, and the Taylor series for $\sin x$ at x_0 converges for all $x \in \mathbb{R}$. The Taylor series for $\sin x$ at $x_0 = 0$ is

$$\sin x = x - \frac{x^3}{3!} + \frac{x^5}{5!} - \cdots. \tag{8.16}$$

For similar reasons, $\cos x$ is also analytic on $\mathbb{R}$ and the Taylor series at x_0 converges for all $x \in \mathbb{R}$. The Taylor series at $x_0 = 0$ is

$$\cos x = 1 - \frac{x^2}{2!} + \frac{x^4}{4!} - \cdots. \tag{8.17}$$

EXAMPLE 8.15: It is tempting to believe that if a function is C^∞, it must be analytic. A well-known example shows that this is not true. The Taylor series may converge, but not to the function. We consider the function

$$f(x) = \begin{cases} e^{-1/x^2} & \text{for } x \neq 0 \\ 0 & \text{for } x = 0 \end{cases}.$$

It can be shown that f is analytic on the open intervals $(-\infty, 0)$ and $(0, \infty)$. We shall focus our attention on the behavior at $x = 0$.

The function f is continuous at $x = 0$ because $\lim_{x \to 0} \exp(-1/x^2) = 0$. Moreover, for each $n \in \mathbb{N}$,

$$\lim_{x \to 0} \frac{e^{-1/x^2}}{x^n} = 0. \tag{8.18}$$

To see why this is true, we replace x with $1/x^2$ in the Taylor series for e^x at $x = 0$. Since all the terms in the series are positive, we deduce that for all $n \geq 1$,

$$e^{1/x^2} \geq \frac{x^{-2n}}{n!}.$$

Consequently,

$$e^{-1/x^2} \leq n! x^{2n}$$

so that,

$$\left| \frac{e^{-1/x^2}}{x^n} \right| \leq n! |x^n|,$$

which implies (8.18).

For $x \neq 0$,

$$f'(x) = \frac{2e^{-1/x^2}}{x^3}$$

so that, using (8.18),

$$\lim_{x \to 0} f'(x) = 0.$$

On the other hand, we can calculate $f'(0)$ using difference quotients and l'Hôpital's rule:

$$f'(0) = \lim_{h \to 0} \frac{f(h) - f(0)}{h}$$

$$= \lim_{h \to 0} \frac{e^{-1/h^2}}{h}$$

$$= \lim_{h \to 0} \frac{2e^{-1/h^2}}{h^3} = 0.$$

Thus f is continuously differentiable at $x = 0$. For general n, an induction argument shows that, for $x \neq 0$,

$$f^{(n)}(x) = x^{-3n} q_n(x) e^{-1/x^2}$$

where $q_n(x)$ has degree $(2n - 1)$. Thus (8.18) implies that

$$\lim_{x \to 0} f^{(n)}(x) = 0$$

for all $n \geq 1$. Again, by an induction argument we can show that for all $n \geq 1$,

$$f^{(n)}(0) = \lim_{h \to 0} \frac{f^{(n-1)}(h)}{h} = 0.$$

Therefore $x \to f^{(n)}(x)$ is continuous on all of $\mathbb{R}$. Because all the derivatives of f are zero at $x = 0$, the Taylor polynomial P_n for f at $x = 0$ is

$$P_n(x) = 0 + 0x + 0\frac{x^2}{2!} + \cdots + 0\frac{x^n}{n!} \equiv 0.$$

The remainder $f(x) - P_n(x) = f(x)$ does not go to zero for any $x \neq 0$. The function f is not analytic at $x = 0$ even though it is C^∞.

It is interesting to understand why this behavior occurs. Near $x = 0$, the derivatives of f must be growing very rapidly, but they do so in such a way that, for any fixed $x_0 \neq 0$, f has a power series expansion at x_0 that converges in some open interval I centered at x_0.

We shall explore this question in the exercises.

The existence of such a function is not just an example of pathology. The class of functions that are C^∞ but not analytic is very important in the study of partial differential equations. In particular, it is useful to be able to construct a C^∞ function f such that

$$f(x) = \begin{cases} 0, & x \leq -\varepsilon \\ 1, & x \geq \varepsilon \end{cases}.$$

These functions, called "cut-off functions," are used to localize the behavior of solutions of partial differential equations.

Exercises for 8.3

1. Consider a power series $\sum_{k=0}^{\infty} a_k x^k$. Suppose that $a_k \neq 0$ for all k and that

$$\lim_{k \to \infty} \frac{|a_{k+1}|}{|a_k|} = L.$$

a) If $L > 0$, show that the radius of convergence of the series is $R = 1/L$.

b) If $L = 0$, show that the series converges for all x.

2. Let

$$f(x) = \sum_{k=1}^{\infty} \frac{x^k}{k^2}.$$

a) What is the radius of convergence R for this series? Does the series converge for $x = \pm R$?

b) What is the series for $f'(x)$? Does the series converge for $x = \pm R$?

c) What is the series for $f''(x)$? Does the series converge for $x = \pm R$?

3. Starting with the power series for $1/(1 + x^2)$ at $x = 0$, find the power series for $\arctan x$ at $x = 0$. What is the radius of convergence for this latter series?

4. Find a power series at $x = 0$ for the error function

$$\text{erf}(x) = \frac{2}{\sqrt{\pi}} \int_0^x e^{-t^2} dt.$$

5. Prove that, for any real α, the binomial expansion

$$(1 + x)^\alpha = 1 + \alpha x + \frac{\alpha(\alpha - 1)}{2!} x^2 + \cdots$$

$$+ \frac{\alpha(\alpha - 1) \cdots (\alpha - n)}{n!} x^n + \cdots$$

converges in $|x| < 1$.

6. Find power series expansions for the following functions at the indicated points

a) $\cosh x$ at $x = 0$

b) xe^{-x^2} at $x = 0$

c) $\dfrac{1}{1 + 2x}$ at $x = 0$

d) $\dfrac{1}{1 + 2x}$ at $x = 1$

7. Determine the radius of convergence of each of the following powers series.

a) $\displaystyle\sum_{k=0}^{\infty} \left(\frac{k^2}{3^k}\right)(x - 1)^k$

b) $\displaystyle\sum_{k=0}^{\infty} \frac{k}{k^2 + 1} x^k$

c) $\displaystyle\sum_{k=0}^{\infty} (k + 1)e^k(x + 3)^k$

d) $\displaystyle\sum_{k=1}^{\infty} \left(\frac{2k + 1}{k}\right)^k x^k$

8. Let $f(x)$ have the power series expansion $\sum_{n=0}^{\infty}(n+1)x^n$. Find the first three terms in the power series expansion at $x = 0$ for the function $g(x) = 1/f(x)$.

9. Compute a power series expansion for the general solution of the ordinary differential equation:

$$y'' + xy' + y = 0.$$

10. In this exercise, we explore further the function $f(x) = \exp(-1/x^2)$ of Example 8.12. Use a software package such as Mathematica, Maple, or MATLAB to symbolically differentiate and plot $f^{(n)}$ for several values of n. Of course, the maximum is becoming much larger, as n increases, but notice also how the peak moves closer to zero. Describe in words how it happens that f has a convergent power series expansion for $x_0 \neq 0$, but not at $x = 0$.

Appendix I

1.1 The Logarithm Functions and Exponential Functions

In this section, we will define and develop the properties of the logarithm and exponential functions. Note that in Chapter 4, we were able to use the inverse function theorem to define x^r for r rational. Using the logarithm and exponential we will be able to define x^r for r irrational as well.

There are various equivalent ways to define the exponential and logarithm functions. The exponential function e^x can be defined by:

(i) the power series expansion $\sum_{k=0}^{\infty} \frac{x^k}{k!}$; or as

(ii) the unique solution of the differential equation $y' = y$ with $y(0) = 1$; or as

(iii) the inverse of the logarithm function.

We choose the third approach which begins with the integral that defines the logarithm.

Definition For $x > 0$, we define

$$\log x = \int_1^x \frac{dt}{t}.$$

Keeping in mind the orientation rule for integrals (7.10), we see that $\log x < 0$ for $0 < x < 1$.

Now we establish the important properties of the logarithm function.

Theorem 1

  a) For $x, y > 0$, $\log xy = \log x + \log y$.
  b) For $x > 0$ and $r \in \mathbb{R}$, $\log(x^r) = r \log x$.
  c) $(\log x)' = 1/x > 0$.
  d) For each real number y, there is a unique x with $\log x = y$.

Proof: To prove a), we write

$$\log(xy) = \int_1^{xy} \frac{dt}{t} = \int_1^{x} \frac{dt}{t} + \int_{x}^{xy} \frac{dt}{t}.$$

We make the change of variable $t = xs$ in the second integral, using the rule (7.15), and we obtain

$$\log(xy) = \int_1^{x} \frac{dt}{t} + \int_1^{y} \frac{ds}{s} = \log(x) + \log(y).$$

At this point, x^r is only defined for $r \in \mathbb{Q}$. We shall establish b) for $r \in \mathbb{Q}$, and return later to show that it is true for r irrational. To establish b), we make the change of variable $t = s^r$ in the integral for $\log x^r$. Since $dt = rs^{r-1}ds$, we find that

$$\log x^r = \int_1^{x^r} \frac{dt}{t}$$

$$= r \int_1^{x} \frac{ds}{s}$$

$$= r \log x.$$

Because $\log x$ has a positive derivative for all $x > 0$, it is one-to-one. Let $y > 0$ be given. We know that $\log(1) = 0$ and $\log x$ is increasing, so $\log(2) > 0$. Then we use b) to deduce that $\log(2^n) = n \log 2$. Thus for n sufficiently large, we have $\log(2^n) > y$. By the intermediate value theorem for continuous functions, there must be a value of $x > 1$ such that $\log x = y$. The argument for $y < 0$ is similar.

Theorem 1 tells us that $x \to \log x$ maps $(0, \infty)$ onto $\mathbb{R}$ in a one-to-one fashion. Hence by the inverse function theorem there exists an inverse function, $E : \mathbb{R} \to (0, \infty)$ that is C^1.

Definition The exponential function $E(x)$ is the inverse of the logarithm function.

We can deduce the following properties of E from those of $\log x$.

Corollary

    a) For all $x, y \in R$, $E(x + y) = E(x)E(y)$.
    b) For $r \in \mathbb{R}$, $E(x)^r = E(rx)$.
    c) $E'(x) = E(x)$.

Proof: To prove a) of the corollory, substitute $x = E(a)$ and $y = E(b)$ in part a) of Theorem 1:

$$\log(E(a)E(b)) = \log(E(a)) + \log(E(b)) = a + b = \log(E(a + b)).$$

Since $\log x$ is one-to-one, it follows that $E(a)E(b) = E(a + b)$. Property b) is proved in a similar fashion for $r \in \mathbb{Q}$. We shall see that b) is true for general r later. To prove c), we use formula (iii) of Theorem 4.8 which relates the derivatives of inverse functions:

$$E'(x) = \frac{1}{((d/dy)\log)(E(x))} = E(x).$$

The corollary is proved.

Now we can extend the definition of x^r to include irrational r. By part b) of Theorem 1, we have

$$x^r = E(\log(x^r)) = E(r \log x)$$

for $r \in \mathbb{Q}$. Then we use this formula to *define* x^r when r is irrational. Thus part b) of Theorem 1 is automatically true for r irrational as well.

Next let e be the unique real number such that $\log e = 1$. Putting $x = e$ in the previous formula, we see that the symbol e^r simply means $E(r)$ where E is the inverse of the logarithm function. We summarize the properties of the exponential function in the usual form.

Theorem 2 The exponential function e^x has the following properties:

    a) $e^{x+y} = e^x e^y$ for all $x, y \in \mathbb{R}$.

b) For all $r \in \mathbb{R}$, $(e^x)^r = e^{rx}$.

c) $(e^x)' = e^x$.

As a consequence of parts b) and c) of Theorem 2, we have the following differentiation rules. For $x > 0$ and $r \in \mathbb{R}$,

$$\frac{d}{dx}x^r = \frac{d}{dx}e^{r\log x} = rx^{r-1}$$

and

$$\frac{d}{dr}x^r = \frac{d}{dr}e^{r\log x} = x^r \log x.$$

Finally, we deduce an interesting connection between a discrete compounding process and exponential growth.

Theorem 3

$$\lim_{y \to \infty} \left(1 + \frac{1}{y}\right)^y = e.$$

Proof: We use Theorem 2, and set $y = 1/x$. We find that

$$\left(1 + \frac{1}{y}\right)^y = \exp\left(y \log\left(1 + \frac{1}{y}\right)\right) = \exp\left(\frac{\log(1 + x)}{x}\right).$$

Now $y \to \infty$ if and only if $x \to 0$. From an application of l'Hôpital's rule we obtain

$$\lim_{x \to 0} \frac{\log(1 + x)}{x} = \lim_{x \to 0} \frac{1}{x + 1} = 1.$$

Because the exponential function is continuous,

$$\lim_{y \to \infty} \left(1 + \frac{1}{y}\right)^y = \lim_{x \to 0} \exp\left(\frac{\log(1 + x)}{x}\right) = e.$$

For applications that involve compound interest, we often consider the case $y = r/n$ where $r > 0$ and $n \in \mathbb{N}$. Then from the formula of Theorem 3, we deduce

$$\lim_{n \to \infty} \left(1 + \frac{r}{n}\right)^n = e^r.$$

I.2 The Trigonometric Functions

As with the exponential function, there are several ways to define and develop the properties of the functions $\sin x$ and $\cos x$. They may be defined by

(i) power series expansions; or as
(ii) the unique solution pair (S, C) of the system of differential equations

$$C' = -S$$
$$S' = C$$

with initial conditions $C(0) = 1, S(0) = 0$; or as
(iii) the coordinates of a point on the unit circle.

The third method has a concrete geometric meaning and that is our choice.

Let γ be a ray from the origin that makes an angle θ with the x axis. When the ray is in the upper half plane, we say that $\theta \geq 0$, and when the ray lies in the lower half plane, we say that $\theta < 0$. Let $P(\theta)$ be the point of intersection of the ray γ with the circle. See Figure I.1.

Definition The function $\cos \theta$ is defined to be the x coordinate of $P(\theta)$, and $\sin \theta$ is defined to be the y coordinate of $P(\theta)$.

We need to put this purely geometric definition into an analytic form. To do this we shall follow the procedure that we used for the exponential function. We define the arcsin function in terms of an integral, and then we deduce the properties of $\sin x$ from those of $\arcsin x$.

Let $l(\theta)$ be the arc length along the circle from $(1, 0)$ to $P(\theta)$. Using radian measure, we can identify θ with $l(\theta)$ when $\theta \geq 0$ and with $-l(\theta)$ when $\theta < 0$. The portion of the circle in $x > 0$ is the graph of $x = g(y) = \sqrt{1 - y^2}$. Using the arc length formula (exercise 8, Section 7.2),

$$\theta = \int_0^{\sin \theta} \sqrt{1 + g'(s)^2}\,ds = \int_0^{\sin \theta} \frac{ds}{\sqrt{1 - s^2}}.$$

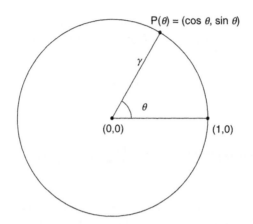

Figure I.1 Angle θ in the unit circle used to define $\sin\theta$ and $\cos\theta$

Setting $y = \sin\theta$, we have

$$\arcsin y = \int_0^y \frac{ds}{\sqrt{1 - s^2}}.$$

Using the sign convention for the oriented integral in one dimension (7.10), we can use this integral to define $\arcsin y$ for $-1 < y < 1$. For $y = \pm 1$, the singularity of $1/\sqrt{1 - s^2}$ at $s = \pm 1$ makes the integral improper. However, we see that

$$\frac{1}{\sqrt{1 - s^2}} = \frac{1}{\sqrt{1 + s}} \frac{1}{\sqrt{1 - s}}.$$

Hence near $s = 1$, the integrand is the product of a continuous (and thus bounded) function times the singular function $(1 - s)^{-1/2}$. Because the integral $\int_0^1 (1 - s)^{-1/2} ds$ converges (Example 7.5 a)), we can define

$$\arcsin 1 = \lim_{y \uparrow 1} \int_0^y \frac{ds}{\sqrt{1 - s^2}}$$

$$= \int_0^1 \frac{ds}{\sqrt{1 - s^2}}.$$

A similar definition is made for $\arcsin(-1)$. The point $(0, 1)$ lies one quarter of the way around the unit circle from the point $(1, 0)$, so we may define the number π by

$$\frac{\pi}{2} = \arcsin 1.$$

Next use the fundamental theorem of calculus to deduce that, for $-1 < y < 1$,

$$\frac{d}{dy} \arcsin y = \frac{d}{dy} \int_0^y \frac{ds}{\sqrt{1 - s^2}}$$

$$= \frac{1}{\sqrt{1 - y^2}} > 0.$$

Thus $\arcsin y$ is C^1 on $(-1, 1)$ and is strictly increasing. Hence by Theorem 4.8, the inverse of $\theta = \arcsin y$ exists on $(-\pi/2, \pi/2)$ and is C^1. This inverse function is $y = \sin \theta$.

We follow a similar procedure for $\cos \theta$. The function $\arccos x$ is defined by the arc length integral:

$$\arccos x = \pi/2 - \int_0^x \frac{ds}{\sqrt{1 - s^2}}.$$

We see that, for $-1 < x < 1$,

$$\frac{d}{dx} \arccos x = -\frac{1}{\sqrt{1 - x^2}} < 0.$$

The function $\arccos x$ maps the interval $(-1, 1)$ onto the interval $(0, \pi)$. Again by Theorem 4.8, the inverse function $x = \cos \theta$ is C^1 and it maps the interval $(0, \pi)$ onto $(-1, 1)$. We can calculate the derivative,

$$\frac{d}{d\theta} \cos \theta = 1 / \left(\frac{d}{dx} \arccos x \right) \Big|_{x = \cos \theta}$$

$$= -\sqrt{1 - x^2} \Big|_{x = \cos \theta}$$

$$= -\sin \theta$$

because $\sin \theta > 0$ for $0 < \theta < \pi$.

Next we exploit the fact that we have established the analytic properties of $\sin\theta$ and $\cos\theta$ on different intervals.   From the geometric
definition, $\cos\theta = \cos(-\theta)$ for $-\pi < \theta \leq 0$ and $\cos\theta$ is differentiable at $\theta = 0$
because $\cos\theta = \sqrt{1 - \sin^2\theta}$ for $-\pi/2 < \theta < \pi/2$. Thus $\cos\theta$ is C^1 on $(-\pi, \pi)$
and $(d/d\theta)\cos\theta = -\sin\theta$. We know already that $\sin\theta$ is C^1 on $(-\pi/2, \pi/2)$, but
now, because

$$
\sin\theta = \begin{cases} \sqrt{1 - \cos^2\theta}, & 0 \leq \theta < \pi \\ -\sqrt{1 - \cos^2\theta}, & -\pi < \theta \leq 0 \end{cases},
$$

we conclude that $\sin\theta$ is also C^1 on $(-\pi, \pi)$ and

$$
\frac{d}{d\theta}\sin\theta = \cos\theta.
$$

Because $(d/d\theta)\sin\theta = \cos\theta$ and $(d/d\theta)\cos\theta = -\sin\theta$, we deduce that both
functions are C^∞ on $(-\pi, \pi)$. Finally we extend the definition of $\cos\theta$ and $\sin\theta$
to all of $\mathbb{R}$ by periodicity:

$$
\sin(\theta + 2\pi) = \sin\theta \quad \text{and} \quad \cos(\theta + 2\pi) = \cos\theta.
$$

We summarize this discussion with

Theorem 4 The functions $\cos\theta$ and $\sin\theta$ are C^∞ and periodic on $\mathbb{R}$ with period
2π. Furthermore,

$$
\frac{d}{d\theta}\sin\theta = \cos\theta \text{ and } \frac{d}{d\theta}\cos\theta = -\sin\theta.
$$

Part II

Part II

Chapter 9

Convergence and Continuity in $\mathbb{R}^n$

Chapter Overview

In this chapter, we develop the language of analysis in higher dimensions. The theorems we prove are extensions to higher dimensions of some of the theorems we saw in earlier chapters. Three different equivalent norms for $\mathbb{R}^n$ are defined and convergence of a sequence of vectors is discussed. Open, closed, and compact sets are introduced. Finally, we prove that continuous functions on a compact set are uniformly continuous and that they attain a minimum and a maximum.

There are many definitions in this section which are very important to understand and to remember. The theorems of analysis are stated in this language, and they cannot be properly applied unless the definitions are understood.

For each $n \in \mathbb{N}$, $\mathbb{R}^n$ will denote the set of all n-tuples of real numbers $\mathbf{x} = (x_1, x_2, \ldots, x_n)$. We shall use boldface letters to denote the elements of $\mathbb{R}^n$, vector-valued functions and matrices. $\mathbb{R}^n$ is a vector space over $\mathbb{R}$ with the usual notions of addition and scalar multiplication. For $\mathbf{x}, \mathbf{y} \in \mathbb{R}^n$ addition is defined to be

$$\mathbf{x} + \mathbf{y} = (x_1 + y_1, x_2 + y_2, \ldots, x_n + y_n),$$

and scalar multiplication by $c \in \mathbb{R}$ is defined to be

$$c\mathbf{x} = (cx_1, cx_2, \ldots, cx_n).$$

9.1 Norms

To discuss the notions of approximation and the convergence of sequences in $\mathbb{R}^n$, we must be able to say when two vector quantities are close together. The concept of a norm will accomplish that; norms are used to measure the distance between points in $\mathbb{R}^n$. They should be thought of as generalizations to higher dimensions of the absolute value of a real number.

Definition 9.1 A *norm* on $\mathbb{R}^n$ is a real-valued function, usually denoted $\|\mathbf{x}\|$, with the following properties:

(Ni) For all $\mathbf{x} \in \mathbb{R}^n$, $\|\mathbf{x}\| \geq 0$ and $\|\mathbf{x}\| = 0$ only when $\mathbf{x} = 0$.
(Nii) For all scalars $c \in \mathbb{R}$ and all $\mathbf{x} \in \mathbb{R}^n$, $\|c\mathbf{x}\| = |c|\|\mathbf{x}\|$.
(Niii) For all $\mathbf{x}, \mathbf{y} \in \mathbb{R}^n$, $\|\mathbf{x} + \mathbf{y}\| \leq \|\mathbf{x}\| + \|\mathbf{y}\|$.

The third property is known as the *triangle inequality*. An immediate consequence of (Niii) is the *reverse triangle inequality*

$$\left| \|\mathbf{x}\| - \|\mathbf{y}\| \right| \leq \|\mathbf{x} - \mathbf{y}\|. \tag{9.1}$$

Here are two norms on $\mathbb{R}^n$ that are very useful.

$$\|\mathbf{x}\|_1 = |x_1| + \cdots + |x_n| \tag{9.2}$$

and

$$\|\mathbf{x}\|_\infty = \max\{|x_1|, \ldots, |x_n|\}. \tag{9.3}$$

The reader can verify that these two functions satisfy the norm properties (Ni), (Nii), and (Niii) in the exercises.

The standard Euclidean norm is

$$\|\mathbf{x}\|_2 = \sqrt{x_1^2 + \cdots + x_n^2}. \tag{9.4}$$

The Euclidean distance between two points $\mathbf{x}$ and $\mathbf{y}$ is $\|\mathbf{x} - \mathbf{y}\|_2$. The Euclidean norm is generated by a scalar product on $\mathbb{R}^n$.

Definition 9.2 A *scalar product* on $\mathbb{R}^n$ is a function from $\mathbb{R}^n \times \mathbb{R}^n$ to $\mathbb{R}$, denoted $\langle x, y \rangle$, with the following properties:

(SPi) (Symmetry) $\langle x, y \rangle = \langle y, x \rangle$.
(SPii) (Nonnegativity) $\langle x, x \rangle \geq 0$, and $\langle x, x \rangle = 0$ only when $x = 0$.
(SPiii) (Linearity) $\langle ax + by, z \rangle = a\langle x, z \rangle + b\langle y, z \rangle$ for all $x, y, z \in \mathbb{R}^n$ and $a, b \in \mathbb{R}$.

Remark (SPi) combined with (SPiii) implies that a scalar product is also linear in the second variable when the first variable is fixed. The scalar function $(x, y) \rightarrow \langle x, y \rangle$ is said to be a bilinear function.

There are many possible scalar products on $\mathbb{R}^n$. The scalar product on $\mathbb{R}^n$ that gives rise to the Euclidean norm $\| \ \|_2$ is

$$\langle x, y \rangle = x_1 y_1 + \ldots + x_n y_n, \tag{9.5}$$

with

$$\|x\|_2 = \langle x, x \rangle^{1/2}.$$

The first two properties (Ni) and (Nii) are obvious for the Euclidean norm. The third property (Niii) follows from the Schwarz inequality which we now state in a lemma.

Lemma 9.1 (Schwarz Inequality) For all $x, y \in \mathbb{R}^n$,

$$|\langle x, y \rangle| \leq \|x\|_2 \|y\|_2 \tag{9.6}$$

with equality holding if and only if x or y is a scalar multiple of the other.

Proof: We give a proof that does not use coordinates. If both $x = 0$ and $y = 0$, there is nothing to prove. We suppose that $y \neq 0$. For $t \in \mathbb{R}$, let $p(t)$ be the quadratic polynomial

$$p(t) = \|x\|_2^2 + 2t\langle x, y \rangle + t^2 \|y\|_2^2 = \|x + ty\|^2.$$

Since $p(t) \geq 0$ for all t, the mininum value of p is also nonnegative. The minimizer of p is $t_* = -\langle x, y \rangle / \|y\|^2$ and the minimum value of p is

$$p(t_*) = -\frac{\langle x, y \rangle^2}{\|y\|^2} + \|x\|^2 \geq 0$$

which implies (9.6).

It is clear that if x is a scalar multiple of y, or if y is a scalar multiple of x, then equality holds in (9.6). Suppose then that $y \neq 0$, and that equality holds in (9.6). Then the minimum value of p is zero so that $0 = p(t_*) = \|x + t_* y\|_2^2$ whence $x = -t_* y$. The proof of the lemma is complete.

Now the triangle inequality for the Euclidean norm follows from the Schwarz inequality. For any $x, y \in \mathbb{R}^n$,

$$
\begin{aligned}
\|x + y\|_2^2 &= \langle x + y, x + y \rangle \\
&= \|x\|_2^2 + 2\langle x, y \rangle + \|y\|_2^2 \\
&\leq \|x\|_2^2 + 2\|x\|_2 \|y\|_2 + \|y\|_2^2 \\
&\leq (\|x\|_2 + \|y\|_2)^2.
\end{aligned}
$$

Thus we see that the Euclidean norm satisfies all of the properties (Ni), (Nii), and (Niii).

An alternate route to the Schwarz inequality is often used in two and three dimensions. In that situation, we use the law of cosines to show that

$$
\langle x, y \rangle = \|x\|_2 \|y\|_2 \cos \theta,
$$

where θ is the smaller angle between the vectors x and y (i.e., $0 \leq \theta \leq \pi$). Then (9.6) follows immediately. In our approach, we derive (9.6) directly from the properties of the scalar product. We then see that

$$
-1 \leq \frac{\langle x, y \rangle}{\|x\|_2 \|y\|_2} \leq 1
$$

which allows us to *define* the angle between two vectors as

$$
\theta = \arccos \left(\frac{\langle x, y \rangle}{\|x\|_2 \|y\|_2} \right).
$$

Definition 9.3 We say that two vectors x and y are *orthogonal* if $\langle x, y \rangle = 0$.

The three norms we have defined on $\mathbb{R}^n$ are related to each other by the following inequalities:

$$
\|x\|_\infty \leq \|x\|_2 \leq \|x\|_1 \leq \sqrt{n}\|x\|_2 \leq n\|x\|_\infty. \tag{9.7}
$$

We leave the proof of these inequalities for the exercises.

Definition 9.4 A sequence of points $\mathbf{x}_k \in \mathbb{R}^n$ *converges* to a point $\mathbf{a}$ in a norm $\| \; \|$ if, for each $\varepsilon > 0$, there is an N (depending on ε) such that, for all $k \geq N$, $\|\mathbf{x}_k - \mathbf{a}\| < \varepsilon$.

The three norms $\| \; \|_1$, $\| \; \|_2$ and $\| \; \|_\infty$ are all equivalent on $\mathbb{R}^n$. By this we mean that convergence in one of the norms implies convergence in the other norms. This equivalence is a consequence of the inequality (9.7).

Convergence in a norm may seem quite abstract, but it can be reduced to familiar one-dimensional convergence. Usually we denote the components of a vector $\mathbf{x}$ with subscripts. However, when we are discussing a sequence of vectors $\mathbf{x}_k$, we denote the components of $\mathbf{x}_k$ with superscripts: $\mathbf{x}_k = (x_k^1, \ldots, x_k^n)$.

Theorem 9.2 A sequence $\mathbf{x}_k \in \mathbb{R}^n$ converges to $\mathbf{a}$ in one of the norms $\| \; \|_1$, $\| \; \|_2$ or $\| \; \|_\infty$ if and only if each component sequence x_k^i converges to a^i in $\mathbb{R}$.

Proof: Suppose that $\mathbf{x}_k$ converges to $\mathbf{a}$ in one (and hence all) of the norms $\| \; \|_1$, $\| \; \|_2$ or $\| \; \|_\infty$. Using the $\| \; \|_\infty$ norm, for each $\varepsilon > 0$, there is an N such that $\|\mathbf{x}_k - \mathbf{a}\|_\infty < \varepsilon$ for all $k \geq N$. Then $|x_k^i - a^i| \leq \|\mathbf{x}_k - \mathbf{a}\|_\infty \leq \varepsilon$ for all $k \geq N$. Hence each component sequence x_k^i converges.

Conversely, let us suppose that each component sequence x_k^i converges to a^i in $\mathbb{R}$ for $i = 1, \ldots, n$. This means that, given an $\varepsilon > 0$, there is an N_i such that for $k \geq N_i$, $|x_k^i - a^i| < \varepsilon$. For some indices i, the N_i may be quite large, and for others, not so large. We want to find an N that will work simultaneously for all i. We let $N = \max N_i$. Then for $k \geq N$, we have $|x_k^i - a^i| < \varepsilon$ for all i, and hence, for $k \geq N$,

$$\|\mathbf{x}_k - \mathbf{a}\|_\infty = \max_{1 \leq i \leq n} |x_k^i - a^i| < \varepsilon.$$

EXAMPLE 9.1:

(i) The sequence $\mathbf{x}_k = (2 + 1/k, \cos(k\pi))$ does not converge, because the second component oscillates between ± 1.

(ii) The sequence $\mathbf{y}_k = (4 + 1/\sqrt{k}, \exp(-1/k))$ converges to $\mathbf{b} = (4, 1)$.

(iii) Let $t_k \to \infty$ be a real sequence. Then the sequence

$$\mathbf{z}_k = e^{-t_k}(\cos(t_k), \sin(t_k), t_k^2)$$

converges to $(0, 0, 0)$.

Theorem 9.3 Let $\| \, \|$ be a norm on $\mathbb{R}^n$ that satisfies (Ni), (Nii), and (Niii). Let the sequence $\mathbf{x}_k$ converge to $\mathbf{a}$, and the sequence $\mathbf{y}_k$ converge to $\mathbf{b}$. Then

    (i) The real sequence $\|\mathbf{x}_k\|$ converges to $\|\mathbf{a}\|$.

    (ii) The sum sequence $\mathbf{x}_k + \mathbf{y}_k$ converges to $\mathbf{a} + \mathbf{b}$.

    (iii) The real sequence $\langle \mathbf{x}_k, \mathbf{y}_k \rangle$ converges to $\langle \mathbf{a}, \mathbf{b} \rangle$.

Proof: (i) is an easy consequence of the reverse triangle inequality. In fact, (9.1) implies that

$$\big| \|\mathbf{x}_k\| - \|\mathbf{a}\| \big| \leq \|\mathbf{x}_k - \mathbf{a}\|.$$

Similarly, (ii) is a simple consequence of the triangle inequality (Niii), using the same proof as in one dimension.

We prove assertion (iii) by using the usual device for dealing with products. We add and subtract a term so that we have to deal with only one variable at a time:

$$|\langle \mathbf{x}_k, \mathbf{y}_k \rangle - \langle \mathbf{a}, \mathbf{b} \rangle| = |\langle \mathbf{x}_k - \mathbf{a}, \mathbf{y}_k \rangle + \langle \mathbf{a}, \mathbf{y}_k - \mathbf{b} \rangle|.$$

Then applying the Schwarz inequality (9.6) to both terms, we have

$$|\langle \mathbf{x}_k, \mathbf{y}_k \rangle - \langle \mathbf{a}, \mathbf{b} \rangle| \leq \|\mathbf{x}_k - \mathbf{a}\|_2 \|\mathbf{y}_k\|_2 + \|\mathbf{a}\|_2 \|\mathbf{y}_k - \mathbf{b}\|_2.$$

By part (i), we know that the real sequence $\|\mathbf{y}_k\|_2$ converges to $\|\mathbf{b}\|_2$ and hence is bounded; there is a constant M_1 such that $\|\mathbf{y}_k\|_2 \leq M_1$ for all k. Letting $M = \max\{M_1, \|\mathbf{a}\|\}$, we have

$$|\langle \mathbf{x}_k, \mathbf{y}_k \rangle - \langle \mathbf{a}, \mathbf{b} \rangle| \leq M(\|\mathbf{x}_k - \mathbf{a}\|_2 + \|\mathbf{y}_k - \mathbf{b}\|_2).$$

For $\varepsilon > 0$ given, we can choose N_1 and N_2 so that

$$\|\mathbf{x}_k - \mathbf{a}\|_2 < \varepsilon \quad \text{for } k \geq N_1$$

and

$$\|\mathbf{y}_k - \mathbf{b}\|_2 < \varepsilon \quad \text{for } k \geq N_2.$$

Finally, we let $N = \max\{N_1, N_2\}$ so that, for $k \geq N$,

$$|\langle \mathbf{x}_k, \mathbf{y}_k \rangle - \langle \mathbf{a}, \mathbf{b} \rangle| \leq 2M\varepsilon.$$

This completes the proof.

Theorem 9.3 shows that the operations of addition and scalar product are continuous with respect to norms.

Definition 9.5 A sequence $\mathbf{x}_k$ is a *Cauchy sequence* if, for each $\varepsilon > 0$, there is an $N \geq 0$ such that for all $k, l \geq N$, $\|\mathbf{x}_k - \mathbf{x}_l\| < \varepsilon$.

It is easy to verify that if a sequence is Cauchy in one of the equivalent norms, then (9.7) implies that it is Cauchy in the other equivalent norms.

Theorem 9.4 A sequence $\mathbf{x}_k$ is Cauchy in $\mathbb{R}^n$ if and only if it converges.

Proof: Suppose that the sequence is a Cauchy sequence. We may suppose that it is Cauchy in the $\| \ \|_\infty$ norm. Because $|x_k^i - x_l^i| \leq \|\mathbf{x}_k - \mathbf{x}_l\|_\infty$ for each i, we conclude that each component sequence x_k^i is a Cauchy sequence of real numbers, and hence that it must converge to some a^i. In view of Theorem 9.2, the sequence $\mathbf{x}_k$ must converge to $\mathbf{a} = (a^1, \ldots, a^n)$.

Suppose now that $\mathbf{x}_k$ is a convergent sequence in $\mathbb{R}^n$. Letting $\| \ \|$ be any one of the equivalent norms, this means that, for each $\varepsilon > 0$, there is an $N > 0$ such that $\|\mathbf{x}_k - \mathbf{a}\| < \varepsilon/2$ for $k \geq N$. Hence for $k, l \geq N$,

$$\|\mathbf{x}_k - \mathbf{x}_l\| \leq \|\mathbf{x}_k - \mathbf{a}\| + \|\mathbf{a} - \mathbf{x}_l\| < \varepsilon/2 + \varepsilon/2 = \varepsilon.$$

As we saw in Chapter 6, the Cauchy criterion is important because it allows us to determine if a sequence, perhaps one generated by some iterative process, is convergent, without knowing its limit.

Exercises for 9.1

1. Show that the norms $\| \ \|_1$ and $\| \ \|_\infty$ satisfy the norm properties (Ni), (Nii), and (Niii).

2. Let a_j, $j = 1, \ldots, n$ be n positive numbers. Define a bilinear function on $\mathbb{R}^n \times \mathbb{R}^n$ by

$$\langle\langle \mathbf{x}, \mathbf{y} \rangle\rangle = a_1 x_1 y_1 + a_2 x_2 y_2 + \ldots + a_n x_n y_n.$$

a) Show that this function $\langle\langle \mathbf{x}, \mathbf{y} \rangle\rangle$ satisfies the properties (SPi), (SPii), and (SPiii).

b) What is the norm that can be derived from this scalar product? Write out the Schwarz inequality for this norm and scalar product in terms of the components and weights a_j.

3. a) By choosing the right vector $\mathbf{y}$ in the Schwarz inequality, show that, for all $\mathbf{x} = (x_1, \ldots, x_n)$,

$$(x_1 + \cdots + x_n)^2 \leq n(x_1^2 + \cdots + x_n^2).$$

b) Show that if equality holds in the previous inequality, then $x_1 = x_2 = \cdots = x_n$.

4. Prove the inequalities (9.7). The inequality of exercise 3 will be useful.
5. Let $\{c_k\}$ be a sequence of real numbers that converges to c in $\mathbb{R}$. Let $\mathbf{x}_k \in \mathbb{R}^n$ be a sequence such that $\mathbf{x}_k$ converges to $\mathbf{x}$. Use the pattern of part (iii) of Theorem 9.3 to show that the sequence $c_k \mathbf{x}_k$ converges to $c\mathbf{x}$.
6. Let the sequence $\mathbf{x}_k$ converge to $\mathbf{a}$. Assuming that $\mathbf{a} \neq 0$, show that there is an N such that $\|\mathbf{x}_k\| \geq \|\mathbf{a}\|/2$ for $k \geq N$.
7. Let $\mathbf{x}_k$ converge to $\mathbf{a}$. Assuming $\mathbf{a} \neq 0$, use Theorem 9.3 and exercises 5 and 6 to show that the unit vectors $\mathbf{u}_k = \mathbf{x}_k/\|\mathbf{x}_k\|$ converge to the unit vector $\mathbf{a}/\|\mathbf{a}\|$.
8. Show that the Euclidean norm has the *Pythagorean property*:

$$\|\mathbf{x} + \mathbf{y}\|_2^2 + \|\mathbf{x} - \mathbf{y}\|_2^2 = 2(\|\mathbf{x}\|_2^2 + \|\mathbf{y}\|_2^2).$$

9. Here is a variation on the Schwarz inequality. For vectors $\mathbf{x}, \mathbf{y} \in \mathbb{R}^n$, show that

$$|\langle \mathbf{x}, \ \mathbf{y}\rangle| \leq \|\mathbf{x}\|_\infty \, \|\mathbf{y}\|_1.$$

9.2 A Little Topology

The terms "open set" and "closed set" are part of a language that allows us to make concise statements of results. They are generalizations of the concepts of open and closed intervals. We first define what we mean by an "open ball." It is the natural extension to several variables of the idea of an open interval. Let $\|\ \|$ be any of the equivalent norms.

Definition 9.6 The *open ball* with center $\mathbf{a}$ and radius $r > 0$ is the set:

$$B(\mathbf{a}, r) = \{\mathbf{x} : \|\mathbf{x} - \mathbf{a}\| < r\}.$$

The *closed ball* with center $\mathbf{a}$ and radius $r > 0$ is:

$$\bar{B}(\mathbf{a}, r) = \{\mathbf{x} : \|\mathbf{x} - \mathbf{a}\| \leq r\}.$$

When we wish to specify the open or closed ball in a particular norm, we shall write $B_1(\mathbf{x}, r)$ for the open ball in the norm $\|\ \|_1$, and so on.

The ball $B_2(0, r)$ really is a disk or a ball. However, in the other norms the shape is rather different. In two dimensions, a sketch reveals that $B_\infty(0, 1)$ is a square of side 2 centered at $(0, 0)$.

Definition 9.7 Let A be a subset of $\mathbb{R}^n$. We say that $\mathbf{x} \in A$ is an *interior point* of A if there is some $r > 0$ such that $B(\mathbf{x}, r) \subset A$. The collection of all interior points of A is called the *interior* of A, sometimes denoted A^o.

Definition 9.8 A set $U \subset R^n$ is an *open* set if each point of U is an interior point of U. This means that $U = U^o$.

Remark Because of the inequalities (9.7), the definition of an open set does not depend on the choice of the norm. For example, in two dimensions, if we can find a small square centered at the point (x, y) that fits inside U, then we can find a circle which fits inside the square.

To make sure that our use of the word "open" is consistent, we must verify that an open ball is indeed an open set in the sense of Definition 9.8.

Let $\mathbf{x} \in B(\mathbf{a}, r)$; this means that $\|\mathbf{x} - \mathbf{a}\| < r$. Let $\delta = r - \|\mathbf{x} - \mathbf{a}\| > 0$. We claim that $B(\mathbf{x}, \delta) \subset B(\mathbf{a}, r)$. For any $\mathbf{y} \in B(\mathbf{x}, \delta)$, we must show that the distance of $\mathbf{y}$ from $\mathbf{a}$ is less than r. This follows easily from the triangle inequality. If $\mathbf{y} \in B(\mathbf{x}, \delta)$,

$$\|\mathbf{y} - \mathbf{a}\| \leq \|\mathbf{y} - \mathbf{x}\| + \|\mathbf{x} - \mathbf{a}\| < \delta + r - \delta = r.$$

Definition 9.9 A point $\mathbf{x}$ is a *boundary point* of a set A if, for each $r > 0$, the open ball $B(\mathbf{x}, r)$ contains a point in A and a point not in A. A boundary point may or may not belong to the set A. The set of all boundary points of A is called the *boundary* of A, denoted ∂A.

Remark Now we can say something about the notion of an open set. It is clear from the definition that an open set contains none of its boundary points. In two dimensions, one might visualize an open set as having a "fuzzy" edge. A second aspect of an open set is that it must have some n-dimensional volume. For example, a plane in $\mathbb{R}^3$ cannot be an open set.

Open sets are useful in defining the domain of a function. If a function f is defined on a open set U, and $\mathbf{x} \in U$, then we can approach the point $\mathbf{x}$ from all directions within U. It is then possible to define the partial derivatives of f at $\mathbf{x}$.

A second important use of open sets is to describe the range of a function. If $\mathbf{f} : \mathbb{R}^n \to \mathbb{R}^n$, and the range of $\mathbf{f}$ contains an open set, then we know that $\mathbf{f}$ does not squash points together into a set of lower dimension. If the range of $\mathbf{f}$ is an open set, and we can solve the equation $\mathbf{f(x) = y}$ for some $\mathbf{y}_0$ in the range of $\mathbf{f}$, then we can solve it for all $\mathbf{y}$ in some open ball centered at $\mathbf{y}_0$.

> **EXAMPLE 9.2:** It often requires a calculation with inequalities to show that a set is open. Let $U = \{(x, y) \in \mathbb{R}^2 : y > x^2\}$. If a point $(x, y) \in U$, can we choose an $r > 0$ so that $B((x, y), r) \subset U$? Since the notion of open set does not depend on the norm, let us use the $\| \ \|_\infty$ norm. The square of side $2r$, centered at (x, y), is:
>
> $$B_\infty((x, y), r) = \{(\xi, \eta) : |x - \xi| < r \text{ and } |y - \eta| < r\}.$$
>
> Now $(\xi, \eta) \in B_\infty((x, y), r)$ implies that $\eta > y - r$ and that
>
> $$|\xi| \leq |\xi - x| + |x| \leq r + |x|.$$
>
> We make the preliminary assumption that $r \leq 1$, which implies that $r^2 \leq r$. Then we have
>
> $$\xi^2 \leq (r + |x|)^2 = r^2 + 2r|x| + x^2 \leq r + 2r|x| + x^2.$$
>
> Now if we choose $r \leq 1$ such that $2r(1 + |x|) < y - x^2$, and $(\xi, \eta) \in B_\infty((x, y), r)$, then
>
> $$\xi^2 \leq r + 2r|x| + x^2 < y - r < \eta.$$
>
> Thus if
>
> $$r = \min\left\{1, \frac{y - x^2}{2(1 + |x|)}\right\},$$
>
> the square $B_\infty((x, y), r) \subset U$.

If a sequence of approximations to some problem belongs to a set F, we want to be able to say that the limit of these approximations also belongs to F. This is one reason for introducing the notion of a closed set.

Definition 9.10 A set F is *closed* if, whenever $\{x_k\}$ is a sequence of points in F and x_k converges to a, then $a \in F$.

Closed sets can be described in terms of open sets and in terms of boundary points.

Theorem 9.5 The following statements are equivalent:

    (i) The set F is closed.
    (ii) The complement of F is open.
    (iii) The set F contains all of its boundary points.

Proof: We show (i) implies (ii). Let U be the complement of F. If $a \in U$, and it is *not* possible to find an $r > 0$ such that $B(a, r) \subset U$, then there must be a sequence of points $x_k \in F$ that converges to a. Since F is assumed closed, we must conclude that $a \in F$ and that contradicts our assumption that $a \in U$.

Next we show that (ii) implies (iii). Let a be a point of U. Because U is assumed to be open, there is an $r > 0$ such that $B(a, r) \subset U$. Hence a cannot be a boundary point of F. Consequently, $\partial F \subset F$.

Finally let us show that (iii) implies (i). Let $x_k \in F$ be a sequence that converges to a. This means that for each $r > 0$, there is an N such that $x_k \in B(a, r)$ for $k \geq N$. If a does not belong to F, then a is a boundary point of F. But F is assumed to contain all of its boundary points, so that we must have $a \in F$.

Yet another definition!

Definition 9.11 The *closure* of a set A is $\bar{A} = A \cup \partial A$.

It is a good exercise to show that $\bar{A}$ is the smallest closed set that contains A.

    EXAMPLE 9.3:

      (i) The empty set ϕ and $\mathbb{R}^n$ are both open and closed subsets of $\mathbb{R}^n$. In fact, we shall see later that they are the only subsets of $\mathbb{R}^n$ that are both open and closed.

      (ii) Let $A = \{(x, y, z) : x^2 + y^2 + z^2 < 1\} \cup \{x^2 + y^2 + z^2 = 1, z \geq 0\}$. The first pair of brackets is the open ball $B_2(0, 1)$ while the second pair is part of the boundary of $B_2(0, 1)$. A is neither open nor closed.

      (iii) Let $A = \{(0, 0), \ldots, (5, 0)\}$. In this example, each point of A is a boundary point of A, and $\partial A = A$. A is closed and has no interior points.

(iv) The closed ball $\bar{B}(\mathbf{a}, r)$ is closed and has a nonempty interior.
(v) We usually think of the boundary of a set as the "skin" of a body, much like the film of soap that encloses a bubble. However, sometimes the boundary of a set can be larger than the set itself. For example, because the rational numbers $\mathbb{Q}$ are dense in $\mathbb{R}$, $\partial\mathbb{Q} = \mathbb{R}$.

We need to know how open and closed sets behave under the operations of intersection and union.

Theorem 9.6

(i) Any union of open sets is again open.
(ii) A finite intersection of open sets is open.
(iii) A finite union of closed sets is closed.
(iv) Any intersection of closed sets is closed.
(v) The only subsets of $\mathbb{R}^n$ which are both open and closed are $\mathbb{R}^n$ itself and the empty set.

Proof: Let $U_\gamma \subset \mathbb{R}^n$ be a collection of open set with index $\gamma \in \Gamma$ and set $U = \cup_{\gamma \in \Gamma} U_\gamma$. If $\mathbf{x} \in U$, then $\mathbf{x} \in U_{\gamma_0}$ for some index γ_0. Since U_{γ_0} is an open set, there is an $r > 0$ such that $B(\mathbf{x}, r) \subset U_{\gamma_0} \subset U$. We have proved (i).

We prove (ii). Suppose $U_1, \ldots U_p$ is a collection of open sets, and let $U = \cap_{k=1}^p U_k$. If $\mathbf{x} \in U$, then because each U_k is open, there is an $r_k > 0$ such that $B(\mathbf{x}, r_k) \subset U_k$. Because there are only a finite number of r_k, $r = \min r_k > 0$. We have $B(\mathbf{x}, r) \subset U_k$ for each k, so $B(\mathbf{x}, r) \subset U$, which proves that U is open.

We can prove (iii) and (iv) using DeMorgan's Laws for sets. Let A' denote the complement of A. Now if $F_1, \ldots, F_p$ are closed sets, and $F = \cup_{k=1}^p F_k$, we show that F is closed by showing that F' is open. For each k, $U_k = F'_k$ is open by Theorem 9.5. By DeMorgan's Law (Chapter 1),

$$F' = (\cup_{k=1}^p F_k)' = \cap_{k=1}^p U_k.$$

Part (ii) tells us that $\cap_{k=1}^p U_k$ is open.

The fourth assertion is proved by showing that the intersection of any collection of closed sets F_γ is the complement of the union of the open sets $U_\gamma = F'_\gamma$.

Finally we prove (v). Suppose to the contrary that there is some nonempty subset U of $\mathbb{R}^n$ with $U \neq \mathbb{R}^n$ that is both open and closed. Then its complement $V = U'$ is nonempty and both open and closed. Let $\mathbf{x} \in U$ and $\mathbf{y} \in V$, and let $\mathcal{L}$ be the line between $\mathbf{x}$ and $\mathbf{y}$. The line $\mathcal{L}$ is parameterized by $\mathbf{z}(t) = t\mathbf{y} + (1 - t)\mathbf{x}$, $0 \leq t \leq 1$. Now let $\tau = \sup\{t \in [0, 1] : \mathbf{z}(t) \in U\}$. There is a sequence $t_k \leq \tau \in [0, 1]$ such that $\mathbf{z}(t_k) \in U$ and t_k converges to τ. The sequence $\mathbf{z}(t_k)$ converges to $\mathbf{z}(\tau)$, and U is closed, so $\mathbf{z}(\tau) \in U$. Therefore we must have $\tau < 1$ because $\mathbf{z}(1) = \mathbf{y} \in V$. On the other hand, U is also open, so there is a $\delta > 0$ such that $\mathbf{z}(t) \in U$ for $|t - \tau| < \delta$. This contradicts the definition of τ.

The class of compact sets is important for the discussion of continuous functions and their extrema, and hence for optimization problems. Finally, here are the last definitions of this section.

Definition 9.12 A set $A \subset \mathbb{R}^n$ is *bounded* if, for any of the equivalent norms, $A \subset B(0, R)$ for some $R > 0$.

Note that a set being bounded says nothing about whether or not it contains its boundary.

Next we make an extension of the Bolzano-Weierstrass theorem to higher dimensions.

Theorem 9.7 Let $\mathbf{x}_k$ be a bounded sequence in $\mathbb{R}^n$. Then there exists a convergent subsequence $\mathbf{x}_{k_l}$.

Proof: Because the sequence $\mathbf{x}_k$ is bounded, there is a constant $R > 0$ such that $\|\mathbf{x}_k\| \leq R$ for all k. This means that, for each i, the component sequence $|x_k^i| \leq R$. We shall use the Bolzano-Weierstrass theorem in one dimension to extract a convergent subsequence from each component. We must take care to make sure that we arrive at a subsequence of indices k_l such that, for each i, $x_{k_l}^i$ converges. Start with $i = 1$. By the Bolzano-Weierstrass theorem in one dimension, we know there exists a subsequence $k_{l'}$ such that $x_{k_{l'}}^1$ converges to a^1. Now we consider the second components, $i = 2$. The subsequence $x_{k_{l'}}^2$ is also bounded. Hence a thinned-out subsequence $x_{k_{l''}}^2$ converges to some number a^2. Clearly $x_{k_{l''}}^1$ still converges to a^1. Thus the subsequence of both components $(x_{k_{l''}}^1, x_{k_{l''}}^2)$ converges to (a^1, a^2). We continue in this way, eventually extracting a subsequence k_l such that, for each $i = 1, \ldots, n$, $x_{k_l}^i$ converges to a^i. According to Theorem 9.2, this means that $\mathbf{x}_{k_l}$ converges to $\mathbf{a}$. The proof is complete.

Definition 9.13 A set K is *compact* if, whenever $\mathbf{x}_k$ is a sequence of points in K, there is a subsequence $\mathbf{x}_{k_j}$ that converges to a point $\mathbf{a} \in K$.

Compact subsets of $\mathbb{R}^n$ are easy to identify.

Theorem 9.8 A set $K \subset \mathbb{R}^n$ is compact if and only if K is closed and bounded.

Proof: To show that compact implies closed and bounded, we prove two assertions:

(i) if K is not closed, then it is not compact; and
(ii) if K is not bounded, then it is not compact.

First suppose K is not closed. Then there is some boundary point $\mathbf{a}$ of K that does not belong to K. However, every ball $B(\mathbf{a}, 1/k)$ contains some point $\mathbf{x}_k \in K$. Clearly $\mathbf{x}_k$ converges to $\mathbf{a}$ and therefore any subsequence of $\mathbf{x}_k$ also converges to $\mathbf{a}$. Because $\mathbf{a} \notin K$, K is not compact. Next suppose that K is not bounded. Then for each $k = 1, 2, 3, \ldots$, there is a point $\mathbf{x}_k \in K$ with $\|\mathbf{x}_k\| \geq k$. The sequence $\mathbf{x}_k$ does not have any convergent subsequence and so K is not compact.

To prove the converse implication, suppose that K is closed and bounded. Let $\mathbf{x}_k$ be a sequence of points in K; $\mathbf{x}_k$ is a bounded sequence. By the Bolzano-Weierstrass theorem, there is a subsequence $\mathbf{x}_{k_l}$ that converges to some point $\mathbf{a}$. Since K is closed, we must have $\mathbf{a} \in K$. This completes the proof.

Theorem 9.9 Properties of compact sets

(i) If F is closed and K is compact, then $F \cap K$ is compact.
(ii) If $K_1, \ldots, K_p \subset R^n$ are compact sets, then $\cup_{k=1}^p K_k$ is compact.
(iii) If $K_\gamma \subset \mathbb{R}^n$, $\gamma \in \Gamma$, is any collection of compact sets, then $\cap_{\gamma \in \Gamma} K_\gamma$ is compact.

Proof: The proofs are left as exercises.

Remark The notion of a *metric* is a generalization of the concept of a norm. A metric is defined as follows:

Let X be a set with a function $d : X \times X \to \mathbb{R}$ such that

(i) $d(x, y) \geq 0$ for all $x, y \in X$ and $d(x, y) = 0$ implies $x = y$;
(ii) $d(x, y) = d(y, x)$ for all $x, y \in X$; and
(iii) $d(x, y) \leq d(x, z) + d(z, y)$ for all $x, y, z \in X$.

The function d is said to be a *metric* on X and the pair (X, d) is said to be a *metric space*.

Any norm on $\mathbb{R}^n$ is an example of a metric on the space $X = \mathbb{R}^n$. We simply take $d(\mathbf{x}, \mathbf{y}) = \|\mathbf{x} - \mathbf{y}\|$. A metric that does not come from a norm arises naturally as the distance between two points as measured on a curved surface, for example the great circle distance between two points on a sphere. All of the topological concepts, including open sets, closed sets, and compact sets, can be defined in a metric space. The book of R. Strichartz [Str] provides an introduction to metric spaces.

Exercises for 9.2

1. For each of the following sets $A \subset \mathbb{R}^2$, (i) sketch the set; (ii) determine whether the set is open, closed, or neither; (iii) describe A°, $\bar{A}$ and ∂A.

   a) $A = \{(x, y) : y > 0 \text{ and } x^2 + y^2 \leq 1\}$.
   b) $A = \{(x, y) : x > -1, y > 0 \text{ and } x + 2y > 0\}$.
   c) $A = \{(x, y) : y \leq e^x \text{ and } y - x \leq 2\}$.
   d) $A = \{(x, y) : y = x - x^2\}$.
   e) $A = \{(x, y) : x, y \in \mathbb{Q} \text{ and } 0 \leq x, y \leq 1\}$.

2. a) Give an example of an infinite collection of closed sets $U_k, k \in \mathbb{N}$, such that $\cap_k U_k$ is not open.
   b) Give an example of an infinite collection of closed sets $F_k, k \in \mathbb{N}$, such that $\cup_k F_k$ is not closed.

3. If $U \subset \mathbb{R}^n$ and $V \subset \mathbb{R}^m$, the Cartesian product $U \times V = \{(\mathbf{x}, \mathbf{y}) : \mathbf{x} \in U, \mathbf{y} \in V\}$ is a subset of $\mathbb{R}^{n+m}$.

   a) Give an example in $\mathbb{R}^2 = \mathbb{R} \times \mathbb{R}$ to show that in general $\partial(U \times V)$ is not equal to $\partial U \times \partial V$.
   b) If U is open in $\mathbb{R}^n$ and V is open in $\mathbb{R}^m$, show that $U \times V$ is an open subset of $\mathbb{R}^{n+m}$. Use balls defined by the norm $\| \ \|_\infty$.
   c) If F is a closed subset of $\mathbb{R}^n$ and H is a closed subset of $\mathbb{R}^m$, show that $F \times H$ is a closed subset of $\mathbb{R}^{n+m}$. Use the definition of closed sets in terms of sequences.

4. Prove Theorem 9.9.

5. Let A be a set in $\mathbb{R}^n$. Show that $\bar{A}$ is the smallest closed set that contains A. Do this by showing (i), $\bar{A}$ is closed, and (ii), if B is any closed set that contains A, then $\bar{A} \subset B$.

6. a) Let K be a *compact* set and let F be a closed set and assume that $K \cap F = \phi$. Show that

$$d = \inf_{\mathbf{x} \in K, \mathbf{y} \in F} \|\mathbf{x} - \mathbf{y}\| > 0.$$

   Hint: assume that $d = 0$ and deduce that $K \cap F \neq \phi$.
   b) Now we see what happens when we relax the hypothesis that K is compact. Let $F = \{(x, 1/x) : x \geq 1\} \subset \mathbb{R}^2$ and let K be the x axis. Both F and K are closed and $K \cap F = \phi$. Verify that

$$d = \inf_{\mathbf{x} \in K, \mathbf{y} \in F} \|\mathbf{x} - \mathbf{y}\| = 0.$$

7. A set $A \subset \mathbb{R}^n$ is said to be *convex* if for each pair of points $\mathbf{a}, \mathbf{b} \in A$, the line segment joining $\mathbf{a}$ and $\mathbf{b}$ is also contained in A. This line segment is easily parameterized by $\mathbf{x}(t) = (1 - t)\mathbf{a} + t\mathbf{b},\ 0 \le t \le 1$.

   a) Show that if A and B are convex sets, then so is $A \cap B$.
   b) Show that if A and B are convex sets, then

$$A + B \equiv \{\mathbf{z} = \mathbf{x} + \mathbf{y} : \mathbf{x} \in A \text{ and } \mathbf{y} \in B\}$$

   is also convex.

8. a) Show that if A is a convex set, then so is the closure $\bar{A}$.
   b) Give an example of a set A that is not convex, but whose closure $\bar{A}$ is convex.

9.3 Continuous Functions of Several Variables

If $G \subset \mathbb{R}^n$, a real-valued function defined on G will be denoted with letters like f, g and h; for example, $f : G \to \mathbb{R}$. Functions with values in $\mathbb{R}^m$, for $m > 1$, will be denoted with boldfaced letters; for example $\mathbf{f} : G \to \mathbb{R}^3$. In terms of components we will write $\mathbf{f}(\mathbf{x}) = (f_1(\mathbf{x}), f_2(\mathbf{x}), \dots, f_m(\mathbf{x}))$. Each component function $f_i : G \to \mathbb{R}$.

The definition of continuity in higher dimensions is the same as it is in one dimension, but with norms substituted for the absolute value.

Definition 9.14 Let $G \subset \mathbb{R}^n$ and let $\mathbf{f} : G \to \mathbb{R}^m$. Let $\mathbf{a} \in G$. We say that $\mathbf{f}$ is *continuous* at $\mathbf{a}$ if, for every sequence of points $\mathbf{x}_k \in G$ that converges to $\mathbf{a}$, the sequence of function values $\mathbf{f}(\mathbf{x}_k)$ converges to $\mathbf{f}(\mathbf{a})$ in $\mathbb{R}^m$. We say that $\mathbf{f}$ is *continuous on G* if $\mathbf{f}$ is continuous at every point of G.

The alternate equivalent formulation of continuity is given in terms of ε and δ.

Alternate(ε, δ) formulation of continuity

The function $\mathbf{f}$ is continuous at $\mathbf{a} \in G$, if for each $\varepsilon > 0$, there is a $\delta > 0$ such that $\|\mathbf{f}(\mathbf{x}) - \mathbf{f}(\mathbf{a})\| < \varepsilon$ for all $\mathbf{x} \in G$ with $\|\mathbf{x} - \mathbf{a}\| < \delta$.

These two formulations are shown to be equivalent in the same manner as in Theorem 3.1.

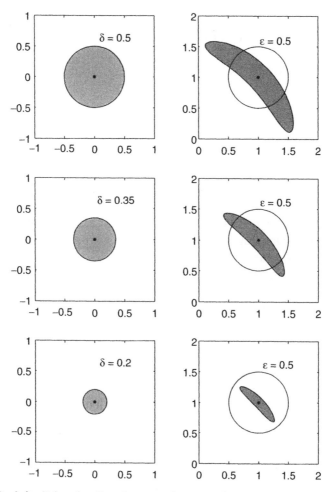

Figure 9.1 On left, disks of radius δ centered at $(0,0)$ for various values of δ. On right, images of the δ disk under a continuous function, and the ε disk of radius $\varepsilon = 1/2$

In Figure 9.1, we consider a continuous function $\mathbf{f} : \mathbb{R}^2 \to \mathbb{R}^2$ such that $\mathbf{f}(0,0) = (1,1)$, and we use the norm $\| \ \|_2$ to measure distances. In the right panels, we see a circle of radius $\varepsilon = 1/2$ centered at the point $(1,1)$. We then try various values of δ to see when the function $\mathbf{f}$ maps the disk of radius δ, centered at $(0,0)$, into the $\varepsilon = 1/2$ disk centered at $(1,1)$. The definition of continuity states that no matter how small the ε disk $B_\varepsilon = B_2((1,1), \varepsilon)$, there is a δ disk $B_\delta = B_2((0,0), \delta)$ such that $\mathbf{f}(B_\delta) \subset B_\varepsilon$.

An immediate consequence of Theorem 9.2 is

Theorem 9.10 A function $\mathbf{f}$ is continuous at $\mathbf{a}$ if and only if each component function f_j is continuous at $\mathbf{a}$.

Because of Theorem 9.10, most of our discussion of continuous functions will focus on the scalar-valued case. It is obvious that if f is continuous, then f is continuous in each variable when the others are held fixed. Is the converse true? It is tempting to think so, but the following counterexample throws cold water on that idea.

EXAMPLE 9.4: Define $f(x, y)$ on $\mathbb{R}^2$ by

$$f(x, y) = \begin{cases} \dfrac{xy}{x^2 + y^2} & \text{for } (x, y) \neq (0, 0) \\ 0 & \text{for } (x, y) = (0, 0) \end{cases}.$$

We see first that f is identically zero along the x axis and along the y axis. This means that $x \to f(x, 0)$ and $y \to f(0, y)$, being constant, are both continuous on $\mathbb{R}$.

However, f is not continuous at $(0, 0)$ in the sense of Definition 9.14. In fact, if $x \neq 0$, then $f(x, mx) = m/(1 + m^2)$. Therefore if we approach $(0, 0)$ with a sequence of points of the form (x_k, mx_k), we have $\lim_{k \to \infty} f(x_k, mx_k) = m/(1 + m^2) \neq 0$.

One can devise more complicated examples in which the function values along each straight line $y = mx$ all converge to the same number as x converges to 0, but the function values as x converges to 0 along a parabola $y = x^2$ converge to a different number. However, we shall not dwell on these examples because this is not a course in pathology.

Continuous functions can be combined in the usual ways to produce new continuous functions.

Theorem 9.11 Let f and g be continuous at **a**. Then,

(i) $\alpha f + \beta g$ is continuous at **a** for any $\alpha, \beta \in \mathbb{R}$;
(ii) fg is continuous at **a**; and
(iii) If $g(\mathbf{a}) \neq 0$, f/g is continuous at **a**.

The proofs are the same as in the one-dimensional case, and we leave them for the exercises.

Theorem 9.12 Let $\mathbf{f} : U \to \mathbb{R}^m$ where $U \subset \mathbb{R}^n$. Suppose that $\mathbf{f}$ is continuous at $\mathbf{x} = \mathbf{a} \in U$. Let $\mathbf{g} : V \to \mathbb{R}^p$ where $V \subset \mathbb{R}^m$ and $\mathbf{f}(U) \subset V$. Suppose that $\mathbf{g}$ is continuous at $\mathbf{f}(\mathbf{a})$. Then the composite function $\mathbf{g} \circ \mathbf{f}$ is continuous at $\mathbf{x} = \mathbf{a}$.

Proof: Let $\mathbf{x}_k$ be a sequence of points in U such that $\mathbf{x}_k$ converges to **a**. Because $\mathbf{f}$ is continuous at **a**, we know that $\mathbf{y}_k = \mathbf{f}(\mathbf{x}_k)$ converges to $\mathbf{f}(\mathbf{a})$ in $\mathbb{R}^m$. Because

g is continuous at $\mathbf{f}(\mathbf{a})$, $\mathbf{g}(\mathbf{f}(\mathbf{x}_k)) = \mathbf{g}(\mathbf{y}_k)$ converges to $\mathbf{g}(\mathbf{f}(\mathbf{a}))$. The proof is complete.

EXAMPLE 9.5:

(i) The coordinate functions $f_j(\mathbf{x}) = x^j$ are continuous on R^n, because $|f_j(\mathbf{x}) - f_j(\mathbf{y})| = |x^j - y^j| \leq \|\mathbf{x} - \mathbf{y}\|$ for any of the norms.

(ii) Because of part (i) and Theorem 9.11, any polynomial in n variables is continuous everywhere on $\mathbb{R}^n$. Rational functions (quotients of polynomials) are continuous at all points where the denominator does not vanish.

(iii) Because of Theorems 9.11 and 9.12, the function

$$\cos\left(\frac{x^2}{1 - (x^2 + y^2)}\right)$$

is continuous on the set $G = \{(x, y) : x^2 + y^2 \neq 1\}$.

(iv) Let $u = x^2 - y^2$ and $v = 2xy$ and let $\mathbf{h}(x, y)$ be the composite function

$$\mathbf{h}(x, y) = (e^u \cos v, e^u \sin v).$$

Theorem 9.12 implies that $\mathbf{h} : \mathbb{R}^2 \to \mathbb{R}^2$ is continuous.

We can now put some ideas of the previous section together with the concept of a continuous function to get our first important, and useful, result.

Theorem 9.13 Let $\mathbf{f}$ be a continuous function on a compact set $K \subset \mathbb{R}^n$ with values in $\mathbb{R}^m$, $\mathbf{f} : K \to \mathbb{R}^m$. Then the image set $\mathbf{f}(K)$ is a compact set in $\mathbb{R}^m$.

Proof: Let $\mathbf{y}_k$ be a sequence of points of $\mathbf{f}(K)$. This means that for each k, there is at least one point $\mathbf{x}_k \in K$ with $\mathbf{f}(\mathbf{x}_k) = \mathbf{y}_k$. Because K is compact, there is a subsequence $\mathbf{x}_{k_j}$ that converges to a point $\mathbf{a} \in K$. We assume that $\mathbf{f}$ is continuous. Hence $\mathbf{y}_{k_j} = f(\mathbf{x}_{k_j})$ converges to $\mathbf{f}(\mathbf{a}) \in \mathbf{f}(K)$. This means that $\mathbf{f}(K)$ is compact.

We recall Definitions 1.10 and 3.8. If f is a real-valued function defined on a set A, we say that f is *bounded* on A if $f(A)$ is a bounded set in $\mathbb{R}$. This means that $\sup_A f$ and $\inf_A f$ are finite.

When there is a point $x^* \in A$ such that $f(x^*) \geq f(x)$ for all $x \in A$, then we say that $f(x^*) = \max_A f$ is the *maximum* of f over A, and that x^* is a *maximizer* for f on A.

When there is a point $x_* \in A$ such that $f(x_*) \le f(x)$ for all $x \in A$, we say that $f(x_*) = \min_A f$ is the *minimum* of f over A and that x_* is a *minimizer* for f over A. Of course, there may be many minimizers and maximizers.

We use the term *extreme point* to refer to both maximizers and minimizers.

Theorem 9.14 Let K be a compact subset of $\mathbb{R}^n$, and let $f : K \to \mathbb{R}$ be continuous on K. Then there is a maximizer $\mathbf{x}^* \in K$ and there is a minimizer $\mathbf{x}_* \in K$: $f(\mathbf{x}_*) \le f(\mathbf{x}) \le f(\mathbf{x}^*)$ for all $\mathbf{x} \in K$.

Proof: We prove the existence of the maximizer. According to Theorem 9.13, $f(K)$ is compact, and hence bounded in $\mathbb{R}$. Thus $M = \sup_K f$ is finite. We must show that $M \in f(K)$. From the definition of the supremum of a set (least upper bound), there is a sequence $y_k \in f(K)$ such that y_k converges upward to M. Since $f(K)$ is compact, it is closed so that $M \in f(K)$. Hence there is a point $\mathbf{a}^* \in K$ such that $f(\mathbf{a}^*) = M \ge f(\mathbf{x})$ for all $\mathbf{x} \in K$. Thus $\mathbf{x}^*$ is a maximizer for f on K. The proof for the existence of the minimizer is the same.

Remark Theorem 9.14 is quite precise in that if either of the hypotheses is relaxed, the result can be false. The set $(0, 1]$ is bounded, but not closed. The function $f(x) = 1/x$ is continuous on $(0, 1]$ but it is unbounded. The set $[0, \infty)$ is closed, but not bounded. The function $f(x) = x$ is continuous on $[0, \infty)$ but it is not bounded.

> **EXAMPLE 9.6:**   Here is an example of the way Theorem 9.14 is used.
> Let K be a compact set in $\mathbb{R}^n$. Suppose that we have a continuous function $f(\mathbf{x})$ on K and that we know that $f(\mathbf{x}) > 0$ for each $\mathbf{x} \in K$. Can we show that there is a constant $c > 0$ such that $f(\mathbf{x}) \ge c$ for all $\mathbf{x} \in K$? In this case we would say that f is bounded away from zero. Theorem 9.14 says that there is a minimizer $\mathbf{x}_* \in K$ such that $f(\mathbf{x}) \ge f(\mathbf{x}_*)$ for all $\mathbf{x} \in K$. Since $\mathbf{x}_* \in K$, we know that we can take $c = f(\mathbf{x}_*) > 0$.

We make a second connection between the notions of continuity and compactness. The next definition extends the concept of uniform continuity to functions of several variables.

Definition 9.15 Let $G \subset \mathbb{R}^n$ and $\mathbf{f} : G \to \mathbb{R}^m$. We say that $\mathbf{f}$ is *uniformly continuous* on G if, for each $\varepsilon > 0$, there is a $\delta > 0$ such that

$$\|\mathbf{f}(\mathbf{x}) - \mathbf{f}(\mathbf{y})\| < \varepsilon \quad \text{for all} \quad \mathbf{x}, \mathbf{y} \in G \text{ with } \|\mathbf{x} - \mathbf{y}\| < \delta.$$

Remark Just as in one dimension, the important difference between uniform continuity and the alternate (ε, δ) version of continuity is that, given an $\varepsilon > 0$, there is a choice of δ that is independent of $\mathbf{x} \in G$. It is also important to remember that uniform continuity only makes sense when we specify a set G.

Theorem 9.15 Let $K \subset \mathbb{R}^n$ be compact, and let $\mathbf{f} : K \to \mathbb{R}^m$ be continuous. Then $\mathbf{f}$ is uniformly continuous on K.

Proof: We show that if $\mathbf{f}$ is not uniformly continuous with these hypotheses, we are led to a contradiction. First, we must carefully state the negation of uniform continuity. For *some* $\varepsilon_0 > 0$, no matter how small we choose $\delta > 0$, there is always some pair $\mathbf{x}, \mathbf{y} \in K$ such that $\|\mathbf{x} - \mathbf{y}\| < \delta$ but $\|\mathbf{f}(\mathbf{x}) - \mathbf{f}(\mathbf{y})\| > \varepsilon_0$. To frame this negation in a more quantitative fashion, we say that, for some $\varepsilon_0 > 0$, there is a pair $\mathbf{x}_k, \mathbf{y}_k \in K$ with $\|\mathbf{x}_k - \mathbf{y}_k\| < 1/k$ but with $\|\mathbf{f}(\mathbf{x}_k) - \mathbf{f}(\mathbf{y}_k)\| > \varepsilon_0$. Now we use the compactness of K and the continuity of $\mathbf{f}$ to produce a contradiction. Because K is compact, we can extract a subsequence $\mathbf{x}_{k_j}$ that converges to a point $\mathbf{a} \in K$. The subsequence $\mathbf{y}_{k_j}$ also converges to $\mathbf{a}$ because $\|\mathbf{x}_{k_j} - \mathbf{y}_{k_j}\| < 1/k_j$. Finally we are ready to draw the noose around our quarry, the contradiction. For all j, we have

$$\|\mathbf{f}(\mathbf{x}_{k_j}) - \mathbf{f}(\mathbf{y}_{k_j})\| \leq \|\mathbf{f}(\mathbf{x}_{k_j}) - \mathbf{f}(\mathbf{a})\| + \|\mathbf{f}(\mathbf{a}) - \mathbf{f}(\mathbf{y}_{k_j})\|.$$

Because $\mathbf{f}$ is continuous, both $\mathbf{f}(\mathbf{x}_{k_j})$ and $\mathbf{f}(\mathbf{y}_{k_j})$ converge to $\mathbf{f}(\mathbf{a})$. The preceding inequality shows that $\|\mathbf{f}(\mathbf{x}_{k_j}) - \mathbf{f}(\mathbf{y}_{k_j})\|$ converges to 0 as $j \to \infty$. This contradicts the assumption that $\|\mathbf{f}(\mathbf{x}_k) - \mathbf{f}(\mathbf{y}_k)\| > \varepsilon_0$ for all k. The proof is complete.

> **EXAMPLE 9.7:** The function $f(x, y) = (x^2 + y^2)^{1/4}$ is continuous on the square $Q = \{0 \leq x, y \leq 1\}$, and hence by Theorem 9.15, it is uniformly continuous on Q.

Finally we consider a last class of sets that are useful for discussing continuous functions.

Definition 9.16 A set $G \subset \mathbb{R}^n$ is *pathwise connected* if, whenever $\mathbf{x}, \mathbf{y} \in G$, there is a continuous function $\gamma : [0, 1] \to G$ such that $\gamma(0) = \mathbf{x}$ and $\gamma(1) = \mathbf{y}$.

In other words, $\mathbf{x}$ and $\mathbf{y}$ can be connected by a continuous path that lies entirely within G.

The next theorem is a generalization of the intermediate value theorem for functions of one variable.

Theorem 9.16 (Intermediate value theorem) Let $G \subset \mathbb{R}^n$ be pathwise connected and let $f : G \to \mathbb{R}$ be continuous. Let $\mathbf{x}, \mathbf{y} \in G$ with $f(\mathbf{x}) < f(\mathbf{y})$. If c is a real number with $f(\mathbf{x}) < c < f(\mathbf{y})$, there is at least one point $\mathbf{z} \in G$ with $f(\mathbf{z}) = c$. A similar conclusion holds if $f(\mathbf{x}) > f(\mathbf{y})$.

Proof: Since G is assumed to be pathwise connected, there is a continuous function $\gamma : [0, 1] \to G$ with $\gamma(0) = \mathbf{x}$ and $\gamma(1) = \mathbf{y}$. The composition $g \equiv f \circ \gamma : [0, 1] \to \mathbb{R}$ is continuous and $g(0) < c < g(1)$. By the intermediate value theorem for continuous functions of one variable, there is a value $t_0 \in (0, 1)$ with $g(t_0) = c$. We take $\mathbf{z} = \gamma(t_0)$.

Corollary Let $K \subset \mathbb{R}^n$ be a compact set that is pathwise connected. Let $f : K \to \mathbb{R}$ be continuous. If c is a real number with

$$\min_K f < c < \max_K f,$$

then there is a point $\mathbf{z} \in K$ with $f(\mathbf{z}) = c$.

Exercises for 9.3

1. Let $A \subset \mathbb{R}^n$ be a subset which is *not* compact. Show that there exists a continuous function $f : A \to \mathbb{R}$ that is unbounded. Hint: Consider the two cases, A not closed, and A not bounded, separately. In each case, construct a function f that is continuous on A and unbounded.

2. Let the function

$$f(x, y) = \frac{|x|^p |y|^q}{\sqrt{x^2 + y^2}} \quad \text{for } (x, y) \neq (0, 0),$$

and set $f(0, 0) = 0$. For what combinations of $p \geq 0$ and $q \geq 0$ is f continuous at $(0, 0)$?

3. Let $\mathbf{f} : \mathbb{R}^n \rightarrow \mathbb{R}^m$. If B is a subset of $\mathbb{R}^m$, we let

$$\mathbf{f}^{-1}(B) = \{\mathbf{x} \in \mathbb{R}^n : \mathbf{f}(\mathbf{x}) \in B\}.$$

$\mathbf{f}^{-1}(B)$ is called the *preimage* of B under $\mathbf{f}$.

a) Show that, if $\mathbf{f}$ is continuous, and $U \subset \mathbb{R}^m$ is open, then the preimage $\mathbf{f}^{-1}(U)$ is open.

b) Show that, if $\mathbf{f}$ is continuous and $F \subset \mathbb{R}^m$ is closed, then the preimage $\mathbf{f}^{-1}(F)$ is closed.

c) Let $\mathbf{v} \in \mathbb{R}^n$ and let $H = \{\mathbf{x} \in R^n : \langle \mathbf{x}, \mathbf{v} \rangle \geq 0\}$. Use part b) of this exercise to show that H is closed and H' is open.

4. Let $f(\mathbf{x})$ be a continuous function on $\mathbb{R}^n$ such that $|f(\mathbf{x})| \leq \|\mathbf{x}\|^{-1}$ for $\|\mathbf{x}\| \geq 10$.

a) Show that f is bounded on $\mathbb{R}^n$.

b) Show that either f has a maximum value on $\mathbb{R}^n$, or that $-f$ has a maximum value on $\mathbb{R}^n$.

5. Let $f(\mathbf{x}) : \mathbb{R}^4 \rightarrow \mathbb{R}$ be continuous. Let S and T be closed subsets of $\mathbb{R}^4$ given by

$$S \subset \{\mathbf{x} : x_1^2 + x_2^2 \leq M_1\} \quad \text{and} \quad T \subset \{\mathbf{x} : x_3^2 + x_4^2 \leq M_2\}.$$

a) Can you show that f has a finite maximum and minimum on each of S and T?

b) Show that f attains a maximum and a minimum on $S \cap T$.

Chapter 10

The Derivative in $\mathbb{R}^n$

Chapter Overview

The concept of differentiability is extended to functions of several variables using the notion of linear approximation. The gradient and directional derivatives are defined. We reprove the rules for differentiating sums, products, and quotients. After introducing the operator norms for matrices, we discuss linear approximation for vector-valued functions. Then we prove the chain rule for vector-valued functions. Finally we prove a version of the mean value theorem for vector-valued functions.

10.1 The Derivative and Approximation in $\mathbb{R}^n$

In Chapter 4, we found that a function of one variable $f(x)$ is differentiable at $x = a$ if and only if there is an affine function,

$$l(x) = f(a) + m(x - a), \tag{10.1}$$

such that the remainder

$$R(h) \equiv f(a + h) - l(a + h) = o(|h|). \tag{10.2}$$

In this case $m = f'(a)$.

Recall from Chapter 4 that a function $g(x) = O(|x|^p)$ if $g(x)/|x|^p$ is bounded as x approaches 0 and $g(x) = o(|x|^p)$ if $\lim_{x \to 0} g(x)/|x|^p = 0$. For a vector-valued function $\mathbf{g}(\mathbf{x})$ of $\mathbf{x} \in \mathbb{R}^n$, $\mathbf{g}(\mathbf{x}) = O(\|\mathbf{x}\|^p)$ means that $\|\mathbf{g}(\mathbf{x})\|/\|\mathbf{x}\|^p$ is bounded as $\|\mathbf{x}\|$ tends to zero, and similiarly for $\mathbf{g}(\mathbf{x}) = o(\|\mathbf{x}\|^p)$.

In higher dimensions, we might expect that a function $f(\mathbf{x})$ would have a linear approximation at $\mathbf{x} = \mathbf{a}$ if each of the partial derivatives $\partial f/\partial x_j(\mathbf{a})$ exists. This turns out to be false. The appropriate way to generalize differentiability to functions of several variables is to use the concept of linear approximation. We begin with scalar-valued functions of several variables.

Definition 10.1 Let $f(\mathbf{x})$ be defined on an open set $G \subset \mathbb{R}^n$. f has a *linear approximation* at the point $\mathbf{a} \in G$ if there is a linear function $m : \mathbb{R}^n \to \mathbb{R}$ such that

$$f(\mathbf{a} + \mathbf{h}) = f(\mathbf{a}) + m(\mathbf{h}) + R(\mathbf{h}), \tag{10.3}$$

and the remainder

$$R(\mathbf{h}) = o(\|\mathbf{h}\|).$$

In other words,

$$\lim_{\|\mathbf{h}\| \to 0} \frac{R(\mathbf{h})}{\|\mathbf{h}\|} = \lim_{\|\mathbf{h}\| \to 0} \frac{f(\mathbf{a} + \mathbf{h}) - f(\mathbf{a}) - m(\mathbf{h})}{\|\mathbf{h}\|} = 0.$$

The linear approximation is $l(\mathbf{x}) = f(\mathbf{a}) + m(\mathbf{x} - \mathbf{a})$.   .

The definition does not depend on the choice of norm. The fact that G is an open set implies that there is $\delta > 0$ such that $\mathbf{a} + \mathbf{h} \in G$ for all $\|\mathbf{h}\| < \delta$. Thus $f(\mathbf{a} + \mathbf{h})$ is defined for all of these $\mathbf{h}$.

Remark For functions of several variables, the term *differentiable* is commonly used in place of the longer phrase "has a linear approximation." However, differentiable in this context means more than just the existence of the partial derivatives. For this reason, we shall use the longer phrase "has a linear approximation."

The linear function m can be thought of as the abstraction of the notion of the derivative of a function of one variable. We shall put it in a more familiar form. The linear function m is uniquely defined by its action on the standard basis vectors $\mathbf{e}_j = (0, \ldots, 0, 1, 0, \ldots, 0)$. Take $\mathbf{h} = h\mathbf{e}_j$ where now h is a scalar.

We assume $\|\mathbf{h}\| = |h| < \delta$. Because $m(\mathbf{h})$ is linear, $m(h\mathbf{e}_j) = hm(\mathbf{e}_j)$. When we substitute $\mathbf{h} = h\mathbf{e}_j$ in (10.3), we have

$$f(\mathbf{a} + h\mathbf{e}_j) - f(\mathbf{a}) = hm(\mathbf{e}_j) + R(h\mathbf{e}_j)$$

or

$$\frac{f(\mathbf{a} + h\mathbf{e}_j) - f(\mathbf{a})}{h} = m(\mathbf{e}_j) + \frac{R(h\mathbf{e}_j)}{h}.$$

Since $R(h\mathbf{e}_j) = o(\|h\mathbf{e}_j)\|) = o(|h|)$, we may take the limit on the left and deduce that

$$\frac{\partial f}{\partial x_j}(\mathbf{a}) = \lim_{h \to 0} \frac{f(\mathbf{a} + h\mathbf{e}_j) - f(\mathbf{a})}{h} = m(\mathbf{e}_j).$$

Hence the elements of the $1 \times n$ matrix for m are

$$m(\mathbf{e}_j) = \frac{\partial f}{\partial x_j}(\mathbf{a}).$$

We have proved

Theorem 10.1 Let $G \subset \mathbb{R}^n$ be an open set with $\mathbf{a} \in G$. Suppose that $f : G \to \mathbb{R}$ has a linear approximation at $\mathbf{x} = \mathbf{a}$. Then the partial derivatives $\partial f / \partial x_j(\mathbf{a})$ exist. The linear function $m(\mathbf{x})$ is unique and is given by $m(\mathbf{x}) = \langle \nabla f(\mathbf{a}), \mathbf{x} \rangle$ where

$$\nabla f(\mathbf{a}) = \left(\frac{\partial f}{\partial x_1}(\mathbf{a}), \ldots, \frac{\partial f}{\partial x_n}(\mathbf{a}) \right)$$

is the gradient vector.

Thus (10.3) becomes

$$f(\mathbf{a} + \mathbf{h}) = f(\mathbf{a}) + \langle \nabla f(\mathbf{a}), \mathbf{h} \rangle + R(\mathbf{h}) \tag{10.4}$$

where $R(\mathbf{h}) = o(\|\mathbf{h}\|)$. We also write (10.4) as

$$f(\mathbf{x}) = l(\mathbf{x}) + R(\mathbf{x}, \mathbf{a})$$

where

$$l(\mathbf{x}) = f(\mathbf{a}) + \langle \nabla f(\mathbf{a}), \mathbf{x} - \mathbf{a} \rangle$$

is the linear approximation and the remainder

$$R(\mathbf{x}, \mathbf{a}) = o(\|\mathbf{x} - \mathbf{a}\|).$$

The graph of $\mathbf{x} \to l(\mathbf{x})$ is the usual tangent plane approximation to the graph of $f(\mathbf{x})$ at the point $(\mathbf{a}, f(\mathbf{a}))$.

To show that a function f has a linear approximation at $\mathbf{x} = \mathbf{a}$, directly from the definition, we must verify that

1) The partial derivatives of f exist at $\mathbf{a}$,

$$AND$$

2) The remainder $R(\mathbf{h}) = o(\|\mathbf{h}\|)$. This means that we must show

$$\lim_{\|\mathbf{h}\| \to 0} \frac{f(\mathbf{a} + \mathbf{h}) - f(\mathbf{a}) - \langle \nabla f(\mathbf{a}), \mathbf{h} \rangle}{\|\mathbf{h}\|} = 0.$$

EXAMPLE 10.1:   Let $f(x, y) = xe^y$. The function f is continuous everywhere and we will verify that it has a linear approximation at $(0, 0)$. We see that $f(0, 0) = 0$ and that the partial derivatives are

$$f_x(0, 0) = 1 \quad \text{and} \quad f_y(0, 0) = 0.$$

Thus the linear approximation, if it exists, must be

$$l(x, y) = x.$$

We let $r = \|(x, y)\|_2 = \sqrt{x^2 + y^2}$. We see that

$$R(x, y) = f(x, y) - l(x, y) = x(e^y - 1).$$

Using the mean value theorem in one variable, $e^y - 1 = ye^\eta$ where η lies between y and 0. Hence

$$\begin{aligned}
|R(x, y)| &= |x||y|e^\eta \\
&\leq (1/2)(x^2 + y^2)e^{|y|} \\
&\leq (1/2)r^2 e^r.
\end{aligned}$$

We conclude that

$$\lim_{r\to 0}\frac{|R(x,y)|}{r} \le \lim_{r\to 0}(1/2)re^r = 0.$$

This means that f has the linear approximation at (0,0), $l(x,y) = x$.

EXAMPLE 10.2: In this example, the remainder does not tend to zero as rapidly as it does in Example 10.1. Let $f(x,y) = (x^2 + y^2)^{1/6}\sin(x + y^2)$. f is clearly continuous everywhere. We want to determine if f has a linear approximation at $(x,y) = (0,0)$. First we check to see if the partial derivatives $f_x(0,0)$ and $f_y(0,0)$ exist. In fact, because $f(0,0) = 0$,

$$f_x(0,0) = \lim_{x\to 0}\frac{f(x,0)}{x} = \lim_{x\to 0}\frac{x^{1/3}\sin x}{x} = 0$$

because $\lim_{x\to 0}\sin x/x = 1$. Similarly,

$$f_y(0,0) = \lim_{y\to 0}\frac{f(0,y)}{y} = \lim_{y\to 0}\frac{y^{1/3}\sin(y^2)}{y} = 0.$$

This means that if f has a linear approximation at $(0,0)$, it must be $l(x,y) \equiv 0$ and the remainder $R(x,y) = f(x,y)$. We see that

$$\begin{aligned}
|R(x,y)| &= |f(x,y)| = r^{1/3}|\sin(x + y^2)|\\
&\le r^{1/3}|x + y^2|\\
&\le r^{1/3}(r + r^2).
\end{aligned}$$

Hence

$$\lim_{r\to 0}\frac{|R(x,y)|}{r} \le \lim_{r\to 0}(r^{1/3} + r^{4/3}) = 0.$$

We conclude that f does have a linear approximation at $(0,0)$. Later we will see that Theorem 10.4 makes it easier to verify when f has a linear approximation.

Definition 10.2 Let $v \in \mathbb{R}^n$ be a vector. The *directional derivative* of f in the direction v at $x = a$ is defined to be the following limit when it exists:

$$D_v f(\mathbf{a}) = \lim_{h \to 0} \frac{f(\mathbf{a} + h\mathbf{v}) - f(\mathbf{a})}{h}$$

when $v = e_j$, $D_v f = \partial f / \partial x_j$.

Theorem 10.2 Let $G \subset \mathbb{R}^n$ be an open set. Assume that $f : G \to \mathbb{R}$ has a linear approximation at $x = a \in G$. Let $v \in \mathbb{R}^n$. Then $D_v f(\mathbf{a})$ exists and

$$D_v f(\mathbf{a}) = \langle \nabla f(\mathbf{a}), \mathbf{v} \rangle. \qquad (10.5)$$

Proof: Substitute $h = h\mathbf{v}$ into (10.4), and divide by h to obtain

$$\frac{f(\mathbf{a} + h\mathbf{v}) - f(\mathbf{a})}{h} = \langle \nabla f(\mathbf{a}), \ \mathbf{v} \rangle + \frac{R(h\mathbf{v})}{h}.$$

We may take the limit to deduce (10.5).

When v is a unit vector, we use the Schwarz inequality to deduce

$$-\|\nabla f(\mathbf{a})\|_2 \le D_v f(\mathbf{a}) \le \|\nabla f(\mathbf{a})\|_2.$$

Equality holds on the right when $v = \nabla f(\mathbf{a}) / \|\nabla f(\mathbf{a})\|_2$; this is the direction of maximum increase of f. Equality holds on the left when $v = -\nabla f(\mathbf{a}) / \|\nabla f(\mathbf{a})\|_2$; this direction is the direction of steepest descent.

Theorem 10.3 If f has a linear approximation at $x = a$, then f is continuous at $x = a$.

Proof: We use the norm $\| \ \|_2$ and the Schwarz inequality in (10.5) to deduce

$$|f(\mathbf{a} + \mathbf{h}) - f(\mathbf{a})| \le |\langle \nabla f(\mathbf{a}), \mathbf{h} \rangle| + |R(\mathbf{h})|$$
$$\le \|\nabla f(\mathbf{a})\|_2 \|\mathbf{h}\|_2 + |R(\mathbf{h})|.$$

Both terms on the right tend to zero as h tends to 0.

We see that the existence of a linear approximation has many consequences. All of the partial derivatives of f exist and all of the directional derivatives exist at $x = a$. However, it is very important to keep in mind that all of the partial derivatives of f may exist at a point, and yet f may not have a linear approximation at that point, as shown by the following example.

EXAMPLE 10.3:   Let

$$
f(x, y) = \begin{cases} \frac{x^2 y}{x^2+y^2} & \text{for } (x, y) \neq (0,0) \\ 0 & \text{for } (x, y) = (0,0). \end{cases}
$$

The function f is continuous at $(0,0)$. Since $f(x,0) = f(0,y) \equiv 0$, it is clear that $f_x(0,0) = f_y(0,0) = 0$. The only possible linear approximation to f at $(0,0)$ is $l(x,y) \equiv 0$. This means that the remainder $R(x,y) = f(x,y)$. Now let $\mathbf{h} = (h, h)$. We see that $R(\mathbf{h}) = f(h,h) = h/2$ so that $R(\mathbf{h})/h = 1/2$ does not tend to zero as h tends to zero. Thus f does *not* possess a linear approximation at $(0,0)$.

We need a simple criterion for f to have a linear approximation at a point **a**.

Theorem 10.4 Let f be defined on an open set $G \subset \mathbb{R}^n$. Suppose that the partial derivatives $\partial f/\partial x_j$, $j = 1, \ldots, n$ exist at each point of G and that for each j, $\mathbf{x} \to \partial f/\partial x_j(\mathbf{x})$ is *continuous* on G. Then f has a linear approximation at each point $\mathbf{a} \in G$.

Proof: We give the proof in two dimensions to avoid the use of many subscripts. Since we are assuming the existence of the partial derivatives of f, we shall trace out a path from **a** to $\mathbf{a} + \mathbf{h}$ that consists of straight line segments that are parallel to the coordinate axes (see Figure 10.1). We can write

$$
f(\mathbf{a} + \mathbf{h}) - f(\mathbf{a}) = f(\mathbf{a} + \mathbf{h}) - f(\mathbf{a}') + f(\mathbf{a}') - f(\mathbf{a})
$$

where we have set $\mathbf{a}' = (a_1, a_2 + h_2)$. Now $t \to f(a_1 + t, a_2 + h_2)$ is differentiable by assumption. Hence we apply the mean value theorem in one dimension to write

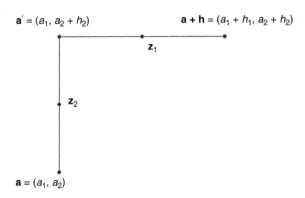

Figure 10.1 Dogleg path from a to a + h

$$f(a_1 + h_1, a_2 + h_2) - f(a_1, a_2 + h_2) = \frac{\partial f}{\partial x}(a_1 + \theta_1, a_2 + h_2)h_1$$

for some θ_1 between h_1 and 0. We can write this more concisely as

$$f(\mathbf{a} + \mathbf{h}) - f(\mathbf{a}') = \frac{\partial f}{\partial x}(\mathbf{z}_1)h_1$$

where $\mathbf{z}_1 = (a_1 + \theta_1, a_2 + h_2)$. In addition $t \rightarrow f(a_1, a_2 + t)$ is differentiable so that

$$f(\mathbf{a}') - f(\mathbf{a}) = \frac{\partial f}{\partial y}(\mathbf{z}_2)h_2$$

where $\mathbf{z}_2 = (a_1, a_2 + \theta_2)$ and where θ_2 lies between h_2 and 0. Thus the remainder can be expressed

$$
\begin{aligned}
R(\mathbf{h}) &= f(\mathbf{a} + \mathbf{h}) - f(\mathbf{a}) - \langle \nabla f(\mathbf{a}), \mathbf{h} \rangle \\
&= f(\mathbf{a} + \mathbf{h}) - f(\mathbf{a}') + f(\mathbf{a}') - f(\mathbf{a}) - \langle \nabla f(\mathbf{a}), \mathbf{h} \rangle \\
&= \left[\frac{\partial f}{\partial x}(\mathbf{z}_1) - \frac{\partial f}{\partial x}(\mathbf{a}) \right] h_1 + \left[\frac{\partial f}{\partial y}(\mathbf{z}_2) - \frac{\partial f}{\partial y}(\mathbf{a}) \right] h_2.
\end{aligned}
$$

Hence

$$|R(\mathbf{h})| \le \max \left\{ \left| \frac{\partial f}{\partial x}(\mathbf{z}_1) - \frac{\partial f}{\partial x}(\mathbf{a}) \right|, \left| \frac{\partial f}{\partial y}(\mathbf{z}_2) - \frac{\partial f}{\partial y}(\mathbf{a}) \right| \right\} \|\mathbf{h}\|_1.$$

As $\mathbf{h}$ tends to 0, both $\mathbf{z}_1$ and $\mathbf{z}_2$ approach $\mathbf{a}$. Since $\partial f/\partial x$ and $\partial f/\partial y$ are assumed to be continuous on G, the quantity in $\{\ \}$ brackets tends to zero as $\|\mathbf{h}\|_1$ tends to zero. Thus $R(\mathbf{h}) = o(\|\mathbf{h}\|_1)$. This completes the proof of Theorem 10.4.

For convenience in stating results, we make the following definition that extends Definition 5.2 to include functions of several variables.

Definition 10.3 Let f be defined on an open set $G \subset \mathbb{R}^n$. We say that $f \in C^1(G)$ if all of the partial derivatives exist and are continuous on G. In Chapter 12 we will define $C^k(G)$ for $k > 1$.

Theorem 10.5 Let $f, g \in C^1(G)$ for some open set G that contains $\mathbf{a}$. Let $\alpha, \beta \in \mathbb{R}$ and let $\mathbf{a}$ be a point in G. Then

(i) $\alpha f + \beta g \in C^1(G)$ and has the linear approximation at $\mathbf{a}$:

$$l(\mathbf{x}) = \alpha f(\mathbf{a}) + \beta g(\mathbf{a}) + \langle [\alpha \nabla f(\mathbf{a}) + \beta \nabla g(\mathbf{a})], \mathbf{x} - \mathbf{a} \rangle.$$

(ii) $fg \in C^1(G)$ and has the linear approximation at $\mathbf{a}$:

$$l(\mathbf{x}) = f(\mathbf{a})g(\mathbf{a}) + g(\mathbf{a})\langle \nabla f(\mathbf{a}), \mathbf{x} - \mathbf{a} \rangle + f(\mathbf{a})\langle \nabla g(\mathbf{a}), \mathbf{x} - \mathbf{a} \rangle.$$

(iii) If $g(\mathbf{x}) \ne 0$ for $\mathbf{x} \in G$, then $f/g \in C^1(G)$ and has the linear approximation at $\mathbf{a}$:

$$l(\mathbf{x}) = \left(\frac{1}{g^2(\mathbf{a})} \right) [g(\mathbf{a})\langle \nabla f(\mathbf{a}), \mathbf{x} - \mathbf{a} \rangle - f(\mathbf{a})\langle \nabla g(\mathbf{a}), \mathbf{x} - \mathbf{a} \rangle].$$

Proof: Using the usual rules for differentiation in one variable we have

$$\nabla(\alpha f + \beta g) = \alpha \nabla f + \beta \nabla g,$$

$$\nabla(fg) = g\nabla f + f\nabla g,$$

and

$$\nabla(f/g) = \frac{g\nabla f - f\nabla g}{g^2}.$$

It can be seen from these formulas that if $f, g \in C^1(G)$, then so are $\alpha f + \beta g$, fg and f/g. The theorem is proved.

It is interesting to see that the formulas for the linear approximations of the sums, products, and quotients of functions arise directly from algebra. For example consider (ii) of Theorem 10.5. The linear approximations for f and g at $\mathbf{x} = \mathbf{a}$ are

$$f(\mathbf{x}) = f(\mathbf{a}) + \langle \nabla f(\mathbf{a}), \mathbf{x} - \mathbf{a} \rangle + R_1(\mathbf{x}, \mathbf{a})$$

and

$$g(\mathbf{x}) = g(\mathbf{a}) + \langle \nabla g(\mathbf{a}), \mathbf{x} - \mathbf{a} \rangle + R_2(\mathbf{x}, \mathbf{a})$$

where both R_1 and R_2 are $o(\|\mathbf{x} - \mathbf{a}\|)$. Hence

$$f(\mathbf{x})g(\mathbf{x}) = f(\mathbf{a})g(\mathbf{a}) + g(\mathbf{a})\langle \nabla f(\mathbf{a}), \mathbf{x} - \mathbf{a} \rangle + f(\mathbf{a})\langle \nabla g(\mathbf{a}), \mathbf{x} - \mathbf{a} \rangle + R(\mathbf{x}, \mathbf{a})$$

where

$$R(\mathbf{x}, \mathbf{a}) = g(\mathbf{a})R_1(\mathbf{x}, \mathbf{a}) + f(\mathbf{a})R_2(\mathbf{x}, \mathbf{a}) + R_1(\mathbf{x}, \mathbf{a})R_2(\mathbf{x}, \mathbf{a}).$$

It is clear that $R(\mathbf{x}, \mathbf{a}) = o(\|\mathbf{x} - \mathbf{a}\|)$.

The next theorem anticipates the chain rule, which is proved in the next section. Let f be defined on an open set G. Let $\mathbf{a} \in G$ and $\mathbf{v} \in \mathbb{R}^n$. Because G is open, there is $\delta > 0$ such that $\mathbf{a} + t\mathbf{v} \in G$ for $|t| < \delta$. Let $g(t) = f(\mathbf{a} + t\mathbf{v})$; g is the restriction of f to the line $\{\mathbf{a} + t\mathbf{v}, -\delta < t < \delta\}$.

Theorem 10.6 Let $f(\mathbf{x}) \in C^1(G)$ where $G \subset \mathbb{R}^n$ is an open set. Let $\mathbf{a} \in G$ and let $\mathbf{v} \in \mathbb{R}^n$. Then for $|t| < \delta$, $t \to g(t)$ is differentiable with

$$\frac{dg}{dt}(t) = \langle \nabla f(\mathbf{a} + t\mathbf{v}), \mathbf{v} \rangle. \tag{10.6}$$

Proof: We use (10.4) with $\mathbf{a}$ replaced by $\mathbf{a} + t\mathbf{v}$ and $\mathbf{h} = h\mathbf{v}$. Then we have

$$g(t + h) - g(t) = f(\mathbf{a} + (t + h)\mathbf{v}) - f(\mathbf{a} + t\mathbf{v})$$
$$= h\langle \nabla f(\mathbf{a} + t\mathbf{v}), \mathbf{v} \rangle + R(h\mathbf{v}).$$

Now $\lim_{h\to 0} R(h\mathbf{v})/h = 0$, so dividing by h and taking the limit, we find

$$\frac{dg}{dt}(t) = \lim_{h\to 0} \frac{g(t+h) - g(t)}{h} = \langle \nabla f(\mathbf{a} + t\mathbf{v}), \mathbf{v}\rangle.$$

We note that when $t = 0$, we get back the formula (10.5) for the directional derivative. This means that we can think of $D_{\mathbf{v}}f(\mathbf{a})$ as the derivative at $t = 0$ of the restriction of f to the line $\{\mathbf{a} + t\mathbf{v}, |t| < \delta\}$.

Theorem 10.7 (Mean value theorem) Let $f \in C^1(G)$ where $G \subset \mathbb{R}^n$ is open. Suppose that $\mathbf{x}, \mathbf{y}$ and the line $\mathcal{L}$ joining them are all contained in G. Then there is a point $\mathbf{z} \in \mathcal{L}$ such that

$$f(\mathbf{x}) - f(\mathbf{y}) = \langle \nabla f(\mathbf{z}), \mathbf{x} - \mathbf{y}\rangle.$$

Proof: The line $\mathcal{L}$ is parameterized by $t\mathbf{x} + (1-t)\mathbf{y} = \mathbf{y} + t(\mathbf{x} - \mathbf{y})$, $0 \le t \le 1$. Let $g(t) = f(t\mathbf{x} + (1-t)\mathbf{y})$. The function g is the restriction of f to the line $\mathcal{L}$ with $g(1) = f(\mathbf{x})$ and $g(0) = f(\mathbf{y})$. By Theorem 10.6 (with $\mathbf{v} = \mathbf{x} - \mathbf{y}$) g is differentiable on $0 < t < 1$ and continuous on the closed interval $0 \le t \le 1$. Hence we can apply the mean value theorem in one dimension to g, and deduce that there is a $0 < \theta < 1$, such that $g(1) - g(0) = g'(\theta)$. Let $\mathbf{z} = \theta\mathbf{x} + (1-\theta)\mathbf{y}$. Then Theorem 10.6 yields

$$\begin{aligned}
f(\mathbf{x}) - f(\mathbf{y}) &= g(1) - g(0) = g'(\theta) \\
&= \langle \nabla f(\theta\mathbf{x} + (1-\theta)\mathbf{y}), \mathbf{x} - \mathbf{y}\rangle \\
&= \langle f(\mathbf{z}), \mathbf{x} - \mathbf{y}\rangle.
\end{aligned}$$

Exercises for 10.1

1. Let $f(x, y) = |x|y$.
   a) For what values of (x, y) does f_y exist?
   b) For what values of (x, y) does f_x exist?
   c) Does f have a linear approximation at $(0, 0)$? If so, what is it?

2. Let $f(x, y) = (x^2y + xy)/\sqrt{x^2 + y^2}$ for $(x, y) \ne (0, 0)$ and $f(0, 0) = 0$. Calculate $f_x(0, 0)$ and $f_y(0, 0)$ from the definitions. Does f have a linear approximation at $(0, 0)$?

3. For each of the following functions, calculate the linear approximation.
   a) $f(x, y) = 2y^2/(x^2 + 3xy)$ at $(1, -1)$.

b) $g(x, y) = x \log(x/y)$ at $(1, 2)$.

c) $h(x, y, z) = \exp(x + y + z^2)$ at $(-1, 2, 3/2)$.

d) Let $\mathbf{v} \in \mathbb{R}^n$ be given. For $\mathbf{x} \in \mathbb{R}^n$, we set $r(\mathbf{x}) = \|\mathbf{x}\|_2^2 + 2\langle \mathbf{x}, \mathbf{v} \rangle - 5$. Calculate the linear approximation at $\mathbf{x} = 0$.

4. For each of the functions in exercise 3, calculate the directional derivative in the direction $\mathbf{u}$ at the point indicated.

a) $\mathbf{u} = (2, 3)/\sqrt{13}$.

b) $\mathbf{u} = (1/\sqrt{2}, -1/\sqrt{2})$.

c) $\mathbf{u} = (2, 1, 2)/3$.

d) $\mathbf{u} = -\mathbf{v}/\|\mathbf{v}\|_2$.

5. Let $f(x, y) = x^2 + y^2/4 + y^3$. Recall that the direction of steepest descent at a point (a, b) is that direction $\mathbf{u} = (u_1, u_2)$ that minimizes $D_\mathbf{u} f(a, b)$. What is the direction of steepest descent for the function f at the points $(a, b) = (1, 2)$ and $(a, b) = (2, 1)$?

6. Suppose that $f(x, y)$ is C^1 on $\mathbb{R}^2$ and that $f(x, y) \geq f(0, 0)$ in the sector $x/2 \leq y \leq 2x$, $x \geq 0$. Show that $D_\mathbf{v} f(0, 0) \geq 0$ for $\mathbf{v} = (1, v_2)$, $1/2 \leq v_2 \leq 2$. Since you do not have enough information to calculate $f_x(0, 0)$ or $f_y(0, 0)$, you must use the definition of $D_\mathbf{v} f$.

7. Let $f : G \to \mathbb{R}$ where G is an open subset of $\mathbb{R}^n$. Assume that $f \in C^1(G)$. Show that for each convex, compact set $K \subset G$, there is a constant C_K such that

$$|f(\mathbf{x}) - f(\mathbf{y})| \leq C_K \|\mathbf{x} - \mathbf{y}\|_\infty \quad \text{for all } \mathbf{x}, \mathbf{y} \in K.$$

10.2 Linear Transformations and Matrix Norms

From linear algebra we know that a *linear transformation* from $\mathbb{R}^n$ to $\mathbb{R}^m$ is a mapping $\mathbf{M}(\mathbf{x})$ such that

$$\mathbf{M}(\alpha \mathbf{x} + \beta \mathbf{y}) = \alpha \mathbf{M}(\mathbf{x}) + \beta \mathbf{M}(\mathbf{y})$$

for all $\mathbf{x}, \mathbf{y} \in \mathbb{R}^n$ and all scalars $\alpha, \beta \in \mathbb{R}$.

We let $\mathbf{e}_j$, $j = 1, \ldots, n$ be the standard basis for $\mathbb{R}^n$ and we let $\mathbf{e}_i$, $i = 1, \ldots, m$ be the standard basis for $\mathbb{R}^m$. The matrix for $\mathbf{M}$ in the standard bases is the array of numbers

$$a_{ij} = \langle \mathbf{M}\mathbf{e}_j, \mathbf{e}_i \rangle.$$

Because we are going to approximate vector-valued mappings by linear transformations, we must be able to measure the size of a matrix, much as we needed to be able to measure the size of a vector. Each of the vector norms on $\mathbb{R}^n$ induces a norm on matrices as follows. Let $\| \ \|$ be any one of the three equivalent norms. We are going to use the same norm on $\mathbb{R}^n$ and on $\mathbb{R}^m$.

Definition 10.4 Let $\mathbf{A}$ be an $m \times n$ matrix. We define

$$\|\mathbf{A}\| = \sup_{\mathbf{x} \neq 0} \frac{\|\mathbf{A}\mathbf{x}\|}{\|\mathbf{x}\|}. \tag{10.7}$$

Since $\|\mathbf{A}\mathbf{x}\|/\|\mathbf{x}\| = \|\mathbf{A}\mathbf{z}\|$ where $\mathbf{z} = \mathbf{x}/\|\mathbf{x}\|$, and $\|\mathbf{z}\| = 1$, we see that an alternate equivalent definition is

$$\|\mathbf{A}\| = \max_{\|\mathbf{z}\|=1} \|\mathbf{A}\mathbf{z}\|.$$

Thus the norm of $\mathbf{A}$ induced by a vector norm is a measure of the size of the image under the mapping $\mathbf{x} \to \mathbf{A}\mathbf{x}$ of the unit ball. We will write $\|\mathbf{A}\|_2$ for the matrix norm induced by the norm $\| \ \|_2$, $\|\mathbf{A}\|_1$ for the norm induced by $\| \ \|_1$, and $\|\mathbf{A}\|_\infty$ for the norm induced by $\| \ \|_\infty$. The matrix norm induced by a vector norm is said to be *subordinate* to that vector norm. These matrix norms are also called "operator norms."

A fourth norm on matrices, which is not induced by a vector norm, is the *Frobenius* norm

$$\|\mathbf{A}\|_F = \sqrt{\sum_{i,j} a_{ij}^2}.$$

The Frobenius norm is simply the Euclidean vector norm of the matrix considered as a vector in $\mathbb{R}^{mn}$.

$\|\mathbf{A}\|_1$ and $\|\mathbf{A}\|_\infty$ are easy to calculate.

Theorem 10.8 The matrix norms

$$\|\mathbf{A}\|_1 = \max_j \sum_i |a_{ij}| \quad \text{and} \tag{10.8}$$

$$\|\mathbf{A}\|_\infty = \max_i \sum_j |a_{ij}| \tag{10.9}$$

We say that $\|\mathbf{A}\|_1$ is the maximum column sum and that $\|\mathbf{A}\|_\infty$ is the maximum row sum.

Proof: We prove the formula for $\|\mathbf{A}\|_1$. For any $\mathbf{x} = (x_1, \ldots, x_n) \in \mathbb{R}^n$, let $\mathbf{y} = \mathbf{A}\mathbf{x}$. Then

$$|y_i| = \left| \sum_j a_{ij} x_j \right| \leq \sum_j |a_{ij}||x_j|$$

which yields

$$\|\mathbf{y}\|_1 = \sum_i |y_i| \leq \sum_{i\,j} |a_{ij}||x_j|$$

$$\leq \sum_j \left(\sum_i |a_{ij}| \right) |x_j|$$

$$\leq \max_j \sum_i |a_{ij}| \sum_j |x_j|$$

$$\leq \max_j \sum_i |a_{ij}| \|\mathbf{x}\|_1.$$

Hence we have

$$\|\mathbf{A}\|_1 = \sup_{\mathbf{x} \neq 0} \frac{\|\mathbf{A}\mathbf{x}\|_1}{\|\mathbf{x}\|_1} \leq \max_j \sum_i |a_{ij}|.$$

To show equality, we must find a vector $\mathbf{x}$ for which $\|\mathbf{A}\mathbf{x}\|_1 = \max_j \sum_i |a_{ij}| \|\mathbf{x}\|_1$. Let k be a column index (there may be more than one) such that

$$\sum_i |a_{ik}| = \max_j \sum_i |a_{ij}|$$

and let $\mathbf{x} = \mathbf{e}_k$. Then $\mathbf{A}\mathbf{e}_k$ is the kth column of $\mathbf{A}$ so that

$$\|\mathbf{A}\mathbf{e}_k\|_1 = \sum_i |a_{ik}| = \max_j \sum_i |a_{ij}| \|\mathbf{e}_k\|_1.$$

The proof of (10.9) is similar and is left as an exercise.

EXAMPLE 10.4: Let

$$A = \begin{bmatrix} 1 & -1 & 2 \\ 2 & 0 & 1 \end{bmatrix}.$$

Then $\|A\|_1 = 3$, $\|A\|_\infty = 4$ and $\|A\|_F = \sqrt{11}$. The norm $\|A\|_2$ is not calculated so easily.

The three operator norms have the property that for all x,

$$\|Ax\| \le \|A\|\|x\|. \tag{10.10}$$

This inequality is an immediate consequence of the definition. In fact, for any $x \ne 0$,

$$\|Ax\| = \frac{\|Ax\|}{\|x\|}\|x\| \le \left(\sup_{x \ne 0} \frac{\|Ax\|}{\|x\|} \right)\|x\| = \|A\|\|x\|.$$

For the Frobenius norm, the Schwarz inequality implies

$$\|Ax\|_2 \le \|A\|_F\|x\|_2.$$

As a consequence of this last inequality, we have

$$\|A\|_2 \le \|A\|_F. \tag{10.11}$$

Theorem 10.9 Let A and B be two matrices of appropriate dimensions so that AB makes sense. Then for all four of the matrix norms,

$$\|AB\| \le \|A\| \|B\|.$$

Proof: We use (10.10) or (10.11) twice. For any $x \in \mathbb{R}^n$,

$$\|ABx\| \le \|A\| \|Bx\| \le \|A\| \|B\| \|x\|.$$

Then

$$\|\mathbf{AB}\| = \sup_{\mathbf{x} \neq 0} \frac{\|\mathbf{ABx}\|}{\|\mathbf{x}\|} \leq \|\mathbf{A}\| \, \|\mathbf{B}\|.$$

Finally, we note that for the operator norms, $\|\mathbf{I}\| = 1$, whereas for the Frobenius norm, $\|\mathbf{I}\|_F = \sqrt{n}$.

Exercises for 10.2

1. Prove the characterization (10.9) of $\|\mathbf{A}\|_\infty$.
2. Show that $\|\mathbf{A}\|_1 = \|\mathbf{A}^T\|_\infty$ and that $\|\mathbf{A}\|_\infty = \|\mathbf{A}^T\|_1$.
3. Let $\mathbf{A}$ be an $m \times n$ matrix.

   a) Let $S = \{\mathbf{x} \in \mathbb{R}^n, \mathbf{x} \neq 0, \mathbf{y} \in \mathbb{R}^m, \mathbf{y} \neq 0\}$. Show that

   $$\|\mathbf{A}\|_2 = \sup_{S} \frac{\langle \mathbf{Ax}, \mathbf{y} \rangle}{\|\mathbf{x}\|_2 \, \|\mathbf{y}\|_2}.$$

   b) Conclude that $\|\mathbf{A}\|_2 = \|\mathbf{A}^T\|_2$.

4. Let the matrix

$$\mathbf{A} = \begin{bmatrix} 1 & 1 \\ 2 & 2 \end{bmatrix}.$$

Show that $\|\mathbf{A}\|_1 < \|\mathbf{A}\|_\infty$ and that if $\mathbf{B} = \mathbf{A}^T$, then $\|\mathbf{B}\|_\infty < \|\mathbf{B}\|_1$. Conclude that there is no set of inequalities like (9.7) that holds for matrix norms.

10.3 Vector-Valued Mappings

Now we are ready to extend the results of Section 10.1 to mappings from $\mathbb{R}^n$ to $\mathbb{R}^m$.

Let $\mathbf{f} : G \to \mathbb{R}^m$ where $G \subset \mathbb{R}^n$. We say that $\mathbf{f} \in C^1(G)$ if each component function $f_i \in C^1(G)$. A linear approximation to $\mathbf{f}$ at the point $\mathbf{a} \in G$ takes the form

$$\mathbf{l}(\mathbf{x}) = \mathbf{f}(\mathbf{a}) + \mathbf{M}(\mathbf{a})(\mathbf{x} - \mathbf{a})$$

where $\mathbf{M}(\mathbf{a})$ is a linear transformation from $\mathbb{R}^n$ to $\mathbb{R}^m$. We can construct $\mathbf{M}$ using the linear approximations to each component of $\mathbf{f}$. For each $i = 1, \ldots m$,

$$f_i(\mathbf{x}) = f_i(\mathbf{a}) + \langle \nabla f_i(\mathbf{a}), \mathbf{x} - \mathbf{a} \rangle + R_i(\mathbf{x} - \mathbf{a})$$

where $R_i(\mathbf{x} - \mathbf{a}) = o(\|\mathbf{x} - \mathbf{a}\|)$. If we arrange $\mathbf{f}(\mathbf{x})$ as a column vector we have

$$\mathbf{f}(\mathbf{x}) = \begin{bmatrix} f_1(\mathbf{x}) \\ f_2(\mathbf{x}) \\ \cdot \\ \cdot \\ f_m(\mathbf{x}) \end{bmatrix} = \begin{bmatrix} f_1(\mathbf{a}) \\ f_2(\mathbf{a}) \\ \cdot \\ \cdot \\ f_m(\mathbf{a}) \end{bmatrix} + \begin{bmatrix} \langle \nabla f_1(\mathbf{a}), \mathbf{x} - \mathbf{a} \rangle \\ \langle \nabla f_2(\mathbf{a}), \mathbf{x} - \mathbf{a} \rangle \\ \cdot \\ \cdot \\ \langle \nabla f_m(\mathbf{a}), \mathbf{x} - \mathbf{a} \rangle \end{bmatrix} + \begin{bmatrix} R_1(\mathbf{x} - \mathbf{a}) \\ R_2(\mathbf{x} - \mathbf{a}) \\ \cdot \\ \cdot \\ R_m(\mathbf{x} - \mathbf{a}) \end{bmatrix}.$$

We let $\mathbf{Df}$ denote the $m \times n$ matrix of partial derivatives

$$\mathbf{Df}(\mathbf{a}) = \left[\frac{\partial f_i}{\partial x_j}(\mathbf{a}) \right].$$

The matrix $\mathbf{Df}$ is the higher-dimensional analogue of the derivative of a real-valued function of one variable. For this reason $\mathbf{Df}$ is often referred to as the *derivative* of $\mathbf{f}$. When $f : \mathbb{R} \to \mathbb{R}$, $x \to f'(x)$ is a real-valued function. When $f : \mathbb{R}^n \to \mathbb{R}$, $\mathbf{x} \to Df(\mathbf{x}) = \nabla f(\mathbf{x})$ is a vector-valued function, and when $\mathbf{f} : \mathbb{R}^n \to \mathbb{R}^m$, $\mathbf{x} \to \mathbf{Df}(\mathbf{x})$ is a matrix-valued function.

The matrix $\mathbf{Df}$ is also called the *Jacobian* matrix of $\mathbf{f}$ and we will use this term. The column vector display of the linear approximations to each f_i shows that we can take the linear transformation $\mathbf{M}(\mathbf{a})$ to be multiplication by the Jacobian matrix $\mathbf{Df}(\mathbf{a})$. Thus the linear approximation to $\mathbf{f}$ at $\mathbf{a}$ can be written more compactly in vector notation as

$$\mathbf{l}(\mathbf{x}) = \mathbf{f}(\mathbf{a}) + \mathbf{Df}(\mathbf{a})(\mathbf{x} - \mathbf{a}). \tag{10.12}$$

The remainder is the vector

$$\mathbf{R}(\mathbf{h}) = \begin{bmatrix} R_1(\mathbf{h}) \\ R_2(\mathbf{h}) \\ \cdot \\ \cdot \\ \cdot \\ R_m(\mathbf{h}) \end{bmatrix}.$$

Now for each i, $1 \le i \le m$,

$$\lim_{\mathbf{h} \to 0} \frac{|R_i(\mathbf{h})|}{\|\mathbf{h}\|} = 0.$$

Hence

$$\lim_{\mathbf{h} \to 0} \frac{\mathbf{R}(\mathbf{h})}{\|\mathbf{h}\|} = 0 \quad \text{in } \mathbb{R}^m.$$

EXAMPLE 10.5:

(i) When $m = 1$, the function is scalar valued. The Jacobian matrix is $1 \times m$, and $\mathbf{Df} = \nabla f$, considered to be a row vector.

(ii) When $n = 1$, the vector-valued function $\mathbf{r}(t) = (x_1(t), \dots, x_m(t))$ describes a curve or path in $\mathbb{R}^m$. The Jacobian matrix in this case is $m \times 1$, usually denoted $\mathbf{r}'(t) = (x_1'(t), \dots, x_m'(t))$, and considered to be a column vector. The vector $\mathbf{r}'(t)$ is a tangent vector to the curve at the point $\mathbf{r}(t)$.

(iii) With $m = n = 3$, we may write $\mathbf{f}(x, y, z) = (u(x, y, z), v(x, y, z), w(x, y, z))$ with, for example,

$$u(x, y, z) = e^{x^2+y^2+z^2}, \quad v(x, y, z) = xy, \quad w(x, y, z) = \frac{x + z}{1 + y^2}.$$

The Jacobian matrix is the 3×3 matrix

$$\mathbf{Df}(x, y, z) = \begin{bmatrix} u_x & u_y & u_z \\ v_x & v_y & v_z \\ w_x & w_y & w_z \end{bmatrix}$$

$$= \begin{bmatrix} 2xe^{x^2+y^2+z^2} & 2ye^{x^2+y^2+z^2} & 2ze^{x^2+y^2+z^2} \\ y & x & 0 \\ \dfrac{1}{1+y^2} & -\dfrac{2y(x+z)}{(1+y^2)^2} & \dfrac{1}{1+y^2} \end{bmatrix}.$$

Now we are ready to prove the chain rule in all its glory.

Theorem 10.10 Let $U \subset \mathbb{R}^n$ be an open set, and let $V \subset \mathbb{R}^m$ also be an open set. Let $\mathbf{f} : U \to \mathbb{R}^m$ with $\mathbf{f}(U) \subset V$. Let $\mathbf{g} : V \to \mathbb{R}^p$. If $\mathbf{f} \in C^1(U)$ and $\mathbf{g} \in C^1(V)$, then the composition $\mathbf{g} \circ \mathbf{f} \in C^1(U)$ with

$$\mathbf{D}(\mathbf{g} \circ \mathbf{f})(\mathbf{x}) = \mathbf{Dg}(\mathbf{f}(\mathbf{x}))\mathbf{Df}(\mathbf{x}). \tag{10.13}$$

For each $\mathbf{a} \in U$, the linear approximation to $\mathbf{g} \circ \mathbf{f}$ is

$$\mathbf{l}(\mathbf{x}) = \mathbf{g}(\mathbf{f}(\mathbf{a})) + \mathbf{Dg}(\mathbf{f}(\mathbf{a}))\mathbf{Df}(\mathbf{a})(\mathbf{x} - \mathbf{a}).$$

Notice that the matrix on the left of (10.13) is $p \times n$; the matrix on the right is the product of the $p \times m$ matrix $\mathbf{Dg}(\mathbf{f}(\mathbf{x}))$ and the $m \times n$ matrix $\mathbf{Df}(\mathbf{x})$.

Proof: It suffices to prove the result for the case in which g is scalar valued. When g is scalar valued, the Jacobian matrix Dg is $1 \times m$, and in fact is just ∇g. Let $\mathbf{a} \in U$. We wish to show that $F(\mathbf{x}) \equiv g(\mathbf{f}(\mathbf{x}))$ is C^1 with $1 \times n$ Jacobian matrix $\nabla F(\mathbf{a}) = \nabla g(\mathbf{f}(\mathbf{a}))\mathbf{Df}(\mathbf{a})$.

Because V is open, there is an $r > 0$ such that $B(\mathbf{f}(\mathbf{a}), r) \subset V$, and because $\mathbf{f}$ is continuous, there is a $\delta > 0$ such that $\mathbf{f}(\mathbf{a} + \mathbf{h}) \in B(\mathbf{f}(\mathbf{a}), r)$ for $\|\mathbf{h}\| < \delta$. We may apply the mean value theorem (Theorem 10.7) to g on $B(\mathbf{f}(\mathbf{a}), r)$. For each $\|\mathbf{h}\| < \delta$, we have

$$\begin{aligned} F(\mathbf{a} + \mathbf{h}) - F(\mathbf{a}) &= g(\mathbf{f}(\mathbf{a} + \mathbf{h})) - g(\mathbf{f}(\mathbf{a})) \\ &= \langle \nabla g(\mathbf{z}), \mathbf{f}(\mathbf{a} + \mathbf{h}) - \mathbf{f}(\mathbf{a}) \rangle \end{aligned} \tag{10.14}$$

where $\mathbf{z}$ lies on the line $\mathcal{L}$ between $\mathbf{f}(\mathbf{a})$ and $\mathbf{f}(\mathbf{a} + \mathbf{h})$. Because $\mathbf{f}$ is C^1, it has a linear approximation at $\mathbf{x} = \mathbf{a}$:

$$\mathbf{f}(\mathbf{a} + \mathbf{h}) - \mathbf{f}(\mathbf{a}) = \mathbf{Df}(\mathbf{a})\mathbf{h} + \mathbf{R}(\mathbf{h}).$$

We substitute this last expression into (10.14) to deduce

$$F(\mathbf{a} + \mathbf{h}) - F(\mathbf{a}) = \langle \nabla g(\mathbf{z}), \mathbf{Df}(\mathbf{a})\mathbf{h} \rangle + \langle \nabla g(\mathbf{z}), \mathbf{R}(\mathbf{h}) \rangle.$$

To calculate $\partial F / \partial x_j$, we specify $\mathbf{h} = h\mathbf{e}_j$ where $\mathbf{e}_j$ is one of the vectors in the standard basis. Dividing this last equation through by h, we have

$$\frac{F(\mathbf{a} + h\mathbf{e}_j) - F(\mathbf{a})}{h} = \langle \nabla g(\mathbf{z}), \mathbf{Df}(\mathbf{a})\mathbf{e}_j \rangle + \langle \nabla g(\mathbf{z}), \mathbf{R}(h\mathbf{e}_j)/h \rangle. \tag{10.15}$$

Now as h tends to 0, $\mathbf{f}(\mathbf{a} + h\mathbf{e}_j)$ converges to $\mathbf{f}(\mathbf{a})$ because $\mathbf{f}$ is continuous. The point $\mathbf{z} \in \mathcal{L}$ lies between $\mathbf{f}(\mathbf{a}+h\mathbf{e}_j)$ and $\mathbf{f}(\mathbf{a})$. Hence $\mathbf{z}$ converges to $\mathbf{f}(\mathbf{a})$ as h tends to 0. Because g has continuous derivatives, $\nabla g(\mathbf{z})$ converges to $\nabla g(\mathbf{f}(\mathbf{a}))$ as h converges to 0. In addition, $R(h\mathbf{e}_j)/h$ converges to 0 as h tends to 0 because $\mathbf{f}$ has a linear approximation at $\mathbf{a}$. Therefore when we take the limit as h tends to 0 in (10.15), we obtain

$$\frac{\partial F}{\partial x_j}(\mathbf{a}) = \lim_{h \to 0} \frac{F(\mathbf{a} + h\mathbf{e}_j) - F(\mathbf{a})}{h}$$

$$= \langle \nabla g(\mathbf{f}(\mathbf{a})), \mathbf{Df}(\mathbf{a})\mathbf{e}_j \rangle.$$

Hence all of the partial derivatives of the composite function F exist at $\mathbf{a}$ and, from the previous equation, we see that they depend continuously on $\mathbf{a}$. In terms of matrix multiplication,

$$D(g \circ \mathbf{f})(\mathbf{a}) = DF(\mathbf{a}) = \nabla g(\mathbf{f}(\mathbf{a}))\mathbf{Df}(\mathbf{a}).$$

The proof of Theorem 10.10 is complete.

EXAMPLE 10.6:

a) Let $\mathbf{r}(t) = (x(t), y(t))$ parameterize a curve γ in $\mathbb{R}^2$. If we have a mapping $(u, v) = \mathbf{f}(x, y) = (f(x, y), g(x, y))$ of $\mathbb{R}^2$ into itself, the curve γ is mapped into the curve $\mathbf{f}(\gamma)$ which is parameterized by $\mathbf{q}(t) = (f(x(t), y(t)), g(x(t), y(t)))$. The chain rule says that the tangent vector $\mathbf{r}'(t) = (x'(t), y'(t))$ at the point $(x(t), y(t)) \in \gamma$ is mapped into the tangent vector $\mathbf{q}'(t)$ at the point $(u(t), v(t)) = (f(x(t), y(t)), g(x(t), y(t))) \in \mathbf{f}(\gamma)$ by the relation

$$\mathbf{q}'(t) = \mathbf{Df}(\mathbf{r}(t))\mathbf{r}'(t)$$

$$= \begin{bmatrix} f_x(x(t), y(t)) & f_y(x(t), y(t)) \\ g_x(x(t), y(t)) & g_y(x(t), y(t)) \end{bmatrix} \begin{bmatrix} x'(t) \\ y'(t) \end{bmatrix}$$

$$= \begin{bmatrix} f_x(x(t), y(t))x'(t) + f_y(x(t), y(t))y'(t) \\ g_x(x(t), y(t))x'(t) + g_y(x(t), y(t))y'(t) \end{bmatrix}.$$

b) If $p(u, v) : \mathbb{R}^2 \to \mathbb{R}$, the gradient vector of the composite function $q(x, y) = p(f(x, y), g(x, y))$ can be calculated by the chain rule,

$$\nabla q(x, y) = (\nabla p)(f(x, y), g(x, y))\mathbf{Df}(x, y),$$

or

$$[q_x(x, y), \ q_y(x, y)]$$

$$= [p_u(f(x, y), g(x, y)), \ p_v(f(x, y), g(x, y))] \begin{bmatrix} f_x(x, y) & f_y(x, y) \\ g_x(x, y) & g_y(x, y) \end{bmatrix}.$$

When we perform the matrix multiplication and equate components, we find

$$q_x(x, y) = p_u(f(x, y), g(x, y))f_x(x, y) + p_v(f(x, y), g(x, y))g_x(x, y),$$

and

$$q_y(x, y) = p_u(f(x, y), g(x, y))f_y(x, y) + p_v(f(x, y), g(x, y))g_y(x, y).$$

These equations are usually written more briefly as

$$q_x = p_u f_x + p_v g_x \quad \text{and} \quad q_y = p_u f_y + p_v g_y.$$

It is important to remember that p_u and p_v are evaluated at $u = f(x, y)$ and $v = g(x, y)$. This becomes crucial when we have to apply the chain rule again to compute higher derivatives of q.

Heretofore we have been careful to use different symbols to distinguish between a function $f(x)$ and its composition with another function, for example, $u(t)$. We write $g(t) = f(u(t))$. However, in many physical applications, one might write $f(t)$ to refer to the composite function. The reason this is done is that the values of f may represent a physical quantity, and we think of f as depending on the variable x (space), or as depending on the variable t (time).

As another example, suppose ρ represents the density in grams/area of some material. ρ is considered to be a function of the physical point which may given in Cartesian coordinates (x, y), or in polar coordinates, (r, θ). The chain rule would be written

$$\frac{\partial \rho}{\partial r} = \frac{\partial \rho}{\partial x} \frac{\partial x}{\partial r} + \frac{\partial \rho}{\partial y} \frac{\partial y}{\partial r}.$$

EXAMPLE 10.7:   We now consider what might be a typical passage from a physics text.

The variables p, T, and V represent pressure, temperature, and volume. It is found that when p is considered to be a function of V and T, p satisfies the differential equation

$$\frac{\partial p}{\partial T} + V\frac{\partial p}{\partial V} + p = 0. \tag{10.16}$$

However, when V is considered to be a function of p and T, equation (10.16) becomes

$$-\frac{\partial V}{\partial T} + V + p\frac{\partial V}{\partial p} = 0. \tag{10.17}$$

How does equation (10.17) follow from (10.16)? In the case of (10.16) it is assumed that $p = f(V, T)$ for some function f, and for (10.17) it is assumed that $V = g(p, T)$ for some function g. To avoid confusion about the meaning of the variables, we will use f_1 to denote differentiation of f with respect to its first argument. f_2 will denote differentiation with respect to its second argument. The same notation will be used for g. Then equation (10.16) can be written

$$f_2 + V f_1 + p = 0. \tag{10.18}$$

If T is held fixed, f and g are inverses:

$$p = f(g(p, T), T).$$

By the chain rule

$$1 = \frac{\partial}{\partial p}f(g(p, T), T) = f_1(g(p, T), T)g_1(p, T) \tag{10.19}$$

and

$$0 = \frac{\partial}{\partial T}f(g(p, T), T) = f_1(g(p, T), T)g_2(p, T) + f_2(g(p, T), T). \tag{10.20}$$

Hence $f_1 = 1/g_1$ and $f_2 = -g_2/g_1$. We substitute these expressions into (10.18) and multiply by g_1 to obtain

$$-g_2 + V + pg_1 = 0.$$

When we write $g_1 = \partial V/\partial p$ and $g_2 = \partial V/\partial T$ we obtain (10.17). Of course, this discussion is not written in the physics book.

The exact analogue of the mean value theorem (Theorem 10.7) is not true for vector-valued mappings. The mean value theorem can be applied to each component of $\mathbf{f} = (f_1, \ldots, f_m)$, thereby producing points $\mathbf{z}_i \in \mathcal{L}$. However the points $\mathbf{z}_i$ are usually not the same. Nevertheless, it is still possible to estimate the difference $\mathbf{f}(\mathbf{x}) - \mathbf{f}(\mathbf{y})$ in terms of the derivative. First we must make an observation about vector-valued mappings of a single variable.

Let $\mathbf{f}(x) = (f_1(x), \ldots, f_m(x))$ be defined on an interval $a \leq x \leq b$ with values in $\mathbb{R}^m$. Assume that each component f_i is continuous. The integral $\int \mathbf{f}(x)dx$ is defined by integrating each component and it produces a vector:

$$\int_a^b \mathbf{f}(x)\,dx = \left(\int_a^b f_1(x)dx, \ldots, \int_a^b f_m(x)dx \right).$$

Theorem 10.11 For any of the equivalent vector norms,

$$\left\| \int_a^b \mathbf{f}(x)\,dx \right\| \leq \int_a^b \|\mathbf{f}(x)\|\,dx. \tag{10.21}$$

Remark Inequality (10.21) is obviously an extension of the inequality $|\int f\,dx| \leq \int |f|\,dx$ for scalar-valued functions.

Proof: We know that the integral of each component is the limit of Riemann sums, and we can take the partition and the selected points to be the same for each component. Let $\Delta x = (b - a)/n$, $x_j = a + j\Delta x$, $j = 0, \ldots n$. Choose $s_j \in [x_{j-1}, x_j]$. Then

$$\int_a^b \mathbf{f}(x)\,dx = \left(\lim_{\Delta x \to 0} \sum_j f_1(s_j)\Delta x, \ldots, \lim_{\Delta x \to 0} \sum_j f_m(s_j)\Delta x \right) = \lim_{\Delta x \to 0} \sum_j \mathbf{f}(s_j)\Delta x.$$

By the triangle inequality,

$$\left\| \sum_j \mathbf{f}(x_j)\Delta x \right\| \leq \sum_j \|\mathbf{f}(s_j)\| \, \Delta x.$$

The sums on the right side are just Riemann sums for the integral $\int \|\mathbf{f}\| \, dx$. Hence

$$\left\| \int_a^b \mathbf{f}(x) \, dx \right\| = \left\| \lim_{\Delta x \to 0} \sum_j \mathbf{f}(s_j) \, \Delta x \right\|$$

$$\leq \lim_{\Delta x \to 0} \sum_j \|\mathbf{f}(s_j)\| \, \Delta x$$

$$\leq \int_a^b \|\mathbf{f}(x)\| \, dx.$$

Now we are ready to prove

Theorem 10.12 Let G be an open set of $\mathbb{R}^n$ and let $\mathbf{f} : G \to \mathbb{R}^m$ be C^1 on G. Suppose that $\mathbf{x}, \mathbf{y}$ and the line $\mathcal{L}$ joining them are all contained in G. Then

$$\|\mathbf{f}(\mathbf{x}) - \mathbf{f}(\mathbf{y})\| \leq \max_{\mathbf{z} \in \mathcal{L}} \|\mathbf{D}\mathbf{f}(\mathbf{z})\| \|\mathbf{x} - \mathbf{y}\|. \qquad (10.22)$$

The norm on $\mathbf{D}\mathbf{f}(\mathbf{z})$ is the matrix norm subordinate to the norm $\| \ \|$.

Proof: The line $\mathcal{L}$ is parameterized by

$$\mathbf{z}(t) = t\mathbf{x} + (1 - t)\mathbf{y}, \quad 0 \leq t \leq 1.$$

By the chain rule, the composite function $t \to \mathbf{g}(t) = \mathbf{f}(\mathbf{z}(t))$ is C^1 with

$$\frac{d\mathbf{g}}{dt} = \mathbf{D}\mathbf{f}(\mathbf{z}(t))\mathbf{z}'(t) = \mathbf{D}\mathbf{f}(\mathbf{z}(t))(\mathbf{x} - \mathbf{y}).$$

Now we have

$$\mathbf{f}(\mathbf{x}) - \mathbf{f}(\mathbf{y}) = \mathbf{g}(1) - \mathbf{g}(0) = \int_0^1 \frac{d\mathbf{g}}{dt}\, dt$$

$$= \int_0^1 \mathbf{Df}(\mathbf{z}(t))(\mathbf{x} - \mathbf{y})\, dt.$$

It follows that

$$\|\mathbf{f}(\mathbf{x}) - \mathbf{f}(\mathbf{y})\| \le \int_0^1 \|\mathbf{Df}(\mathbf{z}(t))(\mathbf{x} - \mathbf{y})\|\, dt$$

$$\le \int_0^1 \|\mathbf{Df}(\mathbf{z}(t))\|\|\mathbf{x} - \mathbf{y}\|\, dt$$

$$\le \max_{\mathbf{z} \in \mathcal{L}} \|\mathbf{Df}(\mathbf{z})\|\|\mathbf{x} - \mathbf{y}\|.$$

The following definition extends the notion of Lipschitz continuity to vector-valued functions. Recall that a set G is convex if, whenever $\mathbf{a} \in G$ and $\mathbf{b} \in G$, the line segment $\mathcal{L}$ between $\mathbf{a}$ and $\mathbf{b}$ is also contained in G.

Definition 10.5 Let $G \subset \mathbb{R}^n$. The function $\mathbf{f} : G \to \mathbb{R}^m$ is *Lipschitz* continuous on G with Lipschitz constant L if

$$\|\mathbf{f}(\mathbf{x}) - \mathbf{f}(\mathbf{y})\| \le L\|\mathbf{x} - \mathbf{y}\| \quad \text{for all } \mathbf{x}, \mathbf{y} \in G.$$

The Lipschitz constant may depend on the choice of norms.

Corollary Let G be an open convex subset of $\mathbb{R}^n$. Suppose that $f : G \to \mathbb{R}^m$ with $f \in C^1(G)$ and that $\|\mathbf{Df}(\mathbf{x})\| \le L$ for all $\mathbf{x} \in G$. Then $\mathbf{f}$ is Lipschitz continuous on G.

Proof: Because G is convex, we may apply Theorem 10.12 and deduce that

$$\|\mathbf{f}(\mathbf{x}) - \mathbf{f}(\mathbf{y})\| \le \max_{\mathbf{z} \in \mathcal{L}} \|\mathbf{Df}(\mathbf{z})\|\ \|\mathbf{x} - \mathbf{y}\| \le L\|\mathbf{x} - \mathbf{y}\|.$$

Exercises for 10.3

1. Calculate the Jacobian matrix for each of the following functions. Give a geometric interpretation of the linear approximation in each case as tangent line, tangent plane, and so on.

   a) $\mathbf{f} : \mathbb{R} \rightarrow \mathbb{R}^3$ given by $\mathbf{f}(t) = (2t, \sqrt{1 + t^2}, e^t)$.
   b) $f : \mathbb{R}^2 \rightarrow \mathbb{R}$ given by $f(x, y) = 2x \sin(x + y^2)$.
   c) $\mathbf{f} : \mathbb{R}^2 \rightarrow \mathbb{R}^3$ given by $\mathbf{f}(x, y) = (x + y^2, \cos(x + y), x^2/y)$.

2. Let $\mathbf{f} : \mathbb{R}^2 \rightarrow \mathbb{R}^2$ be given by

$$\mathbf{f}(x, y) = (x + y + y^2, y - x^2/2).$$

   a) Calculate the Jacobian matrix $\mathbf{Df}(x, y)$.
   b) What is the linear (affine) approximation l to $\mathbf{f}$ at the point $(0, 0)$?
   c) Let Q_h be the square $0 \leq x, y \leq h$. Sketch the image of Q_h under $\mathbf{f}$ and under l. Verify that $\mathbf{f}(Q_h)$ and $\mathbf{l}(Q_h)$ look more and more similar as h gets smaller. In particular, how fast does the difference $\|\mathbf{f}(h, h) - \mathbf{l}(h, h)\|_1$ tend to zero as $h \rightarrow 0$?

3. Let

$$\mathbf{f}(x, y) = \left(xy, \frac{1}{1 + x^2 + y^2} \right).$$

   Let Q_r be the square $0 \leq x, y \leq r$. In view of the corollary to Theorem 10.12, $\mathbf{f}$ is Lipschitz continuous on Q_r. The Lipschitz constant will depend on the norm and on r. Estimate the Lipschitz constant in the norms $\| \ \|_1$ and $\| \ \|_\infty$.

4. Suppose that $\mathbf{f}(x, y) = (x^2 - y^2, 2xy)$ and $\mathbf{g}(x, y) = (x \cos y, x \sin y)$. Note that $\mathbf{f}(0, 0) = \mathbf{g}(0, 0) = (0, 0)$. Show that there is a constant C such that

$$\|\mathbf{g}(x, y) - \mathbf{f}(x, y)\|_1 \leq C \|(x, y)\|_1 \text{ for all } 0 \leq x, y \leq 1.$$

5. Let $f(x, y, t)$ be a C^1 function. Let $x(s, t) = t^2 e^{2s}$ and let $y(s, t) = t + s^2$. The composite function is $F(s, t) = f(x(s, t), y(s, t), t)$. Compute $\partial F/\partial t$ and $\partial F/\partial s$ in terms of the partial derivatives of f. Show how the gradient (row) vector for F is computed from the gradient (row) vector for f and the 3×2 Jacobian matrix of $(s, t) \rightarrow (x, y, t)$.

6. Let $u(x, t) = \psi(x/t)$ where $\psi(s)$ is C^1. Show that u satisfies the partial differential equation $tu_t + xu_x = 0$.

7. Let $f(x, y, t)$ and $g(x, y, t)$ be C^1 functions, and suppose that $u(t)$ and $v(t)$ are C^1 functions such that

$$f(u(t), v(t), t) = 0$$
$$g(u(t), v(t), t) = 0.$$

Differentiate both equations with respect to t and solve for the pair $(u'(t), v'(t))$ in terms of the derivatives of f and g.

8. Let $u(x, y)$ and $v(x, y)$ be two C^1 functions defined on an open set $G \subset \mathbb{R}^2$. We say that u, v is a *Cauchy-Riemann* pair if they satisfy the Cauchy-Riemann equations

$$u_x = v_y, \quad \text{and} \quad u_y = -v_x.$$

Let $g(x, y) = f(u(x, y), v(x, y))$ where f is a C^1 function of u, v. Show that

$$(g_x^2 + g_y^2) = (f_u^2 + f_v^2)(u_x^2 + v_x^2).$$

9. Let u, v be a Cauchy-Riemann pair as in exercise 8, and suppose that $u_x^2 + v_x^2 > 0$. Let f and g be C^2 functions with $g(x, y) = f(u(x, y), v(x, y))$. Show that $f_{uu} + f_{vv} = 0$ if and only if $g_{xx} + g_{yy} = 0$.

10. The Laplace operator is $\Delta u = u_{xx} + u_{yy}$. Use polar coordinates $x = r \cos \theta$, $y = r \sin \theta$, and set $v(r, \theta) = u(r \cos \theta, r \sin \theta)$. Show that $\Delta u = v_{rr} + (1/r)v_r + (1/r^2)v_{\theta\theta}$. This is the Laplace operator in polar coordinates.

Chapter 11

Solving Systems of Equations

Chapter Overview

In this chapter, we will mainly be concerned with the problem of solving systems of equations

$$f_1(x_1, \ldots, x_n) = 0$$
$$f_2(x_1, \ldots, x_n) = 0$$
$$\cdot$$
$$\cdot$$
$$f_n(x_1, \ldots, x_n) = 0. \tag{11.1}$$

In the case of (11.1), the number of equations equals the number of variables, and from our experience with systems of linear equations, we can hope to find conditions that will guarantee the existence of a unique solution. We will also consider systems with more variables than equations,

$$f_1(x_1, \ldots, x_n, y_1, \ldots, y_m) = 0$$
$$f_2(x_1, \ldots, x_n, y_1, \ldots, y_m) = 0$$
$$\cdot$$
$$\cdot$$
$$f_m(x_1, \ldots, x_n, y_1, \ldots y_m) = 0. \tag{11.2}$$

In the case of (11.2), we will find conditions that allow us to solve for the variables $y_1, \ldots, y_m$ in terms of the variables $x_1, \ldots, x_n$.

We begin with a result on perturbations of an invertible matrix. Then we prove the contraction mapping theorem for vector-valued functions exactly as in Chapter 6. In the next section we do not treat Newton's method exactly as in Chapter 6, but rather we prove the Kantorovich existence theorem that is based on Newton's method. The exercises provide opportunities to compute solutions of systems of equations using functional iteration and Newton's method. The inverse function theorem and implicit function theorem are proved. A condition is given for the existence of a global inverse, and this condition is shown to be satisfied for the system of nonlinear equations that resulted from discretizing the boundary value problem of Chapter 5. We also show how Newton's method can be used to compute the points on a curve that is defined implicitly, thereby making a connection between the hypotheses of the implicit function theorem and those of Newton's method. Finally in the last section, we consider the example of a mass suspended by two springs. We show how the inverse and implicit function theorems can be applied in this example to make a complete analysis of the problem.

In this chapter, $\|\mathbf{x}\|$ will denote any one of the equivalent norms on $\mathbb{R}^n$ and $\|\mathbf{A}\|$ will denote the induced matrix (operator) norm.

11.1 Linear Systems

If $\mathbf{A}$ is an $n \times n$ matrix we know that the linear system $\mathbf{Ax} = \mathbf{y}$ is uniquely solvable for each $\mathbf{y} \in \mathbb{R}^n$ if and only if $\det \mathbf{A} \neq 0$. This means that if we change the elements of $\mathbf{A}$ slightly, we will still have $\det \mathbf{A} \neq 0$. We prefer to have a perturbation result of this kind in terms of the norms of matrices. If $\mathbf{A}$ is invertible, and $\mathbf{B}$ is another matrix such that $\|\mathbf{A} - \mathbf{B}\|$ is small, can we say that $\mathbf{B}$ is invertible? First we state an idea that relates $\|\mathbf{A}\|$ and $\|\mathbf{A}^{-1}\|$.

Lemma 11.1 Let $\mathbf{A}$ be an $n \times n$ matrix. If there is a constant $c > 0$ such that $\|\mathbf{Ax}\| \geq c\|\mathbf{x}\|$ for all $\mathbf{x} \in \mathbb{R}^n$, then $\mathbf{A}$ is invertible and

$$\|\mathbf{A}^{-1}\| \leq \frac{1}{c}. \qquad (11.3)$$

Proof: The condition implies that the only solution of $\mathbf{Ax} = \mathbf{0}$ is $\mathbf{x} = \mathbf{0}$. Hence $\mathbf{A}$ is invertible. Letting $\mathbf{y} = \mathbf{Ax}$ so that $\mathbf{x} = \mathbf{A}^{-1}\mathbf{y}$, we have

$$\|\mathbf{A}^{-1}\mathbf{y}\| \leq (1/c)\|\mathbf{y}\|$$

for all $\mathbf{y} \in \mathbb{R}^n$. Therefore

$$\|\mathbf{A}^{-1}\| = \sup_{\|\mathbf{y}\|=1} \|\mathbf{A}^{-1}\mathbf{y}\| \le (1/c).$$

To understand the origin of inequalities for matrices, it sometimes helps to replace the matrices with real numbers. Now suppose that $a \in \mathbb{R}$ and that $a \ne 0$. Then if $|a - b| < |a|$, we must have

$$|b| = |b - a + a| \ge |a| - |a - b| > 0$$

so that $b \ne 0$. Furthermore

$$\frac{1}{b} = \frac{1}{a}\left(\frac{1}{1 - (a - b)/a}\right).$$

Thus $|a - b| < |a| = 1/|a^{-1}|$ implies that $b \ne 0$ and

$$|b^{-1}| \le \frac{|a^{-1}|}{1 - |a^{-1}||a - b|}.$$

Now we prove a theorem about matrices that resembles this inequality.

Theorem 11.2 If $\mathbf{A}$ and $\mathbf{B}$ are $n \times n$ matrices with $\mathbf{A}$ invertible and

$$\|\mathbf{A} - \mathbf{B}\| < 1/\|\mathbf{A}^{-1}\|,$$

then $\mathbf{B}$ is invertible with

$$\|\mathbf{B}^{-1}\| \le \frac{\|\mathbf{A}^{-1}\|}{1 - \|\mathbf{A}^{-1}\| \|\mathbf{A} - \mathbf{B}\|}. \tag{11.4}$$

Proof: For all $\mathbf{x} \in \mathbb{R}^n$,

$$\|\mathbf{B}\mathbf{x}\| = \|\mathbf{B}\mathbf{x} - \mathbf{A}\mathbf{x} + \mathbf{A}\mathbf{x}\| \ge \|\mathbf{A}\mathbf{x}\| - \|(\mathbf{A} - \mathbf{B})\mathbf{x}\|.$$

However, $\|(\mathbf{A} - \mathbf{B})\mathbf{x}\| \le \|\mathbf{A} - \mathbf{B}\| \|\mathbf{x}\|$ and

$$\|\mathbf{x}\| \le \|\mathbf{A}^{-1}\mathbf{A}\mathbf{x}\| \le \|\mathbf{A}^{-1}\| \|\mathbf{A}\mathbf{x}\|$$

so that $\|\mathbf{A}\mathbf{x}\| \geq \|\mathbf{x}\|/\|\mathbf{A}^{-1}\|$. Hence

$$\|\mathbf{B}\mathbf{x}\| \geq \frac{1 - \|\mathbf{A}^{-1}\|\,\|\mathbf{A} - \mathbf{B}\|}{\|\mathbf{A}^{-1}\|}\|\mathbf{x}\|.$$

Then Lemma 4.1 implies that $\mathbf{B}$ is invertible, and $\|\mathbf{B}^{-1}\|$ satisfies (11.4).

Corollary Let $\mathbf{A}$ and $\mathbf{B}$ satisfy the hypotheses of Theorem 11.2. Then

$$\|\mathbf{B}^{-1} - \mathbf{A}^{-1}\| \leq \frac{\|\mathbf{A}^{-1}\|^2\|\mathbf{A} - \mathbf{B}\|}{1 - \|\mathbf{A}^{-1}\|\,\|\mathbf{A} - \mathbf{B}\|}.$$

Proof: We observe that

$$\mathbf{B}^{-1} - \mathbf{A}^{-1} = \mathbf{A}^{-1}(\mathbf{A} - \mathbf{B})\mathbf{B}^{-1}.$$

Thus

$$\|\mathbf{B}^{-1} - \mathbf{A}^{-1}\| \leq \|\mathbf{A}^{-1}\|\,\|\mathbf{A} - \mathbf{B}\|\,\|\mathbf{B}^{-1}\|.$$

Then we use (11.4) to estimate $\|\mathbf{B}^{-1}\|$ on the right side, which yields the corollary.

The corollary says that the solution of $\mathbf{A}\mathbf{x} = \mathbf{b}$ depends continuously on the matrix $\mathbf{A}$. In fact if $\mathbf{A}\mathbf{x} = \mathbf{b}$ and $\mathbf{B}\mathbf{y} = \mathbf{b}$, then

$$\begin{aligned}
\|\mathbf{y} - \mathbf{x}\| &= \|[\mathbf{B}^{-1} - \mathbf{A}^{-1}]\mathbf{b}\| \\
&\leq \|\mathbf{B}^{-1} - \mathbf{A}^{-1}\|\,\|\mathbf{b}\| \\
&\leq \left(\frac{\|\mathbf{A}^{-1}\|^2\|\mathbf{A} - \mathbf{B}\|}{1 - \|\mathbf{A}^{-1}\|\,\|\mathbf{A} - \mathbf{B}\|}\right)\|\mathbf{b}\|.
\end{aligned}$$

11.2 The Contraction Mapping Theorem

If we let $\mathbf{A} = \mathbf{I}$ and $\mathbf{B} = \mathbf{I} - \mathbf{E}$, Theorem 11.2 says that if $\|\mathbf{E}\| < 1$, then the equation $\mathbf{E}\mathbf{x} = \mathbf{x}$ has a unique solution; in this case, it is the trivial solution $\mathbf{x} = 0$. The contraction mapping theorem extends this idea to nonlinear mappings.

In Chapter 6, we considered equations of the form

$$x = g(x) \tag{11.5}$$

where g is a function of one variable. Recall that a solution x_* of equation (11.5) is called a *fixed-point* of the function g. Using the intermediate value theorem, we saw that if g is a continuous map of a closed interval into itself, it has at least one fixed-point. We will not prove the analogue of this kind of theorem in $\mathbb{R}^n$. Instead, we give an n-dimensional version of the contraction mapping theorem that we proved in Chapter 6. The proof is exactly the same, but verifying the hypothesis in higher dimensions is more interesting.

Definition 11.1 Let **g** be a continuous mapping of a set $D \subset \mathbb{R}^n$ into itself. We say that **g** is a *contraction* on D if there is a constant c, $0 \le c < 1$, such that

$$\|\mathbf{g}(\mathbf{x}) - \mathbf{g}(\mathbf{y})\| \le c\|\mathbf{x} - \mathbf{y}\| \tag{11.6}$$

for all $\mathbf{x}, \mathbf{y} \in D$.

Theorem 11.3 (Contraction mapping theorem) Let D be a closed subset of $\mathbb{R}^n$, and let **g** be a contraction on D. Then **g** has a unique fixed-point in D. Furthermore, if $\mathbf{x}_0$ is any point in D, the recursively defined sequence $\mathbf{x}_{n+1} = \mathbf{g}(\mathbf{x}_n)$ converges to the fixed-point $\mathbf{x}_*$.

Remark Recall that the method of constructing the sequence $\mathbf{x}_{n+1} = \mathbf{g}(\mathbf{x}_n)$ is often called *functional iteration*. In the context of differential equations, it is called *Picard iteration*.

Proof of Theorem 11.3: Let $\mathbf{x}_0$ be any starting point in D for the iteration scheme. First we estimate the distance between any two successive iterates. Using (11.6), we have

$$\begin{aligned}
\|\mathbf{x}_{k+1} - \mathbf{x}_k\| &= \|\mathbf{g}(\mathbf{x}_k) - \mathbf{g}(\mathbf{x}_{k-1})\| \\
&\le c\|\mathbf{x}_k - \mathbf{x}_{k-1}\|.
\end{aligned}$$

Applying this estimate over and over again, we arrive at

$$\|\mathbf{x}_{k+1} - \mathbf{x}_k\| \le c^k\|\mathbf{x}_1 - \mathbf{x}_0\|. \tag{11.7}$$

Because $c < 1$, this estimate shows that the difference between successive iterates tends to zero as $k \to \infty$. However, this is not enough to show that the sequence of iterates converges. We must show that the sequence $\mathbf{x}_k$ is a Cauchy sequence. Let k and l be nonnegative integers. Then by (11.7),

$$\|\mathbf{x}_{k+l} - \mathbf{x}_k\| \leq \|\mathbf{x}_{k+l} - \mathbf{x}_{l+k-1}\| + \cdots + \|\mathbf{x}_{k+1} - \mathbf{x}_k\|$$
$$\leq [c^{k+l-1} + \cdots + c^k]\|\mathbf{x}_1 - \mathbf{x}_0\|$$
$$\leq c^k[c^{l-1} + c^{l-2} + \cdots + 1]\|\mathbf{x}_1 - \mathbf{x}_0\|.$$

Now the geometric sum in the square brackets is

$$c^{l-1} + \cdots + 1 = \frac{1 - c^l}{1 - c} \leq \frac{1}{1 - c}.$$

Hence

$$\|\mathbf{x}_{l+k} - \mathbf{x}_k\| \leq \left(\frac{c^k}{1 - c}\right)\|\mathbf{x}_1 - \mathbf{x}_0\|.$$

Since this estimate is independent of l, it shows that $\mathbf{x}_k$ is a Cauchy sequence and hence converges to some point $\mathbf{x}_*$. We must have $\mathbf{x}_* \in D$ because D is assumed to be closed. Finally, taking the limits in both sides of the equation defining the iterates,

$$\mathbf{x}_{k+1} = \mathbf{g}(\mathbf{x}_k),$$

and using the fact that $\mathbf{g}$ is continuous, we deduce that

$$\mathbf{x}_* = \lim_{k \to \infty} \mathbf{x}_{k+1} = \lim_{n \to \infty} \mathbf{g}(\mathbf{x}_k) = \mathbf{g}(\lim_{n \to \infty} \mathbf{x}_k) = \mathbf{g}(\mathbf{x}_*).$$

It remains to be shown that the fixed-point is unique. If $\mathbf{y}_*$ is another fixed-point, we use the contraction property (11.6) of $\mathbf{g}$ to conclude that

$$\|\mathbf{x}_* - \mathbf{y}_*\| = \|\mathbf{g}(\mathbf{x}_*) - \mathbf{g}(\mathbf{y}_*)\| \leq c\|\mathbf{x}_* - \mathbf{y}_*\|.$$

Since $c < 1$, this implies that $\mathbf{x}_* = \mathbf{y}_*$. The proof of Theorem 11.3 is complete.

To prove the n-dimensional analogue of Theorem 6.2, we need to use the vector-valued version of the mean value theorem, Theorem 10.12. Recall that a set A is *convex* if for each pair of points $\mathbf{x}, \mathbf{y} \in A$, the line segment joining $\mathbf{x}$ and $\mathbf{y}$ is also contained in A.

Theorem 11.4 Let D be a closed convex subset of $\mathbb{R}^n$, and let $\mathbf{g}$ be a C^1 mapping of D into itself. Suppose further that there is a constant c, $0 \leq c < 1$, such that

$$\|\mathbf{Dg(x)}\| \leq c \quad \text{for all } \mathbf{x} \in D. \tag{11.8}$$

Then there is a unique fixed point $\mathbf{x}_* \in D$ of $\mathbf{g}$.

Proof: To apply Theorem 10.12, we must be sure that if $\mathbf{x}$ and $\mathbf{y}$ belong to D, then the straight line joining $\mathbf{x}$ and $\mathbf{y}$ lies entirely within D. That is the reason for the additional hypothesis of convexity of D. With the hypothesis (11.8), Theorem 10.12 implies that $\mathbf{g}$ satisfies the contraction condition (11.6). Hence by Theorem 11.3, $\mathbf{g}$ has a unique fixed point in D.

EXAMPLE 11.1: Let $\mathbf{g}(x, y) = (g_1(x, y), g_2(x, y))$ where

$$g_1(x, y) = \sin\left(\frac{x+y}{2}\right) \quad \text{and} \quad g_2(x, y) = 1 - \frac{x^2 + y^2}{4}.$$

You can verify that $\mathbf{g}$ maps the square $D = \{0 \leq x, y \leq 1\}$ into itself. If we can show that $\mathbf{g}$ is a contraction in some norm, Theorem 11.3 guarantees that $\mathbf{g}$ has a unique fixed point in D that can be approximated by functional iteration. Next we calculate

$$\mathbf{Dg}(x, y) = (1/2)\begin{bmatrix} \cos(\frac{x+y}{2}) & \cos(\frac{x+y}{2}) \\ -x & -y \end{bmatrix}.$$

First we try to estimate $\|\mathbf{Dg}\|_\infty$. At each point $(x, y) \in D$,

$$\|\mathbf{Dg}(x, y)\|_\infty = \max\left\{\cos\left(\frac{x+y}{2}\right), \frac{x+y}{2}\right\}.$$

Because $\|\mathbf{Dg}(0, 0)\|_\infty = \|\mathbf{Dg}(1, 1)\|_\infty = 1$, we cannot use Theorem 4.5 to show that $\mathbf{g}$ is a contraction in the norm $\|\ \|_\infty$. Let us try $\|\mathbf{Dg}(x, y)\|_1$ instead.

$$\|\mathbf{Dg}(x, y)\|_1 = (1/2) \max\left\{\cos\left(\frac{x+y}{2}\right) + x, \cos\left(\frac{x+y}{2}\right) + y\right\}.$$

It is easy to verify that

$$(1/2) \max_{D} \left\{ \cos\left(\frac{x+y}{2}\right) + x \right\} = (1/2)(\cos(1/2) + 1) < 1$$

is attained at $(x, y) = (1, 0)$, and similarly

$$(1/2) \max_{D} \left\{ \cos\left(\frac{x+y}{2}\right) + y \right\} = (1/2)(\cos(1/2) + 1) < 1$$

is attained at $(x, y) = (0, 1)$. Consequently, $\|\mathbf{Dg}(x, y)\|_1 \le c = (1/2)$ $(\cos(1/2) + 1) < 1$ for all $(x, y) \in D$. It follows by Theorem 11.4 that g has a unique fixed point in D. The fixed point is the unique solution in D of the system

$$\sin\left(\frac{x+y}{2}\right) = x$$

$$1 - \frac{x^2 + y^2}{4} = y.$$

Starting at the initial point $(x, y) = (1, 1)$, the iterates are

iterate	x	y	$g_1(x, y)$	$g_2(x, y)$
1	1.0000	1.0000	0.8415	0.5000
2	0.8415	0.5000	0.6216	0.7605
3	0.6216	0.7605	0.6373	0.7588
4	0.6373	0.7588	0.6427	0.7545
5	0.6427	0.7545	0.6432	0.7544
6	0.6432	0.7544	0.6433	0.7543
7	0.6433	0.7543	0.6433	0.7543

We see that the iterates have converged (at least to four digits) after seven iterations.

There is a local version of the contraction mapping theorem in $\mathbb{R}^n$ which is the analogue of Theorem 6.4. We leave it as an exercise to formulate and prove the n-dimensional version of Theorem 6.4.

Orders of Convergence

It is important to know that an iterative method converges, but it is equally important to know how fast it converges. Let us make an estimate of how fast the convergence is for functional iteration. Suppose, as in Theorem 11.3, that $\mathbf{g}$ is a contraction satisfying (11.6) with $0 \leq c < 1$. Let $\mathbf{x}_*$ be the fixed point. Then using (11.6), we have

$$\|\mathbf{x}_k - \mathbf{x}_*\| = \|\mathbf{g}(\mathbf{x}_{k-1}) - \mathbf{g}(\mathbf{x}_*)\| \leq c\|\mathbf{x}_{k-1} - \mathbf{x}_*\|.$$

Applying this inequality repeatedly, we find that

$$\|\mathbf{x}_k - \mathbf{x}_*\| \leq c^k\|\mathbf{x}_0 - \mathbf{x}_*\|.$$

The distance between $\mathbf{x}_*$ and $\mathbf{x}_k$ is reduced by the factor c with each iteration. This is an example of what is called *linear convergence*. If c is close to 1, this convergence can be quite slow. The following definition classifies the kinds of convergence.

Definition 11.2 A sequence $\mathbf{x}_k$ converges to $\mathbf{x}_*$ with *order p* if there are constants $C_1 > 0, C_2 > 0$ and K such that

$$C_1 \leq \frac{\|\mathbf{x}_k - \mathbf{x}_*\|}{\|\mathbf{x}_{k-1} - \mathbf{x}_*\|^p} \leq C_2 \quad \text{for } k \geq K.$$

If $p > 1$, the convergence is said to be *superlinear*. When $p = 2$, the convergence is said to be *quadratic*. When $p = 1$, the convergence is *linear*. In this case we must have $C_2 < 1$ to ensure convergence. If only the inequality on the right holds, we say that the convergence is *at least of order p*. When the convergence is quadratic, the number of correct digits approximately doubles with each iteration.

EXAMPLE 11.2:

a) The sequence $x_k = (1/2)^k$ converges linearly to zero with $C_1 = C_2 = 1/2$.

b) If $p > 1$, the sequence $y_k = (1/2)^{p^k}$ converges to zero with order p and $C_1 = C_2 = 1$.

c) Let $\mathbf{A}$ be the matrix

$$\mathbf{A} = \begin{bmatrix} 0 & 1/2 \\ 1/3 & 0 \end{bmatrix}.$$

For any starting value $\mathbf{x}_0$, the sequence $\mathbf{x}_{k+1} = \mathbf{A}\mathbf{x}_k$ converges to zero. The convergence is linear, because for all k,

$$(1/3)\|\mathbf{x}_{k-1}\| \leq \|\mathbf{x}_k\| \leq (1/2)\|\mathbf{x}_{k-1}\|.$$

In particular, if $\mathbf{x}_0 = (1, 0)$,

$$\|\mathbf{x}_k\| = (1/2)\|\mathbf{x}_{k-1}\| \quad \text{for } k \text{ even}$$
$$\|\mathbf{x}_k\| = (1/3)\|\mathbf{x}_{k-1}\| \quad \text{for } k \text{ odd}.$$

As we saw in the case of Newton's method in one dimension, using more information about the function and making a better approximation led to faster convergence. In the next section, we extend Newton's method to higher dimensions.

Exercises for 11.1 and 11.2

1. Construct an example of a matrix $\mathbf{A}$ such that for any of the operator norms,

$$\|\mathbf{A}\|^{-1} < \|\mathbf{A}^{-1}\|.$$

   A 2×2 diagonal example will suffice.
2. Let $\mathbf{E}$ be an $n \times n$ matrix with $\|\mathbf{E}\| < 1$. We know that $\mathbf{I} - \mathbf{E}$ is invertible. By analogy with the geometric series,

$$\frac{1}{1-x} = 1 + x + x^2 + \cdots$$

   which converges for $|x| < 1$, show that we can write $(\mathbf{I} - \mathbf{E})^{-1}$ as a series

$$(\mathbf{I} - \mathbf{E})^{-1} = \mathbf{I} + \mathbf{E} + \mathbf{E}^2 + \cdots$$

   Show that the series converges in any of the operator norms.
3. Let D be a closed subset of $\mathbb{R}^n$, and for each $t \in [a, b]$, let $\mathbf{g}(\mathbf{x}, t)$ be a mapping of D into itself. Suppose that $\mathbf{g} : D \times [a, b] \to D$ is continuous and that there is a constant $0 \leq c < 1$, independent of t, such that

$$\|\mathbf{g}(\mathbf{x}, t) - \mathbf{g}(\mathbf{y}, t)\| \leq c\|\mathbf{x} - \mathbf{y}\|.$$

$\mathbf{x}_*(t)$ be the fixed-point of $g(\mathbf{x}, t)$. Show that $t \to \mathbf{x}_*(t)$ is continuous from $[a, b]$ into D.

4. Show that the function **g** of Example 11.1 is also a contraction on $D = \{0 \le x, y \le 1\}$ in the norm $\| \ \|_2$. Use the fact that $\|\mathbf{Dg}\|_2^2 \le \|\mathbf{Dg}\|_F^2 = \|\nabla g_1\|_2^2 + \|\nabla g_2\|_2^2$.

5. Let

$$g(x, y) = \left(0.6e^{-(x+y)}, \ \sin \left(\frac{y+1}{3} \right) \right).$$

a) Show that the function **g** has a unique fixed-point in the square $D = \{0 \le x, y \le 1\}$. Make estimates $\max_R \|\mathbf{Dg}\|_1 \le c_1$ and $\max_R \|\mathbf{Dg}\|_\infty \le c_\infty$. You only need to show that one of these constants is less than 1 to show that **g** is a contraction on D.

b) Write a code to plot and compute the functional iterates. Start at the point $(1, 1)$. What is the rate of convergence of the iterates?

11.3 Newton's Method

Recall from Chapter 6 that Newton's method is the iterative scheme:

$$x_{k+1} = x_k - \frac{f(x_k)}{f'(x_k)}.$$

We derive Newton's method for a vector-valued mapping $\mathbf{f}(\mathbf{x})$ defined on an open set $G \subset \mathbb{R}^n$ with values in $\mathbb{R}^n$. Recall that at a point $\mathbf{x}$ where each component of $\mathbf{f}$ has a linear approximation, we may write

$$\mathbf{f}(\mathbf{x} + \mathbf{h}) = \mathbf{f}(\mathbf{x}) + \mathbf{Df}(\mathbf{x}) + \mathbf{R}(\mathbf{h}) \tag{11.9}$$

where

$$\lim_{\|\mathbf{h}\| \to 0} \frac{\|\mathbf{R}\|}{\|\mathbf{h}\|} = 0.$$

Now suppose that $\mathbf{x}_*$ is a root of $\mathbf{f}(\mathbf{x}) = 0$ and that $\mathbf{f}$ is C^1 in some neighborhood U of $\mathbf{x}_*$. Let $\mathbf{x}_0 \in U$ be a first approximation to this root. We use (11.9) to make a linear approximation to $\mathbf{f}$ at $\mathbf{x}_0$:

$$\mathbf{f}(\mathbf{x}) = \mathbf{f}(\mathbf{x}_0) + \mathbf{Df}(\mathbf{x}_0)(\mathbf{x} - \mathbf{x}_0) + \mathbf{R}.$$

Instead of trying to solve the equation $\mathbf{f}(\mathbf{x}) = 0$, we consider the linear equation,

$$0 = \mathbf{f}(\mathbf{x}_0) + \mathbf{Df}(\mathbf{x}_0)(\mathbf{x} - \mathbf{x}_0).$$

If $\mathbf{R}$ is small, we hope that the solution of this linear equation will be a good approximation to the solution of $\mathbf{f}(\mathbf{x}) = 0$. We assume that $\mathbf{Df}(\mathbf{x})$ is invertible for $\mathbf{x}$ in U. When we invert $\mathbf{Df}(\mathbf{x}_0)$ and multiply by the inverse, we arrive at the next approximation to the root,

$$\mathbf{x}_1 = \mathbf{x}_0 - \mathbf{Df}(\mathbf{x}_0)^{-1}\mathbf{f}(\mathbf{x}_0).$$

We hope that $\mathbf{x}_1$ is closer to the root than was $\mathbf{x}_0$. More generally, we have the iterative scheme

$$\mathbf{x}_{k+1} = \mathbf{x}_k - \mathbf{Df}(\mathbf{x}_k)^{-1}\mathbf{f}(\mathbf{x}_k). \tag{11.10}$$

It is not difficult to repeat the proof of Theorem 6.5 in higher dimensions. We would assume that we know a root $\mathbf{x}_*$ of $\mathbf{f}(\mathbf{x}) = 0$ and that $\mathbf{Df}(\mathbf{x}_*)$ is invertible. Then we would prove that if our starting point $\mathbf{x}_0$ is close enough to $\mathbf{x}_*$, the Newton iterates converge, at least quadratically, to $\mathbf{x}_*$. We leave it to the reader to provide a proof in higher dimensions in the exercises.

Instead we choose to state and prove a simplified version of the theorem of Kantorovich, which uses Newton's method to prove the *existence* of a root $\mathbf{x}_*$ of $\mathbf{f}(\mathbf{x}) = 0$. For a statement and proof of the Kantorovich theorem, see the book of J. E. Dennis and R. B. Schnabel [De].

Because the existence of the root is not assumed, we shall need to make a careful assumption about the values of the function and its derivatives at the starting point $\mathbf{x}_0$. Newton's method depends on the accuracy of the linear approximation, and we need a lemma to estimate the remainder.

When $\mathbf{f}$ is C^1, we have the linear approximation

$$\mathbf{f}(\mathbf{x}) = \mathbf{f}(\mathbf{x}_0) + \mathbf{Df}(\mathbf{x}_0)(\mathbf{x} - \mathbf{x}_0) + \mathbf{R}(\mathbf{x})$$

where $\|\mathbf{R}(\mathbf{x})\| = o(\|\mathbf{x} - \mathbf{x}_0\|)$ as $\mathbf{x}$ tends to $\mathbf{x}_0$. The following lemma gives a refinement of this result when we make a stronger hypothesis on $\mathbf{f}$.

Lemma 11.5 Let $\mathbf{f}$ be a C^1 function defined on an open set $G \subset \mathbb{R}^n$ with values in $\mathbb{R}^m$. Suppose that $\mathbf{x} \to \mathbf{Df}(\mathbf{x})$ is Lipschitz continuous on G with Lipschitz constant L. Then for each $\mathbf{x}_0 \in G$, there is a $\delta > 0$ such that the remainder $\|\mathbf{R}(\mathbf{x})\| \leq (L/2)\|\mathbf{x} - \mathbf{x}_0\|^2$ for $\|\mathbf{x} - \mathbf{x}_0\| \leq \delta$.

Proof: Notice that $\mathbf{R}(\mathbf{x}_0) = 0$ and that $\mathbf{x} \to \mathbf{R}(\mathbf{x})$ is C^1 because $\mathbf{DR}(\mathbf{x}) = \mathbf{Df}(\mathbf{x}) - \mathbf{Df}(\mathbf{x}_0)$. We could just apply the vector-valued mean value theorem (Theorem 10.12) to $\mathbf{R}$, but that would give us the estimate with a factor of L. However, we need the slightly better result with $L/2$. Thus we start again as we did in the proof of Theorem 10.12. We choose $\delta > 0$ so that the convex set $B(\mathbf{x}_0, \delta) \subset G$ and we suppose that $\|\mathbf{x} - \mathbf{x}_0\| < \delta$. Let $\mathcal{L}$ be the line between $\mathbf{x}_0$ and $\mathbf{x}$. We parameterize this line with

$$\mathbf{g}(t) = t\mathbf{x} + (1 - t)\mathbf{x}_0, \quad 0 \leq t \leq 1.$$

Then

$$\mathbf{R}(\mathbf{x}) = \mathbf{R}(\mathbf{x}) - \mathbf{R}(\mathbf{x}_0) = \mathbf{f}(\mathbf{x}) - \mathbf{f}(\mathbf{x}_0) - \mathbf{Df}(\mathbf{x}_0)(\mathbf{x} - \mathbf{x}_0)$$

$$= \int_0^1 [\mathbf{Df}(\mathbf{g}(t)) - \mathbf{Df}(\mathbf{g}(0))](\mathbf{x} - \mathbf{x}_0) \, dt$$

Now note that $\mathbf{g}(t) - \mathbf{g}(0) = t(\mathbf{x} - \mathbf{x}_0)$. By our hypothesis on $\mathbf{Df}$, we have

$$\|\mathbf{Df}(\mathbf{g}(t)) - \mathbf{Df}(\mathbf{g}(0))\| \leq L\|\mathbf{g}(t) - \mathbf{g}(0)\| = Lt\|\mathbf{x} - \mathbf{x}_0\|.$$

Thus for $\|\mathbf{x} - \mathbf{x}_0\| \leq \delta$,

$$\|\mathbf{R}(\mathbf{x})\| \leq \int_0^1 \|\mathbf{Df}(\mathbf{g}(t)) - \mathbf{Df}(\mathbf{g}(0))\| \, \|\mathbf{x} - \mathbf{x}_0\| \, dt$$

$$\leq L \int_0^1 t\|\mathbf{x} - \mathbf{x}_0\|^2 \, dt$$

$$\leq (L/2)\|\mathbf{x} - \mathbf{x}_0\|^2.$$

Now we are ready to prove

Theorem 11.6 (Kantorovich) Let G be an open subset of $\mathbb{R}^n$, and let $\mathbf{f} : G \to \mathbb{R}^n$ be a C^1 function such that $\mathbf{x} \to \mathbf{Df}(\mathbf{x})$ is Lipschitz continuous with Lipschitz constant L. Let $\mathbf{x}_0 \in G$ and suppose that

(i) $\mathbf{Df}(\mathbf{x}_0)$ is invertible; and
(ii) $r \equiv L\|\mathbf{Df}(\mathbf{x}_0)^{-1}\|^2\|\mathbf{f}(\mathbf{x}_0)\| \leq \frac{1}{3}$.

Let $\mathbf{x}_1 = \mathbf{x}_0 - \mathbf{Df}(\mathbf{x}_0)^{-1}\mathbf{f}(\mathbf{x}_0)$ be the first Newton iterate, and let

$$\delta = \left(\frac{r}{1-r}\right)\|\mathbf{x}_1 - \mathbf{x}_0\|.$$

Suppose that

(iii) $D \equiv \{\mathbf{x} : \|\mathbf{x} - \mathbf{x}_1\| \leq \delta\} \subset G$.

Then the Newton iterates (11.10) remain in D for all $k \in \mathbb{N}$ and converge to a root $\mathbf{x}_* \in D$ of $\mathbf{f}(\mathbf{x}) = 0$. The convergence is at least quadratic.

Remarks How should the condition on r be understood? It is best to look at what it means in one dimension. When f has two derivatives, the Lipschitz constant $L \leq \max |f''(x)|$. Suppose that $f(t)$ is the height above the ground of an elevator, and that t is time. Then $f'(t)$ is the rate of ascent or of descent, and $f''(t)$ is the acceleration. The ratio r is a dimensionless quantity. Indeed, if we let $[x]$ denote the dimension of a quantity, we have

$$[r] = [f''][f']^{-2}[f] = \frac{\text{height}}{\text{time}^2}\frac{\text{time}^2}{\text{height}^2}\text{height} = 1.$$

Assume for the moment that $f(t_0) > 0$ and that the elevator is descending, $f'(t_0) < 0$. Then the elevator will reach the ground provided $f'(t_0)$ is sufficiently negative ($|f'(t_0)|^{-2}$ is sufficiently small) and provided f'' is not too large. The latter condition means that the elevator does not slow down too much.

Note that the ball D is centered at $\mathbf{x}_1$ (not at $\mathbf{x}_0$), and that if r is small, the root lies close to $\mathbf{x}_1$. In particular, if $L = 0$, which implies that $\mathbf{Df}(\mathbf{x})$ is constant, then $r = 0$. In this case, $\mathbf{x}_1$ is the root.

Proof: The plan of the proof is similar to that of Theorem 11.3. We show that the sequence of Newton iterates,

$$\mathbf{x}_{k+1} = \mathbf{x}_k - \mathbf{Df}(\mathbf{x}_k)^{-1}\mathbf{f}(\mathbf{x}_k) \tag{11.11}$$

is a Cauchy sequence and therefore converges.

First we show that, for $\mathbf{x} \in D$, we have $\mathbf{Df}(\mathbf{x})$ invertible and

$$\|\mathbf{Df}(\mathbf{x})^{-1}\| \leq 2\|\mathbf{Df}(\mathbf{x}_0)^{-1}\|. \tag{11.12}$$

In fact, if $\mathbf{x} \in D$, then

$$\|\mathbf{x} - \mathbf{x}_0\| \leq \|\mathbf{x} - \mathbf{x}_1\| + \|\mathbf{x}_1 - \mathbf{x}_0\|$$

$$\leq (\delta + 1)\|\mathbf{x}_1 - \mathbf{x}_0\| = \left(\frac{1}{1-r}\right)\|\mathbf{x}_1 - \mathbf{x}_0\|$$

$$\leq \frac{\|\mathbf{x}_1 - \mathbf{x}_0\|}{2r}$$

because $r \leq 1/3$. Hence

$$\|\mathbf{x} - \mathbf{x}_0\| \leq \frac{\|\mathbf{x}_1 - \mathbf{x}_0\|}{2r} \leq \frac{\|\mathbf{Df}(\mathbf{x}_0)^{-1}\| \, \|\mathbf{f}(\mathbf{x}_0)\|}{2r} = \frac{1}{2L\|\mathbf{Df}(\mathbf{x}_0)^{-1}\|}.$$

To make the notation more compact, for the moment, we let $\mathbf{A} = \mathbf{Df}(\mathbf{x}_0)$ and $\mathbf{B} = \mathbf{Df}(\mathbf{x})$. Thus if $\mathbf{x} \in D$, the previous inequality can be written $\|\mathbf{x} - \mathbf{x}_0\| \leq 1/(2L\|\mathbf{A}^{-1}\|)$. Using the Lipschitz continuity of $\mathbf{x} \to \mathbf{Df}(\mathbf{x})$, we see that for $\mathbf{x} \in D$,

$$\|\mathbf{A} - \mathbf{B}\| \leq L\|\mathbf{x} - \mathbf{x}_0\| \leq \frac{1}{2\|\mathbf{A}^{-1}\|}.$$

Hence

$$\|\mathbf{A}^{-1}\| \, \|\mathbf{A} - \mathbf{B}\| \leq 1/2.$$

Then (11.12) follows directly from an application of Theorem 11.2.

Now we use an induction argument to show that the Newton iterates $\mathbf{x}_k \in D$ for all k and that, for all k,

$$\|\mathbf{x}_k - \mathbf{x}_{k-1}\| \leq r^{k-1}\|\mathbf{x}_1 - \mathbf{x}_0\|. \tag{11.13}$$

We assume that $\mathbf{x}_1, \ldots, \mathbf{x}_k \in D$ and that (11.13) holds for all indices less than or equal to k. We must show that $\mathbf{x}_{k+1} \in D$ and that (11.13) holds with k replaced by $k + 1$.

For our induction argument we will need an estimate of $\|\mathbf{f}(\mathbf{x}_k)\|$. We use the linear approximation to $\mathbf{f}$ at $\mathbf{x}_{k-1}$:

$$\mathbf{f}(\mathbf{x}_k) = \mathbf{f}(\mathbf{x}_{k-1}) + \mathbf{Df}(\mathbf{x}_{k-1})(\mathbf{x}_k - \mathbf{x}_{k-1}) + \mathbf{R}(\mathbf{x}_k, \mathbf{x}_{k-1}).$$

However, x_k is chosen so that $f(x_{k-1}) + Df(x_{k-1})(x_k - x_{k-1}) = 0$. Hence by Lemma 11.5,

$$\|f(x_k)\| = \|R(x_k, x_{k-1})\| \le \frac{L}{2}\|x_k - x_{k-1}\|^2.$$

Assuming that the induction hypothesis (11.13) holds, we have

$$\begin{aligned}
\|f(x_k)\| &\le \frac{L}{2}(r^{k-1}\|x_1 - x_0\|)^2 \\
&\le \frac{r^{2k-2}L}{2}\|Df(x_0)^{-1}\| \, \|f(x_0)\| \, \|x_1 - x_0\|.
\end{aligned} \tag{11.14}$$

Then using (11.12) to estimate $\|Df(x_k)^{-1}\|$, and using (11.14) to estimate $\|f(x_k)\|$, we have

$$\begin{aligned}
\|x_{k+1} - x_k\| &\le \|Df(x_k)^{-1}\| \, \|f(x_k)\| \\
&\le 2\|Df(x_0)^{-1}\| \|f(x_k)\| \\
&= r^{2k-2}L\|Df(x_0)^{-1}\|^2\|f(x_0)\| \|x_1 - x_0\| \\
&\le r^{2k-1}\|x_1 - x_0\| \\
&\le r^k\|x_1 - x_0\|.
\end{aligned}$$

Finally to show that $x_{k+1} \in D$, we use the triangle inequality.

$$\begin{aligned}
\|x_{k+1} - x_1\| &\le \|x_{k+1} - x_k\| + \cdots + \|x_1 - x_1\| \\
&\le (r + \cdots + r^k)\|x_0 - x_0\| \\
&\le \left(\frac{r}{1-r}\right)\|x_1 - x_0\| = \delta.
\end{aligned}$$

Thus $x_{k+1} \in D$. The inequality (11.13) for the difference between successive iterates can be used to show that the sequence x_k is a Cauchy sequence just as in the proof of Theorem 11.3. Hence x_k converges to some point x_*, and $x_* \in D$ because D is closed. The estimate (11.14) implies that $f(x_k)$ tends to zero as $k \to \infty$. Since f is continuous,

$$f(x_*) = \lim_{k \to \infty} f(x_k) = 0.$$

It remains to be shown that $\mathbf{x}_k$ converges at least quadratically to $\mathbf{x}_*$. That is, we must show that there is a constant K such that, for k sufficiently large,

$$\|\mathbf{x}_{k+1} - \mathbf{x}_*\| \le K\|\mathbf{x}_k - \mathbf{x}_*\|^2.$$

The $(k + 1)$st iterate $\mathbf{x}_{k+1}$ is defined by the equation

$$0 = \mathbf{f}(\mathbf{x}_k) + \mathbf{Df}(\mathbf{x}_k)(\mathbf{x}_{k+1} - \mathbf{x}_k). \tag{11.15}$$

Using the linear approximation of $\mathbf{f}$ at $\mathbf{x}_k$, we have

$$0 = \mathbf{f}(\mathbf{x}_*) = \mathbf{f}(\mathbf{x}_k) + \mathbf{Df}(\mathbf{x}_k)(\mathbf{x}_* - \mathbf{x}_k) + \mathbf{R}(\mathbf{x}_*, \mathbf{x}_k).$$

Subtracting (11.15) from this equation yields

$$0 = \mathbf{Df}(\mathbf{x}_k)(\mathbf{x}_* - \mathbf{x}_{k+1}) + \mathbf{R}(\mathbf{x}_*, \mathbf{x}_k)$$

whence

$$\mathbf{x}_{k+1} - \mathbf{x}_* = \mathbf{Df}(\mathbf{x}_k)^{-1}\mathbf{R}(\mathbf{x}_*, \mathbf{x}_k).$$

Now we apply Lemma 11.5 and (11.12) to deduce

$$\|\mathbf{x}_{k+1} - \mathbf{x}_*\| \le L\|\mathbf{Df}(\mathbf{x}_0)^{-1}\| \, \|\mathbf{x}_k - \mathbf{x}_*\|^2.$$

If we take $K = L\|\mathbf{Df}(\mathbf{x}_0)^{-1}\|$, we see that $\mathbf{x}_k$ converges at least quadratically to $\mathbf{x}_*$. The theorem is proved.

In the induction argument of the proof, we could have made sharper estimates and shown that (11.13) can be refined to

$$\|\mathbf{x}_k - \mathbf{x}_{k-1}\| \le r^{2^{k-1}-1}\|\mathbf{x}_1 - \mathbf{x}_0\|.$$

This estimate shows that the iterates are converging more rapidly than is the case for functional iteration and is an indication that the convergence is superlinear.

EXAMPLE 11.3: Let $f(x) = x^2$. The Lipschitz constant $L = 2$. For any $x_0 \neq 0$, the ratio

$$r = 2\frac{f(x_0)}{|f'(x_0)|^2} = \frac{1}{2}$$

so that Theorem 11.6 cannot be used to predict the zero at $x = 0$. This negative result is consistent with the fact that Newton's method converges linearly to this zero.

On the other hand, if we take $f(x) = x^2 - 1$, the ratio is

$$r = 2\frac{|f(x_0)|}{|f'(x_0)|^2} = \frac{|x_0^2 - 1|}{2x_0^2}.$$

Then for $x_0 \geq \sqrt{3/5}$, $r \leq 1/3$. Theorem 11.7 then guarantees that Newton's method, with starting point x_0, will converge quadratically to the root at $x = 1$.

The Lipschitz constant L of the matrix-valued function $\mathbf{x} \to \mathbf{Df}(\mathbf{x})$ plays an important role in the statement and proof of Theorem 11.6. We need a practical way of estimating L. We were able to estimate the Lipschitz constant of the function $\mathbf{f}(\mathbf{x})$ in terms of $\mathbf{Df}$ using the mean value theorem. To estimate the Lipschitz constant of the function $\mathbf{Df}$ using the mean value theorem, we shall need the notion of the second order partial derivatives of a function. This idea is introduced in Chapter 12, but for right now, we assume that the reader is familiar with this concept.

Theorem 11.7 Let G be an open convex subset of $\mathbb{R}^n$, and suppose that $\mathbf{f} : G \to \mathbb{R}^m$ with $\mathbf{f} \in C^1(G)$. Suppose in addition that all of the second order partial derivatives of the component functions,

$$\frac{\partial^2 f_i}{\partial x_j \partial x_k}, \quad i, j, k = 1, \ldots, n$$

exist and are continuous on G. Then for all $\mathbf{x}, \mathbf{y} \in G$,

$$\|\mathbf{Df}(\mathbf{x}) - \mathbf{Df}(\mathbf{y})\|_\infty \leq L\|\mathbf{x} - \mathbf{y}\|_\infty$$

where

$$L \leq \sup_{G} \left\{ \max_{1 \leq i \leq m} \sum_{j,k=1}^{n} \left| \frac{\partial^2 f_i}{\partial x_j \partial x_k} \right| \right\}$$

when this quantity is finite.

Proof: Suppose that $G \subset \mathbb{R}^n$ is an open convex set and suppose that $\mathbf{x} \to \mathbf{A}(\mathbf{x})$ is a matrix-valued function on G that has C^1 entries $a_{ij}(\mathbf{x})$, $i = 1, \ldots, m$, $j = 1, \ldots, n$. How can we estimate the constant L such that

$$\|\mathbf{A}(\mathbf{x}) - \mathbf{A}(\mathbf{y})\| \leq L \|\mathbf{x} - \mathbf{y}\|$$

using the first derivatives of a_{ij}? We will find an estimate in the ∞ norm.

The mean value theorem, Theorem 10.7, says that, for each i, j, there is a point $\mathbf{z}_{ij}$ on the line between $\mathbf{x}$ and $\mathbf{y}$ such that

$$a_{ij}(\mathbf{x}) - a_{ij}(\mathbf{y}) = \langle \nabla a_{ij}(\mathbf{z}_{ij}), \mathbf{x} - \mathbf{y} \rangle.$$

By the variation on the Schwarz inequality (exercise 8, Section 9.2), we have

$$|a_{ij}(\mathbf{x}) - a_{ij}(\mathbf{y})| \leq \|\nabla a_{ij}(\mathbf{z}_{ij})\|_1 \|\mathbf{x} - \mathbf{y}\|_\infty.$$

Then

$$\|\mathbf{A}(\mathbf{x}) - \mathbf{A}(\mathbf{y})\|_\infty = \sup_{1 \leq i \leq m} \sum_{j=1}^{n} |a_{ij}(\mathbf{x}) - a_{ij}(\mathbf{y})|$$

$$\leq \left(\max_{1 \leq i \leq m} \sum_{j=1}^{n} \|\nabla a_{ij}(\mathbf{z}_{ij})\|_1 \right) \|\mathbf{x} - \mathbf{y}\|_\infty.$$

Hence

$$\|\mathbf{A}(\mathbf{x}) - \mathbf{A}(\mathbf{y})\|_\infty \leq L \|\mathbf{x} - \mathbf{y}\|_\infty$$

where

$$L = \sup_{\mathbf{x} \in G} \left\{ \max_{1 \leq i \leq m} \sum_{j=1}^{n} \|\nabla a_{ij}(\mathbf{x})\|_1 \right\}. \tag{11.16}$$

Now we apply this estimate to the case in which $A(x) = Df(x)$ where $f : G \to \mathbb{R}^m$. The elements of $A = Df$ are

$$a_{ij} = \frac{\partial f_i}{\partial x_j}, \quad i = 1, \ldots, m, \; j = 1, \ldots, n$$

and

$$\nabla a_{ij} = \left(\frac{\partial^2 f_i}{\partial x_1 \partial x_j}, \ldots, \frac{\partial^2 f_i}{\partial x_n \partial x_j} \right)$$

so that

$$\|\nabla a_{ij}(x)\|_1 = \sum_{k=1}^{m} \left| \frac{\partial^2 f_i}{\partial x_k \partial x_j}(x) \right|.$$

Hence for all $x, y \in G$,

$$\|Df(x) - Df(y)\|_\infty \leq L \|x - y\|_\infty.$$

Using (11.16), L may be estimated

$$L \leq \max_{x \in G} \left\{ \max_{1 \leq i \leq m} \sum_{j,k=1}^{n} \left| \frac{\partial^2 f_i}{\partial x_j \partial x_k}(x) \right| \right\}. \tag{11.17}$$

Newton's method is implemented on a computer as follows. Enter the initial guess x_0 and iterate with a loop.

```
x = x_0
for k = 1 : N
        solve the linear system Df(x)s = -f(x).
        x = x + s
end the loop
```

As a rule in computing, we do not compute the inverse of a matrix. It is more efficient and more accurate to solve a linear system using a method like Gaussian elimination. The quantity s is called the "Newton step."

EXAMPLE 11.4: Let $f(x, y) = x - \sin(x + y)$, $g(x, y) = y - \cos(x - y)$ and $f(x, y) = (f(x, y), g(x, y))$. A look at the contour lines

$\{(x, y) : f(x, y) = 0\}$ and $\{(x, y) : g(x, y) = 0\}$ shows that there is a root near $(x, y) = (1, 1)$. We check the conditions of Theorem 11.6 to see if Newton's method starting at $(x_0, y_0) = (1, 1)$ will converge.

First we see that $\mathbf{f}(1, 1) = (1 - \sin(2), 0)$ with $\|\mathbf{f}(1, 1)\|_\infty = |1 - \sin(2)| = 0.0907$. Next we see that

$$\mathbf{Df}(x, y) = \begin{bmatrix} 1 - \cos(x + y) & -\cos(x + y) \\ \sin(x - y) & 1 - \sin(x - y) \end{bmatrix}$$

so that

$$\mathbf{Df}(1, 1) = \begin{bmatrix} 1 - \cos(2) & -\cos(2) \\ 0 & 1 \end{bmatrix}.$$

Then

$$\mathbf{Df}(1, 1)^{-1} = \frac{1}{1 - \cos(2)} \begin{bmatrix} 1 & \cos(2) \\ 0 & 1 - \cos(2) \end{bmatrix}$$

and

$$\|\mathbf{Df}(1, 1)^{-1}\|_\infty = \max \left\{ \frac{1 + |\cos(2)|}{|1 - \cos(2)|}, 1 \right\} = 1.$$

We must estimate the Lipschitz constant for $\mathbf{Df}$ using (11.17). We have

$$|f_{xx}| = |f_{xy}| = |f_{yy}| = |\sin(x + y)|$$

and

$$|g_{xx}| = |g_{xy}| = |g_{yy}| = |\cos(x - y)|.$$

Substituting in (11.17), we find that

$$L \leq 4 \max\{|\sin(x + y)|, |\cos(x - y)|\} \leq 4.$$

Thus for our problem, starting at $(x_0, y_0) = (1, 1)$, the ratio

$$r = L\|\mathbf{Df}(1, 1)^{-1}\|_\infty^2 \|\mathbf{f}(1, 1)\|_\infty \leq 4(0.0907) = 0.3628.$$

Thus our estimate for r is slightly larger than the required $1/3$ for Theorem 11.6, but we can try out Newton's method anyway. Remember that $\mathbf{s} = \mathbf{x}_k - \mathbf{x}_{k-1}$ is the Newton step.

iterate	x	y	$\|\mathbf{s}\|_\infty$
1	0.93595115221563	1.00000000000000	0.06404884778437
2	0.93508466508785	0.99802079332802	0.00197920667198
3	0.93508206412647	0.99802005816193	0.00000260096138
4	0.93508206412310	0.99802005816010	0.00000000000336
5	0.93508206412310	0.99802005816010	0

An important practical consideration when using iterative procedures is to know when to stop. How many iterations do we need to achieve a prescribed error tolerance? We must be able to estimate how close we are to the exact solution by looking at the iterates. In the contraction mapping theorem, the error

$$\|\mathbf{x}_k - \mathbf{x}_*\| \approx \frac{1}{1-c}\|\mathbf{x}_k - \mathbf{x}_{k+1}\|.$$

If c is close to 1, the difference $\|\mathbf{x}_k - \mathbf{x}_{k+1}\|$ can underestimate the error.

However, due to the quadratic convergence of Newton's method in the case of a simple root, these differences are a good estimate of the error. For Newton's method, we have

$$\|\mathbf{x}_k - \mathbf{x}_*\| \approx \|\mathbf{x}_k - \mathbf{x}_{k+1}\|.$$

Newton's method yields very rapid convergence when we know the approximate location of the root and can start with a good first guess $\mathbf{x}_0$. However, with a bad first guess, the method may not converge. For this reason, a root-finding procedure usually starts out with a more stable, but slower, root-finding algorithm, and it may finish off with Newton's method.

Exercises for 11.3

1. In this exercise, you are asked to provide a proof of the following theorem that asserts that if a root is known to exist, then Newton's method converges.

   Theorem Let $G \subset \mathbb{R}^n$ be an open set. Assume that $\mathbf{f} : G \to \mathbb{R}^n$ is C^1 and that $\mathbf{x} \to D\mathbf{f}(\mathbf{x})$ is Lipschitz continuous on G with Lipschitz constant L. Suppose that for some $\mathbf{x}_* \in G$, $\mathbf{f}(\mathbf{x}_*) = 0$ and that $D\mathbf{f}(\mathbf{x}_*)$ is invertible. Then there is a $\delta > 0$ such that, if $\|\mathbf{x}_0 - \mathbf{x}_*\| < \delta$, the Newton iterates, given by (11.10), converge at least quadratically to $\mathbf{x}_*$.

a) Use the Lipschitz continuity of $\mathbf{Df}$ and Theorem 11.2 to show that there is a $\delta_0 > 0$ such that $\mathbf{Df}(\mathbf{x})$ is invertible for $\mathbf{x} \in \bar{B}(\mathbf{x}_*, \delta_0)$ with $\|\mathbf{Df}(\mathbf{x})^{-1}\| \le 2\|\mathbf{Df}(\mathbf{x}_*)^{-1}\|$.

b) Suppose that the kth iterate $\mathbf{x}_k \in \bar{B}(\mathbf{x}_*, \delta_0)$. Using the argument following equation (11.5), deduce that

$$\|\mathbf{x}_{k+1} - \mathbf{x}_k\| \le L\|\mathbf{Df}(\mathbf{x}_*)^{-1}\| \, \|\mathbf{x}_k - \mathbf{x}_*\|^2.$$

c) If necessary, choose $\delta \le \delta_0$ so that $c \equiv L\|\mathbf{Df}(\mathbf{x}_*)^{-1}\|\delta < 1$. Show that if $\mathbf{x}_k \in \bar{B}(\mathbf{x}_*, \delta)$, then so is $\mathbf{x}_{k+1}$, and that the sequence $\mathbf{x}_k$ converges to $\mathbf{x}_*$. The inequality of part b) shows that the convergence is at least quadratic.

2. Go through the proof of Theorem 11.6 and show that by taking more care with the book-keeping, we get the estimate

$$\|\mathbf{x}_{k+1} - \mathbf{x}_k\| \le r^{2^k-1}.$$

3. Let $\mathbf{f}(x, y) = (f(x, y), g(x, y))$ where

$$f(x, y) = 2x + y + (1/8)\sin(x - y) - 3$$
$$g(x, y) = -x + 4y - 4.$$

We shall use $\mathbf{p} = (x, y)$.

a) Use Theorem 11.7 to show that

$$\|\mathbf{Df}(\mathbf{p}_1) - \mathbf{Df}(\mathbf{p}_2)\|_\infty \le (1/2)\|\mathbf{p}_1 - \mathbf{p}_2\|_\infty$$

for all $\mathbf{p}_1, \mathbf{p}_2 \in \mathbb{R}^2$.

b) Show that $\|\mathbf{Df}(1, 1)^{-1}\|_\infty \le 2/3$.

c) Show that the hypotheses of Theorem 11.6 are satisfied at $\mathbf{p}_0 = (1, 1)$. Here $G = \mathbb{R}^2$.

d) Use a two-dimensional Newton code, starting at $\mathbf{p}_0 = (1, 1)$, to approximate this root with an error no larger than $\varepsilon = 10^{-6}$.

4. Consider the system of equations

$$x + y^2 + z^2 = a$$
$$y - x^2 = b$$
$$z + x^2 = c.$$

Show that there is an $\varepsilon > 0$ such that when $|a|, |b|, |c| \le \varepsilon$, there is a solution of this system near $(0, 0, 0)$.

5. Let $\mathbf{f}(x, y) = (f(x, y), g(x, y))$ where

$$f(x, y) = \lambda x + y + \alpha(x, y)$$
$$g(x, y) = x + \lambda y + \beta(x, y)$$

where $\alpha, \beta \in C^2(\mathbb{R}^2)$ with all second derivatives bounded on $\mathbb{R}^2$ by a constant $M \geq 0$.

a) Calculate $\|\mathbf{Df}(0, 0)^{-1}\|_\infty$ as a function of λ.

b) Calculate an estimate for the Lipschitz constant L of $(x, y) \to \mathbf{Df}(x, y)$.

c) Show that for $|\lambda|$ sufficiently large, the equation $\mathbf{f}(x, y) = (0, 0)$ has a solution near $(0, 0)$.

6. Write the fixed point problem of exercise 5 of Section 11.2 in the form $\mathbf{f}(\mathbf{x}) = 0$. Use a two-dimensional Newton code to compute numerical approximations to the solution. Start at the point $(1, 1)$. Compare the order of convergence with that of functional iteration in exercise 5 of Section 11.2.

7. Here is a problem from chemical engineering (thanks to William Schiesser of Lehigh). A continuously stirred tank reactor is a vessel through which chemicals flow. While passing through the vessel, they react. We take the volume of the vessel to be V and the flow rate to be Q. We shall consider a reaction between four chemicals, A, B, C, and D. The reaction between these four chemicals takes place according to the following rules:

$$A \to 2B \quad \text{with reaction rate } r_1,$$
$$A \to C \quad \text{with reaction rate } r_2,$$
$$C \to A \quad \text{with reaction rate } r_3,$$
$$B \to C + D \quad \text{with reaction rate } r_4.$$

Let A_0 be the inflow concentration of chemical A. Making an abuse of notation, let A stand for the concentration of chemical A in gram-moles/liter, and similarly for chemicals B, C, and D. The steady state balance equations for the concentrations A, B, C, and D are

$$f_1(A, B, C, D) = -A + A_0 + V(-r_1 - r_2 + r_3)/Q = 0$$
$$f_2(A, B, C, D) = -B + V(2r_1 - r_4)/Q = 0$$
$$f_3(A, B, C, D) = -C + V(r_2 - r_3 + r_4)/Q = 0$$
$$f_4(A, B, C, D) = -D + Vr_4/Q = 0.$$

The reaction rates depend on the amount of chemical present, sometimes in a nonlinear fashion. They are given by

$$r_1 = k_1 A, \quad k_1 = 1.0$$
$$r_2 = k_2 A^{3/2}, \quad k_2 = 0.2$$
$$r_3 = k_3 C^2, \quad k_3 = 0.05$$
$$r_4 = k_4 B^2, \quad k_4 = 0.4.$$

Suppose the volume is $V = 100$ liters, the flow rate is $Q = 50$ liters/second,
and the feed to the reactor is $A_0 = 1.0$ gram mol/liter.

a) Let $\mathbf{f} = (f_1(A, B, C, D), f_2(A, B, C, D), f_3(A, B, C, D), f_4(A, B, C, D))$.
   Show that $\det \mathbf{Df}(A, 0, 0, 0) > 0$ for all $A > 0$.
b) Solve the system $\mathbf{f} = 0$ for $A_0 = 1$ using Newton's method. Take a
   starting point of $(A_0, 0, 0, 0)$.
c) Try different values of A_0. Does the solution (A, B, C, D) appear to
   depend in a continuous fashion on A_0?

11.4 The Inverse Function Theorem

To establish that a function $\mathbf{f}(\mathbf{x})$ has an inverse, we must demonstrate that for
each $\mathbf{y}$, the equation

$$\mathbf{f}(\mathbf{x}) = \mathbf{y}$$

has a unique solution. This defines the inverse function $\mathbf{y} \to \mathbf{g}(\mathbf{y})$. In addition,
we want to know under what conditions the solution $\mathbf{g}(\mathbf{y})$ depends continuously or in a C^1 fashion on $\mathbf{y}$.

   In one dimension, if $f'(x_0) \neq 0$ and f' is continuous, we know that f is either
strictly increasing or strictly decreasing on some interval $I = \{|x - x_0| \leq \delta\}$.
This means that f is one-to-one on I and possesses an inverse there. In higher
dimensions we will make the hypothesis that $\mathbf{Df}(\mathbf{x}_0)$ is invertible.

   We will also need the hypothesis that $\mathbf{Df}(\mathbf{x}_0)$ is invertible to ensure that the
inverse function $\mathbf{g}$ is C^1. Consider $f(x) = x^3$; f is one-to-one on all of $\mathbb{R}$ with
inverse $g(y) = y^{1/3}$. However, the inverse function fails to be differentiable at
$y = 0$, because $f'(0) = 0$.

Theorem 11.8 (Inverse function theorem) Let G be an open set in $\mathbb{R}^n$, and let
$\mathbf{x}_0 \in G$. Let $\mathbf{f} : G \to \mathbb{R}^n$ be a C^1 function such that $\mathbf{Df}(\mathbf{x}_0)$ is invertible. Then

there is an open set $U \subset G$, with $x_0 \in U$, such that the image of U under f is an open subset V of $\mathbb{R}^n$; and there is a C^1 function $g : V \to \mathbb{R}^n$ such that $g(f(x)) = x$ for $x \in U$, and $f(g(y)) = y$ for $y \in V$. Furthermore, if $y = f(x)$, then $Dg(y) = Df(x)^{-1}$.

Here is a plan of the proof. After reducing the problem to a simpler case, we will first show that f is one-to-one on some open ball. This is done in a lemma. Then to show that we can solve the equation for y in an open set containing y_0, we use the contraction mapping theorem.

Proof: Because $Df(x_0)$ is invertible, we can consider the problem of finding the inverse of the modified function

$$f_1(x) \equiv Df(x_0)^{-1}f(x).$$

The modified function f_1 will have a C^1 inverse if and only if this is true of f. Note that $Df_1(x_0) = I$, the identity matrix. Thus we may assume that $Df(x_0) = I$.

Let $y_0 = f(x_0)$. Now we translate the coordinates in both domain and range space. Set $f_2(x) = f(x + x_0) - y_0$. Then $f_2(0) = 0$. The theorem is true for f_2 if and only if it is true for f. Thus we can assume $x_0 = y_0 = 0$. Hence the function f of our reduced problem satisfies

$$f(0) = 0 \quad \text{and} \quad Df(0) = I.$$

Now we show that f is one to one on a ball that contains 0. We place this part of the proof in a lemma. Recall that $B(0, r) = \{\|x\| < r\}$ is the open ball and $\bar{B}(0, r) = \{\|x\| \leq r\}$ is the closed ball with center 0 and radius r.

Lemma 11.9 Let G be an open subset of $\mathbb{R}^n$ with $0 \in G$, and let $f : G \to \mathbb{R}^n$ be a C^1 function. Assume that $f(0) = 0$ and that $Df(0) = I$. Then,

(i) there is an $r > 0$ such that $\bar{B}(0, r) \subset G$, and

$$\|Df(x) - Df(0)\| \leq 1/2 \text{ for } x \in \bar{B}(0, r);$$

(ii) f is one to one on $\bar{B}(0, r)$; and
(iii) the inverse g, defined on $f(\bar{B}(0, r))$, is continuous.

Proof of Lemma (i) Because $Df(x)$ is continuous, we may choose $r > 0$ such that $\|Df(x) - Df(0)\| \leq 1/2$ for $x \in \bar{B}(0, r)$.

To prove (ii), we observe that because of our normalization of the problem, $f(0) = 0$ and $Df(0) = I$. Consequently, the linear approximation to f at $x = 0$ is just $l(x) = x$ so that $f(x) = x + p(x)$ where $p(x)$ is the remainder. Now $Dp(x) = Df(x) - I = Df(x) - Df(0)$ so $\|Dp(x)\| \leq 1/2$ for $\|x\| \leq r$. Let x_1, $x_2 \in \bar{B}(0, r)$, and let $\mathcal{L}$ be the line between x_1 and x_2. Then by the mean value theorem (Theorem 10.12),

$$\|p(x_1) - p(x_2)\| \leq \max_{z \in l} \|Dp(z)\| \|x_1 - x_2\| \leq (1/2)\|x_1 - x_2\|. \qquad (11.18)$$

Now

$$f(x_1) - f(x_2) = x_1 - x_2 + p(x_1) - p(x_2)$$

so that

$$\|f(x_1) - f(x_2)\| \geq \|x_1 - x_2\| - \|p(x_1) - p(x_2)\|$$
$$\geq (1/2)\|x_1 - x_2\|. \qquad (11.19)$$

Hence f is one to one on $\bar{B}(0, r)$. Therefore an inverse function g is defined on the image of $\bar{B}(0, r)$ under f. It follows from (11.19) that if $y_j = f(x_j)$, $j = 1, 2$, then

$$\|g(y_1) - g(y_2)\| \leq 2\|y_1 - y_2\|. \qquad (11.20)$$

This inequality proves both (ii) and (iii).

Proof of Theorem In the next step of the proof, we characterize the image of $\bar{B}(0, r)$ under f. Suppose that $y \in \bar{B}(0, r/2)$. We wish to prove that there exists a solution $x_* \in B(0, r)$ of the equation

$$f(x) = y$$

using the contraction mapping theorem. To do so, we put the equation into the equivalent fixed-point form

$$x = h_y(x) \equiv x - f(x) + y = y - p(x).$$

The equation $f(x) = y$ has a solution in $\bar{B}(0, r)$ if and only if h_y has a fixed-point in $\bar{B}(0, r)$. Recall that in the first part of the proof, we "divided by" the derivative $Df(x_0)$ so as to make $Df_1(x_0) = I$. Then we moved x_0 to 0 by translation. This means that $D_x h = I - Df(x)$ will be small near $x = 0$. The same strategy was used in Example 6.4 to find an interval on which the functional iterates would converge.

First we will show that h_y maps $\bar{B}(0, r)$ into itself. To see this, use (11.18), with $x_1 = x$ and $x_2 = 0$ to deduce

$$\|p(x)\| \leq \|x\|/2 \leq r/2 \quad \text{for } x \in \bar{B}(0, r).$$

Hence, for $y \in \bar{B}(0, r/2)$,

$$\|h_y(x)\| = \|y - p(x)\| \leq r/2 + r/2 = r.$$

Next we verify that h_y is a contraction mapping on $\bar{B}(0, r)$. We have

$$h_y(x_1) - h_y(x_2) = p(x_2) - p(x_1).$$

Again by (11.18),

$$\|h_y(x_1) - h_y(x_2)\| \leq (1/2)\|x_2 - x_1\|.$$

This establishes that h_y is a contraction mapping on the closed set $\bar{B}(0, r)$. Hence h_y has a unique fixed point in $\bar{B}(0, r)$ that is the solution of the equation $f(x) = y$.

Now let V be the open ball $B(0, r/2)$, and let

$$U = B(0, r) \cap \{x : f(x) \in B(0, r/2)\}.$$

U is the intersection of two open sets, and hence is open. We have $f : U \to V$ and $g : V \to U$.

It only remains to be shown that g is C^1 on V. Now for $x \in \bar{B}(0, r)$,

$$\|I - Df(x)\| = \|Df(0) - Df(x)\| \leq 1/2.$$

Hence by Theorem 11.2, $Df(x)$ is invertible for $x \in \bar{B}(0, r)$. Using Cramer's rule, $Df(x)^{-1}$ is constructed from rational functions of the elements of $Df(x)$ which are continuous. Hence $x \to Df(x)^{-1}$ is continuous on $\bar{B}(0, r)$.

Now we show that g has a linear approximation at each point of V. We must find a matrix-valued function $y \to A(y)$ on V such that, if $y_1, y_2 \in V$, then $g(y_2)$ can be approximated

$$g(y_2) = g(y_1) + A(y_1)(y_2 - y_1) + R(y_2, y_1)$$

where

$$\lim_{\|y_1 - y_2\| \to 0} \frac{\|R(y_1, y_2)\|}{\|y_1 - y_2\|} = 0.$$

Let $x_1 = g(y_1)$ and $x_2 = g(y_2)$. As our candidate for $A(y_1)$ we take

$$A(y_1) = Df(x_1)^{-1}.$$

Now with $\mathcal{L}$ being the line between x_1 and x_2, we have

$$
\begin{aligned}
\|R(y_1, y_2)\| &= \|g(y_2) - g(y_1) - A(y_1)(y_2 - y_1)\| \\
&\leq \|A(y_1)\| \|f(x_2) - f(x_1) - Df(x_1)(x_2 - x_1)\| \\
&\leq \|A(y_1)\| \|x_2 - x_1\| \max_{z \in \mathcal{L}} \|Df(z) - Df(x_1)\|.
\end{aligned}
$$

By (11.20), we have

$$\|R(y_1, y_2)\| \leq 2M \|y_2 - y_1\| \max_{z \in \mathcal{L}} \|Df(z) - Df(x_1)\|$$

where

$$M = \max_{x \in B(0,r)} \|Df(x)^{-1}\|.$$

As y_2 tends to y_1, x_2 tends to x_1, so that the line $\mathcal{L}$ shrinks to the single point x_1. Hence the quotient

$$\frac{\|R(y_1, y_2)\|}{\|y_2 - y_1\|} \leq 2M \max_{z \in \mathcal{L}} \|Df(z) - Df(x_1)\| \tag{11.21}$$

as y_2 approaches y_1. This shows that g has a linear approximation at each $y \in V$, with

$$Dg(y) = Df(g(y))^{-1}.$$

Since $x \to Df(x)^{-1}$ and $y \to g(y)$ are continuous, we have $g \in C^1(V)$. The theorem is proved.

Remarks The inverse function theorem is a statement about local inverses. It can be interpreted in terms of systems of equations,

$$
\begin{aligned}
f_1(x_1, \ldots, x_n) &= y_1 \\
f_2(x_1, \ldots, x_n) &= y_2 \\
&\ \vdots \\
f_n(x_1, \ldots, x_n) &= y_n
\end{aligned}
\tag{11.22}
$$

which we write more compactly as $\mathbf{f}(\mathbf{x}) = \mathbf{y}$. Let $\mathbf{x}_0 \in \mathbb{R}^n$, and let $\mathbf{y}_0 = \mathbf{f}(\mathbf{x}_0)$. Then if $\mathbf{f}$ is C^1 and $\mathbf{Df}(\mathbf{x}_0)$ is invertible, we can solve the system for values of $\mathbf{y}$ near $\mathbf{y}_0$. Furthermore, the solution $\mathbf{x}$ depends in C^1 fashion on $\mathbf{y}$. In particular, if we make a small change from $\mathbf{y}_0$ to $\mathbf{y} = \mathbf{y}_0 + \Delta\mathbf{y}$ in any direction, the solution

$$\begin{aligned} \mathbf{x} = \mathbf{g}(\mathbf{y}) &\approx \mathbf{g}(\mathbf{y}_0) + \mathbf{Dg}(\mathbf{y}_0)\Delta\mathbf{y} \\ &= \mathbf{x}_0 + \mathbf{Df}(\mathbf{x}_0)^{-1}\Delta\mathbf{y}. \end{aligned}$$

We can use this approximation to $\mathbf{x}$ as a starting value for an iterative procedure such as Newton's method.

The inverse function theorem does not guarantee the existence of global inverses. For example, even if $\mathbf{Df}(\mathbf{x})$ is invertible for each $\mathbf{x} \in \mathbb{R}^n$, this does not imply that $\mathbf{f} : G \to \mathbb{R}^n$ is one-to-one and onto.

EXAMPLE 11.5:   Consider the function $\mathbf{f} : \mathbb{R}^2 \to \mathbb{R}^2$ given by

$$\mathbf{f}(x, y) = \begin{bmatrix} x^2 - y^2 \\ 2xy \end{bmatrix}.$$

Now $\mathbf{f}$ is C^1, even analytic, and

$$\mathbf{Df}(x, y) = \begin{bmatrix} 2x & -2y \\ 2y & 2x \end{bmatrix}$$

has determinant $4(x^2 + y^2) > 0$ on $G = \mathbb{R}^2 \backslash \{(0,0)\}$. Hence by the inverse function theorem, at each point $(x_0, y_0) \in G$, there is an open set U containing (x_0, y_0) such that $\mathbf{f}$ is one-to-one on U and has a continuous inverse on the open set $V = \mathbf{f}(U)$. However, $\mathbf{f}$ is not globally one-to-one on G. In fact, if $(x_0, y_0) \in G$ then $(-x_0, -y_0) \in G$ as well, and $\mathbf{f}(x_0, y_0) = \mathbf{f}(-x_0, -y_0)$.

You may have already encountered this function in complex variables. The function $\mathbf{f}$ is just the mapping $z \to z^2$ written in terms of real and imaginary parts. Geometrically, we can visualize the action of $\mathbf{f}$ as follows. Think of the xy plane as a rubber sheet and cut it along the negative x axis. Then pull the top edge of this cut around in counterclockwise rotation until it meets the positive axis. Take the lower edge of the cut and pull it around in a clockwise rotation until it too meets the positive x axis. In this way $\mathbf{f}$ covers the set G with itself twice and leaves 0 fixed.

The following theorem gives conditions under which $\mathbf{f}$ is globally one-to-one. The norm used here is the Euclidean norm $\|\mathbf{x}\|_2$.

Theorem 11.10 Let G be an open convex subset of $\mathbb{R}^n$. Suppose that $\mathbf{f} : G \to \mathbb{R}^n$ is C^1 and that there is a constant $c > 0$, independent of $\mathbf{x}$, such that

$$\langle \mathbf{Df}(\mathbf{x})\mathbf{v}, \mathbf{v} \rangle \geq c\|\mathbf{v}\|^2 \quad \text{for all } \mathbf{x} \in \mathbf{G}, \mathbf{v} \in \mathbb{R}^n. \tag{11.23}$$

Then

    (i) $\mathbf{Df}(\mathbf{x})$ is invertible at each $\mathbf{x} \in G$;
    (ii) $\mathbf{f}$ is one-to-one on G; and
    (iii) $W = \mathbf{f}(G)$ is open, and there is a global C^1 inverse $\mathbf{g} : W \to G$.

Proof: First we show (i). If $\mathbf{Df}(\mathbf{x})\mathbf{v} = \mathbf{0}$ for some $\mathbf{v} \in \mathbb{R}^n$, then (11.23) implies that $\mathbf{v} = \mathbf{0}$. This means that $\mathbf{Df}(\mathbf{x})$ is invertible.

Next we demonstrate (ii). Let $\mathbf{x}, \mathbf{y} \in G$. Because G is convex, the line segment $\mathcal{L}$ between $\mathbf{x}$ and $\mathbf{y}$ is contained in G. We parameterize $\mathcal{L}$ by $\gamma(t) = t\mathbf{x} + (1 - t)\mathbf{y}$, $0 \leq t \leq 1$. Using the chain rule, we can write

$$\begin{aligned} \mathbf{f}(\mathbf{x}) - \mathbf{f}(\mathbf{y}) &= \mathbf{f}(\gamma(1)) - \mathbf{f}(\gamma(0)) \\[2mm] &= \int_0^1 \frac{d}{dt}\mathbf{f}(\gamma(t))dt \\[2mm] &= \int_0^1 \mathbf{Df}(\gamma(t))(\mathbf{x} - \mathbf{y})dt \end{aligned}$$

whence

$$\begin{aligned} \langle \mathbf{f}(\mathbf{x}) - \mathbf{f}(\mathbf{y}), \mathbf{x} - \mathbf{y} \rangle &= \int_0^1 \langle \mathbf{Df}(\gamma(t))(\mathbf{x} - \mathbf{y}), \mathbf{x} - \mathbf{y} \rangle dt \\[2mm] &\geq c\int_0^1 \|\mathbf{x} - \mathbf{y}\|^2 dt = c\|\mathbf{x} - \mathbf{y}\|^2. \end{aligned}$$

Then by the Schwarz inequality, we have

$$c\|\mathbf{x} - \mathbf{y}\|^2 \leq \|\mathbf{f}(\mathbf{x}) - \mathbf{f}(\mathbf{y})\|\|\mathbf{x} - \mathbf{y}\|$$

which implies

$$c\|\mathbf{x} - \mathbf{y}\| \leq \|\mathbf{f}(\mathbf{x}) - \mathbf{f}(\mathbf{y})\|. \tag{11.24}$$

Thus $\mathbf{f}(\mathbf{x}) = \mathbf{f}(\mathbf{y})$ forces $\mathbf{x} = \mathbf{y}$.

Finally, we prove (iii). Let $W = \mathbf{f}(G)$. If $\mathbf{y}_0 \in W$, there is a unique $\mathbf{x}_0 \in \mathbb{R}^n$ with $\mathbf{f}(\mathbf{x}_0) = \mathbf{y}_0$. By the inverse function theorem, there is an open set U containing $\mathbf{x}_0$ such that $\mathbf{f}(U) = V$ is an open set containing $\mathbf{y}_0$. However, $V \subset W$, which implies that W is an open set of $\mathbb{R}^n$. The proof of Theorem 11.10 is complete.

Corollary Let $\mathbf{f} : \mathbb{R}^n \to \mathbb{R}^n$ be a C^1 mapping satisfying (11.23) for each $\mathbf{x} \in \mathbb{R}^n$. Then $\mathbf{f}$ is one-to-one and onto $\mathbb{R}^n$ with a C^1 inverse $\mathbf{g}$ defined on all of $\mathbb{R}^n$.

Proof: The only subsets of $\mathbb{R}^n$ that are both open and closed are $\mathbb{R}^n$ itself and the empty set. The set $W = \mathbf{f}(\mathbb{R}^n)$ is clearly nonempty, and it is open by Theorem 11.10. If we can show that W is closed, we can conclude that $\mathbf{f}$ maps onto $\mathbb{R}^n$. Consider a sequence of points $\mathbf{y}_k \in W$, and suppose $\mathbf{y}_k$ converges to $\mathbf{y}_0$. Then $\mathbf{y}_k$ is a Cauchy sequence. The sequence $\mathbf{x}_k = \mathbf{g}(\mathbf{y}_k)$ is also Cauchy because, by (11.24),

$$c\|\mathbf{x}_k - \mathbf{x}_l\| \leq \|\mathbf{y}_k - \mathbf{y}_l\|$$

for all indices k and l. Therefore there is a point $\mathbf{x}_0 \in \mathbb{R}^n$ such that $\mathbf{x}_k$ converges to $\mathbf{x}_0$. However, $\mathbf{f}$ is continuous so

$$\mathbf{y}_0 = \lim_{k \to \infty} \mathbf{y}_k = \lim_{k \to \infty} \mathbf{f}(\mathbf{x}_k) = \mathbf{f}(x_0).$$

We have shown that $\mathbf{y}_0 \in W$, which means that W is closed.

EXAMPLE 11.6: The function $f(x) = \arctan(x)$ has $f'(x) = 1/(1 + x^2) \geq c > 0$ for $x \in I$, where I is any compact subinterval of R. Of course, c depends on I, and there is no $c > 0$ such that (11.23) holds for all $x \in R$. Thus Theorem 11.10 can be applied to f on each bounded open interval, but $f(\mathbb{R}) = (-\pi/2, \pi/2)$. At what point does the argument of the corollary fail when applied to f?

In many important cases the matrix $\mathbf{Df}$ is symmetric. We can use some ideas from linear algebra.

Definition 11.3 An $n \times n$ symmetric matrix $\mathbf{A}$ is *positive definite* if $\langle \mathbf{Ax}, \mathbf{x} \rangle > 0$ for all $\mathbf{x} \in R^n$, $\mathbf{x} \neq 0$.

Lemma 11.11 If the symmetric matrix $\mathbf{A}$ is positive definite, there is a constant $c > 0$ such that $\langle \mathbf{Ax}, \mathbf{x} \rangle \geq c\|\mathbf{x}\|_2^2$ for all $\mathbf{x} \in R^n$.

Proof: This assertion can be proved by showing that c is the smallest eigenvalue of $\mathbf{A}$, which must be positive. Instead we will use a compactness argument. Let $S = \{\mathbf{x} \in \mathbb{R}^n : \|\mathbf{x}\|_2 = 1\}$. S is the unit sphere in $\mathbb{R}^n$. We have $f(\mathbf{x}) \equiv \langle \mathbf{Ax}, \mathbf{x} \rangle > 0$ for all $\mathbf{x} \in S$. f is continuous and S is compact. Hence by Theorem 2.12, there is a minimizer $\mathbf{x}_* \in S$ such that $f(\mathbf{x}) \geq f(\mathbf{x}_*) > 0$ for all $\mathbf{x} \in S$. If we let $c = f(\mathbf{x}_*)$, we have $\langle \mathbf{Ax}, \mathbf{x} \rangle \geq c$ for all $\mathbf{x} \in S$. For any $\mathbf{x} \in \mathbb{R}^n$, $\mathbf{x} \neq 0$, we set $\mathbf{u} = \mathbf{x}/\|\mathbf{x}\|_2 \in S$, and we see that

$$\langle \mathbf{Ax}, \mathbf{x} \rangle = \|\mathbf{x}\|_2^2 \langle \mathbf{Au}, \mathbf{u} \rangle \geq c\|\mathbf{x}\|_2^2.$$

There are several practical tests to determine if a matrix is positive definite. Let $\mathbf{A}_k$ denote the submatrix consisting of the elements a_{ij} with $1 \leq i, j \leq k$.

Theorem The following are equivalent for an $n \times n$ symmetric matrix $\mathbf{A}$:
  (i) $\mathbf{A}$ is positive definite;
  (ii) All of the eigenvalues of $\mathbf{A}$ are strictly positive;
  (iii) $\det \mathbf{A}_k > 0$ for $k = 1, \ldots, n$; and
  (iv) All of the pivots in Gaussian elimination (without row exchanges) are strictly positive.

A proof of this result can be found in the book of G. Strang [St].

To show that condition (11.23) is satisfied when $\mathbf{Df}$ is symmetric, we must show that $\mathbf{Df}(\mathbf{x})$ is uniformly positive definite. This means that the constants that arise when checking (ii), (iii) or (iv) are independent of $\mathbf{x}$.

  EXAMPLE 11.7: In some situations, the function $\mathbf{f}$ arises as the gradient of a scalar potential, $\phi(\mathbf{x})$. In Chapter 12, we will see that if $\mathbf{f}$ is C^1, and $\mathbf{f} = \nabla \phi$ for some scalar function ϕ, then $\mathbf{Df}$ is a symmetric

$n \times n$ matrix. Condition (11.23) will be satisfied if $\mathbf{Df}(\mathbf{x})$ is uniformly positive definite on G. For example, let $\phi(x, y) = (x^2 + y^2)/2 + \cosh(x + y)$. Then

$$\mathbf{f}(x, y) = [\phi_x(x, y), \phi_y(x, y)] = [x + \sinh(x + y), y + \sinh(x + y)],$$

and

$$\mathbf{Df}(x, y) = \begin{bmatrix} 1 + \cosh(x + y) & \cosh(x + y) \\ \cosh(x + y) & 1 + \cosh(x + y) \end{bmatrix}.$$

To see that $\mathbf{Df}$ is positive definite, we appeal to the third criterion of the theorem above. First we check the determinant of the 1×1 submatrix which is the upper left corner element. We note that $1 + \cosh(x + y) \geq 2$. The other determinant to check is $\det \mathbf{Df}(x, y) = 1 + 2\cosh(x + y) \geq 3$. Hence (11.23) holds for $\mathbf{f}$ uniformly on all of $\mathbb{R}^2$. This means that, for each right-hand side (u, v), there is a unique solution to the equations

$$x + \sinh(x + y) = u$$
$$y + \sinh(x + y) = v.$$

An Example with Computation

Here is another example in which condition (11.23) is satisfied. This is also an example in which there can be a very large number of equations.

EXAMPLE 11.8:   Recall the nonlinear boundary value problem for a function $u(x)$ that was discussed in Example 5.5.

$$-u'' + u^3 = q(x), \quad 0 < x < 1 \tag{11.25}$$

$$u(0) = u(1) = 0. \tag{11.26}$$

We used centered difference quotients to approximate the second derivative:

$$u''(x_j) = \frac{u(x_{j+1}) - 2u(x_j) + u(x_{j-1})}{h^2} + O(h^2).$$

After imposing the boundary conditions $u(0) = u(1) = 0$, we arrived at the system of $N - 1$ equations in $N - 1$ unknowns (5.18) which we restate here:

$$2u_1 - u_2 + h^2 u_1^3 = h^2 q_1$$
$$-u_1 + 2u_2 - u_3 + h^2 u_2^3 = h^2 q_2$$
$$-u_2 + 2u_3 - u_4 + h^2 u_3^3 = h^2 q_3$$
$$\cdots\cdots = \cdots$$
$$-u_{N-2} + 2u_{N-1} + h^2 u_{N-1}^3 = h^2 q_{N-1}.$$

We can write this system compactly using vector notation. Let $\mathbf{q} = (q_1, \ldots, q_{N-1})$ and let $\mathbf{T} = (1/h^2)\mathbf{S}$ where $\mathbf{S}$ is the $(N-1) \times (N-1)$ symmetric tridiagonal matrix

$$\mathbf{S} = \begin{bmatrix} 2 & -1 & 0 & \cdots & & & 0 \\ -1 & 2 & -1 & \cdots & & & 0 \\ 0 & -1 & 2 & \cdots & & & 0 \\ \cdot & \cdot & & \cdot & \cdots & & 0 \\ \cdot & \cdot & & & \cdot & \cdots & 0 \\ \cdot & & \cdot & \cdots & & 2 & -1 \\ 0 & 0 & \cdots & & & -1 & 2 \end{bmatrix}. \tag{11.27}$$

Let $\mathbf{N}(\mathbf{u}) = (u_1^3, \ldots, u_{N-1}^3)$. Then the system of equations can be written

$$\mathbf{f}(\mathbf{u}) \equiv \mathbf{T}\mathbf{u} + \mathbf{N}(\mathbf{u}) = \mathbf{q}. \tag{11.28}$$

We claim that $\mathbf{f}(\mathbf{u})$ satisfies condition (11.23). First we show that $\mathbf{S}$ is positive definite. Let $\mathbf{S}_j$, $j = 1, \ldots, N-1$ denote the principal diagonal submatrices. Then $\det \mathbf{S}_1 = 2$, $\det \mathbf{S}_2 = 3$, and in general,

$$\det \mathbf{S}_{j+1} = 2 \det \mathbf{S}_j - \det \mathbf{S}_{j-1}.$$

It follows that $\det \mathbf{S}_j = j + 1$ for $j = 1, \ldots, N-1$. This implies that $\mathbf{S} = \mathbf{S}_{N-1}$ is positive definite.

The Jacobian matrix of $\mathbf{f}$ is

$$\mathbf{Df}(\mathbf{u}) = \mathbf{T} + 3\Lambda(\mathbf{u}^2),$$

where $\Lambda(\mathbf{u}^2)$ is the diagonal matrix with $\lambda_{jj} = u_j^2$, $j = 1, \dots, N - 1$. Now for any vector $\mathbf{v} \in \mathbb{R}^{N-1}$,

$$\langle \Lambda(\mathbf{u}^2)\mathbf{v}, \mathbf{v} \rangle = \sum_{j=1}^{N-1} u_j^2 v_j^2 \geq 0.$$

Hence for any vector $\mathbf{v} \in \mathbb{R}^{N-1}$,

$$\langle \mathbf{Df}(\mathbf{u})\mathbf{v}, \mathbf{v} \rangle = \langle \mathbf{Tv}, \mathbf{v} \rangle + 3\langle \Lambda(\mathbf{u}^2)\mathbf{v}, \mathbf{v} \rangle$$
$$\geq (1/h^2)\langle \mathbf{Sv}, \mathbf{v} \rangle \geq (c/h^2)\|\mathbf{v}\|^2.$$

By the corollary to Theorem 11.10, the system (11.27) has a unique solution for each vector $\mathbf{q} \in \mathbb{R}^{n-1}$.

Now how do we compute the solution of the system? We shall use Newton's method on the function $\mathbf{F}(\mathbf{u}) = \mathbf{f}(\mathbf{u}) - \mathbf{q} = \mathbf{Tu} + \mathbf{Nu} - \mathbf{q}$. Then $\mathbf{DF}(\mathbf{u}) = \mathbf{Df}(\mathbf{u})$. We begin the iterations with $\mathbf{u}_0$ being the solution of the linear system $\mathbf{Tu} - \mathbf{q} = 0$. The Newton iteration scheme for the equation $\mathbf{F}(\mathbf{u}) = \mathbf{0}$ is

solve the linear system $\mathbf{DF}(\mathbf{u}_k)\mathbf{s} = -\mathbf{F}(\mathbf{u}_k)$

which is the same as

solve the linear system $\mathbf{Df}(\mathbf{u}_k)\mathbf{s} = -\mathbf{f}(\mathbf{u}_k) + \mathbf{q}$

update $\mathbf{u}_{k+1} = \mathbf{u}_k + \mathbf{s}$.

We take $q(x) = 100 \exp(-5(x - 0.75)^2)$ in equation (11.25). We take $N = 100$ so that the system (11.27) is 99×99. The first four iterates $\mathbf{u}_0, \mathbf{u}_1, \mathbf{u}_2, \mathbf{u}_3$ are graphed as four functions $u_0(x), u_1(x), u_2(x), u_3(x)$ (see Figure 11.1). The graphs are constructed by joining the point values with straight lines. Since s is the difference between successive iterates, $\|\mathbf{s}\|_\infty$ is a good measure of the convergence of the scheme. The values of $\|\mathbf{s}\|_\infty = \|\mathbf{u}_k - \mathbf{u}_{k-1}\|_\infty$ are listed here.

k	$\|\mathbf{u}_k - \mathbf{u}_{k-1}\|_\infty$
1	2.7424
2	1.4895
3	0.4951
4	0.0511

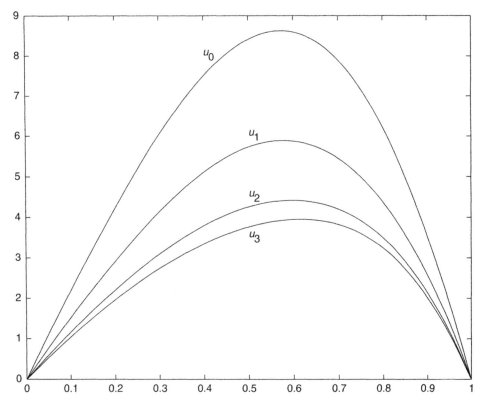

Figure 11.1 Graphs of Newton iterates that approximate the solution of equations (11.25) and (11.26), with $q(x) = 100 \exp(-5(x - 0.75)^2)$

11.5 The Implicit Function Theorem

The setting of the implicit function theorem is underdetermined systems like equations (11.2), in which there are more unknowns than equations. We want to know when it is possible to solve for the variables $y_1, \ldots, y_m$ in terms of the variables $x_1, \ldots, x_n$. When this is possible we can write

$$
\begin{aligned}
y_1 &= g_1(x_1, \ldots, x_n) \\
y_2 &= g_2(x_1, \ldots, x_n) \\
& \quad \cdot \qquad \cdot \\
& \quad \cdot \qquad \cdot \\
y_m &= g_m(x_1, \ldots, x_n),
\end{aligned}
$$

such that the functions $g_1, \ldots, g_m$ satisfy

$$f_1(x_1, \ldots, x_n, g_1(x_1, \ldots, x_n), \ldots, g_m(x_1, \ldots, x_n)) = 0$$
$$f_2(x_1, \ldots, x_n, g_1(x_1, \ldots, x_n), \ldots, g_m(x_1, \ldots, x_n)) = 0$$
$$. = 0$$
$$. = 0$$
$$f_m(x_1, \ldots, x_n, g_1(x_1, \ldots, x_n), \ldots, g_m(x_1, \ldots, x_n)) = 0.$$

This situation can be expressed more compactly in vector notation. Let $\mathbf{x} = (x_1, \ldots, x_n) \in \mathbb{R}^n$, and let $\mathbf{y} = (y_1, \ldots, y_m) \in \mathbb{R}^m$. Then the system (11.2) can be written

$$\mathbf{f}(\mathbf{x}, \mathbf{y}) = \mathbf{0} \tag{11.29}$$

where $\mathbf{f} : \mathbb{R}^{n+m} \rightarrow \mathbb{R}^m$. The function $\mathbf{g} : \mathbb{R}^n \rightarrow \mathbb{R}^m$, and it satisfies

$$\mathbf{f}(\mathbf{x}, \mathbf{g}(\mathbf{x})) = \mathbf{0}. \tag{11.30}$$

We begin the discussion with the case $m = n = 1$ so that we have a scalar equation:

$$f(x, y) = 0.$$

Suppose that $f(a, b) = 0$, and let S be the part of the level set $\{(x, y) : f(x, y) = 0\}$ that contains (a, b). When can S be expressed as the graph of a function $y = g(x)$ with $g(a) = b$? Clearly, not always. If $f(x, y) = x^2 + y^2 - 1$, the level set $S = \{(x, y) : f(x, y) = 0\}$ is the unit circle. Near $(1, 0)$, S is not the graph of a function.

Theorem 11.12 Let $f(x, y)$ be C^1 on an open set G with $(a, b) \in G$. Suppose that $f(a, b) = 0$ and $f_y(a, b) \neq 0$.
Then,

(i) There is an open interval I with $a \in I$ and a C^1 function $g : I \rightarrow \mathbb{R}$ such that $g(a) = b$ and

$$f(x, g(x)) = 0 \quad \text{for } x \in I.$$

(ii) There is an open interval $J \subset \mathbb{R}$ containing b such that, if $(x, y) \in I \times J$ with $f(x, y) = 0$, then $y = g(x)$.
In other words, the part of the level set S that lies over I is the graph of $y = g(x)$. Furthermore, for $x \in I$,

(iii)

$$g'(x) = -\frac{f_x(x, g(x))}{f_y(x, g(x))}. \tag{11.31}$$

Proof: We consider the mapping on G

$$(u, v) = (x, f(x, y)) \equiv \varphi(x, y)$$

which takes G into $\mathbb{R}^2$. This mapping only moves points in the vertical direction, leaving the x coordinate unchanged; see Figure 11.2. In particular, φ maps $S \cap G$

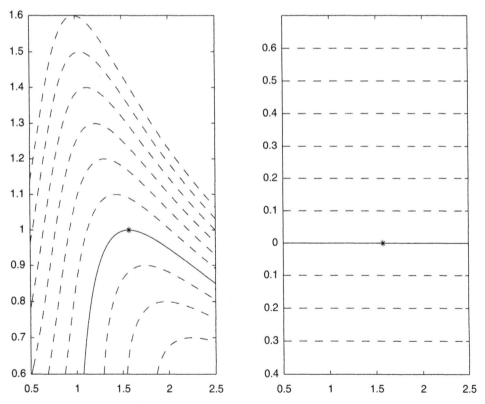

Figure 11.2 On the left, level curves of $f(x, y) = y - \sin(xy)$ with level curve $f(x, y) = 0$ in solid line. On the right, images of these level curves under the mapping $(x, y) \to (x, f(x, y))$

into the u axis. The Jacobian matrix of φ is

$$\mathbf{D}\varphi(x,y) = \begin{bmatrix} 1 & 0 \\ f_x(x,y) & f_y(x,y) \end{bmatrix}$$

and $\det \mathbf{D}\varphi(a,b) = f_y(a,b) \neq 0$. Hence we can apply the inverse function to φ. There exists an open set $U \subset G$ with $(a,b) \in G$, and an open set $V \subset \mathbb{R}^2$ with $(a,0) \in V$, such that $\varphi : U \to V$ is one-to-one and onto with a C^1 inverse $\psi : V \to U$. Because U and V are open, there are open intervals I containing a and J containing b, such that $I \times J \subset U$ and $I \subset V \cap \{0\}$. The mapping φ leaves the x coordinate unchanged so that we can identify the x axis and the u axis. Thus ψ has the form $\psi(x,v) = (x, \sigma(x,v))$. Consequently,

$$(x,v) = \varphi \circ \psi(x,v) = (x, f(x, \sigma(x,v))) \tag{11.32}$$

for $(x,v) \in V$. We set $g(x) = \sigma(x,0)$. From (11.32), we see that

$$f(x, g(x)) = 0.$$

Thus ψ maps the open interval I of the x axis into $S \cap G$, which proves (i).

Now we prove (ii). Because U is an open set containing (a,b), there is an open interval J containing b such that $I \times J \subset U$. For $(x,y) \in I \times J$ with $f(x,y) = 0$, we have $\varphi(x,y) = (x,0)$. Hence $(x,y) = \psi(x,0)$, which means that $y = \sigma(x,0) = g(x)$. We see that $S \cap (I \times J)$ is the graph of $y = g(x)$ over I.

The relation (11.31) follows immediately by differentiating the previous equation. The proof of Theorem 11.11 is complete.

The proof of the higher-dimensional version of the implicit function theorem follows the same pattern. However, we will need some additional notation. For $\mathbf{f}(\mathbf{x}, \mathbf{y}) : \mathbb{R}^{n+m} \to \mathbb{R}^m$, we use $\mathbf{D}_x \mathbf{f}$ to denote the $m \times n$ matrix of partial derivatives

$$\mathbf{D}_x \mathbf{f} = \left[\frac{\partial f_i}{\partial x_j} \right] \quad i = 1, \dots, m \quad j = 1, \dots, n$$

and $\mathbf{D}_y \mathbf{f}$ to denote the $m \times m$ matrix

$$\mathbf{D}_y \mathbf{f} = \left[\frac{\partial f_i}{\partial y_k} \right] \quad i = 1, \dots, m \quad k = 1, \dots, m.$$

Theorem 11.13 Let $\mathbf{a} \in \mathbb{R}^n$, $\mathbf{b} \in \mathbb{R}^m$ and let $G \subset \mathbb{R}^{n+m}$ be an open set with $(\mathbf{a}, \mathbf{b}) \in G$. Let $\mathbf{f}(\mathbf{x}, \mathbf{y}) : G \to \mathbb{R}^m$ be C^1 with $\mathbf{f}(\mathbf{a}, \mathbf{b}) = 0$ and $\mathbf{D}_y\mathbf{f}(\mathbf{a}, \mathbf{b})$ invertible. Then,

   (i) There is an open ball $B(\mathbf{a}, r_1) \subset \mathbb{R}^n$ and a C^1 function $\mathbf{g} : B(\mathbf{a}, r_1) \to \mathbb{R}^m$ with $\mathbf{g}(\mathbf{a}) = \mathbf{b}$ and $\mathbf{f}(\mathbf{x}, \mathbf{g}(\mathbf{x})) = 0$ for $\mathbf{x} \in B(\mathbf{a}, r_1)$.

  (ii) There is an open ball $B(\mathbf{b}, r_2) \subset \mathbb{R}^m$ such that $B(\mathbf{a}, r_1) \times B(\mathbf{b}, r_2) \subset G$ and such that, if $(\mathbf{x}, \mathbf{y}) \in B(\mathbf{a}, r_1) \times B(\mathbf{b}, r_2)$ with $\mathbf{f}(\mathbf{x}, \mathbf{y})) = 0$, then $\mathbf{y} = \mathbf{g}(\mathbf{x})$.

 (iii) For $\mathbf{x} \in B(\mathbf{a}, r_1)$,

$$\mathbf{Dg}(\mathbf{x}) = -[\mathbf{D}_y\mathbf{f}(\mathbf{x}, \mathbf{g}(\mathbf{x}))]^{-1}\mathbf{D}_x\mathbf{f}(\mathbf{x}, \mathbf{g}(\mathbf{x})).$$

Notice how the dimensions match. The matrix on the left is $m \times n$. The matrices on the right are $m \times m$ multiplied by $m \times n$.

Remark It is easy to remember the reason for the hypotheses of Theorem 11.13 if we consider the case of linear systems. Suppose that we have a linear system of m equations in $m + n$ unknowns:

$$\mathbf{f}(\mathbf{x}, \mathbf{y}) = \mathbf{Ax} + \mathbf{By} = \mathbf{b},$$

where $\mathbf{A}$ is an $m \times n$ matrix and $\mathbf{B}$ is an $m \times m$ matrix; $\mathbf{x}$ is an n vector and $\mathbf{y}$ and $\mathbf{b}$ are m vectors. When can we solve the system for $\mathbf{y}$ in terms of the "free" variables $\mathbf{x}$? If we write the system as

$$\mathbf{By} = \mathbf{b} - \mathbf{Ax},$$

it is clear that we need to require that the $m \times m$ matrix $\mathbf{B} = \mathbf{D}_y\mathbf{f}$ be invertible.

Proof: We define the mapping

$$(\mathbf{x}, \mathbf{v}) = \varphi(\mathbf{x}, \mathbf{y}) = (\mathbf{x}, \mathbf{f}(\mathbf{x}, \mathbf{y})).$$

The function φ is C^1 on G, with $\varphi(\mathbf{a}, \mathbf{b}) = (\mathbf{a}, 0)$. Now $\mathbf{D}\varphi$ is the $(n+m) \times (n+m)$ matrix of partial derivatives

$$\mathbf{D}\varphi = \begin{bmatrix} \mathbf{I} & 0 \\ \mathbf{D}_x\mathbf{f} & \mathbf{D}_y\mathbf{f} \end{bmatrix},$$

where $\mathbf{I}$ is the $n \times n$ identity. Taking advantage of the block of zeros in $\mathbf{D}\varphi$, we see that

$$\det \mathbf{D}\varphi(\mathbf{a}, \mathbf{b}) = \det \mathbf{D}_y\mathbf{f}(a, b) \neq 0$$

because $\mathbf{D_y f(a, b)}$ is invertible. Hence we can apply the inverse function theorem to φ. There are open sets $U, V \subset \mathbb{R}^{n+m}$ with $(\mathbf{a, b}) \in U \subset G$, and $(\mathbf{a, 0}) \in V$, and a C^1 mapping $\sigma(\mathbf{x, v}) : V \to \mathbb{R}^m$ such that $\psi(\mathbf{x, v}) = (\mathbf{x}, \sigma(\mathbf{x, v}))$ is the inverse of φ. Consequently,

$$(\mathbf{x, v}) = \varphi \circ \psi(\mathbf{x, v}) = (\mathbf{x}, \mathbf{f}(\mathbf{x}, \sigma(\mathbf{x, v})))$$

for all $(\mathbf{x, v}) \in V$. We set $\mathbf{g(x)} = \sigma(\mathbf{x, 0})$ so that $0 = \mathbf{f}(\mathbf{x, g(x)})$ for all $\mathbf{x} \in V \cap \{\mathbf{v = 0}\}$. Now because U and V are open sets of $\mathbb{R}^{n+m}$, we can choose $r_1, r_2 > 0$ such that $B(\mathbf{a}, r_1) \times B(\mathbf{b}, r_2) \subset U$ and $B(\mathbf{a}, r_1) \times \{\mathbf{0}\} \subset V \cap \{\mathbf{0}\}$. The rest of the proof is the same as that of Theorem 11.12.

We can use the implicit function theorem to give a good description of the tangent plane to a surface.

Definition 11.4 Let $f(\mathbf{x})$ be a C^1 function on an open subset $G \subset \mathbb{R}^n$, and set

$$S = \{\mathbf{x} \in G : f(\mathbf{x}) = 0\}.$$

If $\mathbf{a} \in S$ and $\nabla f(\mathbf{a}) \neq 0$, the *tangent space* to S at $\mathbf{a}$ is the $(n-1)$-dimensional subspace

$$T_\mathbf{a} = \{\mathbf{v} \in \mathbb{R}^n : \langle \nabla f(\mathbf{a}), \mathbf{v} \rangle = 0\}.$$

We now use the implicit function theorem to show that the name "tangent space" is appropriate. $T_\mathbf{a}$ is indeed the set of tangent vectors to curves on S through the point $\mathbf{a}$.

Theorem 11.14 Let G be an open subset of $\mathbb{R}^n$, and let $f : G \to \mathbb{R}$ be a C^1 function. Suppose that $\mathbf{a} \in S = \{\mathbf{x} \in G : f(\mathbf{x}) = 0\}$ and that $\nabla f(\mathbf{a}) \neq 0$. Then $\mathbf{v}$ lies in the tangent space $T_\mathbf{a}$ to S at $\mathbf{a}$ if and only if there is an open interval I with $0 \in I$ and a C^1 function $\mathbf{r} : I \to \mathbb{R}^n$ such that

    (i) $\mathbf{r}(t) \in S$ for $t \in I$;
    (ii) $\mathbf{r}(0) = \mathbf{a}$; and
    (iii) $\mathbf{r}'(0) = \mathbf{v}$.

Proof: If there is a function $\mathbf{r}(t)$ that satisfies (i), then $f(\mathbf{r}(t)) = 0$, and by the chain rule,

$$0 = \frac{d}{dt} f(\mathbf{r}(t)) = \langle \nabla f(\mathbf{r}(t)), \mathbf{r}'(t) \rangle.$$

In particular, putting $t = 0$, (ii) and (iii) imply

$$0 = \langle \nabla f(\mathbf{a}), \mathbf{v} \rangle.$$

To prove the converse implication, we start with a vector $\mathbf{v} \in T_\mathbf{a}$, and we must construct the function $\mathbf{r}$. Without loss of generality, we may assume that $(\partial f/\partial x_n)(\mathbf{a}) \neq 0$. We separate out the last component of $\mathbf{x}$ by writing $\mathbf{x} = (\hat{\mathbf{x}}, x_n)$, $\mathbf{a} = (\hat{\mathbf{a}}, a_n)$ and $\mathbf{v} = (\hat{\mathbf{v}}, v_n)$. According to the implicit function theorem, there is an open set U with $\hat{\mathbf{a}} \in U$ and a C^1 function $g(\hat{\mathbf{x}})$ such that S is the graph of g over U and $g(\hat{\mathbf{a}}) = a_n$. Because U is open, there is a $\delta > 0$ such that $\hat{\mathbf{a}} + t\hat{\mathbf{v}} \in U$ for $|t| < \delta$. For $t \in I \equiv \{|t| < \delta\}$, we define

$$\mathbf{r}(t) = (\hat{\mathbf{a}} + t\hat{\mathbf{v}}, g(\hat{\mathbf{a}} + t\hat{\mathbf{v}})).$$

Because $f(\hat{\mathbf{x}}, g(\hat{\mathbf{x}})) = 0$, it is clear that $\mathbf{r}(t) \in S$ for $t \in I$. Furthermore,

$$\mathbf{r}(0) = (\hat{\mathbf{a}}, g(\hat{\mathbf{a}})) = (\hat{\mathbf{a}}, a_n) = \mathbf{a}.$$

Because g is C^1, $\mathbf{r}$ is C^1 with

$$\mathbf{r}'(0) = (\hat{\mathbf{v}}, \langle \nabla_{\hat{\mathbf{x}}} g(\hat{\mathbf{a}}), \hat{\mathbf{v}} \rangle).$$

It remains to be shown that the last component of $\mathbf{r}'(0)$ is equal to v_n. However, from the implicit function theorem, we have

$$\nabla g(\hat{\mathbf{x}}) = -\frac{\nabla_{\hat{\mathbf{x}}} f(\mathbf{x})}{(\partial f/\partial x_n)(\mathbf{x})}.$$

Hence

$$\langle \nabla g(\hat{\mathbf{a}}), \hat{\mathbf{v}} \rangle = -\frac{\langle \nabla_{\hat{\mathbf{x}}} f(\mathbf{a}), \hat{\mathbf{v}} \rangle}{(\partial f/\partial x_n)(\mathbf{a})} = v_n$$

because $\langle \nabla f(\mathbf{a}), \mathbf{v} \rangle = 0$. The theorem is proved.

Remark Define the mapping $\phi(\hat{\mathbf{x}}) = (\hat{\mathbf{x}}, g(\hat{\mathbf{x}})) : U \rightarrow \mathbb{R}^n$. Then $f(\phi(\hat{\mathbf{x}})) = 0$ for $\hat{\mathbf{x}} \in U$, and $\phi(U)$ is the portion of the surface S that lies above the set U. The calculations in the proof of Theorem 11.14 also show that the tangent space at $\mathbf{a} \in S$ is the image of $\mathbb{R}^{n-1}$ under the linear mapping that has the $n \times (n-1)$ Jacobian matrix $\mathbf{D}\phi(\hat{\mathbf{a}})$: $T_\mathbf{a} = \mathbf{D}\phi(\hat{\mathbf{a}})(\mathbb{R}^{n-1})$. In the exercises, you will extend Theorem 11.14 to the case in which S is the intersection of several surfaces S_i.

Newton's Method and the Implicit Function Theorem

We will use Newton's method to compute the values of the function $y = g(x)$, whose existence is ensured by Theorem 11.13. We begin with the one-dimensional case,

$$f(x, y) = 0.$$

We assume that for some point (a, b), $f(a, b) = 0$. Then for each x near a, we must find the zero of the function $y \rightarrow f(x, y) = 0$. This is like solving a family of problems $f(y) = 0$ with x as a parameter. Let Δx be a small step. The problem is now to compute the solution of $f(a + \Delta x, y) = 0$. The implicit function theorem, Theorem 11.12, tells us that the solution y will be close to b. Hence b will be a good starting value of the Newton iteration scheme:

$$\eta_0 = b$$
$$\eta_{n+1} = \eta_n - \frac{f(a + \Delta x, \eta_n)}{f_y(a + \Delta x, \eta_n)}.$$

We see that the condition $f_y(a, b) \neq 0$ needed for the implicit function theorem is also needed to ensure the quadratic convergence of Newton's method. Let $x_1 = a + \Delta x$ and $y_1 = \lim_{n \to \infty} \eta_n$. Then $f(x_1, y_1) = 0$. In general, we label $x_j = a + j\Delta x$, and we let y_j be the solution of $f(x_j, y) = 0$. We let y_{j-1} be the starting point for the Newton iterates that converge to y_j. Here is a pseudo-code that implements this procedure.

enter a, b and Δx
$y_0 = b$
for $j = 1 : J$
$\qquad x_j = a + j\Delta x$
$\qquad \eta_0 = y_{j-1}$
$\qquad$ for $n = 1 : N$
$\qquad\qquad \eta_n = \eta_{n-1} - f(x_j, \eta_{n-1})/f_y(x_j, \eta_{n-1})$
$\qquad$ end
$\qquad y_j = \eta_N$
end

Here N is the number of iterations needed to get the desired accuracy.

> **EXAMPLE 11.9:** Let $f(x, y) = y - \sin(xy)$. Clearly $f(x, 0) = 0$ for all x. However, $f_y(x, y) = 1 - x\cos(xy)$ whence $f_y(1, 0) = 0$; the implicit function theorem does not hold at $(1, 0)$. This raises the possibility

that there is another branch of solutions of $f(x, y) = 0$ that passes through $(1, 0)$. Some other points where $f(x, y) = 0$ are $(\pi/2^{3/2}, 1/\sqrt{2})$ and $(\pi/2, 1)$. Our computations will indicate that there is a solution curve of the form $y = g(x)$ of $f(x, y) = 0$ with $g(1) = 0$, $g(\pi/2^{3/2}) = 1/\sqrt{2}$ and $g(\pi/2) = 1$.

We take $a = \pi/2$, $b = 1$, and $\Delta x = 0.05$. We move in both directions, to the left and right of a. The solution curve is displayed in Figure 11.3.

We can see that the curve approaches $(1, 0)$ and has a vertical tangent there.

The case of two equations in three variables,

$$f(x, y, z) = 0$$
$$g(x, y, z) = 0 \tag{11.33}$$

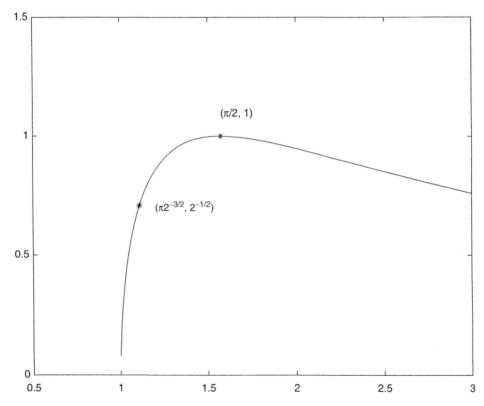

Figure 11.3 Graph of $y = g(x)$, solution of $f(x, y) = y - \sin(xy) = 0$

can be interpreted geometrically as a description of the curve of intersection Γ of the two level surfaces $S_1 = \{(x, y, z) : f(x, y, z) = 0\}$ and $S_2 = \{(x, y, z) : g(x, y, z) = 0\}$. We can use equations (11.33) to provide a parameterization of Γ. If we wish to use x as a parameter, then we must be able to express y as a function $h_1(x)$ and z as a function $h_2(x)$. The implicit function theorem gives us conditions when this can be done. We apply the implicit function theorem with $n = 1$ and $m = 2$, and with y and z playing the roles of y_1 and y_2. Then if (x_0, y_0, z_0) is a solution of (11.33), evaluate the 2×2 determinant

$$\begin{vmatrix} f_y & f_z \\ g_y & g_z \end{vmatrix} = (f_y g_z - f_z g_y)\Big|_{(x_0, y_0, z_0)}.$$

If this determinant is nonzero, there is an open interval I containing x_0 such that $y = h_1(x)$, $z = h_2(x)$ for $x \in I$ with $h_1(x_0) = y_0$ and $h_2(x_0) = z_0$.

EXAMPLE 11.10:   Let $f(x, y, z) = x^2 + y^2 - z^2 - 1$, and let $g(x, y, z) = 4x^2 + y^2 + 4z^2 - 4$. The level surface S_1 on which $f = 0$ is a hyperboloid of one sheet, and the level surface S_2 on which $g = 0$ is an ellipsoid. We note that the points $(0, \pm\sqrt{8/5}, \pm\sqrt{3/5})$ lie on the intersection of S_1 and S_2. Now the 2×2 determinant is

$$f_y g_z - f_z g_y = 20yz,$$

and this determinant is nonzero at these points. Hence there is an open interval I containing zero such that y and z can be expressed as functions of x near each of these four points. The two level surfaces and the two curves of intersection are shown in Figure 11.4.

In fact, we can solve for y and z explicitly in terms of x:

$$y = \pm\sqrt{\frac{8(1 - x^2)}{5}}, \quad z = \sqrt{\frac{3(1 - x^2)}{5}}.$$

We see that in each case, the interval on which we can solve for y and z is $-1 < x < 1$.

Exercises for 11.4 and 11.5

1. Let $\mathbf{f} : \mathbb{R}^2 \to \mathbb{R}^2$ be given by $\mathbf{f}(x, y) = (e^x \cos y, \ e^x \sin y)$.

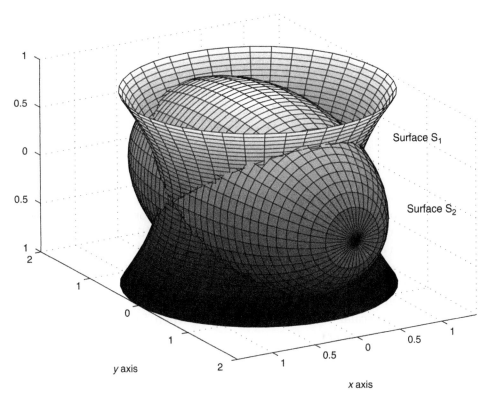

Figure 11.4 Surfaces $f(x, y, z) = x^2 + y^2 - z^2 - 1 = 0$, and $g(x, y, z) = 4x^2 + y^2 + 4z^2 - 4 = 0$ and the curves of intersection

   a) Show that the hypotheses of the inverse function theorem (Theorem 11.8) are satisfied at each $(x, y) \in \mathbb{R}^2$.
   b) Show that $\mathbf{f}$ is not globally one-to-one.
   c) Sketch the images under $\mathbf{f}$ of the vertical lines x = constant and of the horizontal lines y = constant.

2. Let $\mathbf{A}$ be an $n \times n$ real, symmetric, positive definite matrix with smallest eigenvalue $\lambda_0 > 0$. Let $\mathbf{f} : \mathbb{R}^n \to \mathbb{R}^n$ be given by

$$\mathbf{f}(\mathbf{x}) = \mathbf{A}\mathbf{x} + \mathbf{N}(\mathbf{x})$$

where we assume that $\mathbf{N} : \mathbb{R}^n \to \mathbb{R}^n$ is C^1 and that there is a constant $C, 0 \le C < \lambda_0$, such that

$$\|\mathbf{DN}(\mathbf{x})\|_2 \le C \quad \text{for all } \mathbf{x} \in \mathbb{R}^n.$$

a) Show that f satisfies (11.23). The corollary to Theorem 11.10 ensures that, for each $y \in \mathbb{R}^n$, there is a unique solution of $f(x) = y$ and that the inverse function $x = g(y)$ is C^1.

b) Verify that this result may be applied to the system of equations

$$
\begin{aligned}
5x_1 - x_2 + \sin(x_1 + x_2) &= y_1 \\
-x_1 + 5x_2 + \cos(x_1 + x_2) &= y_2
\end{aligned}.
$$

You may use the Frobenius norm $\| \ \|_F$ to estimate $\|DN\|_2$.

3. Suppose that f satisfies the hypotheses of Theorem 11.8 and that in addition, each component of f is C^2. Show that the inverse function g is then also C^2. More generally, show that if f is C^k, then g is also C^k.

4. a) Use the result of Exercise 3 to show that, if f satisfies the hypotheses of Theorem 11.13 (implicit function theorem) and in addition is C^2, then the function g is also C^2.

b) In the scalar-valued case $f : \mathbb{R}^2 \to \mathbb{R}$, when f is assumed to be C^2, calculate $g''(x)$.

5. Consider the pair of equations

$$
\begin{aligned}
x^2 + y^2 + (z - 1)^2 &= 4 \\
-x^2 - y^2 + z^2 &= 1.
\end{aligned}
$$

Each of the equations describes the level surface of a function of three variables.

a) The point $(0, 0, -1)$ lies in this intersection. Is there a curve of intersection through this point? Try the implicit function theorem there.

b) The point $(\sqrt{3}, 0, 2)$ is also in the intersection. Is there a curve of intersection through this point? Does the implicit function theorem say that the curve can be expressed as a function $y \to (x(y), z(y))$?

c) Draw a graph of the two surfaces to explain visually the results of parts a) and b).

6. Let $f(x, y) = (x + y)/2 - \exp(xy - 1)$. Note that $f(1, 1) = 0$.

a) Show that there exists a C^1 function $g(x)$ defined on an interval I, with $1 \in I$, such that $f(x, g(x)) = 0$ for $x \in I$.

b) Use Newton's method to compute points on the graph of $g(x)$ for $0.8 \le x \le 2$ with $\Delta x = 0.05$. Graph the resulting curve.

7. Consider the following system of equations which involves the parameter λ.

$$f(x,y) = (y-x)^2/4 - (x+y)/2 + 1 = 0$$
$$g(x,y) = y + \lambda(x-2)^3 = 0.$$

a) Plot the level curves $f = 0$ and $g = 0$ for various values of λ in the square $\{0 \le x \le 3, \ 0 \le y \le 3\}$. Show that there is a critical value of λ, λ_*, such that

there are no solutions for $\lambda < \lambda_*$;

there is one solution for $\lambda = \lambda_*$;

there are two solutions for $\lambda > \lambda_*$.

b) When $\lambda = 1.5$, the system has two roots. Use a two-dimensional Newton code to compute them.

c) Find the critical value λ_*, and the single root of the system that corresponds to $\lambda = \lambda_*$. Use the following idea. Look for points (x, y, λ) that satisfy the system, but at which the hypothesis of the implicit function theorem is not satisfied. Show that these points must satisfy the three equations

$$f = 0$$
$$g = 0$$
$$f_x g_y - f_y g_x = 0.$$

Solve this system using a three-dimensional version of a Newton code.

8. Let $\mathbf{f}(\mathbf{x}) = (f_1(\mathbf{x}), f_2(\mathbf{x}))$ be a C^1 function on an open subset G of $\mathbb{R}^n$. Let $S_j = \{\mathbf{x} \in G : f_j(\mathbf{x}) = 0\}$, $j = 1, 2$, and let $S = S_1 \cap S_2$. Let $\mathbf{a} \in S$, and assume that $\nabla f_1(\mathbf{a})$ and $\nabla f_2(\mathbf{a})$ are independent vectors. Finally, let

$$M = \{\mathbf{v} \in \mathbb{R}^n : \langle \nabla f_1(\mathbf{a}), \mathbf{v} \rangle = \langle \nabla f_2(\mathbf{a}), \mathbf{v} \rangle = 0\}.$$

Prove an extension of Theorem 11.14. Show that $\mathbf{v} \in M$ if and only if there is a C^1 curve $\mathbf{r}(t) : I \to \mathbb{R}^n$ such that $\mathbf{r}(t) \in S$, $\mathbf{r}(0) = \mathbf{v}$, and $\mathbf{r}'(0) = \mathbf{v}$.

9. Sometimes it is just as important to identify the places where the conditions for the implicit function theorem are not satisfied. This situation

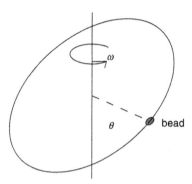

Figure 11.5 Bead on a rotating hoop

may indicate values of parameters that are critical in that they signal a change in the physical problem.

Consider the case of a bead on a rotating hoop. The bead has mass m and can slide without friction on a circular wire or hoop of radius R. The hoop rotates about the z axis with constant angular velocity w (see Figure 11.5). In a steady state, the gravitational force on the bead balances with the centripetal force caused by the rotation of the hoop. The resulting equation is

$$mRw^2 \cos\theta \sin\theta - mg \sin\theta = 0.$$

How does the angle θ depend on the angular velocity w? Clearly $\theta_0 = 0$ is always a solution. Does another solution $\theta(w)$ appear when w is sufficiently large? Look for a point $(w, 0)$ where the hypotheses of the implicit function theorem are not satisfied.

10. Let $G \subset \mathbb{R}^n$ be an open set, and let $f_i : G \to \mathbb{R}$ be C^1 functions, $i = 1, \ldots, k$. Let $S_i = \{\mathbf{x} \in G : f(\mathbf{x}) = 0\}$. Then

$$S = \cap_{i=1}^k S_i = \{\mathbf{x} \in G : \mathbf{f}(\mathbf{x}) = 0\}.$$

where $\mathbf{f}(\mathbf{x}) = (f_1(\mathbf{x}), \ldots, f_k(\mathbf{x}))$. Let $\mathbf{a} \in S$ and assume that the $k \times n$ Jacobian matrix $\mathbf{Df}(\mathbf{a})$ has rank k. We set

$$T_\mathbf{a} = \{\mathbf{v} \in \mathbb{R}^n : \mathbf{Df}(\mathbf{a})\mathbf{v} = 0\}.$$

a) Use the implicit function theorem to show that there is an open set $U \subset \mathbb{R}^{n-k}$ with $\hat{\mathbf{a}} = (a_1, \ldots, a_{n-k}) \in U$ and a C^1 mapping $\phi : U \to \mathbb{R}^n$ such that $\phi(U) \subset S$ and $\phi(\hat{\mathbf{a}}) = \mathbf{a}$.

b) Show that $T_a = \mathbf{D}\phi(\hat{\mathbf{a}})(\mathbb{R}^{n-k})$, which again justifies the use of the term *tangent space* for T_a.

11.6 An Application in Mechanics

In this section, we use both the inverse function theorem and the implicit function to describe the solutions of a problem from mechanics.

Let points $\mathbf{r}_1 = (-1, 0)$ and $\mathbf{r}_2 = (1, 0)$ be located on the x axis. Two springs of unstretched length $l < 1$ are attached at $\mathbf{r}_1$ and $\mathbf{r}_2$. The spring constants are k_1 and k_2, respectively. A body of mass m is attached where their free ends are joined (see Figure 11.6).

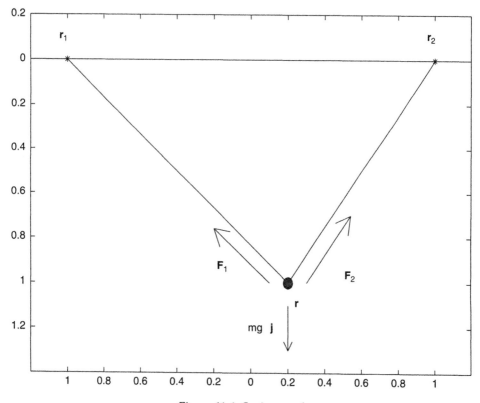

Figure 11.6 Springs and mass

We wish to determine the position of the mass when it is allowed to hang under its weight. The position of the mass is determined by the balance of forces on the mass. When the mass hangs down, the strings are stretched. The restoring force of each spring is proportional to the amount of stretch. Let $\mathbf{r} = (x, y)$ be the position of the mass. Let

$$d_1 = \|\mathbf{r} - \mathbf{r}_1\| = \sqrt{(x+1)^2 + y^2}$$

$$d_2 = \|\mathbf{r} - \mathbf{r}_2\| = \sqrt{(x-1)^2 + y^2},$$

and define the unit vectors

$$\mathbf{u}_1 = \frac{\mathbf{r} - \mathbf{r}_1}{d_1} \quad \text{and} \quad \mathbf{u}_2 = \frac{\mathbf{r} - \mathbf{r}_2}{d_2}.$$

When the spring is stretched (i.e., $d_j > l$), the restoring force is directed in the negative $\mathbf{u}_j$ direction (see Figure 11.6). Thus

$$\mathbf{F}_1 = -k_1(d_1 - l)\mathbf{u}_1 = -k_1\left(\frac{d_1 - l}{d_1}\right)(\mathbf{r} - \mathbf{r}_1)$$

$$\mathbf{F}_2 = -k_2(d_2 - l)\mathbf{u}_2 = -k_2\left(\frac{d_2 - l}{d_2}\right)(\mathbf{r} - \mathbf{r}_2).$$

The balance of forces on the mass (a condition of equilibrium) yields the equation

$$\mathbf{F}_1 + \mathbf{F}_2 - mg\mathbf{j} = 0 \tag{11.34}$$

where we use here the traditional notation $\mathbf{j} = (0, 1)$. In terms of components, these equations can be written

$$p(x, y) \equiv k_1\left(\frac{d_1 - l}{d_1}\right)(x + 1) + k_2\left(\frac{d_2 - l}{d_2}\right)(x - 1) = 0 \tag{11.35}$$

$$q(x, y) \equiv k_1\left(\frac{d_1 - l}{d_1}\right)y + k_2\left(\frac{d_2 - l}{d_2}\right)y = -mg.$$

If we set $\mathbf{f}(x, y) = (p(x, y), q(x, y))$, they can be written

$$\mathbf{f}(x, y) = (p(x, y), q(x, y)) = (0, -mg). \tag{11.36}$$

Instead of requiring the balance of forces when in equilibrium, we could also think of the problem as one of minimizing the potential energy of the system. The potential energy V is chosen so that $-\nabla V = \mathbf{F}_1 + \mathbf{F}_2 - mg\mathbf{j}$, and the equilibrium position of the mass satisfies $\nabla V = 0$. It is not hard to verify that the potential energy of this system is $V(x, y) = \phi(x, y) + mgy$, where

$$\phi(x, y) = (k_1/2)(d_1 - l)^2 + (k_2/2)(d_2 - l)^2.$$

The equations $\nabla V = 0$ are exactly the equations (11.35).

We could prove the existence of solutions of (11.35) by showing that V has a minimizer (x_*, y_*) that would be a solution of (11.35). We prefer to use the inverse and implicit function theorems to analyze this problem. The function $\mathbf{f}$ maps a point in the xy space of positions into a point in the pq plane of forces. Our problem is to find the inverse of this mapping, at least for forces $(p, q) = (0, -mg)$. Because

$$\mathbf{f}(x, y) = \nabla \phi(x, y),$$

which implies $p_y = \phi_{xy} = q_x$, it follows that $\mathbf{Df}$ is symmetric. First we show that $\mathbf{f}$ is locally one-to-one on the set

$$H = \{(x, y) : d_1 > l\} \cap \{(x, y) : d_2 > l\}$$

by showing that $\mathbf{Df}$ is invertible at each point of H. We exploit the symmetry of $\mathbf{Df}$ by showing that $\mathbf{Df}$ is positive definite on H. For a vector $\mathbf{v} = (v_1, v_2) \in \mathbb{R}^2$, we consider

$$\langle \mathbf{Df}(x, y)\mathbf{v}, \ \mathbf{v} \rangle = p_x v_1^2 + 2p_y v_1 v_2 + q_y v_2^2.$$

To clarify the computation we introduce the notation

$$r_1 = k_1 \left(\frac{d_1 - l}{d_1} \right), \quad r_2 = k_2 \left(\frac{d_2 - l}{d_2} \right)$$

and

$$s_1 = \frac{k_1 l}{d_1^3}, \quad s_2 = \frac{k_2 l}{d_2^3}.$$

Then

$$(r_1)_x = s_1(x + 1), \quad (r_2)_x = s_2(x - 1), \quad (r_1)_y = s_1 y, \quad (r_2)_y = s_2 y.$$

Now we can write $p = (x + 1)r_1 + (x - 1)r_2$ so that

$$
\begin{aligned}
p_x &= r_1 + (x + 1)(r_1)_x + r_2 + (x - 1)(r_2)_x \\
&= r_1 + r_2 + s_1(x + 1)^2 + s_2(x - 1)^2
\end{aligned}
\tag{11.37}
$$

and

$$
\begin{aligned}
p_y &= (x + 1)(r_1)_y + (x - 1)(r_2)_y \\
&= (x + 1)ys_1 + (x - 1)ys_2.
\end{aligned}
\tag{11.38}
$$

Similarly, $q = yr_1 + yr_2$, so that

$$
q_y = r_1 + r_2 + y^2(s_1 + s_2).
\tag{11.39}
$$

Using (11.38), we see that

$$
2p_y v_1 v_2 \leq s_1((x + 1)^2 + y^2)v_1^2 + s_2((1 - x)^2 + y^2)v_2^2.
$$

Then (11.37) and (11.39) imply that

$$
\begin{aligned}
\langle \mathbf{Df}(x, y)\mathbf{v}, \ \mathbf{v} \rangle &\geq [p_x - s1((x + 1)^2 + y^2]v_1^2 + [p_y - s_2((x - 1)^2 + y^2)]v_2^2 \\
&\geq (r_1 + r_2)(v_1^2 + v_2^2) \\
&\geq \left[k_1 \left(\frac{d_1 - l}{d_1} \right) + k_2 \left(\frac{d_2 - l}{d_2} \right) \right] (v_1^2 + v_2^2).
\end{aligned}
$$

We see that $\mathbf{Df}$ is positive definite at each point of H and hence invertible there. In physical terms, H is the set of positions of the mass such that both springs are under tension. However, H is not convex, so we cannot apply Theorem 11.10 to deduce that $\mathbf{f}$ is globally one-to-one on H.

To prove the existence and uniqueness of solutions of (11.35), we make a more detailed analysis. We will restrict our attention to the open set

$$
G \equiv H \cap \{(x, y) : |x| < 1\}.
$$

Note that $p < 0$ on the left boundary of G and that $p > 0$ on the right boundary of G. Furthermore, from (11.37), we see that $p_x > 0$ on G. Consequently, for each fixed y, the function $x \rightarrow p(x, y)$ takes on the value zero at exactly one point x between the left boundary of G and the right boundary of G. Hence the level set

$$\{(x, y) \in G : p(x, y) = 0\}$$

is the graph of a function $x = h(y)$, $-\infty < y < \infty$. We can apply the implicit function theorem at each point $(h(y), y)$ to deduce that $h(y)$ is C^1 with

$$h'(y) = -\frac{p_y(h(y), y)}{p_x(x, y)}. \tag{11.40}$$

Now a solution of (11.35) is a point (x_*, y_*) that lies at the intersection of the level curve $\{p = 0\}$ and the level set $\{q = -mg\}$. We will show that, for each $m \in R$, there is a unique point $(x_*(m), y_*(m)) \in G$ where the curve $x = h(y)$ intersects the level set $\{q = -mg\}$. To see this, we consider the restriction of q to the curve $x = h(y)$. This restriction is $y \to q(h(y), y)$. From the definition of q, we have $q(x, y) > 0$ in $G \cap \{y > 0\}$ and $q(x, y) < 0$ in $G \cap \{y < 0\}$. To show that $y \to (q(h(y), y)$ takes on each real value exactly once, we will show that

$$\frac{d}{dy} q(h(y), y) \geq c \tag{11.41}$$

for some constant $c > 0$. However, by (11.37), (11.42), and (11.43),

$$\frac{d}{dy} q(h(y), y) = q_x h_y + q_y$$

$$= q_x \left(-\frac{p_y}{p_x}\right) + q_y$$

$$= \frac{\det \mathbf{Df}(h(y), y)}{p_x(h(y), y)}$$

$$\geq \frac{(r_1 + r_2)(k_1 + k_2)}{r_1 + r_2 + s_1(x + 1)^2 + s_2(x - 1)^2}.$$

Using the definitions of r_1, r_2, s_1, and s_2, we have $r_1 + r_2 \leq k_1 + k_2$ and $s_1(x + 1)^2 + s_2(x - 1)^2 \leq 4(k_1 + k_2)$. Hence

$$\frac{d}{dy} q(h(y), y) \geq (r_2 + r_2)/5.$$

A further computation reveals that for $(x, y) \in G$,

$$r_1 + r_2 = k_1 \left(\frac{d_1 - l}{d_1} \right) + k_2 \left(\frac{d_2 - l}{d_2} \right) \geq C > 0$$

where C depends on k_1 and k_2.

The curve $\{p = 0\}$ is displayed for $l = 0.5$, $k_1 = 1$, $k_2 = 2$ in Figure 11.7. The level curves of the function $q(x, y)$ intersect the curve $p = 0$ at the points where the mass hangs.

Now at each point $(x_*(m), y_*(m))$, we may apply the inverse function theorem (Theorem 11.8). In this context, it says that there is a $\delta > 0$ such that, if $(w_1, w_2) \in \mathbb{R}^2$ with $\|(w_1, w_2)\| < \delta$, then there is a solution of

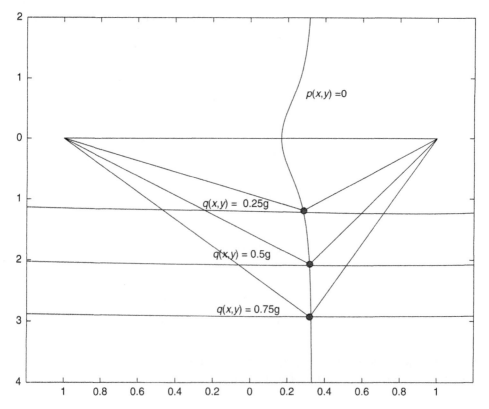

Figure 11.7 Level curve $p(x, y) = 0$ and level curves of q for $k_1 = 1, k_2 = 2$ and $l = 1/2$

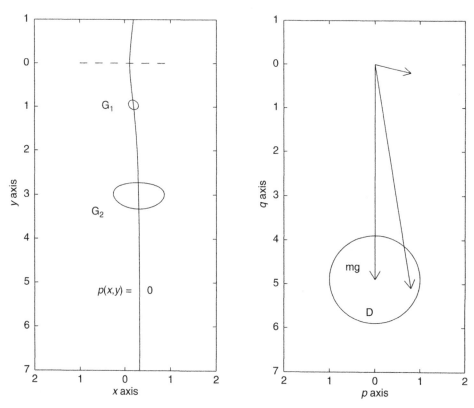

Figure 11.8 On the left, the xy plane of positions, and on the right, the pq plane of forces for the spring mass system

$\mathbf{f}(x, y) = (w_1, w_2 - mg)$ that is locally unique. This means that there exists a solution of the equations for the position of the mass when an additional force is put on the right side of (11.35). In Figure 11.8, we show this relationship. The right half of the figure is the pq plane of forces. D is a disk of radius 1 with center $(0, -mg)$ located at the tip of the gravitational force vector. An additional force, that pushes to the right, is added, yielding a resultant force vector that lies in D. The left side of the figure is the xy plane of positions. The region G_1 is the set of positions (x, y) that corresponds to the forces (p, q) that lie in the disk D when $k_1 = 1$, $k_2 = 2$ and $l = 0.5$. In other words, $\mathbf{f}$ maps G_1 onto D. For the values $k_1 = 0.1$, $k_2 = 0.2$ and $l = 0.5$, $\mathbf{f}$ maps the larger region G_2 on the right onto D. Note that equation (11.35) shows that the level curve $p(x, y) = 0$ remains the same as long as l is fixed, and the ratio k_1/k_2 is fixed.

Exercises for 11.6

1. Remember that we have assumed $l < 1$.

   a) Show that if $m > 0$, the solution of equations (11.35) that lies in $-1 < x < 1$ must satisfy $y < 0$.

   b) If $m = 0$, show that the solution that lies in $-1 < x < 1$ is of the form $(x_0, 0)$. Find x_0.

2. a) Use $\mathbf{D}f(x_0, 0)$ to generate approximate solutions of equations (11.35) when the mass m is small.

   b) The approximations of part a) have $x = x_0$, because $p_y = q_x = 0$ when $y = 0$. How would you generate better approximations in which x is not constantly equal to x_0? What derivatives of p and q would you need to compute?

3. We derive some approximate formulas for the solution of equations (11.35) when the mass m is large.

   a) The curve $p(x, y) = 0$ is asymptotic to a vertical line $x = x_\infty$ as $|y| \to \infty$. What is x_∞?

   b) From (11.44), we know that as $m \to \infty$, the y coordinate of the solution of equations (11.35) tends to $-\infty$. From part a), we know that the x coordinate tends to x_∞. We use these facts to derive an approximate formula for the solution for large m. We write q as

   $$q = k_1 \left(y - \frac{ly}{d_1} \right) + k_2 \left(y - \frac{ly}{d_2} \right).$$

   Show that for $y < 0$ and $|y|$ very large,

   $$q \approx k_1(y + l) + k_2(y + l).$$

   c) Using the approximation for q, deduce an approximate formula for y in terms of m when m is very large.

4. Assign values $l = 0.5$, $k_1 = 1$, and $k_2 = 2$. Take $g = 9.8$.

   a) Use Newton's method to solve the system (11.35) for $m = 0.25, 0.5, 0.75, 0.1$. Use $(x_0, 0)$ as the first guess for the solution when

   $$m = 0.25.$$

   Let the solution of this problem be the first guess for the solution when $m = 0.5$. Continue in this fashion.

   b) Calculate the solution values predicted by approximate formulas of exercise 3 for the case $m = 1$. How well do they compare with the values found by Newton's method?

c) Use Newton's method to compute the solution for $m = 3$, and compare the values with those predicted by the approximate formulas of exercise 3.

5. We continue to assume that $l < 1$. Now let $k_1 = 1$ and $k_2 = 10$, and $m = 0.1$. Plot the level curves $p = 0$ and $q + mgy = 0$. Look carefully in the neighborhood of $(1, 0)$. You should find another solution of the system (11.35), this time with $y > 0$.

a) What are the approximate coordinates of this solution? Does this solution lie in the set H? How do you describe this solution in physical terms?

b) There is still the expected solution of (11.35) in the region $-1 < x < 1$. Does the existence of these two solutions of (11.35) contradict the earlier analysis? Explain.

6. The solution of the springs and mass system of Section 11.6 depends on the choice of the parameters m, k_1, and k_2. Use the implicit function theorem to show that the position (x, y) of the mass depends in a C^1 fashion on the parameters m, k_1, and k_2. Show that $\partial x / \partial k_1 < 0$, $\partial x / \partial k_2 > 0$, and $\partial y / \partial m < 0$.

7. Here is an extension of the problem in Section 11.6. Let $\mathbf{p}_0 = (-1, 0)$ and let $\mathbf{p}_{n+1} = (1, 0)$. Now suppose there are $n + 1$ springs of unstretched length l, and n masses, $m_1, m_2, \ldots, m_n$, arranged as shown in Figure 11.9. Note that in this description, the forces are not the restoring forces used in Section 11.6.

The masses are located at points $\mathbf{p}_j = (x_j, y_j)$, $j = 1, \ldots, n$. For $j = 1, \ldots, n+1$, let

$$d_j = \|\mathbf{p}_j - \mathbf{p}_{j-1}\| = \sqrt{(x_j - x_{j-1})^2 + (y_j - y_{j-1})^2}.$$

We will assume that the springs all have the same spring constant k. The balance of forces on each mass necessary for equilibrium is expressed in the n equations

$$-\mathbf{F}_1 + \mathbf{F}_2 - m_1 g \mathbf{j} = 0$$
$$-\mathbf{F}_2 + \mathbf{F}_3 - m_2 g \mathbf{j} = 0$$
$$. = 0$$
$$. = 0$$
$$-\mathbf{F}_n + \mathbf{F}_{n+1} - m_n g \mathbf{j} = 0.$$

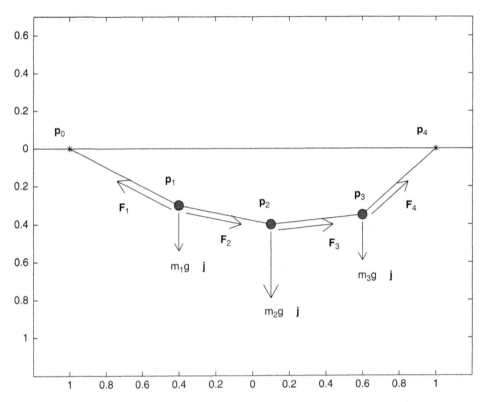

Figure 11.9 Arrangement of springs and masses for Exercise 7

where

$$\mathbf{F}_j = k\left(\frac{d_j - l}{d_j}\right)(\mathbf{p}_j - \mathbf{p}_{j-1}), \quad j = 1, \dots, n+1.$$

Let the coordinates of the jth mass be (x_j, y_j). We group the coordinates as $(\mathbf{x}, \mathbf{y})$ where $\mathbf{x} = (x_1, \dots, x_n)$ and $\mathbf{y} = (y_1, \dots, y_n)$. The potential energy for this system is

$$V = \phi(\mathbf{x}, \mathbf{y}) + \sum_{j=1}^{n} m_j y_j g$$

where

$$\phi(\mathbf{x}, \mathbf{y}) = (k/2) \sum_{j=1}^{n+1} (d_j - l)^2.$$

The balance of forces equations are the same as

$$\mathbf{f}(\mathbf{x}, \mathbf{y}) = \nabla V = 0. \tag{11.42}$$

a) When $n = 2$, there are four scalar equations. Write them out as

$$\mathbf{f}(\mathbf{x}, \mathbf{y}) = (0, 0, -m_1 g, -m_2 g)$$

where $\mathbf{f}(\mathbf{x}, \mathbf{y}) : \mathbb{R}^4 \to \mathbb{R}^4$.

b) What are the equilibrium positions x_1 and x_2 when the masses $m_1 = m_2 = 0$?

c) Let $r_j = k(d_j - l)/d_j$, and let $s_j = kl/d_j^3$, $j = 1, 2, 3$. Using the calcula-

tions of this section as a model, show that if $y_1 = y_2 = 0$, then for any $\mathbf{v} = (v_1, v_2, v_3, v_4) \in \mathbb{R}^4$,

$$\langle \mathbf{Df}(\mathbf{x}, 0)\mathbf{v}, \ \mathbf{v} \rangle \geq (r_1 + r_2)v_1^2 + r_1 v_2^2 + (r_2 + r_3)v_3^2 + r_3 v_4^2.$$

Conclude that if all of the springs are under tension (i.e., $d_j - l > 0$), then $\mathbf{Df}$ is positive definite and hence invertible.

d) Assume that $l < 2/3$. Show that in the equilibrium position with zero masses, the springs are under tension. Use the result of part c) and the inverse function theorem to prove the existence and uniqueness of solutions of the system when $n = 2$ for "small" masses.

e) Let $\mathbf{x}_0 = (-1/3, 1/3)$. Write out the set of linear equations

$$\mathbf{Df}(\mathbf{x}_0, 0) \begin{bmatrix} \mathbf{x} - \mathbf{x}_0 \\ \mathbf{y} \end{bmatrix} = \begin{bmatrix} 0 \\ 0 \\ -m_1 g \\ -m_2 g \end{bmatrix}.$$

Find solutions of these equations. This is the linear approximation to the solution of (11.42), and thus the first step in Newton's method, starting at $(\mathbf{x}_0, 0)$.

e) Write a four-dimensional Newton code to solve the system of springs and masses ($n = 2$) for various values of m_1 and m_2.

Chapter 12

Quadratic Approximation and Optimization

Chapter Overview

The opening section of this chapter deals with quadratic approximation and the second derivative test. Convex functions are discussed in the next section. Then we turn to an application of this material to study the stability of solutions of systems of ordinary differential equations. We return to the example of Section 11.6 of the two springs and mass to discuss the stability of the equilibrium solutions of that system. In the next two sections, we consider the practical question of how to find the minimum of a function without having to calculate the second derivatives. First we present the method of steepest descent. Because this method can converge very slowly, we provide an introduction to the more efficient method of conjugate gradients. Finally, we consider an important class of optimization problems, linear and nonlinear least squares. In particular, we study the problem of fitting a curve to data in which the curve is described by a model that is nonlinear in the parameters. The iterative Gauss-Newton method arises naturally in this context as a way to minimize the error.

12.1 Higher Derivatives and Quadratic Approximation

To make better approximations to a function that will tell us about its behavior near a critical point, we need to know about higher derivatives. We say that a scalar function $f(\mathbf{x})$ defined on an open set $G \subset \mathbb{R}^n$ has second-order partial derivatives on G if the first-order partials $\partial f/\partial x_j$, $j = 1, \ldots, n$ exist at each point of G and the partial derivatives

$$\frac{\partial}{\partial x_i}\left(\frac{\partial f}{\partial x_j}\right),$$

exist at each point of G, $i, j = 1, \ldots n$.

Mathematical life is much simpler because, most of the time, we can calculate the mixed partials in any order. Here is an easily verifiable condition that will guarantee that the order in which we take the second-order partials does not affect the outcome.

Theorem 12.1 Let G be an open subset of $\mathbb{R}^n$, and suppose that $f \in C^2(G)$. Then at each point of G,

$$\frac{\partial}{\partial x_i}\left(\frac{\partial f}{\partial x_j}\right) = \frac{\partial}{\partial x_j}\left(\frac{\partial f}{\partial x_i}\right).$$

Proof: This will be a good exercise in the use of the mean value theorem. To avoid using too many subscripts, we give the proof in two variables, x, y. The proof for more variables is the same. Let $(a, b) \in G$. Because G is an open set, for $h > 0$ sufficiently small, the square $\{a \le x \le a+h, \; b \le y \le b+h\}$ is contained in G. Let $\Delta(h) = f(a+h, b+h) - f(a+h, b) - f(a, b+h) + f(a, b)$. We will show that

$$\frac{\partial}{\partial y}\left(\frac{\partial f}{\partial x}\right)(a, b) = \lim_{h \to 0} \frac{\Delta(h)}{h^2} = \frac{\partial}{\partial x}\left(\frac{\partial f}{\partial y}\right)(a, b).$$

To see this, we first let $\varphi(x) = f(x, b+h) - f(x, b)$ and $\psi(y) = f(a+h, y) - f(a, y)$. Next verify that $\Delta(h) = \varphi(a + h) - \varphi(a)$. Then, using the mean value theorem in one dimension, there is point x_1 between a and $a + h$ such that

$$\Delta(h) = \varphi(a + h) - \varphi(a) = \varphi'(x_1)h.$$

However, $\varphi'(x_1) = f_x(x_1, b + h) - f_x(x_1, b)$, so that by a second application of the mean value theorem, this time in the y variable, we have

$$\Delta(h) = (f_x(x_1, b + h) - f_x(x_1, b))h = \frac{\partial}{\partial y}\left(\frac{\partial f}{\partial x}\right)(x_1, y_1)h^2,$$

where y_1 lies between b and $b + h$. Now we proceed in the same manner with ψ. We can also express

$$\Delta(h) = \psi(b + h) - \psi(b) = \psi'(y_2)h$$

for some y_2 between b and $b + h$. However, $\psi'(y_2) = f_y(a + h, y_2) - f_y(a, y_2)$. Hence we have

$$\Delta(h) = \frac{\partial}{\partial x}\left(\frac{\partial f}{\partial y}\right)(x_2, y_2)h^2$$

for some point x_2 between a and $a + h$. We equate the two expressions for $\Delta(h)$ and divide by h^2 to arrive at

$$\frac{\partial}{\partial y}\left(\frac{\partial f}{\partial x}\right)(x_1, y_1) = \frac{\Delta(h)}{h^2} = \frac{\partial}{\partial x}\left(\frac{\partial f}{\partial y}\right)(x_2, y_2).$$

As h tends to zero, the points (x_1, y_1) and (x_2, y_2) both tend to (a, b). The equality of the partial derivatives at (a, b) follows from the assumed continuity of the second-order partials of f. This completes the proof of Theorem 12.1.

Remark If the C^2 condition is not satisfied, it is possible to cook up an example of a function where the mixed partials depend on the order in which they are taken. Let

$$f(x, y) = \begin{cases} xy(x^2 - y^2)/(x^2 + y^2) & \text{if } (x, y) \neq (0, 0) \\ 0 & \text{if } (x, y) = (0, 0) \end{cases}. \tag{12.1}$$

We leave it to an exercise to show that

$$\frac{\partial}{\partial y}\left(\frac{\partial f}{\partial x}\right)(0, 0) = -1$$

while

$$\frac{\partial}{\partial x}\left(\frac{\partial f}{\partial y}\right)(0,0) = 1.$$

Let G be an open subset of $\mathbb{R}^n$, and let $f : G \to \mathbb{R}$ have second-order partials at the point $\mathbf{a} \in G$. The $n \times n$ matrix of second-order partials is called the *Hessian* of f at $\mathbf{a}$. We write it

$$\mathbf{D}^2 f(\mathbf{a}) = \begin{bmatrix} (f_{x_1})_{x_1}(\mathbf{a}) & (f_{x_1})_{x_2}(\mathbf{a}) & \cdot & \cdot & (f_{x_1})_{x_n}(\mathbf{a}) \\ (f_{x_2})_{x_1}(\mathbf{a}) & (f_{x_2})_{x_2}(\mathbf{a}) & \cdot & \cdot & (f_{x_2})_{x_n}(\mathbf{a}) \\ \cdot & & \cdot & \cdot & \cdot \\ \cdot & & & \cdot & \cdot \\ (f_{x_n})_{x_1}(\mathbf{a}) & (f_{x_n})_{x_2}(\mathbf{a}) & \cdot & \cdot\cdot & (f_{x_n})_{x_n}(\mathbf{a}) \end{bmatrix}. \tag{12.2}$$

We see that $\mathbf{D}^2 f$ is the Jacobian matrix of the vector-valued function $\mathbf{x} \to \nabla f(\mathbf{x})$. In the case in which $f \in C^2(G)$, $\mathbf{D}^2 f$ is a symmetric matrix.

Quadratic Approximation

Recall from Chapter 5 that if f is C^2, we can calculate the Taylor polynomial of degree 2,

$$p_2(x) = f(a) + f'(a)(x - a) + \frac{1}{2}f''(a)(x - a)^2. \tag{12.3}$$

The Taylor polynomial satisfies

$$p_2(a) = f(a), \quad p_2'(a) = f'(a), \quad p_2''(a) = f''(a).$$

We want to construct the analogous quadratic polynomial for functions of n variables, and we want to use it to get information about the critical points.

Let f be a C^2 function on an open set $G \subset \mathbb{R}^n$, and let a $\in G$. There is a $\delta > 0$ such that, if $\mathbf{h} \in \mathbb{R}^n$ with $\|\mathbf{h}\| < \delta$, then $\mathbf{a} + t\mathbf{h} \in G$ for $0 \le t \le 1$.

Let $g(t) = f(\mathbf{a} + t\mathbf{h})$. By the chain rule,

$$g'(t) = \langle \nabla f(\mathbf{a} + t\mathbf{h}), \mathbf{h} \rangle = f_{x_1}(\mathbf{a} + t\mathbf{h})h_1 + \ldots + f_{x_n}(\mathbf{a} + t\mathbf{h})h_n. \tag{12.4}$$

Because $f \in C^2(G)$, we may use the chain rule a second time to compute

$$
\begin{aligned}
g''(t) &= \frac{d}{dt} f_{x_1}(\mathbf{a} + t\mathbf{h})h_1 + \cdots + \frac{d}{dt} f_{x_n}(\mathbf{a} + t\mathbf{h})h_n \\
&= \langle \nabla f_{x_1}(\mathbf{a} + t\mathbf{h}), \mathbf{h} \rangle h_1 + \cdots + \langle \nabla f_{x_n}(\mathbf{a} + t\mathbf{h}), \mathbf{h} \rangle h_n \\
&= \langle \mathbf{D}^2 f(\mathbf{a} + t\mathbf{h})\mathbf{h}, \mathbf{h} \rangle.
\end{aligned}
\tag{12.5}
$$

Note In this chapter we will only use the 2-norm. The symbol $\| \ \|$ will stand for $\| \ \|_2$.

With this preparation, we can state the following:

Theorem 12.2 Let G be an open subset of $\mathbb{R}^n$, and let $f \in C^2(G)$. Let $\mathbf{a} \in G$. For $\|\mathbf{h}\|$ sufficiently small,

$$
f(\mathbf{a} + \mathbf{h}) = f(\mathbf{a}) + \langle \nabla f(\mathbf{a}), \mathbf{h} \rangle + \frac{1}{2} \langle \mathbf{D}^2 f(\mathbf{a})\mathbf{h}, \mathbf{h} \rangle + R_2(\mathbf{h})
\tag{12.6}
$$

where $|R_2(\mathbf{h})| = o(\|\mathbf{h}\|^2)$. The quadratic approximation to f at $\mathbf{x} = \mathbf{a}$ is

$$
q(\mathbf{x}) = f(\mathbf{a}) + \langle \nabla f(\mathbf{a}), \mathbf{x} - \mathbf{a} \rangle + \frac{1}{2} \langle \mathbf{D}^2 f(\mathbf{a})(\mathbf{x} - \mathbf{a}), \mathbf{x} - \mathbf{a} \rangle.
$$

Proof: Let $g(t) = f(\mathbf{a} + t\mathbf{h})$. We apply Taylor's theorem with $n = 1$, and we use the Lagrange form of the remainder to obtain

$$
g(1) = g(0) + g'(0) + \frac{1}{2} g''(\theta)
$$

where $0 < \theta < 1$. However, $g(1) = f(\mathbf{a} + \mathbf{h})$, $g(0) = f(\mathbf{a})$, and using (12.4) and (12.5) we have

$$
f(\mathbf{a} + \mathbf{h}) = f(\mathbf{a}) + \langle \nabla f(\mathbf{a}), \mathbf{h} \rangle + \frac{1}{2} \langle \mathbf{D}^2 f(\mathbf{a} + \theta\mathbf{h})\mathbf{h}, \mathbf{h} \rangle.
$$

We can restate this as saying that there is a point $\mathbf{z}$ on the line between $\mathbf{a}$ and $\mathbf{a} + \mathbf{h}$ such that

$$f(\mathbf{a} + \mathbf{h}) = f(\mathbf{a}) + \langle \nabla f(\mathbf{a}), \mathbf{h} \rangle + \frac{1}{2} \langle \mathbf{D}^2 f(\mathbf{z})\mathbf{h}, \mathbf{h} \rangle. \qquad (12.7)$$

Hence

$$f(\mathbf{a} + \mathbf{h}) = f(\mathbf{a}) + \langle \nabla f(\mathbf{a}), \mathbf{h} \rangle + \frac{1}{2} \langle \mathbf{D}^2 f(\mathbf{a})\mathbf{h}, \mathbf{h} \rangle + R_2(\mathbf{h}),$$

where

$$R_2(\mathbf{h}) = \frac{1}{2} \langle [\mathbf{D}^2 f(\mathbf{z}) - \mathbf{D}^2 f(\mathbf{a})]\mathbf{h}, \mathbf{h} \rangle.$$

Now we estimate $R_2(\mathbf{h})$ using the Schwarz inequality and the fact that $\|\mathbf{A}\mathbf{x}\| \leq \|\mathbf{A}\| \|\mathbf{x}\|$ for any of the norms. We find

$$|R_2(\mathbf{h})| \leq \frac{1}{2} \|[\mathbf{D}^2 f(\mathbf{z}) - \mathbf{D}^2 f(\mathbf{a})]\mathbf{h}\| \, \|\mathbf{h}\|$$

$$\leq \frac{1}{2} \|\mathbf{D}^2 f(\mathbf{z}) - \mathbf{D}^2 f(\mathbf{a})\| \, \|\mathbf{h}\|^2.$$

The presence of the factor $\|\mathbf{h}\|^2$ in this inequality means that $|R_2(\mathbf{h})| = O(\|\mathbf{h}\|^2)$, but in addition, because we assume that $f \in C^2(G)$, we know that the other factor $\|\mathbf{D}^2 f(\mathbf{z}) - \mathbf{D}^2 f(\mathbf{a})\|$ tends to zero as $\|\mathbf{h}\|$ tends to zero, because $\mathbf{z}$ is on the line between $\mathbf{a}$ and $\mathbf{a} + \mathbf{h}$. Together these two facts imply that $R_2(\mathbf{h}) = o(\|\mathbf{h}\|^2)$. The theorem is proved.

In the case of two variables x, y, we can write the quadratic approximation at $(x, y) = (a, b)$ as follows:

$$\begin{aligned} q(x, y) = \ & f(a, b) + f_x(a, b)(x - a) + f_y(a, b)(y - b) \\ & + \frac{1}{2}[f_{xx}(a, b)(x - a)^2 + 2f_{xy}(a, b)(x - a)(y - b) + f_{yy}(a, b)(y - b)^2]. \end{aligned}$$

$$(12.8)$$

The quadratic approximation $q(x, y)$ is the Taylor polynomial of degree 2 in two variables. The polynomial q matches f at (a, b) in the sense that

$$q(a, b) = f(a, b), \quad \nabla q(a, b) = \nabla f(a, b)$$

$$\mathbf{D}^2 q(a, b) = \mathbf{D}^2 f(a, b).$$

EXAMPLE 12.1: Let $f(x, y) = x^2 y \cos y$. Let us find the quadratic approximation to f at $(x, y) = (2, \pi)$. First we have

$$\nabla f(x, y) = (2xy \cos y, \ x^2(\cos y - y \sin y))$$

and

$$\mathbf{D}^2 f(x, y) = \begin{bmatrix} 2y \cos y & 2x(\cos y - y \sin y) \\ 2x(\cos y - y \sin y) & x^2(-2 \sin y - y \cos y) \end{bmatrix}$$

so that

$$\nabla f(2, \pi) = (-4\pi, -4), \quad \mathbf{D}^2 f(2, \pi) = \begin{bmatrix} -2\pi & -4 \\ -4 & 4\pi \end{bmatrix}.$$

Thus the quadratic approximation for f at $(2, \pi)$ is

$$q(x, y) = -4\pi - 4\pi(x - 2) - 4(y - \pi)$$

$$+ \frac{1}{2}[-2\pi(x - 2)^2 - 8(x - 2)(y - \pi) + 4\pi(y - \pi)^2].$$

Definition 12.1 If $G \subset \mathbb{R}^n$ is open and $f \in C^1(G)$, $\mathbf{a} \in G$ is a *critical point* if $\nabla f(\mathbf{a}) = 0$.

The quadratic approximation will provide a criterion to determine when a critical point is a local maximum or a local minimum.

Recall Definition 11.3: An $n \times n$ symmetric matrix $\mathbf{A}$ is positive definite if $\langle \mathbf{Ax}, \mathbf{x} \rangle > 0$ for all $\mathbf{x} \neq 0$. Then Lemma 11.11 asserted that if $\mathbf{A}$ is positive definite, there is a constant $c > 0$ such that $\langle \mathbf{Ax}, \mathbf{x} \rangle \geq c\|\mathbf{x}\|^2$ for all $\mathbf{x}$.

Theorem 12.3 Let $G \subset \mathbb{R}^n$ be an open set and let $f \in C^2(G)$. Suppose that $\mathbf{a} \in G$ is a critical point for f. If the Hessian $\mathbf{D}^2 f(\mathbf{a})$ is positive definite, then f has a local minimum at $\mathbf{x} = \mathbf{a}$. If $\mathbf{D}^2 f(\mathbf{a})$ is negative definite, f has a local maximum at $\mathbf{x} = \mathbf{a}$.

Proof: We prove the assertion when $\mathbf{D}^2 f$ is positive definite. Thus there is a $c > 0$ such that $\langle \mathbf{D}^2 f(\mathbf{a})\mathbf{h}, \mathbf{h} \rangle \geq c\|\mathbf{h}\|^2$. Because G is open, we may choose a $\delta > 0$ such that $\mathbf{a} + \mathbf{h} \in G$ when $\|\mathbf{h}\| < \delta$. According to Theorem 12.2, $|R_2(\mathbf{h})| = o(\|\mathbf{h}\|^2)$. This means that, by choosing $\delta > 0$ that is perhaps smaller, we have $|R_2(\mathbf{h})| \leq (c/4)\|\mathbf{h}\|^2$ for $\|\mathbf{h}\| < \delta$. Now substitute in (12.6):

$$f(\mathbf{a} + \mathbf{h}) \geq f(\mathbf{a}) + (c/2)\|\mathbf{h}\|^2 - |R_2(\mathbf{h})| \geq f(\mathbf{a}) + (c/4)\|\mathbf{h}\|^2$$

for $\|\mathbf{h}\| < \delta$. This shows that f has a local minimum at $\mathbf{x} = \mathbf{a}$. The proof when $\mathbf{D}^2 f$ is negative definite is the same; replace f with $-f$.

Recall from the theorem after Lemma 11.11, that a symmetric matrix $\mathbf{A}$ is positive definite if and only if the determinants of the submatrices $\mathbf{A}_k$ are all strictly positive. The matrix $\mathbf{A}$ is negative definite if $-\mathbf{A}$ is positive definite, or equivalently, $\det(-\mathbf{A}_k) > 0$ for all k. Now, $\mathbf{A}_k$ has odd dimension for k odd and even dimension for k even. Hence $\mathbf{A}$ is negative definite if and only if $\det \mathbf{A}_k < 0$ for k odd, and $\det \mathbf{A}_k > 0$ for k even. As a consequence, we have the n-dimensional version of the second derivative test.

Second Derivative Test

Let $\mathbf{a}$ be a critical point for f, and let $\mathbf{H} = \mathbf{D}^2 f(\mathbf{a})$ be the Hessian matrix for f.

(i) If $\det(\mathbf{H}_k) > 0$ for all k, then f has a local minimum at $\mathbf{a}$.

(ii) If $\det(\mathbf{H}_k) < 0$ for k odd and $\det(\mathbf{H}_k) > 0$ for k even, then f has a local maximum at $\mathbf{a}$.

(iii) If neither of conditions (i) or (ii) is true, but $\det(\mathbf{H}) \neq 0$, then f has a saddle at $\mathbf{a}$.

This yields the usual second derivative test in two dimensions. If $f(\mathbf{x}) = f(x, y)$, then

$$\mathbf{H} = \mathbf{D}^2 f(\mathbf{a}) = \begin{bmatrix} f_{xx}(\mathbf{a}) & f_{xy}(\mathbf{a}) \\ f_{xy}(\mathbf{a}) & f_{yy}(\mathbf{a}) \end{bmatrix},$$

and the conditions for a minimum are:

$$\det H_1 = f_{xx}(\mathbf{a}) > 0 \quad \text{and} \quad \det H_2 = \det H = f_{xx}(\mathbf{a})f_{yy}(\mathbf{a}) - f_{xy}(\mathbf{a})^2 > 0.$$

The conditions for a maximum are:

$$\det H_1 = f_{xx}(\mathbf{a}) < 0 \quad \text{and} \quad \det H_2 = \det H = f_{xx}(\mathbf{a})f_{yy}(\mathbf{a}) - f_{xy}(\mathbf{a})^2 > 0.$$

If $f(\mathbf{x}) = f(x, y, z)$, then

$$\mathbf{H} = \mathbf{D}^2 f(x, y, z) = \begin{bmatrix} f_{xx}(\mathbf{a}) & f_{xy}(\mathbf{a}) & f_{xz}(\mathbf{a}) \\ f_{xy}(\mathbf{a}) & f_{yy}(\mathbf{a}) & f_{yz}(\mathbf{a}) \\ f_{xz}(\mathbf{a}) & f_{yz}(\mathbf{a}) & f_{zz}(\mathbf{a}) \end{bmatrix}.$$

The second derivative conditions for a minimum become

$$\det H_1 = f_{xx}(\mathbf{a}) > 0 \quad \text{and} \quad \det H_2 = f_{xx}(\mathbf{a})f_{yy}(\mathbf{a}) - f_{xy}^2(\mathbf{a}) > 0$$

$$\text{and} \quad \det H_3 = \det H > 0.$$

From our experience in two dimensions, we know that the critical point is a saddle point when $\det \mathbf{D}^2 f(\mathbf{a}) < 0$. This condition is equivalent to $\mathbf{D}^2 f(\mathbf{a})$ having one positive eigenvalue and one negative eigenvalue. The condition for a saddle point in higher dimensions is that $\mathbf{H}$ has eigenvalues of different signs, but no zero eigenvalues. Thus there can be many different kinds of saddle points in higher dimensions.

EXAMPLE 12.2: For the following examples, the origin is the unique critical point. For each example, we let $S_c = \{(x, y, z) : f(x, y, z) = c\}$ denote the level surfaces of f.

(i) Let $f(x, y, z) = x^2 - xz + y^2 + z^2$.

$$\mathbf{D}^2 f(0, 0, 0) = \begin{bmatrix} 2 & 0 & -1 \\ 0 & 2 & 0 \\ -1 & 0 & 2 \end{bmatrix}.$$

It is easy to check that $f(0, 0, 0) = 0$ is the minimum of f. The level surfaces S_c are ellipsoids.

In the next two cases, $(0, 0, 0)$ is a saddle point for f.

(ii) Let $f(x, y, z) = -x^2 + 2y^2 + 4xz - z^2$. We have

$$\mathbf{D}^2 f(0,0,0) = \begin{bmatrix} -1 & 0 & 2 \\ 0 & 2 & 0 \\ 2 & 0 & -1 \end{bmatrix}.$$

The eigenvalues of $\mathbf{D}^2 f(0,0,0)$ are $\lambda = 1,\ 2,\ -3$. The level surfaces S_c are hyperboloids of one sheet for $c > 0$ and hyperboloids of two sheets for $c < 0$.

(iii) $f(x, y, z) = -x^2 - 2y^2 + 4xz - z^2$. In this case,

$$\mathbf{D}^2 f(0,0,0) = \begin{bmatrix} -1 & 0 & 2 \\ 0 & -2 & 0 \\ 2 & 0 & -1 \end{bmatrix}.$$

The eigenvalues of $\mathbf{D}^2 f$ are $\lambda = 1,\ -2,\ -3$. Now S_c is a hyperboloid of one sheet for $c < 0$ and of two sheets for $c > 0$.

Exercises for 12.1

1. Let $f(x, y) = \sqrt{x^2 + 2x + y + 3}$.

a) Calculate the linear approximation $l(x, y)$ to f at $(0, 0)$. Let $E_1(x, y) = |f(x, y) - l(x, y)|$.

b) Calculate the quadratic approximation $q(x, y)$ to f at $(0, 0)$. Let $E_2(x, y) = |f(x, y) - q(x, y)|$.

c) Let Q_h be the square $-h \le x, y \le h$. Compute the maximum of E_1 over Q_h for $h = 1,\ 1/2,\ 1/4$, and $1/8$. By what factor is max E_1 decreasing each time? For what power p is it true that

$$\max_{Q_h} E_1 \approx Ch^p?$$

d) Now compute the maximum of E_2 over Q_h for $h = 1,\ 1/2,\ 1/4$ and $1/8$. By what factor is max E_2 decreasing each time? For what power p is it true that

$$\max_{Q_h} E_2 \approx Ch^p?$$

2. Find the critical points of the following functions. Determine their nature using the second derivative test, Theorem 12.3.

   a) $f(x, y) = x^2 + xy + y^2 - 3x - 9y$.
   b) $f(x, y) = 2x^2 - xy - y^2 - 7x - 4y$.
   c) $f(x, y) = x^2 + 1/(xy) + 2y^2$.
   d) $f(x, y) = x^4 + y^4 - x^2 - 2y^2$.

3. Find the critical points of the following functions. Determine their nature using Theorem 12.3.

   a) $f(x, y, z) = xyz/81 - 1/x + 1/y - 1/z$.
   b) $f(x, y, z) = (2x + y + z) \exp(-4x^2 - y^2 + z^2)$.

4. Verify that $f(x, y) = x^2 + y^4$ has a global minimum at $(0, 0)$ but that Theorem 12.3 cannot be applied.

5. Let $\sigma > 0$ be a parameter. Define

$$f(x, y) = 3x^4 + \sigma x^2 - 4x^2 y + y^2.$$

   a) What is the quadratic (second order) approximation to f at $(0, 0)$?
   b) Is $(0, 0)$ a relative extreme point for f? Which kind? Why?
   c) Now set $\sigma = 0$. What does the second derivative test tell you?
   d) Continue with $\sigma = 0$. Show that, on each line $y = mx$ with $m \neq 0$, $x \to f(x, mx)$ has a minimum value of 0 at $x = 0$.
   e) Now look at the behavior of f along the parabola $y = 2x^2$. Show that, when $\sigma = 0$, there is no $r > 0$ such $f \geq 0$ on $B(0, r)$.

6. Let the vector field $\mathbf{F}(x, y) = (1 - y^2, 0)$ in the square $Q = \{-1 \leq x \leq 1, -1 \leq y \leq 1\}$. Let $\mathbf{p}_0 = (-1/2, -1/2)$ and $\mathbf{p}_1 = (1/2, 1/2)$. Let $\mathbf{q} = (a, b)$ be any point in the square. Define the function

$$W(a, b) = \int_C \mathbf{F} \cdot d\mathbf{r},$$

   where C is any path consisting of a line segment C_1 running from $\mathbf{p}_0$ to $\mathbf{q}$, and a line segment C_2 running from $\mathbf{q}$ to $\mathbf{p}_1$. What is the minimum of W on Q? Does W have any critical points inside Q? What are their nature?

7. Let G be an open subset of $\mathbb{R}^3$. A function $v(x, y, z) \in C^2(G)$ is said to be *harmonic* in G if $\Delta v = v_{xx} + v_{yy} + v_{zz} = 0$ in G.

a) Show that, if v is harmonic, the Hessian $\mathbf{D}^2 v$ is never positive definite and never negative definite.

b) Let $(a, b, c) \in \mathbb{R}^3$, and let $r = \sqrt{(x - a)^2 + (y - b)^2 + (z - c)^2}$. If q is a charge placed at (a, b, c), then

$$v(x, y, z) = \frac{q}{r}$$

is the electrostatic potential of the charge. Verify that v is harmonic for $(x, y, z) \neq (a, b, c)$.

Now let unit positive charges be placed at the points $(-1, 0, 0)$, $(1, 0, 0)$ and $(0, h, 0)$, with

$$
\begin{aligned}
r_1 &= \sqrt{(x + 1)^2 + y^2 + z^2} \\
r_2 &= \sqrt{(x - 1)^2 + y^2 + z^2} \\
r_3 &= \sqrt{x^2 + (y - h)^2 + z^2},
\end{aligned}
$$

and set

$$v = \frac{1}{r_1} + \frac{1}{r_2} + \frac{1}{r_3}.$$

The function v is clearly harmonic by part b).
Verify that $v_z(x, y, 0) = 0$ for all (x, y) except for $(\pm 1, 0)$ and $(h, 0)$, and that $v_x(0, y, 0) = 0$ for $y \neq h$.

c) We shall look for critical points of v on the y axis. Let $f(y, h) = v_y(0, y, 0, h)$. Plot $y \rightarrow f(y, h)$ on the interval $0 \leq y \leq 1$ for values of $h = 1.6$, 1.7, 1.8, 2. Find the roots of $f(y, h) = v_y(0, y, 0) = 0$ using Newton's method when $h = 2$.

d) From the graphs of part c), we can see that there is some value of h between 1.6 and 1.7, where the two roots of $f(y, h) = 0$ coincide. Using the implicit function theorem, show that this occurs for those values of $y = y_*$ and $h = h_*$ such that both $f(y, h) = 0$ and $f_y(y, h) = v_{yy}(0, y, 0, h) = 0$. Use Newton's method in two dimensions to find a numerical approximation for y_* and h_*.

e) Evaluate the Hessian matrix $\mathbf{D}^2 v(0, y, 0, 2)$ for the two critical points that you found in part c). What is the nature of the critical point in each case?

f) Now evaluate the Hessian matrix $\mathbf{D}^2 v$ at the point $(0, y_*, 0, h_*)$ where y_* and h_* are the values found in part d). What is the nature of this critical point?

12.2 Convex Functions

The notion of a convex function (see Section 5.4) is important in optimization problems. Here we give a definition for functions of several variables.

Definition 12.2 Let $A \subset \mathbb{R}^n$ be convex. The function $f : A \to \mathbb{R}$ is *convex* if, for each pair of points $\mathbf{x}_0, \mathbf{x}_1 \in A$,

$$f((1 - t)\mathbf{x}_0 + t\mathbf{x}_1) \leq (1 - t)f(\mathbf{x}_0) + tf(\mathbf{x}_1) \text{ for } 0 \leq t \leq 1.$$

We say that f is *strictly convex* if

$$f((1 - t)\mathbf{x}_0 + t\mathbf{x}_1) < (1 - t)f(\mathbf{x}_0) + tf(\mathbf{x}_1) \text{ for } 0 < t < 1.$$

We will say that f is *concave* if $-f$ is convex.

Recall that in one dimension, we showed that a convex function is always continuous. The same result holds in higher dimensions.

Theorem 12.4 Let G be an open convex set of $\mathbb{R}^n$, and let $f : G \to \mathbb{R}$ be convex. Then f is continuous on G.

Proof: The proof is essentially the same as in one dimension. First we make a change of variable to simplify the notation. Let $\mathbf{a} \in G$. Because G is open, there is an $r > 0$ such that $\{\mathbf{x} : \|\mathbf{x} - \mathbf{a}\|_\infty \leq r\} \subset G$. The image of G under the mapping $\mathbf{x} \to \tilde{\mathbf{x}} = (\mathbf{x} - \mathbf{a})/r$ is an open convex set $\tilde{G}$ such that the closed unit ball $\bar{B} = \{\|\tilde{\mathbf{x}}\|_\infty \leq 1\}$ is contained in $\tilde{G}$. Furthermore, $\tilde{f}(\tilde{\mathbf{x}}) \equiv f(\mathbf{a} + r\tilde{\mathbf{x}})$ is convex on $\tilde{G}$ because f is convex on G. Finally, f will be continuous at $\mathbf{x} = \mathbf{a}$ if and only if $\tilde{f}$ is continuous at $\tilde{\mathbf{x}} = 0$. Thus it is no loss of generality to consider the situation in which f satisfies the conditions of $\tilde{f}$.

Let $\mathbf{v}_k$, $k = 1, \ldots, 2^n$ be the set of vertices of the closed unit ball $\bar{B}$ in the norm $\| \ \|_\infty$, and let $M = \max f(\mathbf{v}_k)$. In exercise 3 of this section, the reader is asked to show that $f(\mathbf{x}) \leq M$ for all $\mathbf{x}$ with $\|\mathbf{x}\|_\infty \leq 1$.

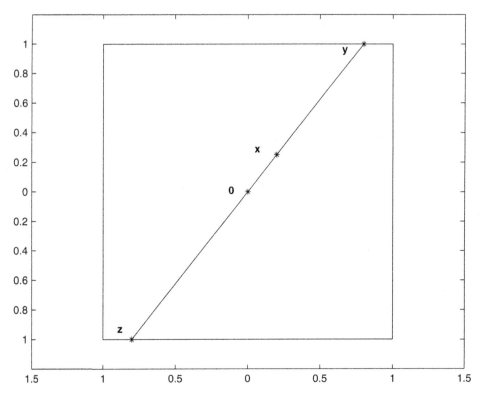

Figure 12.1 Unit ball B in the norm $\|\ \|_\infty$ in two dimensions and intersections y and z of the line through x and 0 with the unit sphere

Now suppose that $\mathbf{x} \in B, \mathbf{x} \neq 0$. Let $y = \mathbf{x}/\|\mathbf{x}\|_\infty$. Then $\mathbf{x}$ lies on the line from 0 to y (see Figure 12.1), and we may write $\mathbf{x} = (1-t)0 + t\mathbf{y}$ with $t = \|\mathbf{x}\|_\infty$. In view of the convexity of f have

$$f(\mathbf{x}) \leq (1-t)f(0) + tf(\mathbf{y}).$$

Hence

$$f(\mathbf{x}) - f(0) \leq \|\mathbf{x}\|_\infty(f(\mathbf{y}) - f(0)) \leq \|\mathbf{x}\|_\infty(M - f(0)).$$

To get an inequality in the opposite sense, we set $\mathbf{z} = -\mathbf{y} = -\mathbf{x}/\|\mathbf{x}\|_\infty$. Then 0 lies on the line from x to z, and we may write $0 = (1-t)\mathbf{x} + t\mathbf{z}$ where $t = \|\mathbf{x}\|_\infty/(1 + \|\mathbf{x}\|_\infty)$. The convexity of f implies

$$f(0) \leq (1-t)f(\mathbf{x}) + tf(\mathbf{z}) = \left(\frac{1}{1 + \|\mathbf{x}\|_\infty}\right)(f(\mathbf{x}) + \|\mathbf{x}\|_\infty f(\mathbf{z})).$$

It follows that

$$f(\mathbf{x}) - f(0) \geq \|\mathbf{x}\|_\infty(f(0) - f(\mathbf{z})) \geq \|\mathbf{x}\|_\infty(f(0) - M).$$

We conclude that

$$|f(\mathbf{x}) - f(0)| \leq \|\mathbf{x}\|_\infty(M - f(0)).$$

This estimate shows that f is continuous at $\mathbf{x} = 0$. The theorem is proved.

The set of minimum points of a convex function is limited in a way that can be very helpful.

Theorem 12.5 Let $A \subset \mathbb{R}^n$ be a convex set, and let $f : A \to \mathbb{R}$ be convex.

(i) If f has a local minimum at $\mathbf{x}_* \in A$, then $f(\mathbf{x}) \geq f(\mathbf{x}_*)$ for all $\mathbf{x} \in A$. In other words, $\mathbf{x}_*$ is a global minimum for f on A.
(ii) If f is strictly convex on A, then there is at most one local minimizer $\mathbf{x}_* \in A$, which by part (i) must be the global minimizer.

Proof: We argue by contradiction.

(i) Suppose that $\mathbf{x}_*$ is a local minimizer for f on A and that there is some other point $\mathbf{y} \in A$ where $f(\mathbf{y}) < f(\mathbf{x}_*)$. Because A is convex, the line $\mathbf{x}(t) = (1 - t)\mathbf{x}_* + t\mathbf{y} \in A$ for $0 \leq t \leq 1$. Now using the convexity of f, we have

$$\begin{aligned} f(\mathbf{x}(t)) &\leq (1 - t)f(\mathbf{x}_*) + tf(\mathbf{y}) \\ &< (1 - t)f(\mathbf{x}_*) + tf(\mathbf{x}_*) = f(\mathbf{x}_*) \end{aligned}$$

for $0 < t < 1$. The sequence $\mathbf{x}_n = \mathbf{x}(1/n)$ converges to $\mathbf{x}_*$, and $f(\mathbf{x}_n) < f(\mathbf{x}_*)$ for all n. This contradicts the assumption that $\mathbf{x}_*$ is a local minimizer.
(ii) We repeat the argument of part (i) using the fact that f is strictly convex. Let $\mathbf{x}_*$ be a local minimizer for f on A, and hence a global minimizer by (i). Suppose that there is a $\mathbf{y} \in A$ such that $f(\mathbf{y}) = f(\mathbf{x}_*)$. Then

$$f(\mathbf{x}(1/2)) < (1/2)f(\mathbf{x}_*) + (1/2)f(\mathbf{y}) = f(\mathbf{x}_*)$$

which is a contradiction. Thus f can have at most one minimizer on A.

In one dimension we saw that if $f \in C^2$, then f is convex if and only if $f'' \geq 0$ (Theorem 5.5). It is not surprising that the analogue for higher dimensions involves the Hessian of f.

Theorem 12.6 Let $G \subset \mathbb{R}^n$ be an open convex set. Suppose that $f : \bar{G} \to \mathbb{R}$ is continuous and that $f \in C^2(G)$.

(i) f is convex on $\bar{G}$ if and only if the Hessian $\mathbf{D}^2 f(\mathbf{x})$ is positive semidefinite for each $\mathbf{x} \in G$.

(ii) If $\mathbf{D}^2 f(\mathbf{x})$ is positive definite for each $\mathbf{x} \in G$, then f is strictly convex on $\bar{G}$.

Proof: We note that G convex implies that $\bar{G}$ is also convex (exercise 6 of Section 9.2). Suppose that f is convex on $\bar{G}$. Let $\mathbf{a} \in G$ and let $\mathbf{v} \in \mathbb{R}^n$ be a unit vector. Because G is open, there is a $\delta > 0$ such that $\mathbf{x}_0 = \mathbf{a} - \delta \mathbf{v}$ and $\mathbf{x}_1 = \mathbf{a} + \delta \mathbf{v}$ both belong to G. Furthermore, $\mathbf{x}(t) = (1 - t)\mathbf{x}_0 + t\mathbf{x}_1 \in G$ for $0 \leq t \leq 1$ with $x(1/2) = \mathbf{a}$. Let $\phi(t) = f(\mathbf{x}(t))$. Now f convex implies that $\phi(t)$ is also convex on the interval $[0, 1]$. It follows from Theorem 5.5 that $\phi''(t) \geq 0$. But

$$\phi''(t) = \langle \mathbf{D}^2 f(\mathbf{x}(t))(\mathbf{x}_1 - \mathbf{x}_0), \mathbf{x}_1 - \mathbf{x}_0 \rangle$$
$$= 4\delta^2 \langle \mathbf{D}^2 f(\mathbf{x}(t))\mathbf{v}, \mathbf{v} \rangle.$$

Consequently,

$$4\delta^2 \langle \mathbf{D}^2 f(\mathbf{a})\mathbf{v}, \mathbf{v} \rangle = \phi''(1/2) \geq 0.$$

Now we prove the other direction in (i). If $\mathbf{x}_0, \mathbf{x}_1 \in G$, we define $\mathbf{x}(t)$ and $\phi(t) = f(\mathbf{x}(t))$ as in part (i). For $0 \leq t \leq 1$, $\phi(t)$ is C^2 and as in part (i),

$$\phi''(t) = \langle \mathbf{D}^2 f(\mathbf{x}(t))(\mathbf{x}_1 - \mathbf{x}_0), \mathbf{x}_1 - \mathbf{x}_0 \rangle.$$

Hence by our assumption on $\mathbf{D}^2 f$, we see that $\phi''(t) \geq 0$ for $0 \leq t \leq 1$. Appealing a second time to Theorem 5.5, we deduce that ϕ is convex on $[0, 1]$. Since $\mathbf{x}_0$ and $\mathbf{x}_1$ are arbitrary points in G, we conclude that f is convex on G. If we make the stronger assumption that $\mathbf{D}^2 f(\mathbf{x})$ is positive definite for each $\mathbf{x} \in G$, we conclude that $\phi'' > 0$, so ϕ is strictly convex. This implies that f is strictly convex on G. Convexity (or strict convexity) of f on G extends to convexity (or strict convexity) of f on $\bar{G}$ by taking appropriate limits.

Corollary Let $G \subset \mathbb{R}^n$ be an open convex set, and suppose that $f \in C^2(G)$ is strictly convex. Then f can have at most one critical point in G, and it must be the global minimizer of f on G.

Proof: Suppose that $\mathbf{x}_* \in G$ is a critical point of f. Because f is strictly convex, Theorem 12.6 tells us that $\mathbf{D}^2 f(\mathbf{x}_*)$ is positive definite. The second derivative test then implies that $\mathbf{x}_*$ is a local minimizer. By Theorem 12.5, we know that $\mathbf{x}_*$ is a global minimizer for f on G, and it must be unique.

The classic reference for convex functions and convex sets is the book of T. Rockafellar [R]. A treatment in a finite dimensional setting is found in the book of K. R. Davidson and A. P. Donsig [Da].

Exercises for 12.2

1. Let A be a convex set, and let $f : A \to \mathbb{R}$ be a convex function. Let $g : f(A) \to \mathbb{R}$ be nondecreasing and convex. Show that the composition $\mathbf{x} \to g(f(\mathbf{x}))$ is convex on A.

2. Let $G \subset \mathbb{R}^n$ be a convex set, and let $\mathbf{A}$ be an $n \times n$ matrix. Let $f : \mathbb{R}^n \to \mathbb{R}$ be convex.

   a) Show that $\mathbf{A}(G)$ is convex.
   b) Show that $g(\mathbf{x}) \equiv f(\mathbf{A}\mathbf{x})$ is convex on G.

3. Let f be a convex function on the closed unit ball $\bar{B} = \{\mathbf{x} \in \mathbb{R}^n : \|\mathbf{x}\|_\infty \leq 1\}$. Let $\mathbf{v}_k$, $k = 1, \ldots, 2^n$ be the vertices of $\bar{B}$, and let $M = \max_k f(\mathbf{v}_k)$. Show that $f(\mathbf{x}) \leq M$ for all $\mathbf{x} \in \bar{B}$.

4. Let G be an open convex subset of $\mathbb{R}^n$ and let $f : \bar{G} \to \mathbb{R}$ be convex.

   a) Show that $F = \{\mathbf{x} \in G : f(\mathbf{x}) \leq 0\}$ is a convex set.
   b) Assume that f is strictly convex and that $f \geq 0$ on ∂G. Show that $\mathbf{x}$ is an interior point of F if and only if $f(\mathbf{x}) < 0$.

5. Let $G \subset \mathbb{R}^n$ be an open convex set. Let $f \in C^1(R)$. Show that f is convex on G if and only if

$$f(\mathbf{y}) \geq f(\mathbf{x}) + \langle \mathbf{y} - \mathbf{x}, \nabla f(\mathbf{x}) \rangle$$

for all $\mathbf{x}, \mathbf{y} \in G$.

12.3 Potentials and Dynamical Systems

In this section, we investigate the relationship between the critical points of a function $V(\mathbf{x})$, $\mathbf{x} \in \mathbb{R}^n$ and the stability of the equilibrium solutions of the second-order system of ordinary differential equations (ODEs),

$$\mathbf{x}'' = -\nabla V(\mathbf{x}). \tag{12.9}$$

This is a system of n second-order ODEs. Systems of this form typically arise in mechanics and come from an application of Newton's second law, $F = ma$. We have taken all of the masses to be 1. We rewrite the system (12.9) as a system of $2n$ first-order ODEs by introducing $\mathbf{y} = \mathbf{x}'$ as a new variable:

$$\mathbf{x}' = \mathbf{y}$$
$$\mathbf{y}' = -\nabla V(\mathbf{x}). \tag{12.10}$$

The solutions of the system (12.10) are paths in the $2n$ dimensional phase space $\mathbb{R}^{2n}$.

> **EXAMPLE 12.3:** We consider a system of two coupled nonlinear oscillators. Let $x_1(t)$ denote the position on the x axis of particle 1, and let x_2 denote the position of particle 2. Newton's law yields two second-order equations:
>
> $$x_1'' = -f(x_1, x_2)$$
> $$x_2'' = -g(x_1, x_2).$$
>
> We assume that there is a scalar function $V(x_1, x_2)$ with $V_{x_1} = f$ and $V_{x_2} = g$. This system is then rewritten in the form (12.10) as
>
> $$x_1' = y_1$$
> $$x_2' = y_2$$
> $$y_1' = -f(x_1, x_2) = -V_{x_1}$$
> $$y_2' = -g(x_1, x_2) = -V_{x_2}. \tag{12.11}$$

Here the phase space is $\mathbb{R}^4$.

Now if $\mathbf{a}$ is a critical point of V ($\nabla V(\mathbf{a}) = 0$), then $\mathbf{x}(t) \equiv \mathbf{a}$, $\mathbf{y}(t) \equiv 0$ is an equilibrium solution of (12.10). For the remainder of this section, $\|\ \|$ will denote the norm $\|\ \|_2$ on $\mathbb{R}^n$, and $\|(\mathbf{x}, \mathbf{y})\| = \sqrt{\|\mathbf{x}\|^2 + \|\mathbf{y}\|^2}$ will be used to denote the two norm on $\mathbb{R}^{2n}$.

Definition 12.3 We say that the equilibrium solution $(\mathbf{a}, \mathbf{b})$ is *stable* for the system (12.10) if the following condition is satisfied. For each $\varepsilon > 0$, there is a $\delta > 0$ such that, if $(\mathbf{x}, \mathbf{y})$ is a solution of (12.10) with $\|(\mathbf{x}(0) - \mathbf{a},\ \mathbf{y}(0) - \mathbf{b})\| < \delta$, then $\|(\mathbf{x}(t) - \mathbf{a}, \mathbf{y}(t) - \mathbf{b})\| < \varepsilon$ for all $t \geq 0$.

The nature of the critical point $\mathbf{a}$ of V will allow us to determine whether the equilibrium solution $(\mathbf{a}, 0)$ is stable or not. First we observe that if $(\mathbf{x}(t), \mathbf{y}(t))$ is a solution of (12.10), then

$$E(t) \equiv \frac{1}{2}\|\mathbf{y}(t)\|^2 + V(\mathbf{x}(t))$$

is constant. To see this, we just need to differentiate E, and use the differential equations (12.10):

$$\frac{d}{dt}E = \frac{1}{2}\frac{d}{dt}\langle \mathbf{y}(t),\ \mathbf{y}(t)\rangle + \frac{d}{dt}V(\mathbf{x}(t))$$

$$= \langle \mathbf{y}'(t),\ \mathbf{y}(t)\rangle + \langle \nabla V(\mathbf{x}(t)),\ \mathbf{x}'(t)\rangle$$

$$= \langle -V(\mathbf{x}(t)),\ \mathbf{y}(t)\rangle + \langle \nabla V(\mathbf{x}(t)),\ \mathbf{y}(t)\rangle$$

$$= 0.$$

Hence a solution curve of (12.10) lies on a level surface of the function

$$H(\mathbf{x}, \mathbf{y}) = \frac{1}{2}\|\mathbf{y}\|^2 + V(\mathbf{x}).$$

H is called the *Hamiltonian* of the system (12.10). In this case, H is just the sum of the kinetic and potential energies.

Now assume that $\mathbf{a}$ is an isolated critical point of V, and that $\mathbf{D}^2 V$ is positive definite. Then

$$\mathbf{D}^2 H(\mathbf{a}, 0) = \begin{bmatrix} \mathbf{D}^2 V(\mathbf{a}) & 0 \\ 0 & \mathbf{I} \end{bmatrix}$$

is also positive definite. Theorem 12.2 (quadratic approximation) implies that, for $(\mathbf{x}, \mathbf{y})$ close to $(\mathbf{a}, 0)$, the level surfaces of H resemble the level surfaces of the quadratic function:

$$q(\mathbf{x}, \mathbf{y}) \equiv V(\mathbf{a}) + \frac{1}{2}\langle D^2 V(\mathbf{a})\mathbf{x}, \mathbf{x}\rangle + \frac{1}{2}\|\mathbf{y}\|^2.$$

The level surfaces of q are ellipsoids. Hence for $(\mathbf{x}, \mathbf{y})$ close to $(\mathbf{a}, 0)$, the level surfaces of H should be closed bounded surfaces that enclose $(\mathbf{a}, 0)$. Since a solution curve of (12.10) must lie on a level surface of H, the solution curves that start near $(\mathbf{a}, 0)$ will not wander away. This suggests that $(\mathbf{a}, 0)$ is a stable equilibrium point for the system (12.10).

A different approach to the stability of equilibrium solutions uses a linear approximation to the system (12.10) around an equilibrium solution:

$$(\mathbf{x}(t), \ \mathbf{y}(t)) \equiv (\mathbf{a}, 0).$$

Let $(\mathbf{x}(t), \ \mathbf{y}(t))$ be a solution of (12.10) that starts near $(\mathbf{a}, \ 0)$. We write this solution in the form

$$\begin{aligned} \mathbf{x}(t) &= \mathbf{a} + \xi(t) \\ \mathbf{y}(t) &= \eta(t). \end{aligned}$$

Substituting in (12.10), we see that

$$\begin{aligned} \frac{d\xi(t)}{dt} &= \eta(t) \\ \frac{d\eta(t)}{dt} &= -\nabla V(\mathbf{a} + \xi(t)). \end{aligned} \tag{12.12}$$

Now assume that $V \in C^2(G)$ for some open set G containing $\mathbf{a}$. Then we can make a linear approximation of ∇V at $\mathbf{a}$:

$$\begin{aligned} \nabla V(\mathbf{x}) &= \nabla V(\mathbf{a}) + D^2 V(\mathbf{a})(\mathbf{x} - \mathbf{a}) + \mathbf{R} \\ &= D^2 V(\mathbf{a})(\mathbf{x} - \mathbf{a}) + \mathbf{R} \end{aligned}$$

where $\mathbf{R} = o(\|\mathbf{x} - \mathbf{a}\|)$. Thus for a small perturbation ξ, we have the approximation

$$\nabla V(\mathbf{a} + \xi) = D^2 V(\mathbf{a})\xi + \mathbf{R}$$

where $\mathbf{R} = o(\|\xi\|)$. We drop the error term and use this approximation in the right side of (12.12) to arrive at the *linearized* system

$$\xi' = \eta$$
$$\eta' = -\mathbf{D}^2 V(\mathbf{a})\xi.$$

This system can be written in matrix form as

$$\frac{d}{dt}\begin{bmatrix} \xi \\ \eta \end{bmatrix} = \mathbf{A}\begin{bmatrix} \xi \\ \eta \end{bmatrix} \tag{12.13}$$

where

$$\mathbf{A} = \begin{bmatrix} 0 & \mathbf{I} \\ -\mathbf{D}^2 V(\mathbf{a}) & 0 \end{bmatrix}.$$

We will say that the system (12.10) is *linearly stable* at $(\mathbf{a},\ 0)$ if the linear system (12.13) is stable at $(0, 0)$. The latter system will be stable at $(0, 0)$ if the eigenvalues λ of $\mathbf{A}$ satisfy $\mathrm{Re}(\lambda) \leq 0$. If there is an eigenvalue λ of $\mathbf{A}$ with $\mathrm{Re}(\lambda) > 0$, then $(\mathbf{a},\ 0)$ is unstable for the system (12.13).

We continue to assume that $\mathbf{a}$ is an isolated critical point of V, but we make no other assumptions about the nature of the critical point. The eigenvalues of $\mathbf{A}$ are determined by the eigenvalues of $\mathbf{D}^2 V(\mathbf{a})$. $\mathbf{D}^2 V(\mathbf{a})$ is symmetric, and hence has real eigenvalues μ_j, $j = 1, \dots, n$ (not necessarily distinct) and an orthogonal basis of eigenvectors $\mathbf{v}_1, \dots, \mathbf{v}_n$, such that $\mathbf{D}^2 V(\mathbf{a})\mathbf{v}_j = \mu_j \mathbf{v}_j$. Let λ be an eigenvalue of $\mathbf{A}$, and suppose that it has an eigenvector in the form $(\mathbf{v}, \mathbf{w})$ where $\mathbf{v}$ is an eigenvector of $\mathbf{D}^2 V(\mathbf{a})$ with eigenvalue $\mu \neq 0$. Then

$$\begin{bmatrix} \lambda \mathbf{v} \\ \lambda \mathbf{w} \end{bmatrix} = \mathbf{A}\begin{bmatrix} \mathbf{v} \\ \mathbf{w} \end{bmatrix} = \begin{bmatrix} \mathbf{w} \\ -\mathbf{D}^2 V(\mathbf{a})\mathbf{v} \end{bmatrix} = \begin{bmatrix} \mathbf{w} \\ -\mu \mathbf{v} \end{bmatrix}.$$

Hence $\lambda \mathbf{v} = \mathbf{w}$ and $\lambda \mathbf{w} = -\mu \mathbf{v}$, which implies that $\lambda^2 \mathbf{v} = -\mu \mathbf{v}$. We deduce that, if $\mathbf{v}$ is an eigenvector of $\mathbf{D}^2 V(\mathbf{a})$, then $(\pm\sqrt{-\mu}\mathbf{v}, \mathbf{v})$ is an eigenvector of $\mathbf{A}$ with eigenvalue $\lambda = \pm\sqrt{-\mu}$. Thus the pairs $(\pm\sqrt{-\mu}\mathbf{v}_j, \mathbf{v}_j)$ are a basis of eigenvectors of $\mathbf{A}$ in $2n$ dimensional complex space. We can conclude that

Theorem 12.7 If $\mathbf{D}^2V(\mathbf{a})$ is positive definite, the point $(\mathbf{a}, 0)$ is linearly stable for (12.10). If $\mathbf{D}^2V(\mathbf{a})$ has a negative eigenvalue, then $(\mathbf{a}, 0)$ is linearly unstable for (12.10).

EXAMPLE 12.4:   We return to the system of two springs and a mass that was described in Section 11.6. Recall that the system (11.35) for the equilibrium solutions is:

$$p(x, y) = k_1(d_1 - l)(1 + x)/d_1 + k_2(d_2 - l)(x - 1)/d_2 = 0 \qquad (12.14)$$

$$q(x, y) = k_1(d_1 - l)y/d_1 + k_2(d_2 - l)y/d_2 = -mg.$$

Recall that $p(x, y) = \phi_x(x, y)$ and $q(x, y) = \phi_y(x, y)$, where

$$\phi(x, y) = (1/2)(d_1 - l)^2 + (1/2)(d_2 - l)^2.$$

The potential energy of the system of two springs and a mass is

$$V(x, y) = \phi(x, y) + mgy.$$

Hence solutions of the system (11.35) are exactly the critical points of the potential energy V.

For the purposes of this example, we will simplify the system somewhat by assuming that the spring constants $k_1 = k_2 = 1$. To be consistent with the notation of this section, we set $x_1 = x$ and $x_2 = y$. Then the potential energy is

$$V(x_1, x_2) = (1/2)(d_1 - l)^2 + (1/2)(d_2 - l)^2 + mgx_2.$$

where $d_1 = \sqrt{(x_1 + 1)^2 + x_2^2}$ and $d_2 = \sqrt{(x_1 - 1)^2 + x_2^2}$.

The time-dependent system of equations is

$$\frac{dx_1}{dt} = y_1$$

$$\frac{dx_2}{dt} = y_2$$

$$m\frac{dy_1}{dt} = -V_{x_1}(x_1, x_2)$$

$$m\frac{dy_2}{dt} = -V_{x_2}(x_1, x_2). \qquad (12.15)$$

According to Theorem 12.7, we can determine the linear stability of the equilibrium points of this system by looking at the critical points of V. In Section 11.6, we assumed that $l < 1$. We found that there was only one critical point $\mathbf{a}$ in $-1 < x_1 < 1$ for each $m > 0$, and that $\mathbf{D}^2 V(\mathbf{a}) = \mathbf{D}^2 \phi(\mathbf{a})$ was positive definite there. This critical point is the minimizer of the potential energy. By Theorem 12.7, it is a linearly stable equilibrium point for the system (12.15).

Now however, we want to make the problem more interesting. We continue to assume that $k_1 = k_2 = 1$, but we do not require $l < 1$. Because of the symmetry of the problem, the critical points of V all lie on the x_2 axis. We take $m = 1$ and $l = 10$. In Figure 12.2, we graph the function

$$x_2 \to V(0, x_2) = (d - l)^2 + mgx_2,$$

where $d = \sqrt{1 + x_2^2}$.

We see that $V_{x_2}(0, x_2) = 0$ at three points, approximately $x_2 = -14.9$, $x_2 = 0.67$, and $x_2 = 4.9$. Now $V_{x_1}(0, x_2) = 0$ for all x_2. Hence $\mathbf{a}_1 = (0, -14.9)$, $\mathbf{a}_2 = (0, .67)$, and $\mathbf{a}_3 = (0, 4.9)$ are each critical points of V, and hence equilibrium points of the system (12.15). We can determine the nature of each of these critical points by calculating $\mathbf{D}^2 V(\mathbf{a})$ for each $\mathbf{a}$. In fact, $\det \mathbf{D}^2 V(\mathbf{a}_1) > 0$ and $V_{x_2 x_2}(\mathbf{a}_1) > 0$ so $\mathbf{a}_1 = (0, -14.9)$ is a minimum. Similarly, $\det \mathbf{D}^2 V(\mathbf{a}_2) > 0$, but $V_{x_2 x_2}(\mathbf{a}_2) < 0$, so $\mathbf{a}_2$ is a local maximum. Finally, $\det \mathbf{D}^2 V(\mathbf{a}_3) < 0$, which means that $\mathbf{a}_3$ is a saddle point for V. The contours of V in a neighborhood of each $\mathbf{a}$ are displayed in Figure 12.3. Hence $\mathbf{a}_1$ is a (linearly) stable equilibrium point for the system (12.15). On the other hand, $\mathbf{a}_2$ and $\mathbf{a}_3$ are (linearly) unstable equilibrium points of the system (12.15). How do we interpret these unstable equilibrium points physically? In both cases, the mass is above the horizontal line joining $\mathbf{r}_1$ and $\mathbf{r}_2$.

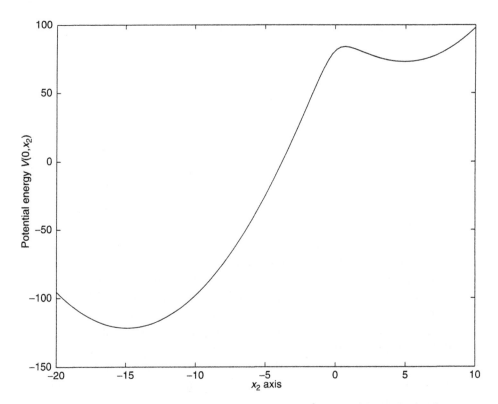

Figure 12.2 Graph of the potential energy $V(0, x_2) = (d-l)^2 + mgy$ of Example 12.4 for $m = 1$, $g = 9.8$, and $l = 10$

Furthermore, because $l = 10$, the springs are compressed ($d - l < 0$) at both of these two critical points. Thus in these two cases, the mass is balanced precariously in an inverted position, and the least perturbation will make it fall.

Exercises for 12.3

1. Consider the single differential equation

$$x'' = 2x + 2x^2 - 4x^3.$$

   a) Introduce the new variable $y = x'$, and write the single second-order equation as a system of two first-order equations.
   b) Find the potential $V(x)$, and graph the contours of $H(x, y) = V(x) + (1/2)y^2$.

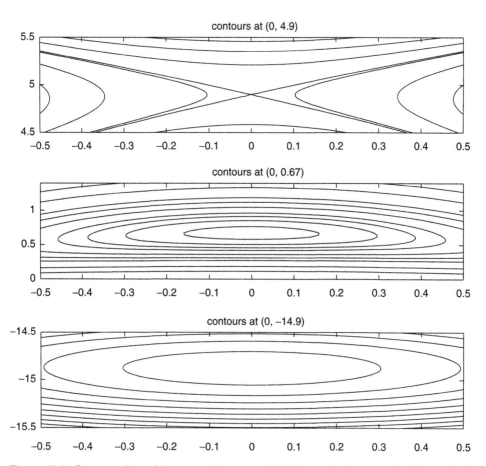

Figure 12.3 Contour plots of the potential energy V of Example 12.4 at each of the critical points

   c) Find the equilibrium points of the system, and determine the linear stability of each of these equilibrium points.

2. Consider the following system of differential equations

$$x_1'' = -x_1 + 4(x_1 - x_2)/(1 + (x_1 - x_2)^2)^2$$
$$x_2'' = -3x_2 + 4(x_2 - x_1)/(1 + (x_2 - x_1)^2)^2.$$

   a) Find the potential $V(x_1, x_2)$ of the system.
   b) Locate the equilibrium points of the system, and determine the linear stability of each of these equilibrium points.

12.4 The Method of Steepest Descent

Now we will address the practical question of how to find the minimizers of f. One approach is to use Newton's method on the system of equations $\nabla f(\mathbf{x}) = 0$, but this requires knowledge of the second derivatives of f. The second derivatives may be difficult or expensive to compute. A second disadvantage of Newton's method is that we must give it a good starting value to ensure convergence. Consequently, much effort has been devoted to finding methods that use only the first derivatives of the function, and that do not require such a good first approximation to the minimizer.

We first develop the method of steepest descent, which can converge very slowly, but uses only the first derivatives of f. We know that at each point $\mathbf{x}$, $\nabla f(\mathbf{x})$ points in the direction of maximum increase of f and that $-\nabla f(\mathbf{x})$ points in the direction of steepest descent. It is natural when seeking a minimum of a function that we move in the direction of steepest descent. It is this simple idea that we want to exploit. We imagine walking down the side of a valley, adjusting our direction at each step so as to be always moving in the direction of steepest descent. Eventually we should arrive at the bottom of the valley, provided we do not reach any smaller hollows along the way. Consider the graph of a function and its level curves displayed in Figure 12.4.

Now we give a more precise description of this procedure. Suppose that f is a C^1 function on an open set $G \subset \mathbb{R}^n$, and let

$$K = \{\mathbf{x} \in G : f(\mathbf{x}) \le f(\hat{\mathbf{x}})\}$$

for some $\hat{\mathbf{x}} \in G$. Assume that K is compact and that $f(\mathbf{x}) < f(\hat{\mathbf{x}})$ for some $\mathbf{x} \in K$. We define a sequence of points $\mathbf{x}_k$ recursively. Assume that the points $\mathbf{x}_0, \ldots, \mathbf{x}_k \in K$ have been defined with $\nabla f(\mathbf{x}_k) \neq 0$. Now let $\tau_k = \max\{\tau\}$ such that the line

$$l_k = \{\mathbf{x}_k - t\nabla f(\mathbf{x}_k), \quad 0 \le t \le \tau\}$$

is contained in K. Then let t_k be a value of t such that

$$f(\mathbf{x}_k - t_k\nabla f(\mathbf{x}_k)) = \min_{0 \le t \le \tau_k} f(\mathbf{x}_k - t\nabla f(\mathbf{x}_k)),$$

and set

$$\mathbf{x}_{k+1} = \mathbf{x}_k - t_k\nabla f(\mathbf{x}_k).$$

We note that $f(\mathbf{x}_{k+1}) < f(\mathbf{x}_k)$ for all k. If at any step $\nabla f(\mathbf{x}_k) = 0$, the procedure stops.

Theorem 12.8 (Steepest descent) Let f, G, and K satisfy the conditions above. Choose the sequence of points $\mathbf{x}_k$ in the manner described above. Then the sequence $\nabla f(\mathbf{x}_k)$ converges to 0. If $\mathbf{z}$ is a cluster point of $\mathbf{x}_k$, then $\nabla f(\mathbf{z}) = 0$.

Proof: First we observe that, at any point $\mathbf{x} \in \partial K$ where $\nabla f(\mathbf{x}) \neq 0$, $\nabla f(\mathbf{x})$ points into the interior of K. This implies that $\tau_k > 0$ for each k.

Suppose the sequence $\nabla f(\mathbf{x}_k)$ does not converge to zero. Then there is an $\varepsilon > 0$ and a subsequence $\mathbf{x}_i$ such that $\|\nabla f(\mathbf{x}_i)\| \geq \varepsilon > 0$. Since K is compact, there is a subsequence of $\mathbf{x}_i$, denoted by $\mathbf{x}_j$, such that $\mathbf{x}_j$ converges to $\mathbf{z} \in K$.

Because $f \in C^1(K)$, the sequence $\nabla f(\mathbf{x}_j)$ converges to $\nabla f(\mathbf{z})$ and, consequently, $\nabla f(\mathbf{z}) \neq 0$. We use this fact to deduce a contradiction. Note that, because f is continuous, $f(\mathbf{x}_j)$ converges to $f(\mathbf{z})$. Furthermore, because $f(\mathbf{x}_j)$ is strictly decreasing, $f(\mathbf{z}) < f(\mathbf{x}_0)$. This means that $\mathbf{z}$ lies in the open set $\{\mathbf{x} : f(\mathbf{x}) < f(\mathbf{x}_0)\} \subset K$. Hence there is an $r > 0$ such that $B(\mathbf{z}, 2r) \subset K$. Now because $\mathbf{x}_j$ converges to $\mathbf{z}$, there is a j_0 such that $\mathbf{x}_j \in B(\mathbf{z}, r)$ for $j \geq j_0$. Thus for $j \geq j_0$, the distance from $\mathbf{x}_j$ to the boundary of K is greater than or equal to r. Let $M = \max_K \|\nabla f(\mathbf{x})\|$. Then for $j \geq j_0$, we have $\tau_j \geq r/M$. Because $\nabla f(\mathbf{z}) \neq 0$, we may choose a $t_0 > 0$ such that we have both $f(\mathbf{z} - t_0 \nabla f(\mathbf{z})) < f(\mathbf{z})$ and $t_0 \leq r/M$, which implies $t_0 \leq \tau_j$ for $j \geq j_0$. Then for $j \geq j_0$, we have

$$
\begin{aligned}
f(\mathbf{x}_{j+1}) &= f(\mathbf{x}_j - t_j \nabla f(\mathbf{x}_j)) \\
&= \min_{0 \leq t \leq \tau_j} f(\mathbf{x}_j - t \nabla f(\mathbf{x}_j)) \\
&\leq f(\mathbf{x}_j - t_0 \nabla f(\mathbf{x}_j)).
\end{aligned}
$$

Remembering that both f and ∇f are continuous, we may take the limit as $j \to \infty$ to deduce

$$f(\mathbf{z}) \leq f(\mathbf{z} - t_0 \nabla f(\mathbf{z})) < f(\mathbf{z}),$$

which is impossible. The theorem is proved.

Corollary If f has a single critical point $\mathbf{z} \in K$, then $\mathbf{x}_k$ converges to $\mathbf{z}$ and $f(\mathbf{z}) = \min_K f$.

Remark Let l be the half line, $l = \{\mathbf{x}_k - t\nabla f(\mathbf{x}_k), \ t \geq 0\}$, and set

$$\Delta(\mathbf{x}_k, t) = f(\mathbf{x}_k - t\nabla f(\mathbf{x}_k)).$$

In the procedure of Theorem 12.8, we choose t_k to minimize $\Delta(\mathbf{x}_k, t)$. This means that we choose t_k so that $(d/dt)\Delta(\mathbf{x}_k, t_k) = 0$. Now

$$\frac{d}{dt}\Delta(\mathbf{x}_k, t) = -\langle \nabla f(\mathbf{x}_k - t\nabla f(\mathbf{x}_k)), \nabla f(\mathbf{x}_k)\rangle$$

so that we are choosing the point $\mathbf{x}_{k+1} \in l$ such that $\langle \nabla f(\mathbf{x}_{k+1}), \nabla f(\mathbf{x}_k)\rangle = 0$. Since $\nabla f(\mathbf{x})$ is always normal to the level surfaces of f, the line l is tangent to a level surface of f at $\mathbf{x}_{k+1}$. The next step, from $\mathbf{x}_{k+1}$ to $\mathbf{x}_{k+2}$, is taken in a direction orthogonal to that level surface and, of course, orthogonal to l. The path of steepest descent therefore consists of orthogonal straight line segments, as seen in the following examples.

EXAMPLE 12.5:   As in Figure 12.4, we let

$$f(x, y) = ((x - 1)^2 + y^2 + 0.2)((x + 1)^2 + y^2).$$

The path on the left begins at $(-.5, 1.5)$, and the path on the right begins at $(1, 1.5)$; see Figure 12.5. In each case, five iterations were made. The numerical results are displayed here.

x	y	$f(x, y)$
1.00000000000000	1.50000000000000	15.31250000000000
0.43578587297164	0.06331817410339	1.07890217274122
0.85144126245456	-0.07399408912869	0.78123188997631
0.88361130287183	0.00131099810580	0.75766703313241
0.88742874745332	0.00073402740305	0.75762308294995

-0.50000000000000	1.50000000000000	11.75000000000000
-0.33029953221041	0.03378795829798	0.88616848571048
-0.99214457798086	-0.05175709865049	0.01143151972298
-0.99838720688360	0.00009182937089	0.00001094321619
-1.00000259729227	-0.00000482857801	0.00000000012626

As can be seen in this example, the method of steepest descent may not converge to the global minimum but rather to a local minimum. The sequence of iterates may even converge to the saddle point.

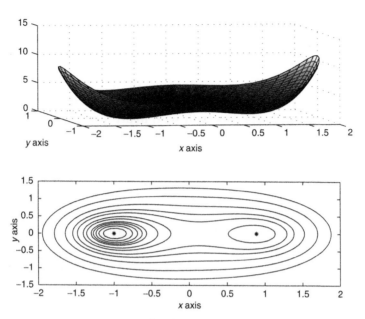

Figure 12.4 Graph of $f(x, y) = ((x - 1)^2 + y^2 + 0.2)((x + 1)^2 + y^2)$ above, and level curves of f below. There is a global minimum at $(-1, 0)$, a local miminum at $(0.8873, 0)$ and a saddle point near $(0, 0)$

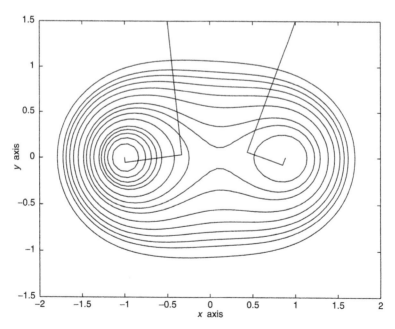

Figure 12.5 Two steepest descent paths. The one on the left goes to the global minimum at $(-1, 0)$, and the one on the right goes to the local minimum at $(0.8873, 0)$

The method of steepest descent can be very inefficient in the case of a "long narrow valley." A long narrow valley can be modeled by an ellipse with large eccentricity (see Figure 12.6).

> **EXAMPLE 12.6:**   Let $f(x, y) = .4x^2 - .9xy + .7y^2$. Starting from a point near the major axis yields a path with many steps. The numerical results for 20 iterations are displayed here.

x	y	$f(x, y)$
3.00000000000000	2.50000000000000	1.22500000000000
2.85765000000000	1.74080000000000	0.91059724900000
2.22997350000000	1.85838246250000	0.67689320831443
2.12422208002125	1.29395588917250	0.50316836496848
1.65759393013912	1.38143415888498	0.37402810063847

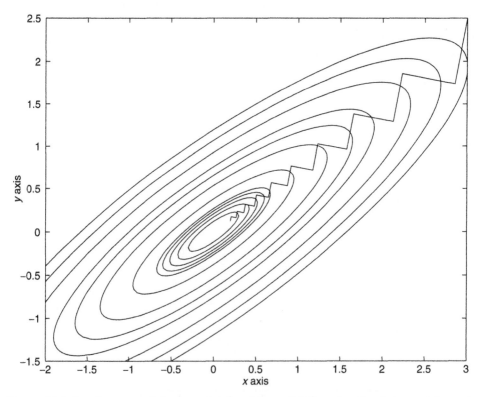

Figure 12.6 Level curves of $f(x, y) = 0.4x^2 - 0.9xy + 0.7y^2$, and path of steepest descent starting at $(3, 2.5)$

1.57907292568172	0.96203279776488	0.27803554451961
1.23211756358947	1.02691391004477	0.20668120287822
1.17381181564750	0.71507208705447	0.15363927585875
0.91600152468507	0.76334726095463	0.11420831644625
0.87254806272552	0.53152099228382	0.08489639346961
0.68088637413398	0.56743646646434	0.06310761984193
0.64860492489866	0.39508542926765	0.04691116727401
0.50620117417860	0.42179028227817	0.03487111852123
0.48211897315070	0.29361121077091	0.02592077997384
0.37627957823820	0.31352618910007	0.01926744629469
0.35837199045773	0.21825539699063	0.01432202729582
0.27970365934020	0.23305105783788	0.01064590138954
0.26638753269304	0.16223977964937	0.00791336015528
0.20791491606744	0.17323208576696	0.00588221269157
0.19801301318693	0.12060066493456	0.00437237467397

There are various ways of accelerating this process. One way is to not follow the path of steepest descent at each point. Surprisingly, this can produce faster convergence. One method of this type is known as the method of *conjugate gradients*. In the case of Example 12.3, this method will converge to the minimum in no more than two steps, no matter where we start. We will examine this method in the next section.

Exercises for 12.4

1. Let $f(x, y) = x^2/4 + y^2/9$. Derive the following recursive formula for the points $p_k = (x_k, y_k)$ in the method of steepest descent applied to f. Let $u_k = -f_x(x_k) = -x_k/2$, and let $v_k = -f_y(x_k) = -2y_k/9$.

a) Show that

$$t_k = -\frac{u_k x_k + (4/9)v_k y_k}{u_k^2 + (4/9)v_k^2},$$

and then

$$x_{k+1} = x_k + t_k u_k, \quad y_{k+1} = y_k + t_k v_k.$$

b) Starting with $p_0 = (x_0, y_0) = (2, 2)$, calculate the first two steps p_1 and p_2.

2. Consider the quadratic function

$$f(\mathbf{x}) = \frac{1}{2}\langle \mathbf{A}\mathbf{x}, \mathbf{x} \rangle - \langle \mathbf{b}, \mathbf{x} \rangle,$$

where $\mathbf{A}$ is an $n \times n$, symmetric, positive definite matrix and $\mathbf{b} \in \mathbb{R}^n$ is a given vector.

a) Let $\mathbf{x}_* = \mathbf{A}^{-1}\mathbf{b}$. Show that f can be rewritten

$$f(\mathbf{x}) = \frac{1}{2}\langle \mathbf{A}(\mathbf{x} - \mathbf{x}_*), \mathbf{x} - \mathbf{x}_* \rangle - \frac{1}{2}\langle \mathbf{b}, \mathbf{x}_* \rangle.$$

b) Show that f has a unique minimum at $\mathbf{x}_*$.

3. Starting at an arbitrary point $\mathbf{x}_0$, and using the first form of the function $f(\mathbf{x})$ in exercise 2, implement the method of steepest descent. Verify the following formulas:

a) The descent directions

$$\mathbf{d}_k \equiv -\nabla f(\mathbf{x}_k) = \mathbf{b} - \mathbf{A}\mathbf{x}_k.$$

The descent directions are thus the *residuals*.

b) The minimizing value of t is

$$t_k = \frac{\langle \mathbf{d}_k, \mathbf{d}_k \rangle}{\langle \mathbf{A}\mathbf{d}_k, \mathbf{d}_k \rangle}.$$

c) $\mathbf{x}_{k+1} = \mathbf{x}_k + t_k \mathbf{x}_k$.

4. Write a code to implement the method of steepest descent in two dimensions for a general function $f(x, y)$. Set a maximum of 50 iterations. Try it out on the function of exercise 1, starting at different places to see how it works.

5. Let $f(x, y) = x^2/25 + (y - x^2/4)^2$. Graph the level curve $f(x, y) = 1$ to see why this function is called a "banana function." Now try the method of steepest descent, starting at the point $\mathbf{p}_0 = (5, 4)$. How does the method work? Describe what happens using graphs and pictures.

12.5 Conjugate Gradient Methods

In the method of steepest descent, at each step, we took the descending direction to be $-\nabla f(\mathbf{x})$. As we saw in Example 12.6, this way of choosing the directions of descent can produce very slow convergence. In this section, we explore a method that chooses its directions differently.

We will develop the method first in the context of a quadratic problem. Let

$$f(\mathbf{x}) = \frac{1}{2}\langle \mathbf{A}\mathbf{x}, \mathbf{x} \rangle - \langle \mathbf{b}, \mathbf{x} \rangle \qquad (12.16)$$

where $\mathbf{A}$ is an $n \times n$ symmetric positive definite matrix and $\mathbf{b}$ is a given vector. We can rewrite the expression for f in the form

$$f(\mathbf{x}) = \frac{1}{2}\langle \mathbf{A}(\mathbf{x} - \mathbf{x}_*), \, \mathbf{x} - \mathbf{x}_* \rangle - \frac{1}{2}\langle \mathbf{A}\mathbf{x}_*, \, \mathbf{x}_* \rangle,$$

where $\mathbf{x}_* = \mathbf{A}^{-1}\mathbf{b}$ is the unique solution of $\mathbf{A}\mathbf{x} = \mathbf{b}$. Thus f attains its global minimum at the point $\mathbf{x}_*$. Consequently, a procedure to minimize f is also a method of solving $\mathbf{A}\mathbf{x} = \mathbf{b}$. The gradient of f is easily calculated to be

$$\nabla f(\mathbf{x}) = \mathbf{A}\mathbf{x} - \mathbf{b} = \mathbf{A}(\mathbf{x} - \mathbf{x}_*). \qquad (12.17)$$

Let $\mathbf{d}_0$ be a descent direction at the point $\mathbf{x}_0$, by which we mean,

$$\langle \nabla f(\mathbf{x}_0), \mathbf{d}_0 \rangle < 0.$$

Then, in any kind of descent procedure, we would seek to find t_0 such that

$$f(\mathbf{x}_0 + t_0\mathbf{d}_0) = \min_t f(\mathbf{x}_0 + t\mathbf{d}_0).$$

At the minimizing value t_0, we have

$$0 = \frac{d}{dt}f(\mathbf{x}_0 + t\mathbf{d}_0)\bigg|_{t=t_0} = \langle \nabla f(\mathbf{x}_0 + t_0\mathbf{d}_0), \, \mathbf{d}_0 \rangle.$$

Let $\mathbf{x}_1 = \mathbf{x}_0 + t_0\mathbf{d}_0$. In the method of steepest descent, the next direction of descent would be $\mathbf{d}_1 = -\nabla f(\mathbf{x}_1) = -\mathbf{A}(\mathbf{x}_1 - \mathbf{x}_*)$, and we have the orthogonality relation seen before:

$$\langle \mathbf{d}_1, \mathbf{d}_0 \rangle = \langle \mathbf{A}(\mathbf{x}_* - \mathbf{x}_1), \mathbf{d}_0 \rangle = 0.$$

However, we see that the optimal direction of descent is $\tilde{\mathbf{d}}_1 = \mathbf{x}_* - \mathbf{x}_1$, because, if we descend in the direction $\tilde{\mathbf{d}}_1$, we will pass through the minimizer $\mathbf{x}_*$. This optimal descent direction is not orthogonal to $\mathbf{d}_0$ in the usual sense, but as we see from the previous equation, $\langle \mathbf{A}\tilde{\mathbf{d}}_1, \mathbf{d}_0 \rangle = 0$. This suggests that we try to find descent directions that satisfy this kind of orthogonality condition.

Definition 12.4 Two vectors $\mathbf{u}, \mathbf{v} \in \mathbb{R}^n$ are $\mathbf{A}$ *orthogonal* if $\langle \mathbf{A}\mathbf{u}, \mathbf{v} \rangle = 0$. A set of vectors $\mathbf{v}_j$ is an $\mathbf{A}$ orthogonal set if $\langle \mathbf{A}\mathbf{v}_i, \mathbf{v}_j \rangle = 0$ for $i \neq j$.

In fact, $(\mathbf{x}, \mathbf{y}) \to \langle \mathbf{A}\mathbf{x}, \mathbf{y} \rangle$ defines a new scalar product on $\mathbb{R}^n$ and a corresponding norm:

$$\|\mathbf{x}\|_\mathbf{A} = \sqrt{\langle \mathbf{A}\mathbf{x}, \mathbf{x} \rangle}.$$

The level curve surfaces of f given by (12.16) are just the the sets

$$\{\mathbf{x} : \|\mathbf{x} - \mathbf{x}_*\|_\mathbf{A}\} = c.$$

Before we attempt to devise an algorithm using descent directions that are $\mathbf{A}$ orthogonal, we investigate some of the useful properties of a set of $\mathbf{A}$ orthogonal vectors. The eigenvectors of $\mathbf{A}$ are an orthogonal set in the usual sense and are also an $\mathbf{A}$ orthogonal set. In general, vectors that are orthogonal in the usual sense are not $\mathbf{A}$ orthogonal. For example, the standard basis vectors $\mathbf{e}_j$ may not be $\mathbf{A}$ orthogonal because $\langle \mathbf{A}\mathbf{e}_i, \mathbf{e}_j \rangle = a_{ij}$. However, starting with the standard basis $\mathbf{e}_j$, we can construct a set of vectors $\mathbf{v}_j$ that are $\mathbf{A}$ orthogonal using the Gram-Schmidt process.

Now suppose that we have an $\mathbf{A}$ orthogonal set of vectors, $\mathbf{d}_j$, $j = 0, \ldots,$ $n - 1$. These vectors will be independent and thus form a basis for $\mathbb{R}^n$. Hence the solution $\mathbf{x}_*$ of $\mathbf{A}\mathbf{x} = \mathbf{b}$ can be expanded (uniquely)

$$\mathbf{x}_* = \alpha_0 \mathbf{d}_0 + \cdots + \alpha_{n-1} \mathbf{d}_{n-1}. \tag{12.18}$$

We wish to determine the coefficients α_j without knowledge of $\mathbf{x}_*$. Apply $\mathbf{A}$ to both sides of (12.18) to find

$$\mathbf{b} = \mathbf{A}\mathbf{x}_* = \alpha_0 \mathbf{A}\mathbf{d}_0 + \cdots + \alpha_{n-1} \mathbf{A}\mathbf{d}_{n-1}.$$

Finally, take the usual scalar product of $\mathbf{b}$ with each $\mathbf{d}_j$. We arrive at

$$\langle \mathbf{b}, \mathbf{d}_j \rangle = \sum_{i=0}^{n-1} \alpha_i \langle \mathbf{A}\mathbf{d}_i, \mathbf{d}_j \rangle$$

$$= \alpha_j \langle \mathbf{A}\mathbf{d}_j, \mathbf{d}_j \rangle,$$

so that

$$\alpha_j = \frac{\langle \mathbf{b}, \mathbf{d}_j \rangle}{\langle \mathbf{A}\mathbf{d}_j, \mathbf{d}_j \rangle}. \tag{12.19}$$

In this formula for the α_j, $\mathbf{x}_*$ does not appear. Hence we have found a method of computing the solution $\mathbf{x}_*$ by substituting the α_j, as computed in (12.19), into (12.18).

Next we reformulate this method to express the solution as the limit of a finite sequence of points $\mathbf{x}_k$ generated by an iterative process.

Theorem 12.9 Let $\mathbf{d}_j$, $j = 0, \ldots, n - 1$ be an $\mathbf{A}$ orthogonal set of vectors. Let $\mathbf{x}_0 \in \mathbb{R}^n$ be arbitrary. We generate a sequence $\mathbf{x}_k$ as follows:

$$\mathbf{x}_{k+1} = \mathbf{x}_k + \alpha_k \mathbf{d}_k, \quad k \geq 0,$$

where

$$\alpha_k = -\frac{\langle \mathbf{g}_k, \mathbf{d}_k \rangle}{\langle \mathbf{A}\mathbf{d}_k, \mathbf{d}_k \rangle} \tag{12.20}$$

and

$$g_k = \nabla f(\mathbf{x}_k) = \mathbf{A}\mathbf{x}_k - \mathbf{b}.$$

The sequence $\mathbf{x}_k$ converges to $\mathbf{x}_*$ in n steps or less.

Proof: Using the fact that the $\mathbf{d}_j$ form a basis, we can write

$$\mathbf{x}_* - \mathbf{x}_0 = \alpha_0 \mathbf{d}_0 + \cdots + \alpha_{n-1} \mathbf{d}_{n-1}.$$

As before, we deduce that

$$\langle \mathbf{A}(\mathbf{x}_* - \mathbf{x}_0), \mathbf{d}_j \rangle = \alpha_j \langle \mathbf{A}\mathbf{d}_j, \mathbf{d}_j \rangle$$

so that

$$\alpha_j = \frac{\langle \mathbf{A}(\mathbf{x}_* - \mathbf{x}_0), \mathbf{d}_j \rangle}{\langle \mathbf{A}\mathbf{d}_j, \mathbf{d}_j \rangle}. \tag{12.21}$$

The sequence $\mathbf{x}_k$ is simply

$$\mathbf{x}_k = \mathbf{x}_0 + \alpha_0 \mathbf{d}_0 + \cdots + \alpha_{k-1} \mathbf{d}_{k-1}.$$

The only thing left is to show that the two formulas (12.20) and (12.21) are the same. However,

$$\begin{aligned}
\mathbf{A}(\mathbf{x}_* - \mathbf{x}_0) &= \mathbf{b} - \mathbf{A}\mathbf{x}_0 \\
&= \mathbf{b} - \mathbf{A}\mathbf{x}_k + \mathbf{A}\mathbf{x}_k - \mathbf{A}\mathbf{x}_0 \\
&= \mathbf{A}(\mathbf{x}_k - \mathbf{x}_0) - g_k.
\end{aligned}$$

Finally we see that

$$\begin{aligned}
\langle \mathbf{A}(\mathbf{x}_* - \mathbf{x}_0), \mathbf{d}_k \rangle &= \langle \mathbf{A}(\mathbf{x}_k - \mathbf{x}_0), \mathbf{d}_k \rangle - \langle g_k, \mathbf{d}_k \rangle \\
&= -\langle g_k, \mathbf{d}_k \rangle
\end{aligned}$$

because $\langle \mathbf{A}\mathbf{d}_j, \mathbf{d}_k \rangle = 0$ for $j < k$. Thus formula (12.21) agrees with formula (12.20).

Next we remark that the formula (12.20) for α_k can be expressed as the result of a line minimization. Indeed, suppose that $\mathbf{x}_k$ and $\mathbf{d}_k$ have been chosen.

We claim that α_k given by (12.20) is that value α_* that minimizes the function $\alpha \to f(\mathbf{x}_k + \alpha \mathbf{d}_k)$. The minimizer α_* must satisfy

$$0 = \frac{d}{d\alpha} f(\mathbf{x}_k + \alpha \mathbf{d}_k)\bigg|_{\alpha=\alpha_*} = \langle \nabla f(\mathbf{x}_k + \alpha_* \mathbf{d}_k), \mathbf{d}_k \rangle$$

$$= \langle \mathbf{A}\mathbf{x}_k + \alpha_* \mathbf{A}\mathbf{d}_k - \mathbf{b}, \mathbf{d}_k \rangle$$
$$= \langle \mathbf{g}_k, \mathbf{d}_k \rangle + \alpha_* \langle \mathbf{A}\mathbf{d}_k, \mathbf{k}_k \rangle.$$

Consequently,

$$\alpha_* = -\frac{\langle \mathbf{g}_k, \mathbf{d}_k \rangle}{\langle \mathbf{A}\mathbf{d}_k, \mathbf{d}_k \rangle}.$$

Thus the algorithm for finding the minimum of $f(\mathbf{x}) = (1/2)\langle \mathbf{A}\mathbf{x}, \mathbf{x} \rangle - \langle \mathbf{b}, \mathbf{x} \rangle$ may be written as follows:

Given an $\mathbf{A}$ orthogonal basis $\mathbf{d}_j$, $j = 0, \ldots, n-1$, start with any point $\mathbf{x}_0 \in \mathbb{R}^n$. Then define the sequence $\mathbf{x}_k$ by

$$\mathbf{x}_{k+1} = \mathbf{x}_k + \alpha_k \mathbf{d}_k, \quad k \geq 0$$

where α_k is the minimizer of $\alpha \to f(\mathbf{x}_k + \alpha \mathbf{d}_k)$.

Up to this point, we have assumed the prior knowledge of an $\mathbf{A}$ orthogonal basis of vectors $\mathbf{d}_j$. This, of course, requires knowledge of $\mathbf{A}$. If we want to extend this procedure to minimize f that are not of the form (12.16), we must find a way to generate the descent directions $\mathbf{d}_j$ as we go along. However, it seems plausible that this can be done, because the $\mathbf{d}_j$ are generated by the Gram-Schmidt process, which is an iterative procedure.

The final form of the conjugate gradient algorithm is an iterative procedure that produces a sequence of points $\mathbf{x}_k$ and a sequence of descent directions $\mathbf{d}_k$.

Theorem 12.10 (The conjugate gradient algorithm)

Let $f(\mathbf{x}) = (1/2)\langle \mathbf{A}\mathbf{x}, \mathbf{x} \rangle - \langle \mathbf{b}, \mathbf{x} \rangle$ where $\mathbf{A}$ is an $n \times n$ symmetric positive definite matrix. Let $\mathbf{x}_*$ be the minimizer of f. Starting at an arbitrary point $\mathbf{x}_0$, define

$$\mathbf{d}_0 = -\mathbf{g}_0 = \mathbf{b} - \mathbf{A}\mathbf{x}_0 = -\nabla f(\mathbf{x}_0) \tag{12.22}$$

and

$$\mathbf{x}_{k+1} = \mathbf{x}_k + \alpha_k \mathbf{d}_k, \quad k \geq 0, \tag{12.23}$$

where α_k is the minimizer of $\alpha \to f(\mathbf{x}_k + \alpha \mathbf{d}_k)$. The directions $\mathbf{d}_k$ are determined by

$$\mathbf{d}_{k+1} = -\mathbf{g}_{k+1} + \beta_k \mathbf{d}_k, \quad k \geq 0, \tag{12.24}$$

where $\mathbf{g}_k = \mathbf{A}\mathbf{x}_k - \mathbf{b} = \nabla f(\mathbf{x}_k)$ and

$$\beta_k = \frac{\langle \mathbf{A}\mathbf{g}_{k+1}, \mathbf{d}_k \rangle}{\langle \mathbf{A}\mathbf{d}_k, \mathbf{d}_k \rangle}. \tag{12.25}$$

An alternate formula for β_k is

$$\beta_k = \frac{\langle \mathbf{g}_{k+1}, \mathbf{g}_{k+1} \rangle}{\langle \mathbf{g}_k, \mathbf{g}_k \rangle} = \frac{\|\mathbf{g}_{k+1}\|^2}{\|\mathbf{g}_k\|^2}. \tag{12.26}$$

Then the sequence $\mathbf{x}_k$ converges to $\mathbf{x}_*$ in n steps or less.

The proof of this theorem requires more linear algebra than we have time for here. However, we shall make some comments about the algorithm and illustrate it with a simple example.

The set of $\mathbf{d}_j$ constructed using the algorithm is an $\mathbf{A}$ orthogonal set. Furthermore, the $\mathbf{g}_j$ are a set of orthogonal vectors in the usual sense: $\langle \mathbf{g}_j, \mathbf{g}_k \rangle = 0$ for $j \neq k$. The first descent direction $\mathbf{d}_0$ is just the negative of the gradient $\mathbf{g}_0 = -\nabla f(\mathbf{x}_0)$. This algorithm starts out the same way as the steepest descent algorithm. At the next step, we see from (12.24) that $\mathbf{d}_1$ is not equal to the negative gradient $-\mathbf{g}_1$. The direction $\mathbf{d}_1$ is obtained by subtracting off the projection (in the $\mathbf{A}$ orthogonal sense) of $-\mathbf{g}_1$ on $\mathbf{d}_0$. This operation automatically makes $\mathbf{d}_1$ $\mathbf{A}$ orthogonal to $\mathbf{d}_0$. Luckily, from (12.26) we see that β_0 can be calculated without knowing the matrix $\mathbf{A}$, but just from a knowledge of $\mathbf{g}_1$. In this way, $\mathbf{d}_{k+1}$ is chosen so that it is $\mathbf{A}$ orthogonal to $\mathbf{d}_k$.

EXAMPLE 12.7: In a two-dimensional problem, the procedure ends in, at most, two steps. Let $f(x, y) = x^2/9 + y^2$. The minimum is obviously $(0, 0)$, and the matrix

$$A = \begin{bmatrix} 2/9 & 0 \\ 0 & 2 \end{bmatrix}.$$

Two vectors $\mathbf{u} = (u_1, u_1)$ and $\mathbf{v} = (v_1, v_2)$ are $\mathbf{A}$ orthogonal if $u_1 v_1 + 9 u_2 v_2 = 0$. We take $(x_0, y_0) = (2, \sqrt{5}/3)$ on the ellipse $f(x, y) = 1$. In Figure 12.7, we have displayed the starting point (x_0, y_0) and the next point (x_1, y_1). Notice that the steepest descent direction at (x_1, y_1) does not go directly toward the minimum at $(0, 0)$, whereas the conjugate gradient direction does.

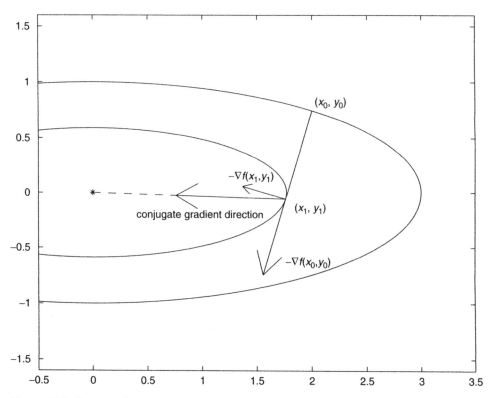

Figure 12.7 Steepest descent directions and conjugate gradient direction for the quadratic function $f(x, y) = x^2/9 + y^2$

Extension to Nonquadratic Functions

If f is a C^2 function, we see by Theorem 12.2 that, at a critical point $\mathbf{a}$, we can write

$$f(\mathbf{x}) = f(\mathbf{a}) + \frac{1}{2}\langle \mathbf{A}(\mathbf{x} - \mathbf{a}), \mathbf{x} - \mathbf{a}\rangle + \mathbf{R}(\mathbf{x} - \mathbf{a})$$

where $\mathbf{A} = \mathbf{D}^2 f(\mathbf{a})$ and $\|\mathbf{R}(\mathbf{x} - \mathbf{a})\| = o(\|\mathbf{x} - \mathbf{a}\|^2)$. If $\mathbf{A}$ is positive definite, f has a strict minimum at $\mathbf{a}$, and it is plausible that the conjugate gradient method would work to find $\mathbf{a}$. However, we do not know $\mathbf{A}$. The point of the reformulation of Theorem 12.10 is to state the conjugate gradient algorithm in a way that does not involve $\mathbf{A}$ explicitly. In particular, the alternate formula (12.26) for β_k is stated in terms of the vectors $\mathbf{g}_k = \nabla f(\mathbf{x}_k)$. The Fletcher-Reeves method is the conjugate gradient algorithm as stated in Theorem 12.10 applied to a nonquadratic function.

Fletcher-Reeves Method

Step 1. Given $\mathbf{x}_0$, compute $\mathbf{g}_0 = \nabla f(\mathbf{x}_0)$, and set $\mathbf{d}_0 = -\mathbf{g}_0$.
Step 2. For $k = 0, \ldots, n - 1$,

    a) Find the minimizer α_k of the function $\alpha \rightarrow f(\mathbf{x}_k + \alpha \mathbf{d}_k)$.
    b) Set $\mathbf{x}_{k+1} = \mathbf{x}_k + \alpha \mathbf{d}_k$.
    c) Compute $\mathbf{g}_{k+1} = \nabla f(\mathbf{x}_{k+1})$.
    d) Unless $k = n - 1$, set $\mathbf{d}_{k+1} = -\mathbf{g}_{k+1} + \beta_k \mathbf{d}_k$ where

$$\beta_k = \frac{\langle \mathbf{g}_{k+1}, \mathbf{g}_{k+1}\rangle}{\langle \mathbf{g}_k, \mathbf{g}_k\rangle}.$$

Step 3. Replace $\mathbf{x}_0$ by $\mathbf{x}_n$ and go back to Step 1.

The Fletcher-Reeves method is convergent because every $n - 1$ steps, we take a steepest descent step. The steps in between do not lead to larger values of f and, hopefully, lead to smaller values. In the case of two dimensions, we start with a steepest descent step, and then alternate a conjugate gradient step with a steepest descent step. In three dimensions we do one steepest descent step followed by two conjugate gradient steps.

The methods of steepest descent and conjugate gradient enable us to locate minima without having to compute the Hessian matrix of the function. On the other hand, at each step we must conduct a line search to find the minimum of a function of one variable. The work is shifted from computing second derivatives to minimizing functions of one variable, perhaps many times.

EXAMPLE 12.8: We have seen that the method of steepest descent required many steps to reach the minimum of the function f in Example 12.6. The method of conjugate gradients finds the minimum in just two steps because f is quadratic. In this example, we modify f by adding a cubic term, and we use the Fletcher-Reeves method. Let

$$f(x,y) = 0.4x^2 - 0.9xy + 0.7y^2 + 0.1y^3.$$

The function f still has a minimum at $(0,0)$. The contours of f are displayed in Figure 12.8. We use the Fletcher-Reeves method starting at $p_0 = (1.5, 1.5)$. The first step is the steepest descent step, arriving at p_1. The next step, from p_1 to p_2, is a conjugate gradient step. From p_2 to p_3 is a steepest descent step, and from p_3 to p_4, another conjugate gradient step.

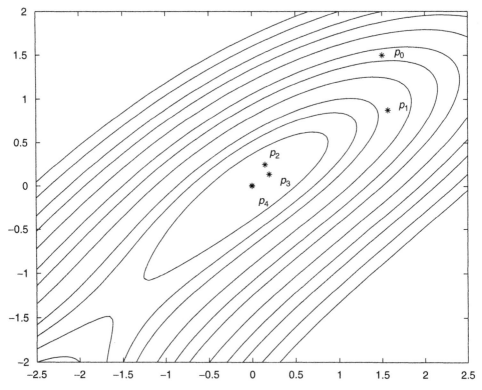

Figure 12.8 Contours of $f(x,y) = 0.4x^2 - 0.9xy + 0.7y^2 + 0.1y^3$ with steps calculated by the Fletcher-Reeves method, starting at $p_0 = (1.5, 1.5)$

Here are the values as computed by the Fletcher-Reeves method. If you use a pure steepest descent method on this particular function, it will take many more iterations to reach the same degree of accuracy.

x	y	$f(x, y)$
1.50000000000000	1.50000000000000	0.78750000000000
1.56561540232877	0.87665367787665	0.35054628926051
0.14946628158749	0.24525065658422	0.01952368441218
0.19844390399303	0.13539848502813	0.00465103712321
-0.00045933430099	0.00485549750020	0.00001860620853
0.00185573714125	0.00132813524573	0.00000039429576
-0.00000069435054	0.00000107930295	0.00000000000168

A more detailed discussion of the method of steepest descent, and the method of conjugate gradients can be found in the book of S. Nash and A. Sofer [N]. The Fletcher-Reeves method is described in the book of D. G. Luenberger, [Lu].

Exercises for 12.5

1. Suppose that $\mathbf{A}$ is a real symmetric positive definite matrix. Let $\mathbf{u}$ and $\mathbf{v}$ be eigenvectors of $\mathbf{A}$ with different eigenvalues. Show that $\mathbf{u}$ and $\mathbf{v}$ are $\mathbf{A}$ orthogonal.

2. Let $\mathbf{A}$ be an $n \times n$ symmetric positive definite matrix, and define

$$\langle \mathbf{x}, \mathbf{y} \rangle_{\mathbf{A}} \equiv \langle \mathbf{A}\mathbf{x}, \ \mathbf{x} \rangle.$$

   a) Show that $\langle \mathbf{x}, \mathbf{y} \rangle_{\mathbf{A}}$ is a scalar product on $\mathbb{R}^n$.
   b) It follows that

$$\|\mathbf{x}\|_{\mathbf{A}} \equiv \sqrt{\langle \mathbf{x}, \ \mathbf{x} \rangle_{\mathbf{A}}}$$

   is a norm on $\mathbb{R}^n$. What is the Schwarz inequality for this scalar product and norm?

3. Let f be defined as in (12.16). Let $\mathbf{d}_0, \dots, \mathbf{d}_{n-1}$ be a basis of $\mathbf{A}$ orthogonal vectors, and let L_k be the subspace spanned by $\mathbf{d}_0, \dots, \mathbf{d}_{k-1}$. Starting with any $\mathbf{x}_0 \in \mathbb{R}^n$, define the sequence

$$\mathbf{x}_k = \mathbf{x}_{k-1} + \alpha_{k-1}\mathbf{d}_{k-1}$$

where α_{k-1} minimizes the function $\alpha \to f(\mathbf{x}_{k-1} + \alpha \mathbf{d}_{k-1})$. Use an induction argument to show that

$$f(\mathbf{x}_k) = \min\{f(\mathbf{x}) : \mathbf{x} \in M_k\}$$

where $M_k = \mathbf{x}_0 + L_k$ is the translate of the subspace L_k. Conclude that $\langle \nabla f(\mathbf{x}_k), \mathbf{d}_j \rangle = 0$ for $j = 0, 1, \ldots, k-1$.

4. Let f be given as in (12.16), with $\mathbf{b} = 0$ and

$$\mathbf{A} = \begin{bmatrix} 1 & 1 \\ 1 & 4 \end{bmatrix}.$$

Let $\mathbf{x}_0 = (1, 1)$. Calculate two steps in the Fletcher-Reeves method. You should find $\mathbf{x}_2 = (0, 0)$.

5. Write a program to implement the Fletcher-Reeves algorithm. Try it out on a simple problem to minimize $f(\mathbf{x}) = x_1^2 + 2x_2^2 - x_1 + x_2$. By completing the square, we know that the minimum is attained at $(1/2, -1)$.

12.6 Some Optimization Problems

Linear and Nonlinear Models

Given a collection of data pairs (t_i, y_i), $i = 1, \ldots, m$, we often try to fit a curve $t \to f(\mathbf{c}, t)$ as closely as possible to these points. The function $f(\mathbf{c}, t)$ is called a *model* for the data. The variables $\mathbf{c} = (c_1, \ldots, c_2)$ are called the *parameters* of the model. The choice of model is usually based on some knowledge of the origin of the data. For example, if the data come from a certain chemical process, a model may be devised that involves physical and chemical principles that, it is hoped, give a good description of the process. The parameters are then adjusted to give the best fit of the model to the data.

The fit of the model to the data may be measured in many ways, but one common measure is the sum of the squares of the differences $(f(\mathbf{c}, t_i) - y_i)^2$.

The sum

$$E^2(\mathbf{c}) = \sum_{i=1}^{m} (f(\mathbf{c}, t_i) - y_i)^2$$

is called the *squared error*. The optimization problem is to find that choice of parameters **c** that minimizes the squared error.

If the model is a linear function of the parameters the problem is called a *linear* least squares problem. If the model is nonlinear in one or more of the parameters, the problem is a *nonlinear* least squares problem.

Linear Least Squares

The most general linear model with n parameters is of the form

$$f(\mathbf{c}, t) = c_1 f_1(t) + c_2 f_2(t) + \cdots + c_n f_n(t).$$

In this case the squared error is

$$E^2(\mathbf{c}) = \sum_{i=1}^{m} \left(\sum_{j=1}^{n} c_j f_j(t_i) - y_i \right)^2.$$

To minimize E, we look for the critical points of E. The critical points are the solutions of the n linear equations

$$\frac{\partial E^2}{\partial c_k} = 0, \quad k = 1, \ldots, n.$$

We write the equations out for the case $n = 2$. We find

$$\frac{\partial E^2}{\partial c_1} = 2 \sum_{i=1}^{m} (f_1(t_i)c_1 + f_2(t_i)c_2 - y_i) f_1(t_i) = 0$$

$$\frac{\partial E^2}{\partial c_2} = 2 \sum_{i=1}^{m} (f_1(t_i)c_1 + f_2(t_i)c_2 - y_i) f_2(t_i) = 0.$$

These equations can be rewritten

$$\left(\sum_{i=1}^{m} f_1^2(t_i)\right) c_1 + \left(\sum_{i=1}^{m} f_1(t_i)f_2(t_i)\right) c_2 = \sum_{i=1}^{m} y_i f_1(t_i) \qquad (12.27)$$

$$\left(\sum_{i=1}^{m} f_1(t_i)f_2(t_i)\right) c_1 + \left(\sum_{i=1}^{m} f_2^2(t_i)\right) c_2 = \sum_{i=1}^{m} y_i f_2(t_i). \qquad (12.28)$$

The equations (12.27) and (12.28) can also be derived in a more geometric fashion. Let $\mathbf{A}$ be the $m \times 2$ matrix

$$\mathbf{A} = \begin{bmatrix} f_1(t_1) & f_2(t_1) \\ f_1(t_2) & f_2(t_2) \\ \cdot & \cdot \\ \cdot & \cdot \\ f_1(t_m) & f_2(t_m) \end{bmatrix}.$$

Let $\mathbf{y} = (y_1, \ldots, y_m)$. The m values taken on by the model function $f(\mathbf{c}, t_i)$, $i = 1, \ldots, m$, are given by the matrix product $\mathbf{Ac}$. The squared error is

$$E^2(\mathbf{c}) = \|\mathbf{Ac} - \mathbf{y}\|_2^2.$$

Thus the squared error is just the square of the distance, as measured in the Euclidean norm, from the point $\mathbf{y} = (y_1, \ldots, y_m)$ to the point $\mathbf{Ac}$ in the range of $\mathbf{A}$. Assuming that the rank of $\mathbf{A}$ is 2, the range of $\mathbf{A}$ is a two-dimensional subspace M of $\mathbb{R}^m$. The question of minimizing the squared error can be seen as the problem of finding that point $\mathbf{z} \in M$ that lies closest to $\mathbf{y}$. The vector $\mathbf{z}$ is the orthogonal projection of $\mathbf{y}$ onto M and hence, $\mathbf{z} - \mathbf{y}$ is orthogonal to all the elements of M. That is to say,

$$\langle \mathbf{z} - \mathbf{y}, \mathbf{Ac} \rangle = 0$$

for all $\mathbf{c} \in \mathbb{R}^2$. However, this means that $\mathbf{z} - \mathbf{y}$ lies in the nullspace of $\mathbf{A}^T$, so

$$\mathbf{A}^T\mathbf{z} = \mathbf{A}^T\mathbf{y}.$$

Keeping in mind that $\mathbf{z} = \mathbf{Ac}$ for some $\mathbf{c} \in \mathbb{R}^n$, we see that the pair $\mathbf{c} = (c_1, c_2)$ that minimizes the squared error must satisfy the *normal equation*:

$$\mathbf{A}^T\mathbf{Ac} = \mathbf{A}^T\mathbf{y}. \qquad (12.29)$$

If we multiply out $A^T A$, we will get exactly the equations (12.27) and (12.28). If the matrix A has rank 2, the 2×2 matrix $A^T A$ is invertible, and the normal equations have a unique solution.

> **EXAMPLE 12.9:** A very common choice of linear model with two parameters is $f(t) = c_1 + c_2 t$. That is, $f_1(t) \equiv 1$ and $f_2(t) = t$. The model in this case is a straight line. If two of the points t_i are distinct, the matrix A has rank 2, in which case the normal equations have a unique solution.
>
> Another example with two parameters is $f(t) = c_1 e^{-t} + c_2 e^{-2t}$. We might use this model if the data come from some experiment that produced signals that decay exponentially.
>
> An example with three parameters in the model is
>
> $$f(c_1, c_2, c_3, t) = c_1/(1+t^2) + c_2 t/(1+t^2) + c_3 t^2/(1+t^2).$$

Nonlinear Least Squares

The linear least squares problem can be reduced to a problem in linear algebra. When the model is nonlinear in the parameters, we must use more general optimization techniques. In a few cases, the nonlinear problem can be converted to a linear one.

> **EXAMPLE 12.10:** Suppose that our model is of the form
>
> $$f(a, b, t) = a e^{bt}.$$
>
> Let the data points be $(t_1, y_1) = (0, 1)$, $(t_2, y_2) = (1, 2)$, and $(t_3, y_3) = (2, 5)$. The total squared error is
>
> $$E^2(a, b) = (f(a, b, 0) - y_1)^2 + (f(a, b, 1) - y_2)^2 + (f(a, b, 2) - y_3)^2$$
> $$= (a - 1)^2 + (ae^b - 2)^2 + (ae^{2b} - 5)^2.$$
>
> The problem is to find the choice of parameters a and b that will minimize E^2. One way to find these values of a and b is to seek

solutions of the nonlinear system of equations:

$$\frac{\partial E^2}{\partial a} = 2(a-1) + 2(ae^b - 2)e^b + 2(ae^{2b} - 5)e^{2b} = 0$$

$$\frac{\partial E^2}{\partial b} = 2(ae^b - 2)ae^b + 2(ae^{2b} - 5)2ae^{2b} = 0.$$

We can compute a numerical solution of this system.

$$a_* = 0.8839 \quad b_* = 0.8644 \quad E(a_*, b_*) = 0.0235.$$

The Hessian matrix $\mathbf{D}^2 E^2(a_*, b_*)$ is positive definite, so this is a (local) minimum.

On the other hand, we can make a change of the dependent variable to transform the problem into one that is linear in the parameters. Let $c = \log a$. Then take the logarithm of both sides of

$$f(a, b, t_i) \approx y_i$$

to get

$$c + bt_i \approx \log y_i.$$

We are now using a model $t \to c + bt$ that is linear in the parameters c and b. We form the new squared error function

$$\hat{E}^2(c, b) = (c + bt_1 - \log y_1)^2 + (c + bt_2 - \log y_2)^2 + (c + bt_3 - \log y_3)^3.$$

To minimize this squared error, we set the partial derivatives of $\hat{E}$ equal to zero, thus obtaining the linear system,

$$\frac{\partial \hat{E}^2}{\partial c} = 2c + 2(c + b - \log 2) + 2(c + 2b - \log 5) = 0$$

$$\frac{\partial \hat{E}^2}{\partial b} = 2c + 2(c + b - \ln 2) + 4(c + 2b - \ln 5) = 0.$$

This system has the solution

$$c^* = -0.0372 \quad b^* = 0.8047 \quad \hat{E}(c^*, b^*) = 0.0083.$$

However, the problem of minimizing $E^2(a, b)$ is not equivalent to the problem of minimizing $\hat{E}^2(c, b)$ because

$$\log(ae^b t - y) \neq c + bt - \log y.$$

Indeed, we see that

$$0.8839 = a_* \neq e^{c_*} = e^{-0.0372} = 0.9635$$

and $b_* \neq b^*$. If it is important to minimize the squared error in the form of E^2, we must use the nonlinear formulation.

The Gauss-Newton Method

More generally, let us suppose that there are data points (t_i, y_i) with $t_1 < t_2 < \ldots t_m$. We set $\mathbf{y} = (y_1, \ldots, y_m)$. Assume a model function $f(\mathbf{c}, t)$ where $\mathbf{c} = (c_1, \ldots, c_n) \in \mathbb{R}^n$, $n < m$. We define a vector-valued function

$$\mathbf{f}(\mathbf{c}) = (f(\mathbf{c}, t_1), \ldots, f(\mathbf{c}, t_m))$$

from $\mathbb{R}^n$ into $\mathbb{R}^m$. The squared error is now

$$E^2(\mathbf{c}) = \sum_{i=1}^{m} (f(\mathbf{c}, t_i) - y_i)^2 = \|\mathbf{f}(\mathbf{c}) - \mathbf{y}\|_2^2. \tag{12.30}$$

One approach to the problem of minimizing E^2 is to set the partial derivatives of E^2 equal to zero and to search for a critical point. This means we must attempt to solve the $n \times n$ possibly nonlinear system

$$\frac{\partial E^2}{\partial c_1} = 2 \sum_i (f(\mathbf{c}, t_i) - y_i) \frac{\partial f}{\partial c_1}(\mathbf{c}, t_i) = 0$$

$$. = \ldots$$

$$. = \ldots$$

$$\frac{\partial E^2}{\partial c_n} = 2 \sum_i (f(\mathbf{c}, t_i) - y_i) \frac{\partial f}{\partial c_n}(\mathbf{c}, t_i) = 0.$$

We can apply Newton's method to this system, but this will involve calculating the second derivatives of the model function f.

The simplest Gauss-Newton method, which we describe below, is a way of numerically minimizing the squared error using only the first derivatives of f. Unfortunately there are situations where the method does not converge, no matter how close we start to the minimizer of the squared error. The method may be improved by testing each iterate and modifying the step taken if it is too big. This procedure leads to a Gauss-Newton method that is convergent if the initial guess is close enough to a minimizer. It is widely used in the statistical analysis of data. We present the simple Gauss-Newton method here because it is easy to understand.

Again we look for a more geometric way of analyzing this problem. We recognize that the squared error (12.30) is just the distance from the point y to a point $f(c)$ in the range of f. Of course, when f is nonlinear in c, the range of f need not be a subspace of $\mathbb{R}^n$. We assume that f is C^1 and that the $m \times n$ Jacobian matrix

$$\mathbf{Df}(\mathbf{c}) = \left[\frac{\partial f(\mathbf{c}, t_i)}{\partial c_j}\right], \quad i = 1, \ldots, m, \quad j = 1, \ldots, n,$$

has rank n. With this assumption on f, the range of f is an n-dimensional smooth surface lying in $\mathbb{R}^m$. The nonlinear least squares problem consists of finding the point z in the range of f that is closest to y. Then c_* is the point in $\mathbb{R}^n$ such that $f(c_*) = z$.

Now it is no longer possible to find z by projection, but we can use projections by making linear approximations to the range of f. Indeed, let c_0 be a first guess for the minimizer. We make the following linear approximation to f at the point c_0:

$$\mathbf{f}(\mathbf{c}) \approx \mathbf{f}(\mathbf{c}_0) + \mathbf{Df}(\mathbf{c}_0)(\mathbf{c} - \mathbf{c}_0). \tag{12.31}$$

Then we find the solution c_1 in the linear least squares sense of the linear equation:

$$\mathbf{f}(\mathbf{c}_0) + \mathbf{Df}(\mathbf{c}_0)(\mathbf{c} - \mathbf{c}_0) = \mathbf{y}. \tag{12.32}$$

In geometric terms, this procedure can be described as follows: Let M_0 be the image of $\mathbb{R}^n$ under the affine mapping

$$\mathbf{c} \to \mathbf{f}(\mathbf{c}_0) + \mathbf{Df}(\mathbf{c}_0)(\mathbf{c} - \mathbf{c}_0).$$

M_0 is the translate of a linear subspace. Let z_1 be the orthogonal projection of y onto M_0. Then $c_1 \in \mathbb{R}^n$ is the unique solution of (12.32) when y is replaced by z_1. At the next step we let M_1 be the image of the affine mapping

$$c \rightarrow f(c_1) + Df(c_1)(c - c_1).$$

Then z_2 is the projection of y onto M_1 and c_2 is the unique solution of (12.32) with y replaced by z_2 and c_0 replaced by c_1. This procedure can be used to generate a sequence of points c_k that we hope converges to the solution c_* that minimizes the squared error (12.30).

It can be shown that if the surface that is the range of f is "flat enough" near the point $f(c_*)$, then the procedure will converge, and the convergence will usually be linear. However, unlike the case of linear least squares, there may be more than one minimizing point, and the method may converge to a local minimum rather than a global minimum.

In Figure 12.9 we sketch the geometry of this procedure in the case that there is only one parameter c ($n = 1$) and two data points (t_1, y_1), (t_2, y_2) ($m = 2$). The range of $f(c) = (f(c, t_1), t(c, t_2))$ is a curve in $\mathbb{R}^2$ parameterized by c, and the linear approximation (12.31) is just the tangent line to the curve at $f(c_0) = (f(c_0, t_1), f(c_0, t_2))$.

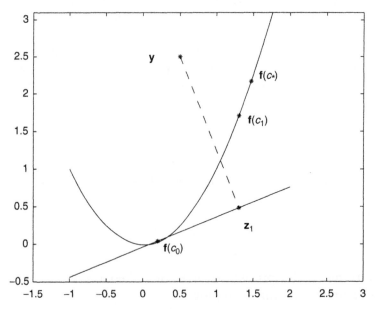

Figure 12.9 Linear approximation of the range of the model function, and orthogonal projection of y onto the tangent line

The Gauss-Newton algorithm can be viewed as a modification of the usual Newton algorithm. Suppose we let

$$g(c) = f(c) - y.$$

Then the problem of minimizing the squared error (12.30) is the same as that of finding a point c_* that minimizes $\|g(c)\|_2^2$. Since $Df = Dg$, equation (12.32) can be written

$$Dg(c_0)s = -g(c_0),$$

where $s = c - c_0$. The next iterate c_1 is found by solving this equation for s in the least squares sense. Then $c_1 = c_0 + s$. Thus the Gauss-Newton algorithm is the same as the Newton algorithm, except that at each iteration, we solve the linear system in the sense of least squares. The scheme can be described as follows:

Enter the initial guess c_0. Then iterate with a loop

$$c = c_0$$
$$\text{for } k = 1 : N$$
$$\quad \text{solve the linear system } Dg(c_{k-1})s = -g(c_{k-1})$$
$$\quad \text{in the sense of least squares}$$
$$\quad c_k = c_{k-1} + s$$
$$\text{end the loop}$$

The Gauss-Newton method is discussed in some detail in the books [N], and [De]. In the exercises to follow we take examples from sociology and biology. Some sources for these kind of problems are the books of J. S. Coleman [Col] and, D. M. Bates and D. B. Watts [Ba].

Exercises for 12.6

1. The Gauss-Newton method. Let A be an $m \times n$ matrix with rank n. Recall that the least squares solution of a linear equation $Ax = y$ is the solution of the *normal equations*

$$A^T Ax = A^T y.$$

Since A has rank n, $A^T A$ is invertible, and the least squares solution is given by

$$x_* = [A^T A]^{-1} A^T y.$$

For the purposes of analysis (not actual computation), show that the Gauss-Newton scheme can be written

$$\mathbf{c}_k = \mathbf{c}_{k-1} - [\mathbf{Dg}^T(\mathbf{c}_{k-1})\mathbf{Dg}(\mathbf{c}_{k-1})]^{-1}\mathbf{Dg}^T(\mathbf{c}_{k-1})\mathbf{g}(\mathbf{c}_{k-1}). \qquad (12.33)$$

2. The Gauss-Newton method. We take a simple case with $n = 1$ and $m = 2$. Suppose that $\mathbf{f}(c) = (c, Ac^2 + 1)$. We take $\mathbf{y} = (0, 0)$ so the least squares problem is simply the problem of finding c_* that minimizes $\|\mathbf{f}(c)\|_2^2 = c^2 + (Ac^2 + 1)^2$. The minimizer is obviously $c_* = 0$.

a) Use exercise 1 to show that for this $\mathbf{f}$ $(= \mathbf{g})$, the Gauss-Newton iteration scheme can be viewed as the iteration scheme for the fixed-point problem $c = h(c)$ where

$$h(c) = \frac{2A^2c^3 - 2Ac}{1 + 4A^2c^2}.$$

b) Check that $h'(0) = -2A$. Use the results of Section 6.1 to show that if $|A| < 1/2$, there is an interval $|c| < \delta$ such that, if $|c_0| < \delta$, the iterates $c_k = h(c_{k-1})$ will converge to $c_* = 0$.

c) Show by computation or analysis that when $A = 1$, the fixed-point iterates do not converge, no matter how close we start to the minimizer $c_* = 0$.

3. Here is an example in which a nonlinear model is used to fit data. In sociology, the spread of information through a population is often modeled by a logistic differential equation. Let $y(t)$ be the fraction of the population that has received some information (such as a rumor) at time t. Then $1 - y(t)$ is the fraction that has not received the information. It is assumed that the rate of change of $y(t)$ is given by the differential equation

$$y'(t) = ry(t)(1 - y(t)),$$

where r is a growth rate. The solution with $y(0) = 1/(1 + c)$ can be written in the form

$$f(c, r, t) = \frac{1}{1 + ce^{-rt}}.$$

To test the model and make predictions, we try to find the parameter values c and r that give the best fit to data in the least squares sense.

a) Suppose that the data are

t	y
1	0.1670
3	0.1799
5	0.2623
10	0.4752
15	0.7351
20	0.8966

Construct the squared error function

$$E^2(c, r) = \sum_i (f(c, r, t_j) - y_i)^2,$$

and make a contour plot over the rectangle $\{6 \leq c \leq 10, 0.15 \leq r \leq 0.25\}$. From this contour plot, you can make a good estimate of where the minimum is attained.

b) Use the Gauss-Newton method to find the values of c_* and r_* that minimize the error function. Plot $t \to f(c_*, r_*, t)$ on the interval $0 \leq t \leq 21$ and superimpose the data points to see how good your fit is.

c) With the values c_* and r_* in the model function $f(c, r, t)$, predict at what time 95% of the population will have received the information.

4. In physical chemistry, the Michaelis-Menton model for enzyme kinetics relates the "velocity" of an enzymatic reaction to the concentration x of the enzyme. The model function is

$$y = f(x, \beta, \gamma) = \frac{\beta x}{\gamma + x}.$$

A typical data set is

t	y
.02	75
.05	94
.11	130
.20	154
.54	193
1.09	205

a) Construct the squared error function $E^2(\beta, \gamma)$, and plot contours on the rectangle $206 \leq \beta \leq 210$, $0.05 \leq \gamma \leq 0.07$. Use level curve values in the range 630 to 700. From the contour plot, make an estimate of where the minimum value of the error is attained.

b) Use the Gauss-Newton method to find the values β_* and γ_* that minimize the squared error. Plot $x \to f(x, \beta_*, \gamma_*)$ on the interval $0 \leq x \leq 1.2$ together with the data to see how well your curve fits.

Chapter 13

Constrained Optimization

Chapter Overview

In Chapter 12, we studied the problem of finding the extreme points of functions defined on open sets of $\mathbb{R}^n$, usually all of $\mathbb{R}^n$. However, in many problems in physics and economics, we wish to find the extreme points of a function subject to some constraint. We introduce the technique of Lagrange multipliers to find necessary conditions for the existence of extreme points of constrained problems. An example, that of a three bar linkage, is used to illustrate how constraints arise in a physical problem, and how the number of constraints depends on the choice of variables used to describe the problem. When we investigate the dependence of an extreme value on a parameter in the problem, we find that the Lagrange multiplier is a measure of the sensitivity of the solution to changes in the parameter. We then turn to the question of finding sufficient conditions for a local maximum or minimum of a constrained problem. Constraints involving inequalities are treated next, leading to the KKT equations. Finally, an important example from economics, the Averch-Johnson effect, is discussed in detail.

13.1 Lagrange Multipliers

We will pose constrained optimization problems in terms of a function $f(\mathbf{x})$, which is the *objective* function, and one or more functions $g(\mathbf{x})$, which are the *constraint* functions. Let G be an open set of $\mathbb{R}^n$. We suppose that $f, g : G \to \mathbb{R}$ are C^1 functions. When there is only one constraint function g, the *constraint set*

$$S = \{\mathbf{x} \in G : g(\mathbf{x}) = 0\}.$$

The optimization problem that we seek to solve is

$$\text{Find the extreme points of } f\Big|_S .$$

EXAMPLE 13.1:   In a typical problem from calculus, there is a rectangular yard with dimensions x and y. Then for a given area A, what choice of x and y will yield the shortest perimeter? That is, we wish to minimize the objective function

$$f(x, y) = 2(x + y)$$

subject to the constraint

$$g(x, y) = xy - A = 0.$$

In this simple problem, we use the constraint to solve for $y = A/x$. We thereby reduce the problem to that of minimizing the function of one variable

$$f(x, A/x) = 2(x + A/x)$$

without a constraint.

In Example 13.1, we saw that the constraint effectively reduced the number of variables in the problem, commonly referred to as the "number of degrees of freedom." In most cases, however, we will not be able to solve the constraint equation for one of the variables globally. The constraint set may be a surface that is not the graph of a single function. Consequently, we will pursue a more indirect, local approach which does not reduce the number of variables, but rather adds another variable: the Lagrange multiplier. We first prove some sample theorems in three dimensions to illustrate the method. We then prove a general result in n dimensions with multiple constraints.

Theorem 13.1 Suppose that $G \subset \mathbb{R}^3$ is an open set and that $f, g : G \to \mathbb{R}$ are both C^1 functions. Let $S = \{(x, y, z) \in G : g(x, y, z) = 0\}$, and suppose that

$\mathbf{a} = (a_1, a_2, a_3) \in S$. We assume that $\nabla g(\mathbf{a}) \neq 0$. If $f|_S$ has a local extreme point at $\mathbf{a}$, there is a constant λ^* such that the pair $(\mathbf{a}, \lambda^*)$ satisfies the equation

$$\nabla f(\mathbf{a}) = \lambda^* \nabla g(\mathbf{a}). \tag{13.1}$$

Remark Theorem 13.1 provides a necessary condition that a local extreme point of $f|_S$ must satisfy. The function

$$L(x, y, z, \lambda) = f(x, y, z) - \lambda g(x, y, z)$$

is the Lagrange function. We will refer to equations (13.1), together with the constraint equation $g = 0$, as the *Lagrange equations*. The Lagrange equations are a system of four equations in four unknowns for the quantities a_1, a_2, a_3 and λ^*. A solution $(\mathbf{a}, \lambda^*)$ of this system is a critical point of L. The number λ^* is the *Lagrange multiplier*.

We also remark that this result uses an abstract existence theorem, the implicit function theorem, to derive a system of equations that we can attempt to solve by whatever analytical or numerical means we have at our disposal.

Proof: Since $\nabla g(\mathbf{a}) \neq 0$, at least one component of $\nabla g(\mathbf{a})$ is nonzero. Without loss of generality, we can assume that $\partial g / \partial z(\mathbf{a}) \neq 0$. By the implicit function theorem (Section 11.5), there is an open set $U \subset \mathbb{R}^2$ containing (a_1, a_2) and a C^1 function $p(x, y)$ such that $p(a_1, a_2) = a_3$ and

$$g(x, y, p(x, y)) = 0 \quad \text{for} \quad (x, y) \in U. \tag{13.2}$$

In other words, S is the graph of $z = p(x, y)$ near the point $\mathbf{a}$. Because $\mathbf{a}$ is an extreme point of $f|_S$, it follows that (a_1, a_2) is an extreme point of the function $(x, y) \to \psi(x, y) = f(x, y, p(x, y))$. Hence

$$\frac{\partial \psi}{\partial x}(a_1, a_2) = \frac{\partial \psi}{\partial y}(a_1, a_2) = 0.$$

Writing out these derivatives using the chain rule, we have

$$0 = \frac{\partial f}{\partial x}(\mathbf{a}) + \frac{\partial f}{\partial z}(\mathbf{a})\frac{\partial p}{\partial x}(a_1, a_2) \tag{13.3}$$

$$0 = \frac{\partial f}{\partial y}(\mathbf{a}) + \frac{\partial f}{\partial z}(\mathbf{a})\frac{\partial p}{\partial y}(a_1, a_2). \tag{13.4}$$

However, equation (13.2) implies that for $(x, y) \in U$,

$$0 = \frac{\partial g}{\partial x}(x, y, p(x, y)) + \frac{\partial g}{\partial z}(x, y, p(x, y))\frac{\partial p}{\partial x}(x, y)$$

$$0 = \frac{\partial g}{\partial y}(x, y, p(x, y)) + \frac{\partial g}{\partial z}(x, y, p(x, y))\frac{\partial p}{\partial y}(x, y),$$

and in particular for $(x, y) = (a_1, a_2)$. Therefore

$$\frac{\partial p}{\partial x}(a_1, a_2) = -\frac{\partial g/\partial x}{\partial g/\partial z}(\mathbf{a})$$

$$\frac{\partial p}{\partial y}(a_1, a_2) = -\frac{\partial g/\partial y}{\partial g/\partial z}(\mathbf{a}).$$

We substitute these expressions into (13.3) and (13.4) to deduce

$$\frac{\partial f}{\partial x}(\mathbf{a}) = \lambda^*\frac{\partial g}{\partial x}(\mathbf{a})$$

$$\frac{\partial f}{\partial y}(\mathbf{a}) = \lambda^*\frac{\partial g}{\partial y}(\mathbf{a}).$$

The Lagrange multiplier is

$$\lambda^* = \frac{\partial f/\partial z}{\partial g/\partial z}(\mathbf{a}).$$

This completes the proof of Theorem 13.1.

Before giving examples of the use of Theorem 13.1, we first consider an example in two dimensions that highlights the importance of the condition $\nabla g \neq 0$.

EXAMPLE 13.2:   Let the objective function $f(x, y) = x^2 + y^2$, and let the constraint function $g(x, y) = x^2 - (y - 1)^3$.   We want to find

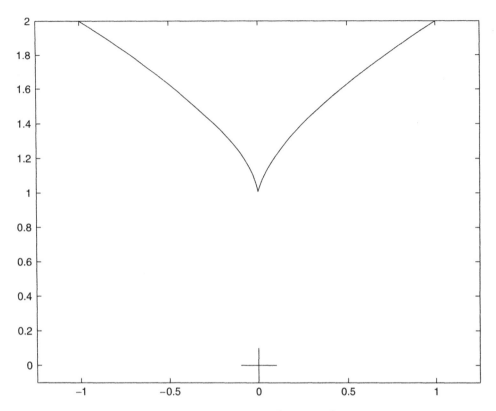

Figure 13.1 Constraint set $g(x, y) = x^2 - (y - 1)^3 = 0$ for Example 13.2

the minimum of f, subject to the constraint that $g(x, y) = 0$. From Figure 13.1, we see that the minimizer is obviously the point $(0, 1)$. The system of Lagrange equations is

$$2x = 2\lambda x$$
$$2y = -3\lambda(y - 1)^2$$
$$x^2 = (y - 1)^3.$$

However, $(0, 1)$ is not a solution of this system because $y = 1$ does not satisfy the second equation. What has gone wrong? The answer is that $\nabla g = (0, 0)$ at the point $(0, 1)$.

We usually think of the constraint surface given by a smooth function, for example $g = x^2 - (y - 1)^3$, as being smooth, with a well-defined normal everywhere. However, such surfaces can have cusps and sharp edges. It is important to remember that, if the maximum or minimum of f on the constraint set S occurs at a point **a** where $\nabla g(\mathbf{a}) = 0$, **a** may not be a solution of the Lagrange equations.

EXAMPLE 13.3:   In quantum mechanics, the ground state energy of a particle in a box of dimensions x, y, and z is

$$E = \frac{h^2}{8m} \left(\frac{1}{x^2} + \frac{1}{y^2} + \frac{1}{z^2} \right),$$

where h is Planck's constant and m is the mass of the particle. If the box has volume V, which set of dimensions will minimize the ground state energy?

In our terms, we want to minimize E subject to the constraint $xyz = V$. The constraint function is $g(x, y, z) = xyz - V$, and $\nabla g = (yz, xz, xy)$ is never equal to zero when $g(x, y, z) = 0$. Hence the minimizer of E, subject to the constraint $g = 0$, must satisfy the Lagrange equations. The Lagrange function is $L(x, y, z) = E(x, y, z) - \lambda xyz$, and the Lagrange equations are

$$-\frac{2h^2}{8mx^3} = \lambda yz$$

$$-\frac{2h^2}{8my^3} = \lambda xz$$

$$-\frac{2h^2}{8mz^3} = \lambda xy$$

$$xyz = V.$$

Multiply the first equation by x, the second by y and the third by z. It follows that a solution of the first three equations satisfies $x = y = z$. From the fourth equation, we deduce that $x = y = z = V^{1/3}$. The ground state energy is then $E = 3(h^2/8m)V^{-2/3}$.

In many circumstances, there are several constraints. We consider the following situation. Suppose that we have two constraint functions g_1 and g_2. Set

$S_1 = \{(x, y, z) : g_1(x, y, z) = 0\}$ and $S_2 = \{(x, y, z) : g_2(x, y, z) = 0\}$. Then we may wish to optimize f over the set $S = S_1 \cap S_2$. The general case is described in the following manner.

Let the number of constraints be $k < n$. Let G be an open subset of $\mathbb{R}^n$, and suppose that

$$f, g_1, g_2, \ldots, g_k : G \to \mathbb{R} \quad \text{are } C^1 \text{ functions.} \tag{13.5}$$

The constraint set is $S = \cap_{i=1}^k S_i$, where

$$S_i = \{\mathbf{x} \in \mathbb{R}^n : g_i(\mathbf{x}) = 0\}.$$

Our problem is again that of finding the extreme points of $f|_S$. The following theorem deals with the case of multiple constraints and includes Theorem 13.1.

Theorem 13.2 Assume that (13.5) holds. Let $\mathbf{a} \in S$ be a local extreme point of $f|_S$, and suppose that the vectors

$$\nabla g_i(\mathbf{a}), \quad i = 1, \ldots, k \quad \text{are linearly independent.} \tag{13.6}$$

Then there are Lagrange multipliers λ_i^*, $i = 1, \ldots, k$, such that

$$\nabla f(\mathbf{a}) = \lambda_1^* \nabla g_1(\mathbf{a}) + \cdots + \lambda_k^* \nabla g_k(\mathbf{a}). \tag{13.7}$$

Remark Let λ stand for the vector $(\lambda_1, \ldots, \lambda_k)$. The function

$$L(\mathbf{x}, \lambda) = f(\mathbf{x}) - \sum_{i=1}^k \lambda_i g_i(\mathbf{x})$$

is the Lagrange function for this problem and the numbers λ_i^* are the Lagrange multipliers. The system of $n + k$ equations in $n + k$ unknowns, consisting of equations (13.7) and the k constraint equations $g_i = 0$, form the Lagrange equations for this problem.

Proof: We first give a proof in the case $n = 3$ with $k = 2$ constraints because it can be visualized. Then we give a sketch of the proof for the general case.

To make the notation more concise, we let $\mathbf{g}(x, y, z) : \mathbb{R}^3 \to \mathbb{R}^2$ denote the function

$$\mathbf{g}(x, y, z) = \begin{bmatrix} g_1(x, y, z) \\ g_2(x, y, z) \end{bmatrix}$$

and

$$\mathbf{Dg(a)} = \begin{bmatrix} \partial_x g_1(\mathbf{a}) & \partial_y g_1(\mathbf{a}) & \partial_z g_1(\mathbf{a}) \\ \partial_x g_2(\mathbf{a}) & \partial_y g_2(\mathbf{a}) & \partial_z g_2(\mathbf{a}) \end{bmatrix}.$$

The condition (13.6) says that the rows of $\mathbf{Dg(a)}$ are independent. Hence at least one of the 2×2 submatrices must be invertible. By relabeling the coordinates if necessary, we can assume that

$$\mathbf{D}_{yz}\mathbf{g(a)} \equiv \begin{bmatrix} \partial_y g_1(\mathbf{a}) & \partial_z g_1(\mathbf{a}) \\ \partial_y g_2(\mathbf{a}) & \partial_z g_2(\mathbf{a}) \end{bmatrix}$$

has a nonzero determinant. Then by the implicit function theorem, there is an open interval I containing a_1, and C^1 functions $y = p(x)$ and $z = q(x)$ defined on I such that $p(a_1) = a_2$, $q(a_1) = a_3$, and such that

$$g_1(x, p(x), q(x)) = 0$$

$$g_2(x, p(x), q(x)) = 0, \quad \text{for } x \in I. \tag{13.8}$$

The constraint set S is a curve in $\mathbb{R}^3$. Equations (13.8) say that S is parameterized by $x \to (x, p(x), q(x))$ and has tangent vector $(1, p'(x), q'(x))$.

We have assumed that $f|_S$ has a local extreme point at $\mathbf{a}$. Hence $x \to f(x, p(x), q(x))$ has a local extreme point at $x = a_1$. Therefore

$$0 = f_x(\mathbf{a}) + f_y(\mathbf{a})p'(a_1) + f_z(\mathbf{a})q'(a_1). \tag{13.9}$$

If we differentiate equations (13.8) with respect to x, and set $x = a_1$, we see that

$$0 = \partial_x g_1(\mathbf{a}) + \partial_y g_1(\mathbf{a})p'(a_1) + \partial_z g_1(\mathbf{a})q'(a_1) \tag{13.10}$$

$$0 = \partial_x g_2(\mathbf{a}) + \partial_y g_2(\mathbf{a})p'(a_1) + \partial_z g_2(\mathbf{a})q'(a_1). \tag{13.11}$$

Equations (13.9), (13.10), and (13.11) say that $\nabla f(\mathbf{a})$, $\nabla g_1(\mathbf{a})$, and $\nabla g_2(\mathbf{a})$ are each orthogonal to the tangent vector $(1, p'(a_1), q'(a_1))$. Since $\nabla g_1(\mathbf{a})$ and $\nabla g_2(\mathbf{a})$ are assumed independent, they span the two-dimensional orthogonal complement of the tangent vector. Hence $\nabla f(\mathbf{a})$ can be written as a linear combination of these two vectors, and this is exactly the assertion of the theorem.

Now we sketch the proof of the general case. We write the coordinates as $\mathbf{x} = (\mathbf{u}, \mathbf{w})$ where $\mathbf{u} = (x_1, \ldots, x_{n-k})$ and $\mathbf{w} = (x_{n-k+1}, \ldots, x_n)$. We take the functions g_i as the components of a mapping $\mathbf{g} : G \to \mathbb{R}^k$ which we write $\mathbf{g}(\mathbf{u}, \mathbf{w})$. Our hypothesis is that the gradient vectors $\nabla g_i(\mathbf{a})$, $i = 1, \ldots, k$ are independent. Relabeling the coordinates if necessary, we can assume that the $k \times k$ Jacobian matrix $\mathbf{D_w g(a)}$ is invertible. In view of the implicit function theorem, we can assert that there is an open set $U \subset \mathbb{R}^{n-k}$ that contains $\hat{\mathbf{a}} = (a_1, \ldots, a_{n-k})$ and a C^1 function $\mathbf{w} = \mathbf{p}(\mathbf{u})$ defined on U such that $(\hat{\mathbf{a}}, \mathbf{p}(\hat{\mathbf{a}})) = \mathbf{a}$ and $\mathbf{g}(\mathbf{u}, \mathbf{p}(\mathbf{u})) = 0$ for $\mathbf{u} \in U$. To simplify the notation, we set $\phi(\mathbf{u}) = (\mathbf{u}, \mathbf{p}(\mathbf{u}))$. In terms of ϕ, we have $\phi(\hat{\mathbf{a}}) = \mathbf{a}$ and

$$\mathbf{g}(\phi(\mathbf{u})) = 0 \quad \text{for all } \mathbf{u} \in U. \tag{13.12}$$

The function ϕ is a *parameterization* of the surface S near $\mathbf{a}$. The $n \times (n - k)$ Jacobian matrix

$$\mathbf{D}\phi(\hat{\mathbf{a}}) = [\mathbf{I}, \mathbf{Dp}(\hat{\mathbf{a}})]$$

has rank $n - k$. Recall that the *tangent space* at $\mathbf{a} \in S$ is

$$T_{\mathbf{a}} = \{\mathbf{v} \in \mathbb{R}^n : \mathbf{Dg(a)v} = 0\}.$$

As we saw in exercise 10 of Section 11.5,

$$T_{\mathbf{a}} = \mathbf{D}\phi(\hat{\mathbf{a}})(\mathbb{R}^{n-k}). \tag{13.13}$$

Equation (13.13) implies that each of the gradient vectors $\nabla g_i(\mathbf{a})$ is orthogonal to $T_{\mathbf{a}}$ and thus belongs to N, which is the orthogonal complement in $\mathbb{R}^n$ of $T_{\mathbf{a}}$. Since $T_{\mathbf{a}}$ has dimension $n - k$, N must have dimension k. Because the k gradient vectors $\nabla g_i(\mathbf{a})$ are assumed independent, they span N. Finally, if $\mathbf{a}$ is a local extreme point for $f|_S$, then $\hat{\mathbf{a}}$ is a local extreme point for the function $\mathbf{u} \to f(\phi(\mathbf{u}))$. Therefore, using the chain rule,

$$\nabla f(\mathbf{a})\mathbf{D}\phi(\hat{\mathbf{a}}) = 0. \tag{13.14}$$

This last equation means that $\nabla f(\mathbf{a})$ must also lie in N. Consequently $\nabla f(\mathbf{a})$ can be written as a linear combination of the gradient vectors $\nabla g_i(\mathbf{a})$:

$$\nabla f(\mathbf{a}) = \lambda_1^* \nabla g_1(\mathbf{a}) + \cdots + \lambda_k^* \nabla g_k(\mathbf{a}).$$

The proof is complete.

An important part of the analysis of a problem is to determine how many constraints must be imposed. This number can depend on the choice of variables used to describe the problem.

> **EXAMPLE 13.4:**   We will study a three bar linkage consisting of three bars and two masses (see Figure 13.2). We assume that the bars each have length 1 and that the bar on the left can pivot about the point $(0,0)$, while the bar on the right pivots about the point $(2,0)$. The masses m_1 and m_2 are located at the point where the bars join in the middle. We neglect the mass of the bars. Our goal is to determine the configuration of the bars and masses when hanging under the force of gravity. To do this, we will seek to minimize the potential energy of the system.
>
> We first describe this system in terms of Cartesian coordinates. Let (x_1, y_1) denote the position of m_1 and (x_2, y_2) the position of m_2.

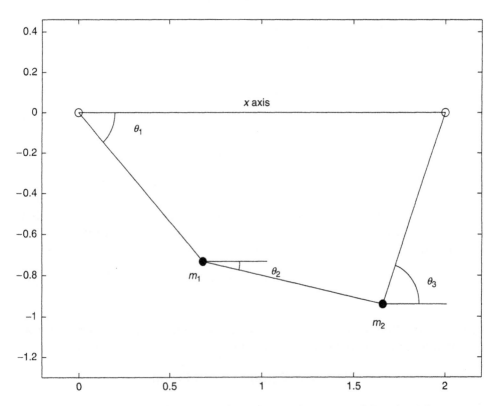

Figure 13.2   Three bar linkage with angles θ_1, θ_2, and θ_3 indicated

How should we express the potential energy of the system in these coordinates?

We note that it is possible for the masses to rise above the x axis. When mass m_1 is raised as far as possible, the bars and the line segment from $(0,0)$ to $(2,0)$ form an isosceles triangle, with the left-most bar making a positive angle $\theta_1 = \cos^{-1}(1/4)$ with the x axis. The height to which the mass m_1 can be raised is then $d = \sin(\theta_1) = \sqrt{15/16}$; the same is true for the mass m_2. One way of assigning the potential energy of the system is

$$V(x_1, y_1, x_2, y_2) = m_1 g(d + y_1) + m_2 g(d + y_2).$$

When the mass m_1 is in its lowest position ($y_1 = -d$), the contribution to the potential energy of that mass is zero, and similarly for the mass m_2.

What are the constraints? They are determined by the geometry of the linkage. If we fix the position of one of the masses, the position of the other is also fixed. The first mass can only move in the circle of radius 1 about the point $(0,0)$, and this circle can be parameterized by a single variable. Thus the system has only one degree of freedom. Since the two positions depend on four variables, we would expect that we need three constraints to describe the problem. The constraints come from the fact that the bars always have length 1. In this example, we use h for the constraint functions to avoid confusion with the gravitational constant g. The constraints are:

$$
\begin{aligned}
h_1(x_1, y_1, x_2, y_2) &\equiv x_1^2 + y_1^2 - 1 = 0 \\
h_2(x_1, y_1, x_2, y_2) &\equiv (x_2 - x_1)^2 + (y_2 - y_1)^2 - 1 = 0 \\
h_3(x_1, y_1, x_2, y_2) &\equiv (x_2 - 2)^2 + y_2^2 - 1 = 0.
\end{aligned}
$$

Viewing the problem this way, the Lagrange function is $L = V - \lambda_1 h_1 - \lambda_2 h_2 - \lambda_3 h_3$, and the Lagrange equations are 7 equations in 7 unknowns: $x_1, y_1, x_2, y_2, \lambda_1, \lambda_2, \lambda_3$:

$$
\begin{aligned}
0 &= V_{x_1} = 2\lambda_1 x_1 + 2\lambda_2(x_1 - x_2) \\
m_1 g &= V_{y_1} = 2\lambda_1 y_1 + 2\lambda_2(y_1 - y_2) \\
0 &= V_{x_2} = 2\lambda_2(x_2 - x_1) + 2\lambda_3(x_2 - 2) \\
m_2 g &= V_{y_2} = 2\lambda_2(y_2 - y_1) + 2\lambda_3 y_2 \\
h_1 &= 0 \\
h_2 &= 0 \\
h_3 &= 0.
\end{aligned}
$$

It is possible to find a solution of this system of equations using a numerical technique such as Newton's method.

The problem can also be posed in terms of the angular variables θ_1, θ_2 and θ_3 (see Figure 13.2). Since $y_1 = \sin \theta_1$ and $y_2 = -\sin \theta_3$, V can be expressed in terms of these variables as

$$V(\theta_1, \theta_2, \theta_3) = m_1 g(d + \sin \theta_1) + m_2 g(d - \sin \theta_3).$$

Now that we have used three variables to describe the position of the linkage, and there is still only one degree of freedom, we expect to see two constraints. From Figure 13.2, we see that they are (with different hs):

$$h_1(\theta_1, \theta_2, \theta_3) \equiv \cos \theta_1 + \cos \theta_2 + \cos \theta_3 - 2 = 0 \tag{13.15}$$

$$h_2(\theta_1, \theta_2, \theta_3) \equiv \sin \theta_1 + \sin \theta_2 + \sin \theta_3 = 0. \tag{13.16}$$

A convenient way to plot the various configurations of the linkages is to use θ_2 as a parameter, and to determine θ_1 and θ_3 from equations (13.15) and (13.16). The reason that we can do this is given by the implicit function theorem. The appropriate Jacobian matrix to check is

$$\det J(\theta_1, \theta_3) = \det \begin{bmatrix} \dfrac{\partial h_1}{\partial \theta_1} & \dfrac{\partial h_1}{\partial \theta_3} \\ \dfrac{\partial h_2}{\partial \theta_1} & \dfrac{\partial h_2}{\partial \theta_3} \end{bmatrix} = \sin(\theta_3 - \theta_1).$$

The reader must verify that $-\pi/3 \le \theta_2 \le \pi/3$ and $\theta_3 - \theta_1 \ne 0$.

Our problem is now to minimize V as a function of θ_1, θ_2 and θ_3, subject to the constraints (13.15) and (13.16). The Lagrange function for this problem is

$$L(\theta_1, \theta_2, \theta_3) = V(\theta_1, \theta_2, \theta_3) - \lambda h_1(\theta_1, \theta_2, \theta_3) - \mu h_2(\theta_1, \theta_2, \theta_3),$$

and the Lagrange equations are

$$\nabla V = \lambda \nabla h_1 + \mu \nabla h_2$$
$$h_1 = 0$$
$$h_2 = 0.$$

In terms of the components, we have

$$m_1 g \cos \theta_1 = -\lambda \sin \theta_1 + \mu \cos \theta_1$$
$$0 = -\lambda \sin \theta_2 + \mu \cos \theta_2$$
$$-m_2 g \cos \theta_3 = -\lambda \sin \theta_3 + \mu \cos \theta_3$$
$$h_1 = 0$$
$$h_2 = 0. \tag{13.17}$$

It can be verified that ∇h_1 and ∇h_2 are linearly independent for all θ_1, θ_2, and θ_3 that satisfy the constraints $h_1 = h_1 = 0$. Hence the minimizer of the potential energy must satisfy the Lagrange equations (13.17). We will construct numerical solutions of this system in the exercises. We should also note the presence of the parameters m_1 and m_2 in the problem. How does the solution of this system depend on these parameters? This question is investigated in the next section.

Exercises for 13.1

1. Let $f(x, y) = x^2 y^2$ and $g(x, y) = x^2/4 + y^2/9 - 1$.

   a) Verify that $\nabla g \neq 0$ for all (x, y) such that $g(x, y) = 0$. Write out the Lagrange equations, and find the extreme points of f restricted to $S = \{g = 0\}$.

   b) We usually think that, at a critical point of f restricted to $g = 0$, the level curves of f and g are tangent. Does this happen at the minimum point of f restricted to $g = 0$? What is the value of the Lagrange multiplier there?

   c) At the minimum point, check that the conditions of Theorem 13.1 are satisfied.

2. Let $f(x, y) = x^3 + x^2 + y^3/3$, and let $g(x, y) = x^2 + y^2 - 36$.

   a) Write out the Lagrange equations, and find all of the candidates for the extreme points of f restricted to $\{g = 0\}$.

   b) Which of the solutions of the Lagrange equations are local maxima, and which are global maxima? Where is the minimum?

3. Let $f(x, y, z) = x+y-z$, and let the constraints be $g_1(x, y, z) = x^2+y^2-1$ and $g_2(x, y, z) = x + z$.

a) Verify that ∇g_1 and ∇g_2 are linearly independent.

b) Use the Lagrange equations to find the extreme points of f subject to the constraints $g_1 = 0$ and $g_2 = 0$.

4. Let the functions

$$d(x_1, y_1, x_2, y_2) = (x_1 - x_2)^2 + (y_1 - y_2)^2$$
$$g_1(x_1, y_1) = x_1 y_1 - 4$$
$$g_2(x_2, y_2) = (x_2 - 1)^2 + y_2^2 - 1.$$

The function d is, of course, the distance squared between the point (x_1, y_1) which lies on the curve $g_1 = 0$, and the point (x_2, y_2) which lies on the curve $g_2 = 0$.

a) Plot the two constraint curves in the xy plane. Verify that ∇g_1 and ∇g_2 are linearly independent for all points (x_1, y_1, x_2, y_2) that satisfy the constraints.

b) Write out the Lagrange equations (six equations in six unknowns) for the problem

$$\text{minimize} \quad d$$
$$\text{subject to} \quad g_1(x_1, y_1) = 0$$
$$g_2(x_2, y_2) = 0,$$

and solve them. There are four real solutions. You may want to use a software system to find all of these roots.

c) Determine which solution is the global minimum and which are the local minimums.

d) Are there any maxima for d subject to $g_1 = 0$ and $g_2 = 0$? Explain.

5. Consider the link-spring system in Figure 13.3. The link on the left is attached to a pivot at the point $(0, 0)$, is assumed to have negligible mass, and has length 1. The spring has length l when unstretched and it is attached at the point $(2, 0)$. A body of mass m hangs at the joint of the link and the spring, located at the point (x, y). The potential energy of this contraption is

$$V(x, y) = \frac{k}{2}(d - l)^2 + mgy,$$

where $d = \sqrt{(x - 2)^2 + y^2}$ and k is the spring constant.

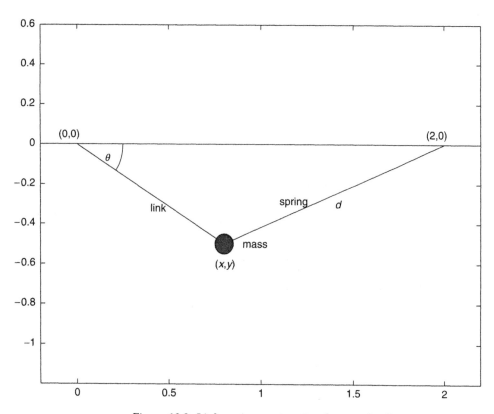

Figure 13.3 Link-spring contraption for exercise 7

a) To find the position of the mass in equilibrium, we must minimize
   the potential energy, subject to a constraint. What is the constraint?
b) What is the system of Lagrange equations?
c) This problem has one degree of freedom. What variable can you use
   to write this problem as an unconstrained minimization problem?
   What is the function we want to minimize in this case?
d) Set $m = 1, k = 1, l = 0.5$ and $g = 9.8$. Find the minimum potential
   energy and the coordinates (x, y) of the position of the mass.

6. Recall the three bar linkage of Example 13.4 where the constraint equations were

$$h_1(\theta_1, \theta_2, \theta_3) = \cos\theta_1 + \cos\theta_2 + \cos\theta_3 - 2 = 0$$
$$h_2(\theta_1, \theta_2, \theta_3) = \sin\theta_1 + \sin\theta_2 + \sin\theta_3 = 0.$$

Verify that ∇h_1 and ∇h_2 are linearly independent on the set of $(\theta_1, \theta_2, \theta_3)$ that satisfy the constraints $h_1 = h_2 = 0$.

7. We continue with Example 13.4.

   a) Write a program that uses Newton's method to solve the equations (13.15) and (13.16) for θ_1 and θ_3 as a function of θ_2 for 11 values of θ_2 in the range $-\pi/3 \le \theta_2 \le \pi/3$. Plot the position of the bars in the linkage for each of these values of θ_2.

   b) Let the numerically determined functions of part a) be denoted $\theta_1 = \varphi(\theta_2)$ and $\theta_3 = \psi(\theta_2)$. Set $m_1 = 1$, $m_2 = 2$, and $g = 9.8$. Then substitute $\varphi(\theta_2)$ and $\psi(\theta_2)$ into the expression for the potential energy $V(\theta_1, \theta_3)$ to arrive at a numerical function $V = V(\theta_2)$. Plot this function on $\pi/3 \le \theta_2 \le \pi/3$, and estimate the value of θ_2 that minimizes V.

8. We continue with the three bar linkage and consider the constraint equations in Cartesian coordinates.

   a) Show that if y_1 is fixed (which determines x_1), we can solve explicitly for y_2 in terms of x_1 and y_1:

   $$y_2 = \frac{y_1}{2} - \frac{(2-x_1)\sqrt{4x_1-1}}{2\sqrt{5-4x_1}}.$$

   b) Thus we can substitute for y_2 in the expression $V = m_1 g(d + y_1) + m_2 g(d + y_2)$ and obtain V as a function of y_1 alone. Set $m_1 = 1$, $m_2 = 2$, and $g = 9.8$. Graph this function, and use a root finder to determine at what point $V'(y_1) = 0$. Find the minimum potential energy and the corresponding configuration of the bars. Compare this with your results from Exercise 6.

9. Assume the hypotheses of Theorem 13.2. Show that if $\mathbf{v} \in T_a$, then $\mathbf{v}$ is a tangent vector to S at the point $\mathbf{a} \in S$.

13.2 Dependence on Parameters and Second-Order Conditions

In many optimization problems, there are parameters present. The extreme values of the objective function, and the location of the extreme points, will also depend on the parameters. In this case, the Lagrange multiplier λ associated with an extreme value may have an important interpretation, and it actually may be more important than the extreme value itself. We first state and prove a theorem in this regard when the parameter appears in the constraint.

Theorem 13.3 Let f and g be C^1 functions on an open set $G \subset \mathbb{R}^n$. Let $I \subset \mathbb{R}$ be an open interval and for $c \in I$, let $S_c = \{\mathbf{x} : g(\mathbf{x}) = c\}$. Let $\mathbf{x}(c) \in S_c$ and $\lambda(c)$ be a solution of the Lagrange equations,

$$\nabla f = \lambda \nabla g \tag{13.18}$$

$$g = c. \tag{13.19}$$

We assume that $c \to \mathbf{x}(c)$ is a C^1 function on I, and that $c \to \lambda(c)$ is continuous on I. Then for $c \in I$,

$$\frac{d}{dc} f(\mathbf{x}(c)) = \lambda(c). \tag{13.20}$$

Proof: We note that, because $g(\mathbf{x}(c)) = c$, we have

$$1 = \frac{d}{dc} g(\mathbf{x}(c)) = \langle \nabla g(\mathbf{x}(c)), \frac{d\mathbf{x}}{dc} \rangle.$$

Hence by (13.18)

$$\frac{d}{dc} f(\mathbf{x}(c)) = \langle \nabla f(\mathbf{x}(c)), \frac{d}{dc} \mathbf{x}(c) \rangle$$

$$= \lambda(c) \langle \nabla g(\mathbf{x}(c)), \frac{d}{dc} \mathbf{x}(c) \rangle$$

$$= \lambda(c).$$

The proof is complete.

Remark It is important to note that solutions of the Lagrange functions do not necessarily correspond to extreme points of $f|_{S_c}$. However, in many situations we can deduce by other means that a particular solution of the Lagrange equations does correspond to an extreme point. This is the case in Example 13.5, which is given next. Furthermore, we do not yet have general conditions that will ensure that a family of solutions $c \to (\mathbf{x}(c), \lambda(c))$ is continuous or C^1. We will address this question shortly.

Theorem 13.3 shows that the Lagrange multiplier can provide a measure of the sensitivity of the extreme point to changes in the constraint level. Often, the data in an applied problem contain errors, in particular the value of c in the constraint. A large multiplier λ^* indicates that the solution may not be too reliable.

EXAMPLE 13.5:   In a very simplified model of consumer preference, we assume that a consumer has a choice of two products, e.g. CDs and beer. If he buys x_1 dollars of CDs and x_2 dollars of beer, his "satisfaction" with these purchases is modeled by a *utility function* $u(x_1, x_2)$. We take our utility function to be

$$u(x_1, x_2) = 2x_1^{1/2} + 3x_2^{1/3}.$$

The consumer's budget allows only c dollars to be spent on these two items, and we assume that he spends all of it, so that the constraint is

$$x_1 + x_2 = c.$$

Of course, we assume that $x_1, x_2 \geq 0$. How does the (rational) consumer maximize his satisfaction subject to this constraint, and how will his maximal satisfaction change as his budget for CDs and beer is allowed to increase? The optimization problem is

$$\text{maximize } u \quad \text{subject to } x_1 + x_2 = c.$$

The Lagrange equations for this problem are

$$x_1^{-1/2} = \lambda$$
$$x_2^{-2/3} = \lambda$$
$$x_1 + x_2 = c.$$

The utility function u is strictly concave. The restriction of u to the portion of the line $x_1 + x_2 = c$ that lies in the first quadrant attains a unique maximum in the interior of this line segment. Furthermore, the gradient $\nabla g(x_1, x_2) = (1, 1) \neq 0$ everywhere. Hence the maximizer $(x_1^*(c), x_2^*(c))$ must be a solution of the Lagrange equations

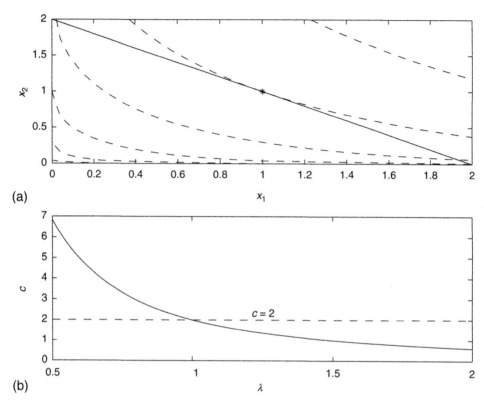

(a)

(b)

Figure 13.4 Figure 13.4a: Level curves of the utility function of Example 13.5 are shown in dashed curves. The constraint set $S = \{x_1 + x_2 = c = 2\}$ is indicated by the solid line. The maximum of $u|_S$ is attained at $(1, 1)$. Figure 13.4b: The graph of $q(\lambda)$ indicating where $q(\lambda) = c = 2$

(see Figure 13.4a). Solving for x_1 and x_2 in the first two equations and substituting in the constraint yields an equation for λ,

$$\lambda^{-2} + \lambda^{-3/2} = c.$$

The function $q(\lambda) \equiv \lambda^{-2} + \lambda^{-3/2}$ is monotone decreasing so that there is a unique solution $\lambda^*(c)$ for the Lagrange multiplier. According to Theorem 13.3, $\lambda^*(c)$ is the rate of increase of the maximal value of utility: $u^*(c) = u(x_1^*(c), x_2^*(c))$. In this context $\lambda^*(c)$ is called the *marginal utility*. In particular, when $c = 2$, $\lambda^* = 1$. As c increases, the marginal utility decreases (see Figure 13.4b). Although it is not difficult to compute $u^*(c)$ for various values of c, Theorem 13.3 allows us to determine the behavior of $(d/dc)u^*(c)$ without this computation.

The following result deals with the more general situation in which a parameter c appears in both the objective function and the constraint. Theorem 13.3 is a special case of

Theorem 13.4 Let $f(\mathbf{x}, c)$ and $g(\mathbf{x}, c)$ be C^1 functions on $G \times I$ where G is an open subset of $\mathbb{R}^n$ and I is an open interval. Let

$$S_c = \{\mathbf{x} : g(\mathbf{x}, c) = 0\} \subset G.$$

We suppose that, for each $c \in I$, $\mathbf{x}(c) \in S_c$, and $\lambda(c)$ is a solution pair of the Lagrange equations:

$$\nabla_{\mathbf{x}} f(\mathbf{x}, c) = \lambda(c) \nabla_{\mathbf{x}} g(\mathbf{x}(c), c)$$

$$g(\mathbf{x}(c), c) = 0.$$

Assume that $\mathbf{x}(c)$ is a C^1 function on I and that $\lambda(c)$ is continuous on I. Then

$$\frac{d}{dc} f(\mathbf{x}(c), c) = \frac{\partial L}{\partial c}(x(c), \lambda(c)). \tag{13.21}$$

Proof: Because $g(\mathbf{x}(c), c) = 0$ for $c \in I$, it follows that

$$0 = \frac{d}{dc} g(\mathbf{x}(c), c) = \left\langle \nabla g(\mathbf{x}(c), c), \frac{d\mathbf{x}}{dc} \right\rangle + \frac{\partial g}{\partial c}(\mathbf{x}(c), c).$$

Therefore

$$\frac{d}{dc} f(\mathbf{x}(c), c) = \left\langle \nabla f(\mathbf{x}(c), c), \frac{d\mathbf{x}}{dc} \right\rangle + \frac{\partial f}{\partial c}(\mathbf{x}(c), c)$$

$$= \left\langle \lambda(c) \nabla g(\mathbf{x}(c), c), \frac{d\mathbf{x}}{dc} \right\rangle + \frac{\partial f}{\partial c}$$

$$= -\lambda(c) \frac{\partial g}{\partial c} + \frac{\partial f}{\partial c}$$

$$= \frac{\partial L}{\partial c}.$$

Remark The importance of Theorem 13.4 is that we can calculate the rate of change of the extreme value of f by calculating the partial derivative $\partial L / \partial c$ first, and then evaluating it at $\mathbf{x}^*(c)$.

EXAMPLE 13.6: We return to the three bar linkage of Example 13.4. We assume that for each m_1 and m_2, the system (13.17) of five equations in five unknowns $\theta_1, \theta_2, \theta_2, \lambda, \mu$ has a unique solution that corresponds to a minimum of V. We denote this solution

$$\theta_1^*(m_1, m_2), \quad \theta_2^*(m_1, m_2), \quad \theta_3^*(m_1, m_2).$$

The minimum value of the potential energy is

$$V^*(m_1, m_2) = V(\theta_1^*, \theta_2^*, \theta_3^*, m_1, m_2).$$

Now we invoke Theorem 13.4 to see how the minimum potential energy changes as a function of the masses m_1 and m_2. It is not difficult to extend Theorem 13.4 to the case in which the objective function f depends on several parameters. Applying this extension of Theorem 13.4 to V^*, we see that

$$\begin{aligned} \partial V^*/\partial m_1 &= \partial V/\partial m_1(\theta_1^*, \theta_2^*, \theta_3^*, m_1, m_2) \\ &= g(d + \sin\theta_1^*(m_1, m_1)) \end{aligned}$$

and

$$\begin{aligned} \partial V^*/\partial m_2 &= \partial V/\partial m_2(\theta_1^*, \theta_2^*, \theta_3^*, m_1, m_2) \\ &= g(d - \sin\theta_3^*(m_2, m_2)). \end{aligned}$$

For any position (except the extreme lower positions) when $d = -\sin\theta_1$ or $d = \sin\theta_3$, we always have $d + \sin\theta_1 > 0$ and $d - \sin\theta_3 > 0$. Hence

$$\partial V^*/\partial m_1 > 0 \quad \text{and} \quad \partial V^*/\partial m_2 > 0.$$

This result may seem a bit surprising because, when m_1 is increased while m_2 remains fixed, the equilibrium position shifts, with the mass m_1 hanging lower. This would appear to decrease the potential energy of the first mass. However at the same time, the second mass m_2 moves higher, thereby increasing the potential energy of the second mass. The net result is to increase the potential energy V.

Continuous Dependence on Parameters

In Theorems 13.3 and 13.4, we assumed that the critical point $\mathbf{x}(c)$ is a well-defined C^1 function. Now we address the matter of providing a condition when this is true. First we state a simple lemma.

Lemma 13.5 Let $\mathbf{A}$ be an $n \times n$ real symmetric matrix, and let $\mathbf{b}$ be a nonzero vector in $\mathbb{R}^n$, considered as an $n \times 1$ matrix. Let T be the orthogonal complement of $\mathbf{b}$,

$$T = \{\mathbf{x} \in \mathbb{R}^n : \langle \mathbf{x},\ \mathbf{b} \rangle = 0\}.$$

Form the $(n + 1) \times (n + 1)$ matrix

$$\mathbf{C} = \begin{bmatrix} \mathbf{A} & \mathbf{b} \\ \mathbf{b}^T & 0 \end{bmatrix}.$$

Then, if $\langle \mathbf{Ax}, \mathbf{x} \rangle \neq 0$ for all $\mathbf{x} \in T$, $\mathbf{x} \neq 0$, the matrix $\mathbf{C}$ is invertible.

Proof: Write a vector $\mathbf{z} \in \mathbb{R}^{n+1}$ as $\mathbf{z} = (\mathbf{x}, y)$ where $\mathbf{x} \in \mathbb{R}^n$. If $\mathbf{Cz} = 0$, then $\mathbf{x}$ and y satisfy the two equations

$$\mathbf{Ax} + y\mathbf{b} = 0$$
$$\langle \mathbf{b},\ \mathbf{x} \rangle = 0.$$

The second equation says that $\mathbf{x} \in T$. If we take the scalar product of $\mathbf{x}$ with the first equation, we deduce

$$0 = \langle \mathbf{Ax},\ \mathbf{x} \rangle + y\langle \mathbf{b}, \mathbf{x} \rangle = \langle \mathbf{Ax},\ \mathbf{x} \rangle.$$

In view of the hypothesis on $\mathbf{A}$, we conclude that $\mathbf{x} = 0$. Returning to the first equation, $\mathbf{b} \neq 0$ implies $y = 0$.

We will apply the implicit function theorem to the Lagrange equations, and for this reason we will need to calculate the Hessian of the Lagrange function $L(\mathbf{x}, \lambda, c) = f(\mathbf{x}, c) - \lambda g(\mathbf{x}, c)$, which depends on the parameter c in both the objective function f and in the constraint function g. The Hessian with respect to both $\mathbf{x}$ and λ is the $(n + 1) \times (n + 1)$ matrix

$$\mathbf{D}^2_{\mathbf{x}, \lambda} L = \begin{bmatrix} D^2_{\mathbf{x}} L & -\nabla g^T \\ -\nabla g & 0 \end{bmatrix}. \tag{13.22}$$

The Hessian of L with respect to $\mathbf{x}$ is

$$\mathbf{D}_{\mathbf{x}}^2 L = \begin{bmatrix} \frac{\partial^2 (f - \lambda g)}{\partial x_1 \partial x_1} & \cdots & \frac{\partial^2 (f - \lambda g)}{\partial x_1 \partial x_n} \\ \cdot & \cdots & \cdot \\ \cdot & \cdots & \cdot \\ \frac{\partial^2 (f - \lambda g)}{\partial x_1 \partial x_n} & \cdots & \frac{\partial^2 (f - \lambda g)}{\partial x_n \partial x_n} \end{bmatrix}. \tag{13.23}$$

Theorem 13.6 Let $G \subset \mathbb{R}^n$ be an open subset, and let $J \subset \mathbb{R}$ be an open interval. Suppose that $f, g : G \times J \to \mathbb{R}$ are C^2 functions. The Lagrange equations with a parameter $c \in J$ are

$$\nabla_{\mathbf{x}} f(\mathbf{x}(c), c) = \lambda(c) \nabla_{\mathbf{x}} g(\mathbf{x}(c), c) \tag{13.24}$$

$$g(\mathbf{x}(c), c) = 0. \tag{13.25}$$

Let $\mathbf{x}_0 = \mathbf{x}(c_0)$ and let $\lambda_0 = \lambda(c_0)$ be a solution pair of (13.24) and (13.25) for the parameter value $c = c_0 \in J$. Assume that $\mathbf{b} = \nabla_{\mathbf{x}} g(\mathbf{x}_0, c_0) \neq 0$, and let $T_0 = \{\mathbf{x} \in \mathbb{R}^n : \langle \mathbf{b}, \mathbf{x} \rangle = 0\}$ be the tangent space to S_{c_0} at the point $\mathbf{x}_0$. Suppose further that the Hessian matrix (13.23) satisfies

$$\langle \mathbf{D}_{\mathbf{x}}^2 L(x_0, \lambda_0, c_0) \mathbf{v}, \mathbf{v} \rangle \neq 0$$

for all nonzero $\mathbf{v} \in T_0$. Then there is an open interval $I \subset J$ with $c_0 \in I$ and a family of solutions $\mathbf{x}(c), \lambda(c)$ defined for $c \in I$ such that

(a) $\mathbf{x}(c)$ and $\lambda(c)$ are both C^1 functions on I; and
(b) $\mathbf{x}(c_0) = \mathbf{x}_0$ and $\lambda(c_0) = \lambda_0$.

Proof: Because $\mathbf{b} = \nabla g_{\mathbf{x}}(\mathbf{x}_0, c_0) \neq 0$, and $\mathbf{A} = \mathbf{D}_{\mathbf{x}}^2 L(x_0, \lambda_0, c_0)$ satisfies the conditions of Lemma 13.5, we conclude that the Hessian matrix (13.22) is also nonsingular at $(\mathbf{x}_0, \lambda_0, c_0)$. Now we apply the implicit function theorem to the function $\mathbf{F} : \mathbb{R}^{n+2} \to \mathbb{R}^{n+1}$ defined by

$$\mathbf{F}(\mathbf{x}, \lambda, c) = \begin{bmatrix} \nabla_{\mathbf{x}} f(\mathbf{x}, c) - \lambda \nabla_{\mathbf{x}} g(\mathbf{x}, c) \\ -g(\mathbf{x}, c) \end{bmatrix}.$$

The Jacobian matrix of $\mathbf{F}$ with respect to $\mathbf{x}$ and λ is exactly the Hessian matrix (13.22). The result follows immediately.

Second-Order Conditions

Finally, we consider the question of providing a sufficient condition for the existence of a maximizer or minimizer.

We consider a problem with k constraints, $k < n$,

$$
\begin{array}{ll}
\text{minimize} & f(\mathbf{x}) \\
\text{subject to} & g_i(\mathbf{x}) = 0, \quad i = 1, \dots, k.
\end{array}
\tag{13.26}
$$

We let $\mathbf{g}(\mathbf{x}) = (g_1(\mathbf{x}), \dots, g_k(\mathbf{x}))$. The constraint set is

$$
S = \{\mathbf{x} : \mathbf{g}(\mathbf{x}) = 0\}.
$$

We may also state the problem as

$$
\text{minimize } f|S.
$$

When we minimize f without a constraint, a sufficient condition for a critical point $\mathbf{a}$ to be a local minimum is that the Hessian $\mathbf{D}^2 f(\mathbf{a})$ be positive definite; that is,

$$
\langle \mathbf{D}^2 f(\mathbf{a})\xi, \xi \rangle > 0
$$

for all $\xi \in \mathbb{R}^n$, $\xi \neq 0$. We seek a similar sufficient condition for the problem (13.26).

We suppose that, near the point $\mathbf{a} \in S$, the surface S is the image of a C^2 function $\phi : U \to \mathbb{R}^n$ where U is an open subset of $\mathbb{R}^{n-k}$ that contains $\hat{\mathbf{a}} = (a_1, \dots, a_{n-k})$ and that $\phi(\hat{\mathbf{a}}) = \mathbf{a}$. Such a function was constructed in the proof of Theorem 3.2 using the implicit function theorem. Then near $\mathbf{a}$, the restriction of f to S is given by the function

$$
\psi(\mathbf{u}) = f(\phi(\mathbf{u})), \quad \mathbf{u} \in U.
$$

Let us compute the Hessian of ψ. First we use the chain rule and find that

$$
\frac{\partial \psi}{\partial u_j} = \sum_{m=1}^{n} \frac{\partial f}{\partial x_m} \frac{\partial \phi_m}{\partial u_j}.
$$

Next we push on and compute the second derivatives:

$$
\frac{\partial^2 \psi}{\partial u_i \partial u_j} = \sum_{l=1}^{n} \sum_{m=1}^{n} \frac{\partial^2 f}{\partial x_l \partial x_m} \frac{\partial \phi_l}{\partial u_i} \frac{\partial \phi_m}{\partial u_j} + \sum_{m=1}^{n} \frac{\partial f}{\partial x_m} \frac{\partial^2 \phi_m}{\partial u_i \partial u_j}.
$$

We can simplify this expression somewhat by using the following notation. We will use $\partial\phi/\partial u_j$ to denote the jth column of the Jacobian matrix $\mathbf{D}\phi$. Thus

$$\mathbf{D}\phi = \left[\frac{\partial\phi}{\partial u_1}, \ldots, \frac{\partial\phi}{\partial u_{n-k}} \right].$$

Then for a vector $\xi \in \mathbb{R}^{n-k}$,

$$\mathbf{D}\phi\xi = \sum_{i=1}^{n-k} \xi_i \frac{\partial\phi}{\partial u_i}.$$

Hence

$$\frac{\partial^2\psi}{\partial u_i \partial u_j} = \left\langle \mathbf{D}^2 f \frac{\partial\phi}{\partial u_i}, \frac{\partial\phi}{\partial u_j} \right\rangle + \sum_{m=1}^{n} \frac{\partial f}{\partial x_m} \frac{\partial^2\phi_m}{\partial u_i \partial u_j}.$$

Consequently for vectors $\xi, \eta \in \mathbb{R}^{n-k}$,

$$
\begin{aligned}
\langle \mathbf{D}^2\psi\xi, \eta \rangle &= \sum_{i,j=1}^{n-k} \frac{\partial^2\psi}{\partial u_i \partial u_j} \xi_i \eta_j \\
&= \langle \mathbf{D}^2 f \mathbf{D}\phi\xi, \mathbf{D}\phi\eta \rangle + \sum_{m=1}^{n} \frac{\partial f}{\partial x_m} \langle \mathbf{D}^2\phi_m\xi, \eta \rangle.
\end{aligned}
\tag{13.27}
$$

This can be expressed in a matrix equation,

$$\mathbf{D}^2\psi = [\mathbf{D}\phi]^T \mathbf{D}^2 f \mathbf{D}\phi + \sum_{m=1}^{n} \frac{\partial f}{\partial x_m} \mathbf{D}^2\phi_m.$$

In equation (13.27), we see that to verify that $\mathbf{D}^2\psi$ is positive definite, we must have information about a sum of the Hessians $\mathbf{D}^2\phi_m$. This is not a practical requirement. We will see in the next theorem that there is a verifiable sufficient condition for a minimum that can be stated in terms of the Lagrange function.

Theorem 13.7 Let the functions f and g_i be C^2 functions on an open set $G \subset \mathbb{R}^n$ and let $L(\mathbf{x}, \lambda) = f(x) - \sum_{i=1}^{k} \lambda_i g_i(\mathbf{x})$ be the Lagrange function for the problem

(13.26). We suppose that $\mathbf{a} \in S$ and $\lambda_i^*, i = 1, \ldots, k$ is a solution of the Lagrange system:

$$\nabla_{\mathbf{x}} L = \nabla f - \sum_{i=1}^{k} \lambda_i \nabla g_i = 0 \tag{13.28}$$

$$g_i = 0, \ i = 1, \ldots, k. \tag{13.29}$$

We also suppose that the gradient vectors $\nabla g_i(\mathbf{a})$ are linearly independent. If $\langle \mathbf{D}_{\mathbf{x}}^2 L(\mathbf{a}, \lambda^*) \mathbf{v}, \mathbf{v} \rangle > 0$ for all $\mathbf{v}$ in the tangent space $T_{\mathbf{a}}$ to S at $\mathbf{a}$, then $\mathbf{a}$ is a (strict) local minimizer of the problem (13.26).

Proof: As in the proof of Theorem 13.2, we can use the assumption that the gradient vectors $\nabla g_i(\mathbf{a})$ are independent, together with the implicit function theorem, to construct the function ϕ on an open set $U \subset \mathbb{R}^{n-k}$. In addition, because we assume that f and the constraint functions g_i are C^2, it can be shown that ϕ is also C^2. The key observation to make here is that, because each $g_i = 0$ on the constraint set S,

$$\psi(\mathbf{u}) = f(\phi(\mathbf{u})) = L(\phi(\mathbf{u}), \lambda) \tag{13.30}$$

for all $\mathbf{u} \in U$. Instead of f, we may use L in the right side of equation (13.27). Evaluating at $\mathbf{u} = \hat{\mathbf{a}}$ and $\lambda^* = (\lambda_1^*, \ldots, \lambda_k^*)$, we find that

$$\langle \mathbf{D}^2 \psi(\hat{\mathbf{a}})\xi, \ \xi \rangle = \langle \mathbf{D}_{\mathbf{x}}^2 L(\mathbf{a}, \lambda^*) \mathbf{D}\phi(\hat{\mathbf{a}})\xi, \ \mathbf{D}\phi(\hat{\mathbf{a}})\xi \rangle + \sum_{m=1}^{n} \frac{\partial L}{\partial x_m}(\mathbf{a}, \lambda^*)\langle \mathbf{D}^2 \phi_m(\hat{\mathbf{a}})\xi, \ \xi \rangle.$$

However, because $\mathbf{a}$ and λ^* solve the Lagrange equation, $\partial L / \partial x_m(\mathbf{a}, \lambda^*) = 0$ for each m. Furthermore $\mathbf{D}\phi(\hat{\mathbf{a}})\xi$ lies in the tangent space $T_{\mathbf{a}}$ for each $\xi \in \mathbb{R}^{n-k}$. Hence

$$\langle \mathbf{D}^2 \psi(\hat{\mathbf{a}})\xi, \ \xi \rangle = \langle \mathbf{D}_{\mathbf{x}}^2 L(\mathbf{a}, \lambda^*) \mathbf{D}\phi(\hat{\mathbf{a}})\xi, \ \mathbf{D}\phi(\hat{\mathbf{a}})\xi \rangle > 0$$

for all $\xi \in \mathbb{R}^{n-k}$. The theorem is proved.

EXAMPLE 13.7:   Consider the trivial minimization problem:

$$\text{minimize } f(x, y) = x^2 + y^2$$
$$\text{subject to } g(x, y) = x^2/4 + y^2 - 1 = 0.$$

The Lagrange function is $L(x, y, \lambda) = x^2 + y^2 - \lambda(x^2/4 + y^2 - 1)$. The minimizers are $(x, y) = (0, \pm 1)$, with $\lambda = 1$. If we compute the Hessians $\mathbf{D}^2_{x,y}L$ and $\mathbf{D}^2_{x,y,\lambda}L$ and evaluate at $(x, y, \lambda) = (0, \pm 1, 1)$, we get

$$\mathbf{D}^2_{x,y}L = \begin{bmatrix} 3/2 & 0 \\ 0 & 0 \end{bmatrix}$$

and

$$\mathbf{D}^2_{x,y,\lambda}L = \begin{bmatrix} 3/2 & 0 & -1/2 \\ 0 & 0 & -2 \\ -1/2 & -2 & 0 \end{bmatrix}.$$

The Hessian $\mathbf{D}^2_{x,y}L$ is positive definite on the one-dimensional tangent space T (the same at both minimum points) that is the subspace of vectors $(v_1, 0)$. However, the points $(0, \pm 1, 1)$ are saddle points for the Lagrange function considered as a function of three variables, because the eigenvalues for $\mathbf{D}^2_{x,y,\lambda}L$ are 1.3462, 2.1895 and -2.0357. In an exercise, we will see that a solution of the Lagrange equations is a saddle point of $L(\mathbf{x}, \lambda)$, considered as a function of the $n+1$ variables $(\mathbf{x}, \lambda)$, whenever $\mathbf{D}^2_{\mathbf{x}}L$ is positive definite on the tangent space or negative definite on the tangent space.

Theorem 13.7 also provides a criterion for the existence of a C^1 family of local maxima or minima of the problem (13.6).

Corollary Assume the hypotheses of Theorem 13.6, and suppose that the Hessian $\mathbf{D}^2_{\mathbf{x}}L(\mathbf{x}_0, \lambda_0)$ is positive definite on the tangent space T_0. Then there is a $\delta > 0$ such that the solutions $\mathbf{x}(c)$ of the Lagrange equations (13.24) and (13.25) are local minimizers for $f|_{S_c}$ for $|c - c_0| < \delta$. If $\mathbf{D}^2_{\mathbf{x}}L(\mathbf{x}_0, \lambda_0)$ is negative definite on T_0, the points $\mathbf{x}(c)$ are local maximizers.

Proof: We combine Theorem 13.6 with Theorem 13.7 in the case of one constraint ($k = 1$). We can construct a family of parameterizations $\phi(\mathbf{u}, c)$ and define the restriction of $f(\mathbf{x}, c)$ to S_c, $\psi(\mathbf{u}, c) = f(\phi(\mathbf{u}, c), c)$. The hypothesis that $\mathbf{D}^2_{\mathbf{x}}L(\mathbf{x}_0, \lambda_0)$ is positive definite on the tangent space T_0 implies that $\mathbf{D}^2\psi(\hat{\mathbf{x}}_0, c_0)$

is positive definite. Because f and g are C^2 functions, there is a $\delta > 0$ such that $\mathbf{D}^2\psi$ remains positive definite for $|c - c_0| < \delta$. Hence $\mathbf{x}(c)$ is a local minimizer for this range of c. The same argument applies when $\mathbf{x}_0$ is a local maximizer.

Exercises for 13.2

1. Consider the minimization problem with two constraints,

$$\begin{array}{ll} \text{minimize} & f \\ \text{subject to} & g_1(\mathbf{x}) = 0 \\ & g_2(\mathbf{x}) = 0. \end{array}$$

Let $S_j = \{\mathbf{x} \in \mathbb{R}^n : g_j(\mathbf{x}) = 0\}$, $j = 1, 2$, and let the constraint set be $S = S_1 \cap S_2$.

a) Assume that $\mathbf{a} \in S$ with $\nabla g_1(\mathbf{a})$ and $\nabla g_2(\mathbf{a})$ linearly independent. Let T_j be the tangent space to S_j at the point $\mathbf{a}$. What is the dimension of T_j? Let T be the tangent space to S at $\mathbf{a}$. Show that $T = T_1 \cap T_2$. What is the dimension of T?

b) Let $\mathbf{Dg(a)}$ be the $2 \times n$ Jacobian matrix of $\mathbf{g(x)} = (g_1(\mathbf{x}), g_2(\mathbf{x}))$. Show that T is the nullspace of $\mathbf{Dg(a)}$.

2. Let $f(x, y, z) = z - x^2 - y^2$, $g_1(x, y, z) = x + y + z - 1$ and $g_2(x, y, z) = x^2 + y^2 + z^2 - 4$. We wish to find the maximum and minimum of f subject to the constraints $g_1 = 0$ and $g_2 = 0$.

a) What is the constraint set $S = \{(x, y, z) : g_1(x, y, z) = 0\} \cap \{(x, y, z) : g_2(x, y, z) = 0\}$?

b) Write out the Lagrange equations (five equations in five unknowns: $x, y, z, \lambda, \mu,$) and find the solutions.

c) What is the tangent space T to S at the point $\mathbf{a} \in S$ where $\min f|_S$ is attained? Verify that Theorem 13.7 is satisfied.

3. Let the functions

$$\begin{aligned} d(x_1, y_1, x_2, y_2) &= (x_1 - x_2)^2 + (y_1 - y_2)^2 \\ g_1(x_1, y_1) &= 5x_1^2 - 6x_1 y_1 + 5y_1^2 - 64 \\ g_2(x_2, y_2) &= (x_2 + 1/2)^2 + (y_2 - 1/2)^2 - 2. \end{aligned}$$

a) Plot the two constraint curves in the xy plane. One is a circle, and one is an ellipse, and both are symmetric with respect to the line $x + y = 0$.

We wish to find the solutions of the problem

$$\text{minimize } d$$
$$\text{subject to} \quad g_1(x_1, y_1) = 0$$
$$g_2(x_2, y_2) = 0.$$

There are eight real and four complex solutions of the Lagrange equations. The real solutions of the Lagrange system consist of one global minimum, three local minima, two saddle points and two points where the global maximum is attained.

b) Four of the solutions to the Lagrange system lie on the line $x+y = 0$. Find them, determine which is the global minimum, and which is the local minimum. What are the other solutions?

c) Verify that the second-order conditions of Theorem 13.7 are satisfied at the minima. Note that in this case of four variables, the Hessian $\mathbf{D}^2 L$ is a 4×4 matrix. The tangent space will be a two-dimensional subspace of $\mathbb{R}^4$.

4. Suppose that f, g are C^2, and let $S = \{x : g(x) = 0\}$. Suppose that $f|_S$ has a local minimum at $\mathbf{a} \in S$ and that $\nabla g(\mathbf{a}) \neq 0$. Let $T_\mathbf{a}$ be the tangent space to S at $\mathbf{a}$. Fill in the steps of the following argument to provide a short, simple proof that

$$\langle \mathbf{D}_\mathbf{x}^2 L(\mathbf{a}, \lambda^*)\mathbf{v}, \mathbf{v} \rangle \geq 0$$

for all $\mathbf{v} \in T_\mathbf{a}$.

a) If $\mathbf{v} \in T_\mathbf{a}$, there is C^2 function $\mathbf{r}(t) : (-1, 1) \rightarrow S$ with $\mathbf{r}(0) = \mathbf{a}$ and $\mathbf{r}'(0) = \mathbf{v}$. Why?

b) Show that $\frac{d^2}{dt^2} L(\mathbf{r}(t), \lambda^*)|_{t=0} \geq 0$.

c) Show that

$$\frac{d^2}{dt^2} L(\mathbf{r}(t), \lambda^*)|_{t=0} = \langle \mathbf{D}_\mathbf{x}^2 L(\mathbf{a}, \lambda^*)\mathbf{v}, \mathbf{v} \rangle.$$

5. Suppose that f and g are C^2, and let $L(\mathbf{x}, \lambda) = f(\mathbf{x}) - \lambda g(\mathbf{x})$. Suppose that $(\mathbf{a}, \lambda^*)$ is a solution of the Lagrange equations such that $\nabla g(\mathbf{a}) \neq 0$ and such that $\mathbf{D}_\mathbf{x}^2 L(\mathbf{a}, \lambda^*)$ is positive definite on the tangent space $T_\mathbf{a}$. Use Lemma 13.5 to show that the function of $n + 1$ variables $(\mathbf{x}, \lambda) \rightarrow L(\mathbf{x}, \lambda)$ has a saddle point at $(\mathbf{a}, \lambda^*)$. Show that we can draw the same conclusion if $\mathbf{D}_\mathbf{x}^2 L(\mathbf{a}, \lambda^*)$ is negative definite on $T_\mathbf{a}$.

13.3 Constrained Optimization with Inequalities

In many optimization problems, the constraints are a mixture of equalities and inequalities. This situation often occurs in business and economics. When the objective function and the constraint functions are linear, these optimization problems are tackled with the techniques of linear programming. Some of these techniques can be modified to treat problems with nonlinear constraints.

Up to now we have not been concerned with the sign of the multiplier. However, in the context of inequality constraints, it becomes important. To begin, we consider the simplest case of a single inequality constraint:

$$\text{maximize} \quad f(x, y)$$
$$\text{subject to} \quad h(x, y) \leq 0.$$

Let $Q = \{\mathbf{p} = (x, y) : h(\mathbf{p}) \leq 0\}$. Assume that Q has a nonempty interior. We distinguish between two cases.

(i) The extreme point $\mathbf{p}^*$ occurs on ∂Q. We say that the constraint $h \leq 0$ is *active*.

(ii) The extreme point $\mathbf{p}^*$ occurs in the interior of Q. We say that the constraint $h \leq 0$ is *inactive*.

Notice in case (i) that both $\nabla f(\mathbf{p}^*)$ and $\nabla h(\mathbf{p}^*)$ point out of Q. Thus $\nabla f(\mathbf{p}^*)$ will be a positive multiple μ^* of $\nabla h(\mathbf{p}^*)$. In case (ii), $\nabla f(\mathbf{p}^*) = 0$, so $\nabla f(\mathbf{p}^*) = \mu^* \nabla h(\mathbf{p}^*)$ with $\mu^* = 0$. We can combine both cases into a single set of necessary conditions.

$$\nabla f(\mathbf{p}^*) = \mu^* \nabla h(\mathbf{p}^*)$$
$$h(\mathbf{p}^*) \leq 0$$
$$\mu^* \geq 0.$$

This is the simplest of the Karush-Kuhn-Tucker (KKT) conditions.

When we combine equality constraints with inequality constraints, the situation can be more complicated. We begin with an example involving one constraint equality and one inequality. We consider the problem

$$\text{maximize} \quad f(x, y) \tag{13.31}$$
$$\text{subject to} \quad g(x, y) = 0 \tag{13.32}$$
$$h(x, y) \leq 0. \tag{13.33}$$

We let $\mathbf{p} = (x, y)$, $S = \{\mathbf{p} : g(\mathbf{p}) = 0\}$ and $Q = \{\mathbf{p} : h(\mathbf{p}) \leq 0\}$. The constraint set is $S \cap Q$. In optimization theory, the constraint set is often called the *feasible* set.

Again we must distinguish between the two cases.

(i) When the constraint $h \leq 0$ is active, the extreme point $\mathbf{p}^* \in \partial Q \cap S$. The gradient $\nabla h(\mathbf{p}^*)$ points out of Q. However, when there is an equality constraint, the gradient $\nabla f(\mathbf{p}^*)$ need not point out of Q. We are comparing the values of f restricted to S with the values of h restricted to S. Thus it is the component of $\nabla f(\mathbf{p}^*)$ that is tangent to S that must be a positive multiple $\mu*$ of the component of $\nabla h(\mathbf{p}^*)$ tangent to S. We illustrate the possibilities in Figure 13.5. In Figure 13.5a, we see displayed the situation in which both $\nabla f(\mathbf{p}^*)$ and $\nabla h(\mathbf{p}^*)$ point out of Q. In Figure 13.5b, we see a situation in which $\nabla h(\mathbf{p}^*)$ points out of Q, but $\nabla f(\mathbf{p}^*)$ points into Q. Nevertheless, the components of $\nabla f(\mathbf{p}^*)$ and $\nabla h(\mathbf{p}^*)$ that are perpendicular to $\nabla g(\mathbf{p}^*)$ (i.e., tangent to S) point in the same direction.

(ii) When the constraint $h \leq 0$ is inactive, $\nabla f(\mathbf{p}^*)$ will be a multiple λ^* of $\nabla g(\mathbf{p}^*)$.

These various possibilities are combined in an elegant set of necessary conditions that a solution of the problem (13.31)–(13.33) must satisfy. They are called the KKT conditions.

Theorem 13.8 Assume that $f, g,$ and h are C^1. Let $\mathbf{p}^*$ be a local maximum of the problem (13.31)–(13.33). Assume that $\nabla g(\mathbf{p}^*) \neq 0$ and that, if $h(\mathbf{p}^*) = 0$ (the constraint is active) then $\nabla h(\mathbf{p}^*)$ and $\nabla g(\mathbf{p}^*)$ are linearly independent. Then there are multipliers λ^*, and μ^* such that $\mathbf{p}^*$, λ^*, and μ^* satisfy the system (13.34)–(13.38):

$$\nabla f(\mathbf{p}^*) = \lambda^* \nabla g(\mathbf{p}^*) + \mu^* \nabla h(\mathbf{p}^*) \tag{13.34}$$

$$g(\mathbf{p}^*) = 0 \tag{13.35}$$

$$\mu^* h(\mathbf{p}^*) = 0 \tag{13.36}$$

$$\mu^* \geq 0 \tag{13.37}$$

$$h(\mathbf{p}^*) \leq 0 \tag{13.38}$$

Proof: Assume that the constraint $h \leq 0$ is inactive (i.e., $h(p^*) < 0$). Then $\mathbf{p}^*$ lies in the interior of Q. In this case, Theorem 13.1 implies that there is a

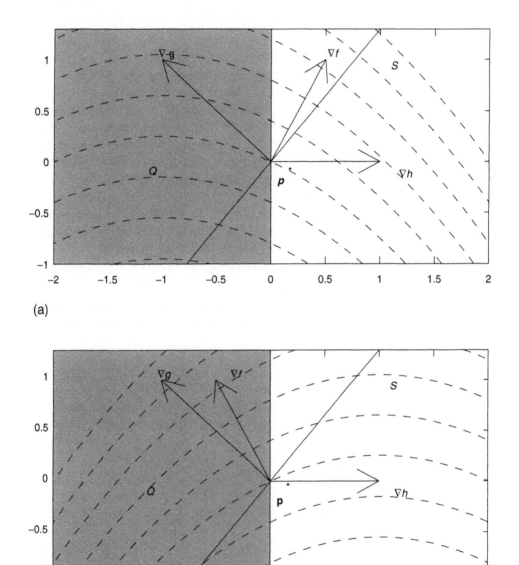

(a)

(b)

Figure 13.5 Level curves of the objective function f (dashed lines) and constraint sets Q (shaded) and S (solid line). In Figure 13.5a, $\nabla f(\mathbf{p}^*)$ points out of Q, while in Figure 13.5b, $\nabla f(\mathbf{p}^*)$ points into Q

multiplier λ^* such that $\nabla f(p^*) = \lambda^* \nabla g(p^*)$. It follows that p^*, λ^*, and $\mu^* = 0$ satisfy (13.34)–(13.38).

When the constraint is active, $h(\mathbf{p}^*) = 0$. Since $\nabla g(\mathbf{p}^*)$ and $\nabla h(\mathbf{p}^*)$ are independent, Theorem 13.2 (multiple constraints) implies that there are multipliers λ^* and μ^* such that λ^*, μ^* and $\mathbf{p}^*$ satisfy (13.34)-(13.36) and (13.38). It remains for us to show that $\mu^* \geq 0$. Suppose to the contrary that $\mu^* < 0$. Here again we use the assumption that $\nabla g(\mathbf{p}^*)$ and $\nabla h(\mathbf{p}^*)$ are independent. This means that there is some vector $\mathbf{v}$ with $\langle \nabla g(\mathbf{p}^*), \mathbf{v} \rangle = 0$ and $\langle \nabla h(\mathbf{p}^*), \mathbf{v} \rangle > 0$. Because $\mathbf{v}$ lies in the tangent plane to S at $\mathbf{p}^*$, we can use Theorem 11.14 to assert that there is a curve $\mathbf{p}(t) \in S$ such that $\mathbf{p}(0) = \mathbf{p}^*$ and $\mathbf{p}'(0) = \mathbf{v}$. Hence

$$\frac{d}{dt} f(\mathbf{p}(t))|_{t=0} = \langle \nabla f(\mathbf{p}^*), \mathbf{p}'(0) \rangle$$

$$= \lambda^* \langle \nabla g(\mathbf{p}^*), \mathbf{v} \rangle + \mu * \langle \nabla h(\mathbf{p}^*), \mathbf{v} \rangle < 0.$$

At the same time,

$$\frac{d}{dt} h(\mathbf{p}(t))|_{t=0} = \langle \nabla h(\mathbf{p}^*), \mathbf{v} \rangle > 0.$$

Hence there is a $\delta > 0$ such that, for $-\delta < t < 0$, $f(\mathbf{p}(t)) > f(\mathbf{p}^*)$, $g(\mathbf{p}(t)) = 0$, and $h(\mathbf{p}(t)) < 0$. This contradicts the assumption that $\mathbf{p}^*$ is a local maximizer of f subject to the constraints (13.32) and (13.33). The theorem is proved.

When we formulate a minimization problem, the inequality constraints are usually stated $h \geq 0$. Thus if (x^*, y^*) is a solution of the optimization problem

$$\text{minimize } f$$
$$\text{subject to } g = 0$$
$$h \geq 0,$$

and ∇g and ∇h are independent whenever the constraint $h \geq 0$ is active, then there are multipliers $\lambda = \lambda^*$ and $\mu = \mu^*$ such that (x^*, y^*), λ^*, and μ^* solve the system

$$\nabla f = \lambda \nabla g + \mu \nabla h$$
$$g = 0$$
$$\mu h = 0$$
$$\mu \geq 0$$
$$h \geq 0.$$

The systems (13.34)–(13.38) or (13.42)–(13.46) are known as the KKT conditions. We give an example of such a problem and its solution.

EXAMPLE 13.8: Let $f(x,y) = (x-2)^2 + (y+1)^2$, and let $g(x,y) = x^2 + y^2 - 1$. Our problem will be to minimize f subject to the equality constraint $g(x,y) = 0$ and two inequality constraints, $h_1(x,y) = x \geq 0$ and $h_2(x,y) = y \geq 0$. In other words, we wish to minimize f over the portion of the unit circle that lies in the first quadrant. The level curves of f are circles centered at the point $(x,y) = (2,-1)$, so it is clear by inspection that the minimum point occurs at $(1,0)$, where $f = 2$. However, we will work through the KKT conditions to arrive at this conclusion. Notice that $\nabla g(x,y) \neq 0$ on the set $S = \{(x,y) : g(x,y) = 0\}$. At the points on S at which one of the constraints $h_j \geq 0$ is active, ∇g and the gradient of the active constraint function are independent. The KKT equations are

$$
\begin{aligned}
2(x-2) &= 2\lambda x + \mu_1 \\
2(y+1) &= 2\lambda y + \mu_2 \\
x^2 + y^2 &= 1 \\
\mu_1 x &= 0 \\
\mu_2 y &= 0 \\
\mu_1, \mu_2 &\geq 0 \\
x, y &\geq 0.
\end{aligned}
$$

First we consider the possibility that the minimizer occurs at a point on the circle where both $x > 0$ and $y > 0$ (both of the constraints $x \geq 0$ and $y \geq 0$ are inactive). Because the minimizer must satisfy the KKT equations, we must have $\mu_1 = \mu_2 = 0$. In this case, the minimizer must solve the smaller system

$$
\begin{aligned}
2(x-2) &= 2\lambda x \\
2(y+1) &= 2\lambda y \\
x^2 + y^2 &= 1.
\end{aligned}
$$

However, the solutions to this system are $(x,y) = \pm(2,-2)/\sqrt{5}$, both of which fall outside of the first quadrant.

To see the effect of the constraints $x, y \geq 0$, we return to the original KKT equations. If $x = 0$, so that this constraint is active, from the first equation we have $\mu_1 = -4 < 0$, which violates the sixth line of the KKT conditions. Hence we must assume $x > 0$ and $\mu_1 = 0$.

Next we assume that the constraint $y = 0$ is active. From the third equation, we deduce that $x = \pm 1$. We can only accept $x = 1$ because of the constraint $x \geq 0$. From the second equation, because $y = 0$, we see that $\mu_2 = 2$. The first equation, with $x = 1$ and $\mu_1 = 0$, implies that $\lambda = -1$. Thus the complete solution is $(x^*, y^*) = (1, 0)$, $\lambda^* = -1$, $\mu_1^* = 0$, $\mu_2^* = 2$.

The KKT conditions provide the most information when the constraint set and the objective function are convex. In this case, the KKT conditions can be sufficient for the existence of a minimizer. More information on this question is found in the book of Nash and Sofer [N].

Exercises for 13.3

1. Let $f(x, y) = x + 2y$, and consider the problem of maximizing f subject to the contraints $g(x, y) = x^2 - y^2 = 0$ and $h(x, y) = x \leq 0$.

   a) Sketch the constraint set and the level lines of f, and find the maximizer by inspection.

   b) Does the maximizer satisfy the KKT conditions? Why or why not?

2. Let $f(x, y) = x + y$. Consider the problem of finding the minimum of f subject to the constraints $x^2 + y^4/4 = 1$ and $x \geq 0$.

   a) Formulate the KKT conditions for this problem.

   b) Find the minimizer of this problem, and verify that the KKT conditions are satisfied at the minimizer.

3. The KKT conditions are easily formulated for problems that involve more variables. Consider the problem of finding the maximizer of $f(x, y, z) = z$ subject to the constraints:

$$g(x, y, z) = z + x^2 + y^2 - 1 = 0$$
$$h_1(x, y, z) = (1/2) - x - z \leq 0$$
$$h_2(x, y, z) = x \leq 0.$$

   a) Formulate the KKT conditions for this problem.

   b) Show that there is a unique maximizer. Which of the constraint inequalities is active and which is inactive at the maximizer?

4. We return to the the spring-link system of exercise 5 of Section 13.1. Now impose the additional constraint that the spring cannot be stretched more than a certain distance. Specifically, we require that $d \leq 1.5$. Formulate the KKT conditions for this problem and solve it.

13.4 Applications in Economics

Economists use constrained optimization extensively. However, they often do not specify precisely the functions used to model a given situation. They are less concerned with determining the optimal numerical values than they are with understanding the qualitative behavior of solutions of optimization problems. Their goal is to elucidate principles of economics.

We first make a short list of some terms.

Let x denote the amount of capital a firm invests to produce a certain product, or to provide a service. This can be money used to build a plant and buy materials.

Let y be the amount of labor employed in production (in some units like man-hours).

The *production function* is the level of output when the firm uses x units of capital and y units of labor, $z = f(x, y)$. It is natural to assume that $f(0, y) = f(x, 0) = 0$. A typical production function used by economists is the Cobb-Douglas function:

$$f(x, y) = Ax^\alpha y^\beta, \quad \alpha + \beta = 1, \quad \alpha, \beta \geq 0. \tag{13.47}$$

It has the feature that it is homogeneous of degree 1, that is, $f(\lambda x, \lambda y) = \lambda f(x, y)$ for $\lambda \geq 0$. Economists say that the production function increases according to scale. If we double x and double y, the production level z will be doubled.

If we assume that the firm is the sole supplier of this product or service (a monopolist), the price it can charge for the product depends on how much is produced. Usually, as more is produced, the price will go down until the market is saturated and the firm can sell no more at any price. This process is described by a *demand function* $p(z)$, where $p(z)$ is the price that can be charged at production level z.

The *revenue* that is realized at production level z is $R(z) = p(z)z$. In terms of capital and labor, R can be expressed as

$$R(x, y) = p(f(x, y))f(x, y). \tag{13.48}$$

The *cost* of using x units of capital and y units of labor is $C(x, y) = rx + wy$, where r is the interest rate the firm pays to borrow the capital and w is the wage rate.

The *profit* is obviously

$$\pi(x, y) = R(x, y) - C(x, y) = R(x, y) - rx - wy. \tag{13.49}$$

The *rate of return* on the investment of x units of capital is

$$q(x, y) = \frac{R(x, y) - wy}{x}. \tag{13.50}$$

The Averch-Johnson Effect

This example is due to H. Averch and L. Johnson who analyzed the effect of government regulation of the telephone and telegraph industry [A].

We suppose there is a monopolist firm that is regulated by government. The form of the regulation is to limit the rate of return of the firm on its investment. We want to see how this form of regulation (constraint) affects the maximum profit and how the firm uses labor and capital to achieve the maximum profit. The optimization problem is

$$\text{maximize} \quad \pi(x, y)$$
$$\text{subject to} \quad q(x, y) \leq s$$

where s is the percentage return allowed by the regulation.

We consider only values of $s > r$, for otherwise there will be no profit. In fact if $q(x, y) \leq s \leq r$, then

$$\pi(x, y) = R(x, y) - rx - wy = xq(x, y) - rx$$
$$\leq (s - r)x$$
$$\leq 0.$$

Furthermore, we assume that $R(x, 0) = R(0, y) = 0$. Consequently, we can only have $\pi(x, y) > 0$ for $x, y > 0$. We do not need to include $x \geq 0$ and $y \geq 0$ as constraints. We assume that $\pi(x_1, y_1) > 0$ for some point (x_1, y_1).

We set

$$h(x, y, s) = R(x, y) - sx - wy = x(q(x, y) - s).$$

Then when $x > 0$, $h(x, y, s) \leq 0$ if and only if $q(x, y) \leq s$. Our problem is now

$$\text{maximize} \quad \pi(x, y) \tag{13.51}$$

$$\text{subject to} \quad h(x, y, s) \leq 0. \tag{13.52}$$

Our analysis of this problem is done in two parts. In Part I, we make a basic assumption about the revenue function $R(x, y)$, but we do not specify the production function $z(x, y)$, or how R depends on the production level z. We are able to derive considerable information about the dependence of the constrained maximizer on the allowed rate of return, s. These results are stated in Theorem 13.11.

In Part II we consider the case in which z is a Cobb-Douglas production function, and we are able to get a more complete picture of how the firm uses capital and labor to maximize its profits under the constraint of the limitation of its rate of return on investment. The more detailed results are stated in Theorem 13.12.

Part I

Let $E = \{(x, y) : x \geq 0 \text{ and } y \geq 0\}$, and let $E^o = \{(x, y) : x > 0 \text{ and } y > 0\}$.

We assume that R is continuous on E and C^2 on the interior E^o. Furthermore we assume

$$R(x, 0) = 0 \text{ for } x \geq 0, \quad R(0, y) = 0 \text{ for } y \geq 0 \quad \text{and}$$

$$0 \leq R(x, y) \leq 1. \tag{13.53}$$

In addition, we make the key assumption that the Hessian $\mathbf{D}^2 R$ is negative definite for $(x, y) \in E^o$. We assume that

$$R_{yy} < 0 \text{ and } R_{xx} R_{yy} - R_{xy}^2 > 0 \text{ for } (x, y) \in E^o. \tag{13.54}$$

Of course, this means that R is strictly concave ($-R$ is strictly convex).
  The next two lemmas explore the consequences of the hypotheses on R.

Lemma 13.9 The profit function π attains a global maximum at a unique point $(x_0, y_0) \in E^o$.

Proof: Let $K = \{(x, y) : rx + wy \leq 1\} \cap E$. K is a compact set. Since $R \leq 1$, we see that $\{(x, y) : \pi(x, y) > 0\} \subset K$. Hence π must attain a positive maximum at a point $(x_0, y_0) \in K \cap E^o$. However, π is strictly concave on E^o, and thus, by Theorem 12.5, (x_0, y_0) is the unique global maximizer of π on E^o.

  The rate of return at the unconstrained maximum is

$$s_0 = \frac{R(x_0, y_0) - rx_0 - wy_0}{x_0}.$$

From the definition of $h(x, y, s)$, we see that $h(x_0, y_0, s_0) = 0$.
  Now we turn our attention to the constrained maximization problem (13.51) and (13.52).

Lemma 13.10 For $r < s \leq s_0$, there is a unique maximizer $(x^*(x), y^*(s))$ of problem (13.51) and (13.52) that satisfies

  (i) $h(x^*(s), y^*(s), s) = 0$;
  (ii) $h_y(x^*(s), y^*(s), s) = 0$; and
  (iii) $h_x(x^*(s), y^*(s), s) < 0$.

Proof: First we show that a maximizer of (13.51) and (13.52) must satisfy (i). Suppose on the contrary that a maximizer (x^*, y^*) satisfied $h(x^*, y^*, s) < 0$. This would mean that (x^*, y^*) is a local maximizer of π in E^o. However, π has a unique local (global) maximizer in $(x_0, y_0) \in E^o$, so $(x^*, y^*) = (x_0, y_0)$. Consequently,

$$h(x^*, y^*, s) = h(x_0, y_0, s) = h(x_0, y_0, s_0) + (s_0 - s)x_0$$
$$= (s_0 - s)x_0 \geq 0,$$

which is a contradiction.
  Thus if (x^*, y^*) is a maximizer of (13.51) and (13.52), it must lie on the curve $H_s = \{(x, y) : h(x, y, s) = 0\}$, where the constraint is active. Furthermore,

$$\pi(x, y) = h(x, y, s) + (s - r)x,$$

and we are assuming that $s > r$. Thus the problem (13.51) and (13.52) is equivalent to the problem

$$\text{find } \max\{x : (x, y) \in H_s\}.   \tag{13.55}$$

Our next goal is to prove the existence of a unique maximizer of the problem (13.55). Because $h(0, 0, s) = 0$ and $h(x, y, s) < 0$ for $sx+yw > 1$, H_s is nonempty, closed, and contained in $E \cap \{(x, y) : sx + wy \le 1\}$. Thus H_s is compact. Consequently, there exists a point $(x^*, y^*) \in H_s$ with $x^* \ge x$ for all $(x, y) \in H_s$. To prove uniqueness of the maximizer, we suppose that there is a $\tilde{y} \ne y^*$ such that $(x^*, \tilde{y}) \in H_s$. Now h is also strictly concave. Hence

$$h\left(x^*, \frac{y^* + \tilde{y}}{2}, s\right) > (1/2)h(x^*, y^*, s) + (1/2)h(x^*, \tilde{y}, s) = 0.$$

However, for x sufficiently large, $h(x, \frac{y^*+\tilde{y}}{2}, s) < 0$. By the intermediate value theorem, there must be an $\tilde{x} > x^*$ such that $h(\tilde{x}, \frac{y^*+\tilde{y}}{2}, s) = 0$. This contradicts the maximality of x^*. Hence the maximizer (x^*, y^*) must be unique.

Finally we consider (ii) and (iii). First suppose that $h_y(x^*, y^*, s) \ne 0$. Because $h \in C^1$, there is a value y such that $h(x^*, y, s) > h(x^*, y^*, s) = 0$. Then, arguing as we did in the previous paragraph, there is a value $x > x^*$ such that $h(x, y, s) = 0$. However, this contradicts the assumption that $x^* \ge x$ for all $(x, y) \in H_s$. We conclude that $h_y(x^*, y^*, s) = 0$. A similar argument shows that we must have $h_x(x^*, y^*, s) \le 0$. It remains to be shown that $h_x(x^*, y^*, s) < 0$. Suppose to the contrary that $h_x(x^*, y^*, s) = 0$. Then (x^*, y^*) would be a critical point of h in E^o, and because h is strictly concave on E^o, it would be a local maximum. By Theorem 12.5, (x^*, y^*) would be the unique global maximizer of h on E^o. However, this is not possible, because $h(x^*, y^*, s) = 0$ and

$$h(x_0, y_0, s) = h(x_0, y_0, s_0) + (s - s_0)x_0 = (s - s_0)x_0 \ge 0.$$

The proof is complete.

We are now ready to prove the following theorem.

Theorem 13.11 For $r < s \le s_0$, there is a unique point $(x^*(s), y^*(s))$ on the curve $h(x, y, s) = 0$ where the solution of (13.51) and (13.52) is attained. Furthermore the function $s \to (x^*(s), y^*(s))$ is C^1, with $dx^*/ds < 0$.

Proof: We saw in Lemma 13.10 that there is a unique point $(x^*, y^*) = (x^*(s), y^*(s)) \in H_s$ at which this maximum is attained. Furthermore, Lemma 13.10 says that this point lies at the intersection of the curves

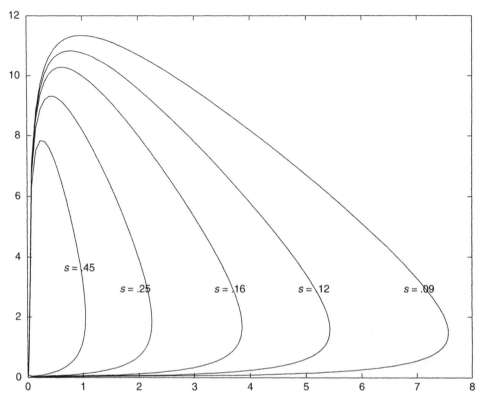

Figure 13.6 Typical curves $h(x, y, s) = 0$ for several values of s. In each case, the set H_s lies inside the curve $h = 0$ and the set G_s lies outside the curve $h = 0$

$$h_y(x, y, s) = 0 \tag{13.56}$$

$$h(x, y, s) = 0. \tag{13.57}$$

With the goal of applying the implicit function theorem to equations (13.56) and (13.57), we compute the Jacobian matrix J with respect to x and y:

$$J = \begin{bmatrix} h_{xy} & h_{yy} \\ h_x & h_y \end{bmatrix} = \begin{bmatrix} R_{xy} & R_{yy} \\ h_x & 0 \end{bmatrix}.$$

J is invertible provided $R_{yy} h_x$ is not zero at $(x^*(s), y^*(s))$. However, $R_{yy} < 0$ everywhere, because $\mathbf{D}^2 R$ is negative definite, and $h_x(x^*(s), y^*(s)) < 0$ by Lemma 13.10. The implicit function theorem implies that $s \to (x^*(s), y^*(s))$ is C^1 and

$$\frac{d}{ds}\begin{bmatrix} x^*(s) \\ y^*(s) \end{bmatrix} = -J^{-1}\begin{bmatrix} h_{ys} \\ h_s \end{bmatrix}$$

$$= \frac{1}{h_x R_{yy}}\begin{bmatrix} 0 & -R_{yy} \\ -h_x & R_{xy} \end{bmatrix}\begin{bmatrix} 0 \\ -x \end{bmatrix}.$$

Hence

$$\frac{dx^*}{ds} = \frac{x^*(s)}{h_x(x^*(s), y^*(s))} < 0.$$

The theorem is proved.

Notice that we did not use the implicit function theorem to prove the existence of the function $s \to (x^*(s), y^*(s))$, because that was already established. We used the implicit function theorem to deduce the regularity of the function and the expression for dx^*/ds.

Next, we can use the KKT equations to determine how the profit π changes along the curve of maximizers.

Corollary The maximized profit $\pi(x^*(s), y^*(s))$ decreases as s decreases.

Proof: The same argument as was used in the proof of Theorem 13.8 shows that the maximizer of the problem (13.51) and (13.52) satisfies the KKT equations (with $g = 0$):

$$\nabla\pi = \mu^*\nabla h$$
$$\mu^*h = 0$$
$$\mu^* \geq 0$$
$$h \leq 0.$$

When we calculate $(d/ds)\pi$ and use the first of the KKT equations, we find

$$\frac{d}{ds}\pi(x^*(s), y^*(s)) = \pi_x\frac{dx^*}{ds} + \pi_y\frac{dy^*}{ds}$$

$$= \mu^*\left(h_x\frac{dx^*}{ds} + h_y\frac{dy^*}{ds}\right).$$

However, because $h(x^*(s), y^*(s), s) = 0$ for $r < s \leq s_0$, we deduce that

$$h_x\frac{dx^*}{ds} + h_y\frac{dy^*}{ds} = -h_s = x^*(s).$$

Thus

$$\frac{d}{ds}\pi(x^*(s), y^*(s)) = \mu^*(s)x^*(s) \geq 0$$

because $\mu^* \geq 0$. The corollary is proved.

Part II

Our assumptions (13.53) and (13.54) on the revenue function $R(x, y)$ allowed us to deduce that as the regulation becomes more restrictive, that is, as s decreases toward r, the amount of capital used by the firm to maximize its profit will increase. What happens to the amount of labor? The expression for dy^*/ds is

$$\begin{aligned}\frac{dy^*}{ds} &= \frac{-x^* R_{xy}}{h_x R_{yy}} \\ &= -\frac{R_{xy}}{R_{yy}}\frac{dx^*}{ds}.\end{aligned} \tag{13.58}$$

In this equation, R_{xy}, R_{yy} and h_x are evaluated at $(x^*(s), y^*(x))$. We see that the rate of change of labor depends on the sign of $R_{xy}(x^*(s), y^*(s))$. To draw some conclusions about this issue, we must make some more specific assumptions about the form of R.

We assume that the production function $z = f(x, y)$ is a Cobb-Douglas function $z = x^\alpha y^\beta$ with $\alpha + \beta = 1$ and $\alpha, \beta > 0$. Then we assume that the revenue function $R(z)$ is continuous on $[0, \infty)$, C^2 on $(0, \infty)$ and that it satisfies the following conditions:

$$0 \leq R(z) \leq 1 \text{ for } z \geq 0 \text{ with } R(0) = 0, \tag{13.59}$$

$$R'(z) > 0 \text{ for } z > 0, \tag{13.60}$$

$$R''(z) < 0 \text{ for } z > 0. \tag{13.61}$$

These assumptions describe a market that becomes saturated as the production level $z \to \infty$.

First we must verify that if R satisfies (13.59)–(13.61), then $R(x, y) = R(x^\alpha y^\beta)$ satisfies (13.53) and (13.54). The condition $R(0) = 0$, and the fact that $z(x, 0) =$

$z(0, y) = 0$, implies that $R(x, 0) = R(0, y) = 0$. Next we compute the partial derivatives:

$$R_{xx} = R''(z_x)^2 + R' z_{xx} \tag{13.62}$$

$$R_{yy} = R''(z_y)^2 + R' z_{yy} \tag{13.63}$$

$$R_{xy} = R'' z_x z_y + R' z_{xy}. \tag{13.64}$$

The partial derivatives of the Cobb-Douglas production function are

$$z_x = \alpha x^{\alpha-1} y^\beta = \alpha \frac{z}{x} \quad \text{and} \quad z_y = \beta x^\alpha y^{\beta-1} = \beta \frac{z}{y}$$

$$z_{xx} = \alpha(\alpha - 1)\frac{z}{x^2}, \quad z_{yy} = \beta(\beta - 1)\frac{z}{y^2}, \quad z_{xy} = \alpha\beta\frac{z}{xy}.$$

Substitution in (13.62)–(13.64) yields

$$R_{xx} = \frac{\alpha^2 z}{x^2}\left(z R'' + \left(1 - \frac{1}{\alpha}\right) R'\right), \tag{13.65}$$

$$R_{yy} = \frac{\beta^2 z}{y^2}\left(z R'' + \left(1 - \frac{1}{\beta}\right) R'\right), \tag{13.66}$$

$$R_{xy} = \frac{\alpha\beta z}{xy}(z R'' + R'). \tag{13.67}$$

Because $\alpha < 1$ and $\beta < 1$, we see from (13.65) and (13.66) that $R_{xx} < 0$ and $R_{yy} < 0$. Furthermore, using the fact that $\alpha + \beta = 1$, we have

$$R_{xx} R_{yy} - R_{xy}^2 = -\frac{\alpha\beta z^3 R' R''}{x^2 y^2} > 0$$

by assumptions (13.60) and (13.61). Therefore, the conclusions of Lemma 13.9 and 13.10 and Theorem 13.11 hold for $R(x, y) = R(x^\alpha y^\beta)$ when R satisfies (13.59)–(13.61). In particular, the profit function $\pi(x, y)$ achieves a unique maximum at a point $(x_0, y_0) \in E$. Let $z_0 = x_0^\alpha y_0^\beta$ be the production level, and let s_0 be the rate of return at (x_0, y_0).

Theorem 13.12 Let $R(z)$ be continuous on $[0, \infty)$, C^2 on $(0, \infty)$ and satisfy conditions (13.59)–(13.61). Let $z = x^\alpha y^\beta$ be a Cobb-Douglas production function with $\alpha, \beta > 0$ and $\alpha + \beta = 1$.

Then for each s, $r < s \le s_0$, there is a unique point $(x^*(s), y^*(s))$ on the curve $h(x, y, s) = 0$ where the solution of (13.51) and (13.52) is attained. Let $z^*(s) = (x^*(s))^\alpha (y^*(s))^\beta$ be the corresponding level of production. Then the functions $s \to (x^*(s), y^*(s))$ and $s \to z^*(s)$ are C^1 with $dx^*/ds < 0$ and $dz^*/ds < 0$. If in addition

$$zR''(z) + R'(z) < 0 \quad \text{for } z \ge z_0, \tag{13.68}$$

then $dy^*/ds > 0$ for $r < s \le s_0$.

Proof: The assertion about $x^*(s)$ is already proved in Theorem 13.11. To prove that $z^*(s)$ is strictly decreasing, we first calculate

$$\frac{dz^*}{ds} = z_x \frac{dx^*}{ds} + z_y \frac{dy^*}{ds}$$

$$= \frac{\alpha z^*}{x^*} \frac{dx^*}{ds} + \frac{\beta z^*}{y^*} \frac{dy^*}{ds}.$$

We then substitute the expression (13.58) for dy^*/ds, and we simplify the result to obtain

$$\frac{dz^*}{ds} = \left(\frac{z^*}{x^* y^* R_{yy}} \right) \left(\frac{dx^*}{ds} \right) (\alpha y^* R_{yy} - \beta x^* R_{xy}). \tag{13.69}$$

Now using (13.66) and (13.67) the quantity

$$\alpha y^* R_{yy} - \beta x^* R_{xy} = -\frac{\alpha \beta z^* R'(z^*)}{y^*}.$$

Hence (13.69) becomes

$$\frac{dz^*}{ds} = -\left(\frac{\alpha \beta (z^*)^2 R'(z^*)}{x^* (y^*)^2 R_{yy}} \right) \left(\frac{dx^*}{ds} \right).$$

Because $R' > 0$, $R_{yy} < 0$ and $dx^*/ds < 0$, we conclude that $dz^*/ds < 0$.

Finally we prove the assertion about the sign of dy^*/ds. We substitute (13.67) into (13.58) to obtain

$$\frac{dy^*}{ds} = \frac{\alpha\beta z}{x^* y^*} \left(\frac{-dx^*/ds}{R_{yy}} \right) (z^* R''(z^*) + R'(z^*)). \qquad (13.70)$$

Since $R_{yy} < 0$ everywhere, and it was shown in Theorem 13.11 that $dx^*/ds < 0$, it follows that $dy^*/ds > 0$ if and only if $z^* R''(z^*) + R'(z^*) < 0$. Using the hypothesis (13.68), we see that $dy^*/ds > 0$ as long as $z^*(s) \geq z_0$. However, $z^*(s_0) = z_0$, and we have just shown that $dz^*/ds < 0$. Hence we will have $z^*(s) \geq z_0$ for $r < s \leq s_0$. The theorem is proved.

As a matter of public policy, one can argue that imposing this kind of government regulation has a bad social consequence. Firms will react by increasing capital and decreasing the use of labor, thereby costing jobs. Of course, there are many other factors, for example the tax laws, to be taken into account.

EXAMPLE 13.9:   In this example, we use the Cobb-Douglas production function $z = x^{0.5} y^{0.5}$ and the revenue function $R(z) = z/(1 + z)$. As a cost function we take $C(x, y) = 0.04x + 0.06y$.

First we verify the hypotheses (13.59)–(13.61). Clearly $R(0) = 0$ and $0 \leq R(z) \leq 1$, with $R(z)$ tending to 1 as $z \to \infty$. Next we see that

$$R'(z) = \frac{1}{(1 + z)^2} > 0$$

and

$$R''(z) = -\frac{2z}{(1 + z)^4} < 0.$$

To prove that (13.68) is satisfied, we must find the maximizer of the unregulated profit function $\pi(x, y)$. A critical point for π is a point where

$$\pi_x = R_x - 0.04 = 0$$
$$\pi_y = R_y - 0.06 = 0.$$

Now $R_x = R'(z) z_x$, and $z_x = z/(2x)$ and $z_y = z/(2y)$. Hence the equations for the critical point of π become

$$zR'(z) = 0.08x$$
$$zR'(z) = 0.12y$$

(13.71)

Therefore the critical point lies on the line $y = 2x/3$. The production function z restricted to this line is $z(x) = x\sqrt{2/3}$. We substitute this expression into (13.71), and we find that x must satisfy

$$R'(x\sqrt{2/3}) = \sqrt{6}/25 = 0.097970589.$$

In our case, with $R(z) = z/(1+z)$, we have $R'(z) = 1/(1+z)^2$, which is strictly decreasing, and $R'(0) = 1$. Hence there is exactly one solution to this equation at $x_0 = 2.6880$, whence $y_0 = 1.7920$ and $z_0 = 2.1947$. Finally we calculate

$$zR''(z) + R'(z) = \frac{1-z}{(1+z)^3} < 0 \text{ for } z > 1.$$

Thus the assumption (13.68) holds for $z \geq z_0 = 2.1947$. The location of the maximized profit for various values of s for Example 13.9 is shown in Figure 13.7.

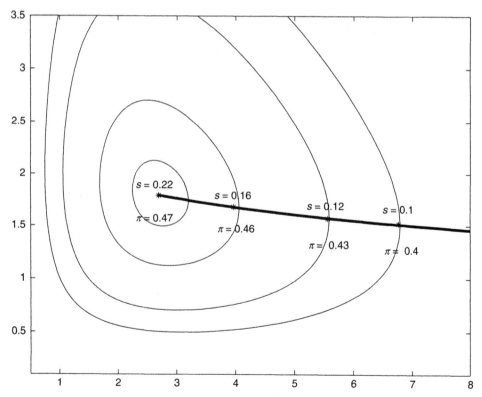

Figure 13.7 Example 13.9. Level curves of the profit function π and the curve $\pi_y = 0$ in heavier line with location of constrained maximum of profit for various values of s

Many examples of constrained optimization problems in the field of economics are provided in the book of L. Blume and C. P. Simon [Bl].

Exercises for 13.4

1. Let a revenue function $R(z) = z/\sqrt{1+z^2}$. Assume we have a Cobb-Douglas production function $z = x^\alpha y^\beta$, and a profit function $\pi(x,y) = R(z(x,y)) - rx - wy$, $r, w > 0$.

   a) Verify that R satisfies (13.59)–(13.61).

   b) Find the maximizer of the unconstrained profit function when $r = 0.04$, $w = 0.06$, and $\alpha = 0.25, \beta = 0.75$.

   c) Does this combination of R and the production function $z = x^{1/4}y^{3/4}$ satisfy (13.68)?

   In the next three exercises, we look for a general class of production functions $z(x,y)$ such that $R(x,y) = R(z(x,y))$ satisfies (13.53) and (13.54).

2. Show that for any vector $\mathbf{v} = (v_1, v_2) \in \mathbb{R}^2$,

$$\langle \mathbf{D}^2_{x,y} R(x,y)\mathbf{v}, \ \mathbf{v}\rangle$$

$$= R''(z(x,y))(v_1 z_x + v_2 z_y)^2 + R'(z(x,y))\langle \mathbf{D}_{x,y} z^2 \mathbf{v}, \ \mathbf{v}\rangle. \qquad (13.72)$$

3. Suppose that the production function $z(x,y)$ is a general function that is homogeneous of degree 1. As a function of the polar coordinates (t, θ), $t = \sqrt{x^2 + y^2}$, z has the form $z = tg(\theta)$. We assume that g is a C^2 function of θ, $0 \le \theta \le \pi/2$. To use the formula (13.72), you need to calculate the partial derivatives z_x, z_y, z_{xx}, z_{xy} and z_{yy} in terms of $t, g(\theta), g'(\theta)$ and $g''(\theta)$.
   Using (13.72), show that

$$\langle \mathbf{D}^2_{x,y} R(x,y)\mathbf{v}, \ \mathbf{v}\rangle = \frac{R''(z(x,y))}{t^2}[(xv_1 + yv_2)g(\theta) + (xv_2 - yv_1)g'(\theta)]^2$$

$$+ \frac{R'(z(x,y))}{t^3}(g(\theta) + g''(\theta))(xv_2 - yv_1)^2.$$

4. Now assume that $z \to R(z)$ satisfies (13.59)–(13.61), i.e., $0 \leq R(z) \leq 1$, $R(0) = 0$, $R'(z) > 0$, and $R''(z) < 0$. We suppose that g satisfies

$$g(0) = g(\pi/2) = 0 \quad \text{and} \quad g(\theta) + g''(\theta) < 0. \tag{13.73}$$

(It is a consequence of (13.73) that $g(\theta) > 0$ for $0 < \theta < \pi/2$.) With these assumptions and the formula of exercise 3, show that $z(t, \theta) = tg(\theta)$ is a production function such that $R(x, y) = R(z(x, y))$ satisfies (13.53) and (13.54).

5. a) The Cobb-Douglas production function is of the form discussed in exercise 3, with $g(\theta) = (\cos \theta)^\alpha (\sin \theta)^\beta$. Verify that (13.73) is satisfied.
   b) Verify that for any $A > 0$, $g(\theta) = A\theta(\pi/2 - \theta)$ satisfies (13.73). What is the corresponding production function?
   c) Another commonly used production function has the form

$$z(x, y) = xy/(\alpha x + \beta y)$$

where $\alpha, \beta > 0$. In polar coordinates, this kind of z may be written $z = tg(\theta)$ where

$$g(\theta) = \frac{\cos \theta \sin \theta}{\alpha \cos \theta + \beta \sin \theta}.$$

Show that this g also satisfies (13.73).

6. Suppose that a firm has a manager whose goal is to maximize revenue by advertising, without letting profits drop below a certain level m. Let $a \geq 0$ be the cost of advertising. Suppose that $R(y, a)$ is the revenue generated at production level y, with advertising cost a. Let $C(y)$ be the cost of producing y units of output. We assume that C and R are both C^1 functions, with $C' > 0$ and $\partial R/\partial a > 0$. The profit is

$$\pi(y, a) = R(y, a) - C(y) - a.$$

The manager wishes to maximize R with constraints

$$\pi \geq m, \quad \text{and} \quad a, y \geq 0.$$

a) Formulate this problem in terms of the KKT conditions.
b) Assuming that there is an optimal solution (y^*, a^*) with $y^* > 0$, and that the appropriate gradients are independent, show that $\pi(y^*, a^*) = m$; the profit realized is the minimum permissible.
c) Show that $R_y(y^*, a^*) > 0$ and that $\pi_y(y^*, a^*) < 0$.

7. Consider a production function of three variables x, y, z. The variables represent the amount of certain materials needed in a manufacturing process. Suppose that the production function is

$$f(x, y, z) = \left(\frac{1}{x} + \frac{4}{y} + \frac{2}{z} \right)^{-1}.$$

The price of x is w_1, the price of y is w_2 and the price of z is w_3. The cost function for the materials is

$$C(x, y, z) = w_1 x + w_2 y + w_3 z.$$

We will assume that $w_1 = 12$, $w_2 = 5$, and $w_3 = 2$. The firm wishes to minimize the cost of the materials needed to produce at least 2 units of output. The optimization problem is

$$\begin{aligned} \text{minimize} \quad & C \\ \text{subject to} \quad & f \geq 2. \end{aligned}$$

Of course, we have the added constraints $x \geq 0$, $y \geq 0$, and $z \geq 0$.

a) What are the KKT conditions for this problem?
b) Find the minimizer.
c) Suppose that the suppliers of material x charge the stated price $w_1 = 12$ only for a certain minimum purchase. Instead of the side conditions $x \geq 0$, $y \geq 0$, and $z \geq 0$, we have

$$x \geq 10 \quad \text{and} \quad y \geq 0 \quad z \geq 0.$$

Now find the minimizer.

Chapter 14

Integration in $\mathbb{R}^n$

Chapter Overview

Section 14.1 provides a brief treatment of Riemann integration over generalized rectangles in $\mathbb{R}^n$, including the Fubini theorem. It relies on the more detailed treatment of integrals in one dimension given in Chapter 7. In Section 14.2, we discuss integration over Jordan domains. Numerical methods on rectangles in $\mathbb{R}^2$ are the next topic. The most detailed treatment in this chapter is given to the change of variable theorem for integrals in n variables. This result is then combined with numerical methods to provide numerical evaluation of various integrals in two and three dimensions that fall outside the usual range of examples found in calculus books. Improper integrals on $\mathbb{R}^n$ are discussed, and some results are applied to the calculation of probability densities of several random variables.

14.1 Integration Over Generalized Rectangles

We begin with integration over a generalized rectangle where the development closely parallels that of Chapter 7.

A *generalized rectangle* $R \subset \mathbb{R}^n$ is the Cartesian product of one-dimensional closed intervals

$$R = [a_1, b_1] \times \cdots \times [a_i, b_i] \times \cdots \times [a_n, b_n]$$

with $a_i \leq b_i$. The *n-dimensional volume* of a generalized rectangle R is $\mathrm{vol}(R) = (b_1 - a_1)(b_2 - a_2) \ldots (b_n - a_n)$. When R is a rectangle in $\mathbb{R}^2$, we will use the term *area*.

We will say that a collection of generalized rectangles is *nonoverlapping* if they intersect only on faces, edges, or corners. If $F = \cup_j R_j$ is a finite union of generalized rectangles, it is clear that F may be decomposed into a finite union of nonoverlapping rectangles R_i. Then we define

$$\mathrm{vol}(F) = \sum_i \mathrm{vol}(R_i).$$

The notion of a partition of a generalized rectangle R and of a refinement of a partition are essentially the same as in one dimension. Because the notation can become cumbersome, we give the definitions in two dimensions. The reader can readily extend these ideas to higher dimensions.

Let $R = [a, b] \times [c, d]$ be a rectangle. Let P_x be a one-dimensional partition (see Chapter 7) of the interval $[a, b]$, $a = x_0 < x_1 < \cdots < x_m = b$, and let P_y be a partition of the interval $[c, d]$, $c = y_0 < y_1 < \cdots < y_n = d$. We then form the subrectangles

$$R_{ij} = \{(x, y) : x_{i-1} \leq x \leq x_i, \; y_{j-1} \leq y \leq y_j\}.$$

The collection of rectangles R_{ij}, $i = 1, \ldots, m$, $j = 1, \ldots, n$, forms a *partition P* of the rectangle R.

A *refinement* of the partition P is obtained by constructing refinements P_x' of P_x and P_y' of P_y. The collection P' of the resulting subrectangles R_{kl}' is a refinement of the partition P. Each R_{kl}' is a subrectangle of one of the subrectangles R_{ij} of the original partition P. If P and Q are both partitions of the rectangle R, then $P \cup Q$ is the common refinement of both P and Q that is formed by using the common refinements $P_x \cup Q_x$ and $P_y \cup Q_y$.

When we are dealing with the partition P of a generalized rectangle $R \subset \mathbb{R}^n$, we will denote the generalized subrectangles of the partition by R_j, using a single index.

Now we turn to the definition of the Riemann integral. Let R be a generalized rectangle, $R \subset \mathbb{R}^n$, and let $f : R \to \mathbb{R}$ be a bounded function. We form a partition P of R into generalized subrectangles R_j. Let

$$M_j = \sup_{R_j} f \quad \text{and} \quad m_j = \inf_{R_j} f.$$

We form the upper sum

$$U(f, P) = \sum_{j} M_j \text{vol}(R_j),$$

and the lower sum

$$L(f, P) = \sum_{j} m_j \text{vol}(R_j).$$

The analogues of Lemmas 7.1 and 7.2 are proved in the same way in higher dimensions. Thus for any two partitions P and Q of I,

$$L(f, P) \leq U(f, Q). \tag{14.1}$$

Therefore every upper sum is an upper bound for all of the lower sums and every lower sum is a lower bound for all of the upper sums. We define the *lower integral* of f over R as

$$\underline{\int}_R f = \sup_P L(f, P),$$

where the supremum is taken over all partitions P. The *upper integral* of f over R as

$$\overline{\int}_R f = \inf_Q U(f, Q),$$

where the infimum is taken over all partitions Q. We have

$$\underline{\int}_R f \leq \overline{\int}_R f.$$

Definition 14.1 A bounded function f is *Riemann integrable* over the generalized rectangle R if

$$\underline{\int}_R f = \overline{\int}_R f.$$

When this equality holds, we set

$$\int_R f = \int_{\underline{R}} f = \overline{\int}_R f.$$

The analogue of Theorem 7.3 is also true, and the proof is the same.

Lemma 14.1 Let R be a generalized rectangle, and let $f : R \to \mathbb{R}$ be a bounded function. Then f is Riemann integrable on R if and only if, for each $\varepsilon > 0$, there is a partition P of R such that $|U(P, f) - L(P, f)| < \varepsilon$.

As we saw in Chapter 7, another useful way of looking at the notion of the integral is to form the *Riemann sum*

$$S(f, P) = \sum_j f(\mathbf{x}_j)\mathrm{vol}(R_j),$$

where $\mathbf{x}_j$ is picked from R_j. The norm $\|P\|$ of the partition P is the maximum length of the edges of the R_j.

It is clear that for each partition P, and any choice of the points $\mathbf{x}_j$,

$$L(f, P) \le S(f, P) \le U(f, P).$$

The theorem that corresponds to Theorem 7.8 is

Theorem 14.2 Let R be a generalized rectangle, and let $f : R \to \mathbb{R}$ be bounded. The following assertions are equivalent.

(i) The function f is Riemann integrable over R.
(ii) There is number I such that, for each $\varepsilon > 0$, there is $\delta > 0$ such that $|S(f, P) - I| < \varepsilon$ for all partitions P of R with $\|P\| < \delta$. In this case, $\int_R f = I$.

The following theorems are proved exactly as they were in one dimension.

Theorem 14.3 Let $R \subset \mathbb{R}^n$ be a generalized rectangle and suppose that $f, g : R \to \mathbb{R}$ are both integrable. Then

(i) For all $a, b \in \mathbb{R}$, $af + bg$ is integrable on R and

$$\int_R (af + bg) = a \int_R f + b \int_R g.$$

(ii) If $f(\mathbf{x}) \le g(\mathbf{x})$ for all $\mathbf{x} \in R$, then

$$\int_R f \le \int_R g.$$

Theorem 14.4 Let $R \subset \mathbb{R}^n$, and suppose that $f : R \to \mathbb{R}$ is bounded and integrable on R. Then $\mathbf{x} \to |f(\mathbf{x})|$ is integrable on R, with

$$\left| \int_R f \right| \le \int_R |f|.$$

Theorem 14.5 Let $R \subset \mathbb{R}^n$ be a generalized rectangle, and suppose that $f : R \to \mathbb{R}$ is continuous. Then f is integrable over R.

In the next section, we make an important extension of Theorem 14.5 to allow f to be discontinuous on a "small" set.

To be able to use the techniques of one-dimensional integration, we must be able to express integrals over rectangles in terms of integrals in one variable at a time. This is the point of the theorem of Fubini. We will write points in $\mathbb{R}^n$ as $(\mathbf{x}, \mathbf{y})$, where $\mathbf{x} \in \mathbb{R}^k$ and $\mathbf{y} \in \mathbb{R}^{n-k}$. A generalized rectangle $R \subset \mathbb{R}^n$ will be written as the Cartesian product $R = R_1 \times R_2$, with $R_1 \subset \mathbb{R}^k$ and $R_2 \subset \mathbb{R}^{n-k}$.

Theorem 14.6 (Fubini) Let f be bounded and integrable over the generalized rectangle $R \subset \mathbb{R}^n$. For each $\mathbf{x} \in R_1$, we assume that $\mathbf{y} \to f(\mathbf{x}, \mathbf{y})$ is integrable on R_2, and we define

$$F(\mathbf{x}) = \int_{R_2} f(\mathbf{x}, \mathbf{y}) \, d\mathbf{y}.$$

Then $F(\mathbf{x})$ is integrable over R_1, and

$$\int_R f = \int_{R_1} F(\mathbf{x}) \, d\mathbf{x} = \int_{R_1} \left[\int_{R_2} f(\mathbf{x}, \mathbf{y}) \, d\mathbf{y} \right] d\mathbf{x}.$$

Proof: To make the notation simpler, we give the proof in two variables x and y. First we note that $F(x)$ is bounded because f is bounded on R, and for all $a \le x \le b$,

$$|F(x)| \le \int_c^d |f(x, y)| dy \le (d - c) \sup_R |f|.$$

We appeal to the integrability criterion Lemma 14.1 to show that $F(x)$ is integrable. Let the rectangle $R = [a, b] \times [c, d]$, and let P be a partition of R into sub-rectangles $R_{ij} = \{x_{i-1} \leq x \leq x_i, \, y_{j-1} \leq y \leq y_j\}$ where $a = x_0 < \cdots < x_m = b$ is a partition P_x of $[a, b]$ and $c = y_0 < \cdots < y_n = d$ is a partition P_y of $[c, d]$. We let

$$M_{ij} = \sup_{R_{ij}} f \quad \text{and} \quad m_{ij} = \inf_{R_{ij}} f$$

with

$$N_i = \sup_{[x_{i-1}, x_i]} F \quad \text{and} \quad n_i = \inf_{[x_{i-1}, x_i]} F.$$

Now for $x \in [x_{i-1}, x_i]$,

$$m_{ij}(y_j - y_{j-1}) \leq \int_{y_{j-1}}^{y_j} f(x, y) dy \leq M_{ij}(y_j - y_{j-1}).$$

Then summing over j, we have

$$\sum_{j=1}^{n} m_{ij}(y_j - y_{j-1}) \leq F(x) \leq \sum_{j=1}^{n} M_{ij}(y_j - y_{j-1}).$$

Since this is true for all $x \in [x_{i-1}, x_i]$, we deduce that

$$\sum_{j=1}^{n} m_{ij}(y_j - y_{j-1}) \leq n_i \leq N_i \leq \sum_{j=1}^{n} M_{ij}(y_j - y_{j-1}).$$

Next we multiply each of these inequalities, $i = 1, \ldots, m$, by $(x_i - x_{i-1})$, and we sum over i to obtain

$$\sum_{i,j} m_{ij}\text{area}(R_{ij}) \leq \sum_{i=1}^{m} n_i(x_i - x_{i-1})$$

$$\leq \sum_{i=1}^{m} N_i(x_i - x_{i-1}) \leq \sum_{i,j} M_{ij}\text{area}(R_{ij}).$$

Thus for any partition P of the rectangle R, we have

$$L(f, P) \leq L(F, P_x) \leq U(F, P_x) \leq U(f, P). \qquad (14.2)$$

Let $\varepsilon > 0$ be given. Because f is integrable over R, Lemma 14.1 tells us that we may choose P such that $U(f, P) - L(f, P) < \varepsilon$. Then the inequality (14.2) implies

$$U(F, P_x) - L(F, P_x) \leq U(f, P) - L(f, P) < \varepsilon.$$

A second application of Lemma 14.1 tells us that F is integrable on $[a, b]$. Furthermore, we may use (14.2) to conclude that

$$\int_R f = \sup_P L(f, P) \leq \int_a^b F(x)dx \leq \inf_P U(f, P) = \int_R f,$$

which is to say,

$$\int_R f = \int_a^b \left[\int_c^d f(x, y)dy \right] dx.$$

The theorem is proved.

Corollary Let $R \subset \mathbb{R}^n$ be a generalized rectangle, and let $f : R \to \mathbb{R}$ be continuous. Then $\int_R f$ may be calculated using iterated integrals in either order:

$$\int_R f = \int_a^b \int_c^d f(x, y)dy \, dx = \int_c^d \int_a^b f(x, y)dx \, dy.$$

Exercises for 14.1

1. Let $R \subset \mathbb{R}^2$ be a rectangle, and suppose that $f : R \to \mathbb{R}$ is integrable. Suppose that $f(x, y) \geq 0$ for all points $(x, y) \in R$ with x and y rational. Show that $\int_R f \geq 0$.

2. Let $R \subset \mathbb{R}^n$ be a generalized rectangle, and suppose that $f(\mathbf{x}) = g(\mathbf{x})h(\mathbf{x})$ where $g, h : R \to \mathbb{R}$ are both continuous. For a partition P of R into subrectangles R_j, we construct the sums

$$T(f, P) = \sum_j g(\mathbf{x}_j)h(\mathbf{y}_j)\mathrm{vol}(R_j).$$

where $\mathbf{x}_j$ and $\mathbf{y}_j$ are chosen in R_j. If it were the case that $\mathbf{x}_j = \mathbf{y}_j$, these would be Riemann sums, but we do not assume that. Show that the sums $T(f, P)$ converge to $\int_R f$ as $\|P\|$ tends to zero.

3. Let $R = [a, b] \times [a, b]$ be a square in $\mathbb{R}^2$. Suppose that $f : R \to \mathbb{R}$ is bounded and that f is continuous on the triangles $\{a \le x < y \le b\}$ and $\{a \le y < x \le b\}$. Show that f is integrable on R by appealing to the integrability criterion of Lemma 14.1.

4. Suppose that $R = [a, b] \times [a, b]$, that $f(x, y)$ is continuous on $\{a \le x \le y \le b\}$ and that $f = 0$ for $x > y$. In view of exercise 3, f is integrable on R.

a) For each $x \in [a, b]$, verify that $y \to f(x, y)$ is integrable on $[a, b]$, and set $F(x) = \int_a^b f(x, y)dy$.

b) Use the Fubini theorem to deduce that

$$\int_R f = \int_a^b \left[\int_a^x f(x, y)dy \right] dx.$$

5. Let $f(x, y)$ be a continuous function on $R = [a, b] \times [a, b]$. Show carefully how the Fubini theorem can be used to justify the formula

$$\int_a^b \int_a^x f(x, y)dy\,dx = \int_a^b \int_y^b f(x, y)dx\,dy.$$

6. Let $u(x)$ be continuous on $\{x \ge 0\}$. Show that for $x \ge 0$,

$$\int_0^x \int_0^t u(s)ds\,dt = \int_0^x (x - s)u(s)ds.$$

14.2 Integration Over Jordan Domains

Up to this point, the development of the Riemann integral in several dimensions follows the development in one dimension exactly. However, it should be noted that we have not stated an analogue to Theorem 7.9, which says that

$\int_a^b f = \int_a^c f + \int_c^b f$ for $a < c < b$. Now we must introduce a new notion to deal with the fact that the sets we integrate over may be much more complicated than generalized rectangles. This new notion is *Jordan content*.

Definition 14.2 Let $F \subset \mathbb{R}^n$ be a bounded set. F has *Jordan content zero*, if for each $\varepsilon > 0$, there is a finite collection of generalized rectangles R_j, $j = 1, \ldots, N$, such that $F \subset \cup_{j=1}^N R_j$ and

$$\sum_{j=1}^N \text{vol}(R_j) < \varepsilon.$$

We say that a bounded set $D \subset \mathbb{R}^n$ is a *Jordan domain* if its boundary ∂D is a set of zero Jordan content.

> **EXAMPLE 14.1:**   The graph of a continuous function $y = f(x)$, $a \leq x \leq b$, is a set with zero Jordan content in $\mathbb{R}^2$. Let $\varepsilon > 0$ be given. Using the fact that f is uniformly continuous on $[a, b]$, there is a $\delta > 0$ such that $|f(x) - f(y)| < \varepsilon$ whenever $x, y \in [a, b]$ with $|x - y| < \delta$. Choose N so large that $h = (b - a)/N < \delta$, and make a partition of $[a, b]$ with $x_j = a + jh$, $j = 0, 1, \ldots N$. Let
>
> $$R_j = \{(x, y) : x_{j-1} \leq x \leq x_j,\ |y - f(x_j)| \leq \varepsilon\}.$$
>
> Then $(x, f(x)) \in R_j$ for $x_{j-1} \leq x \leq x_j$, so the graph of f is contained in $\cup_{j=1}^N R_j$. Because area $(R_j) = 2\varepsilon(x_j - x_{j-1})$, we have $\sum_{j=1}^N$ area $(R_j) = 2\varepsilon(b - a)$.
>   In Figure 14.1, we see the graph of $y = f(x) = x^2$ on the interval $0 \leq x \leq 1$ covered by small rectangles.

In a similar fashion, we can show that the graph of a continuous function $z = f(x, y)$ over a rectangle $R = \{a \leq x \leq b,\ c \leq y \leq d\}$ has zero Jordan content in $\mathbb{R}^3$.

It is an obvious consequence of the definition that, if $A \subset B$ and B has zero Jordan content, then A also has zero Jordan content. We leave it to the reader in the exercises for this section to prove that finite intersections and unions of sets of zero Jordan content again have zero Jordan content.

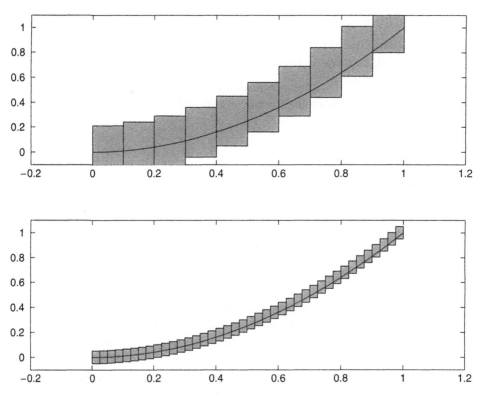

Figure 14.1 Graph of $y = f(x) = x^2$. In the upper figure, the graph is covered by 10 rectangles of total area $\varepsilon = 0.4$ In the lower figure, it is covered by 40 rectangles of total area $\varepsilon = 0.1$

The basic theorem for Riemann integration in higher dimensions is

Theorem 14.7 Let f be a bounded function that is continuous on a generalized rectangle R except possibly on a set F of Jordan content zero. Then f is Riemann integrable over R.

Proof: To make the proof more readable, we will give it in two dimensions. The proof in higher dimensions is similar.

We assume that the rectangle $R = [a, b] \times [c, d]$. Let $\varepsilon > 0$ be given. According to Definition 14.2, there is a finite collection of rectangles R_j such that $F \subset \cup_{j=1}^{N} R_j$ and $\sum_{j=1}^{N} \mathrm{vol}(R_j) < \varepsilon$. Subdividing some of the R_j if necessary, we can assume that the R_j are nonoverlapping. Thus $R_j = [a_{j-1}, a_j] \times [c_{j-1}, c_j]$ with the numbers a_j forming a partition P_x of $[a, b]$ and the numbers c_j forming a partition P_y of $[c, d]$. Using P_x and P_y, we form a partition of R that includes the R_j as subrectangles. We can write $R = S \cup T$ where $S = \cup_{j=1}^{N} R_j$ and where T is the union of the remaining subrectangles in P. The set T is compact and f is continuous on T because $F \subset S$. Consequently f is uniformly continuous on T. This means that there is a $\delta > 0$ such that, if $\mathbf{p}_1 = (x_1, y_1)$ and $\mathbf{p}_2 = (x_2, y_2) \in T$ with $\|\mathbf{p}_1 - \mathbf{p}_2\|_\infty < \delta$, then $|f(\mathbf{p}_1) - f(\mathbf{p}_2)| < \varepsilon$. Make a refinement P' of P such that $\|P'\| < \delta$. We will show that f is Riemann integrable on R by an appeal to Lemma 14.1. We denote the subrectangles of P' by R'_k. They may be divided into two mutually exclusive categories: those that are contained in S and those that are contained in T. Let $M_k = \sup_{R'_k} f$, $m_k = \inf_{R'_k} f$, and let $M = \sup_R |f|$. We have

$$U(f, P') - L(f, P') = \sum_k (M_k - m_k)\text{area}(R'_k)$$

$$= \sum_{R'_k \subset S} (M_k - m_k)\text{area}(R'_k) + \sum_{R'_k \subset T} (M_k - m_k)\text{area}(R'_k).$$

The first sum on the right is bounded by

$$2M \sum_{R'_k \subset S} \text{area}(R'_k) = 2M \text{ area}(S) < 2M\varepsilon.$$

The second sum on the right is bounded by

$$\varepsilon \sum_{R'_k \subset T} \text{area}(R'_k) \le \varepsilon \text{ area}(R).$$

We conclude that

$$U(f, P') - L(f, P') < (2M + \text{area}(R))\varepsilon,$$

which implies that f is Riemann integrable on R.

The following result shows that the values of f on a set of zero Jordan content do not affect the value of the integral $\int_R f$.

Theorem 14.8 Let $R \subset \mathbb{R}^n$ be a generalized rectangle, and let $f, g : R \to \mathbb{R}$ be bounded integrable functions. If $f(\mathbf{x}) = g(\mathbf{x})$ except for $\mathbf{x}$ in a set F of zero Jordan content, then $\int_R f = \int_R g$.

Proof: The function $f - g$ is bounded and integrable on R and is zero except for $\mathbf{x} \in F$. For $\varepsilon > 0$, there is a finite collection of generalized rectangles R_j

such that $F \subset \cup R_j$ and $\sum_j \text{vol}(R_j) < \varepsilon$. As in the proof of Theorem 14.7, we embed this collection of generalized rectangles in a partition P of R. If M is a bound for $|f - g|$ on R, we have

$$-\varepsilon M \le L(f - g, P) \le U(f - g, P) \le \varepsilon M.$$

Since ε is arbitrary, we conclude that $\int_R - \int_R g = \int_R f - g = 0$. The theorem is proved.

Theorem 14.7 allows us to define the integral of a continuous function f over a Jordan domain D. We first enclose D in a generalized rectangle R. Next we define an extension of f to R:

$$\tilde{f}(\mathbf{x}) = \begin{cases} f(\mathbf{x}) & \text{for } \mathbf{x} \in D \\ 0 & \text{for } \mathbf{x} \notin D \end{cases}.$$

We note that $\tilde{f}$ is continuous on R except possibly on the boundary of D, which is a set of Jordan content zero.

Definition 14.3 Let $D \subset \mathbb{R}^n$ be contained in a generalized rectangle R, and let $f : D \to \mathbb{R}$ be continuous on the interior of D. The integral of $\tilde{f}$ over R is defined, and we set

$$\int_D f(\mathbf{x}) \, d\mathbf{x} = \int_R \tilde{f}(\mathbf{x}) \, d\mathbf{x}.$$

Of course, for this definition to be unambiguous, it must be shown that the value of the integral is independent of the enclosing rectangle R. This point is left to the exercises.

We use this approach to define the volume of a Jordan domain. Let the set $D \subset \mathbb{R}^n$. We assume that D is bounded and can be enclosed in some generalized rectangle R. The *characteristic function* of D is

$$\chi_D(\mathbf{x}) = \begin{cases} 1 & \text{for } \mathbf{x} \in D \\ 0 & \text{for } \mathbf{x} \notin D \end{cases}.$$

Definition 14.4 We define the n-dimensional *volume* of a Jordan domain to be

$$\text{vol}(D) = \int_R \chi_D(\mathbf{x}) \, d\mathbf{x}$$

where R is any generalized rectangle that contains D.

A comment is in order regarding this definition. For consistency, we need to know that a set D has zero Jordan content if and only if D is a Jordan domain with $\text{vol}(D) = 0$. This is a rather technical exercise that we leave to the reader.

EXAMPLE 14.2:

a) In view of Example 14.1, we deduce that a bounded set $D \subset \mathbb{R}^2$ is a Jordan domain, and hence has area, if the boundary of D consists of a finite set of the graphs of continuous functions $y = f(x)$ or $x = g(y)$.

b) A bounded set $D \subset \mathbb{R}^3$ is a Jordan domain, and hence has volume, if the boundary of D consists of a finite set of graphs of continuous functions $z = f(x, y)$, $y = g(x, z)$, or $x = h(y, z)$.

The next theorem records the properties of Jordan domains that we will need.

Theorem 14.9

(i) If D is a Jordan domain, then, for each $\varepsilon > 0$, there is a collection of nonoverlapping generalized rectangles $R_j, j = 1, \ldots, N$ and a subcollection $j = 1, \ldots, M$, $M < N$, such that

$$\cup_{j=1}^{M} R_j \subset D \subset \cup_{j=1}^{N} R_j$$

and such that

$$\text{vol}(D) - \varepsilon \leq \sum_{j=1}^{M} \text{vol}(R_j) \leq \sum_{j=1}^{N} \text{vol}(R_j) \leq \text{vol}(D) + \varepsilon.$$

(ii) If A and B are Jordan domains in $\mathbb{R}^n$ with $A \subset B$, then $\text{vol}(A) \leq \text{vol}(B)$.

(iii) If A and B are Jordan domains in $\mathbb{R}^n$ such that $A \cap B$ is a set of Jordan content zero, then $A \cup B$ is a Jordan domain, and

$$\text{vol}(A \cup B) = \text{vol}(A) + \text{vol}(B).$$

(iv) Suppose that A and B are Jordan domains in $\mathbb{R}^n$ satisfying the hypotheses of part (iii). If $f : A \cup B \to \mathbb{R}$ is continuous on the interior of $A \cup B$, then

$$\int_{A \cup B} f = \int_A f + \int_B f.$$

Proof: (i) The proof is very similar to that of Theorem 14.8. We leave the proof of parts (i) and (ii) to the exercises. We will see that (iii) is a consequence of (iv).

We now prove (iv). Let R be a generalized rectangle containing both A and B. Next note that $\partial(A \cup B) \subset \partial A \cup \partial B$. Because ∂A and ∂B both have zero Jordan content, so does $\partial(A \cup B)$. Hence $A \cup B$ is a Jordan domain. We define

$$f_1(\mathbf{x}) = \begin{cases} f(\mathbf{x}) & \text{for } \mathbf{x} \in A \\ 0 & \text{for } \mathbf{x} \notin A \end{cases},$$

and

$$f_2(\mathbf{x}) = \begin{cases} 0 & \text{for } \mathbf{x} \notin B \\ f(\mathbf{x}) & \text{for } \mathbf{x} \in B \end{cases}.$$

Let $\tilde{f}$ be the extension by zero outside $A \cup B$. The function f_1 is continuous on the interior of A, and f_2 is continuous on the interior of B. Hence both f_1 and f_2 are integrable with

$$\int_R f_1 = \int_A f \quad \text{and} \quad \int_R f_2 = \int_B f.$$

The functions $\tilde{f}$ and $f_1 + f_2$ agree except possibly on the set $A \cap B$. Since this set is assumed to have zero Jordan content, we conclude from Theorem 14.8 that

$$\int_{A \cup B} f = \int_R \tilde{f} = \int_R (f_1 + f_2) = \int_A f + \int_B f.$$

The proof of (iv) is complete. The assertion (iii) follows from (iv) by taking $f = \chi_{A \cup B}$, whence $f_1 = \chi_A$ and $f_2 = \chi_B$.

The following mean value theorem for integrals is the higher-dimensional analogue of Theorem 7.12. It is useful whenever we want to relate the pointwise behavior of a function to its integral.

Theorem 14.10 Let $K \subset \mathbb{R}^n$ be a compact Jordan domain that is pathwise connected, and let $f : K \to \mathbb{R}$ be continuous. Let $g : K \to \mathbb{R}$ be a bounded function that satisfies $g(\mathbf{x}) \geq 0$ for all $\mathbf{x} \in K$, and suppose that g is continuous except possibly on a set of Jordan content zero. Then there is a point $\mathbf{z} \in K$ such that

$$\int_K f(\mathbf{x})g(\mathbf{x})\, d\mathbf{x} = f(\mathbf{z}) \int_K g(\mathbf{x})\, d\mathbf{x}.$$

Proof: Because K is compact and f is continuous, f attains a maximum at some point $\mathbf{v} \in K$ and attains a minimum at some point $\mathbf{u} \in K$. Because $g \geq 0$, we have $\int_K g \geq 0$ and

$$f(\mathbf{u})g(\mathbf{x}) \leq f(\mathbf{x})g(\mathbf{x}) \leq f(\mathbf{v})g(\mathbf{x})$$

for all $\mathbf{x} \in K$. This immediately implies the integral inequalities

$$f(\mathbf{u}) \int_K g(\mathbf{x})\, d\mathbf{x} \leq \int_K f(\mathbf{x})g(\mathbf{x})\, d\mathbf{x} \leq f(\mathbf{v}) \int_K g(\mathbf{x})\, d\mathbf{x}.$$

If $\int_K g = 0$, then $\int_K fg = 0$ so the result is true for all $\mathbf{z} \in K$. If $\int_K g > 0$, we may divide to find

$$f(\mathbf{u}) \leq \frac{\int_K f(\mathbf{x})g(\mathbf{x})\, d\mathbf{x}}{\int_K g(\mathbf{x})\, d\mathbf{x}} \leq f(\mathbf{v}).$$

By the intermediate value theorem for continuous functions (Theorem 9.16), there is a point $\mathbf{z} \in K$ such that

$$f(\mathbf{z}) = \frac{\int_K f(\mathbf{x})g(\mathbf{x})\, d\mathbf{x}}{\int_K g(\mathbf{x})\, d\mathbf{x}}.$$

This concludes the proof of the theorem.

When $g(\mathbf{x}) \equiv 1$, the right-hand side of this equation is what we call the average of the function over the set K. The average of a continuous function over a set K converges to $f(\mathbf{x}_0)$ if the set shrinks to the point $\mathbf{x}_0$.

Corollary Let $f : G \to \mathbb{R}$ be continuous where G is some open set of $\mathbb{R}^n$. Let $\mathbf{x}_0 \in G$ and recall that $B(\mathbf{x}_0, \delta) = \{\mathbf{x} : \|\mathbf{x} - \mathbf{x}_0\| < \delta\}$. The norm can be any of the equivalent norms. Then

$$\lim_{\delta \to 0} \frac{\int_{B(\mathbf{x}_0, \delta)} f(\mathbf{x})\, d\mathbf{x}}{\mathrm{vol}(B(\mathbf{x}_0, \delta))} = f(\mathbf{x}_0). \tag{14.3}$$

Exercises for 14.2

1. Let R be a rectangle in $\mathbb{R}^2$, and let $f : R \to \mathbb{R}$ be continuous. Following the procedure of Example 14.1, show that the graph of f is a set of zero content in $\mathbb{R}^3$.
2. Show that if $F_i, i = 1, \ldots, N$ is a collection of sets of zero Jordan content, then $\cup_i F_i$ and $\cap_i F_i$ are both sets of zero Jordan content.
3. Show that in Definition 14.3, the value of $\int_D f$ is independent of the enclosing generalized rectangle R.
4. Recall that it was desirable to show that the usage of the terms *content* and *volume* is consistent. To do this, show that a bounded set $F \subset \mathbb{R}^n$ has zero Jordan content if and only if F is a Jordan domain with $\mathrm{vol}(F) = 0$.
5. For any two sets $A, B \subset \mathbb{R}^n$, show that $\partial(A \cup B) \subset \partial A \cup \partial B$.
6. Show that if A and B are Jordan domains with $A \subset B$, then $\mathrm{vol}(A) \le \mathrm{vol}(B)$.
7. Let $D_0 \subset \mathbb{R}^n$ be a compact Jordan domain, and let $u, v : D_0 \to \mathbb{R}$ be continuous functions with $u(\mathbf{x}) \le v(\mathbf{x})$ for all $\mathbf{x} \in D_0$. Define $D \subset \mathbb{R}^{n+1}$ to be

$$D = \{(\mathbf{x}, y) : \mathbf{x} \in D_0, \text{ and } u(\mathbf{x}) \le y \le v(\mathbf{x})\}.$$

   a) Show that D is a bounded Jordan domain.
   b) Use the Fubini theorem to show that if f is continuous on D, then

$$\int_D f = \int_{D_0} \left[\int_{u(\mathbf{x})}^{v(\mathbf{x})} f(\mathbf{x}, y) dy \right] d\mathbf{x}.$$

8. Give an example in which the mean value theorem for integrals (Theorem 14.10) fails when K is not pathwise connected.

14.3 Numerical Methods

In Section 7.4 we discussed some of the simpler, but effective, methods of making numerical estimates of the values of integrals that cannot be done analytically. In this section, we see how these methods can be used to estimate integrals in two and three dimensions. We recall the three quadrature rules from Section 7.4 and the expressions for the error. Let the interval $[a, b]$ be divided into n equal subintervals of length $h = (b - a)/n$. We denote the partition points by $x_0 = a$, $x_i = a + ih$, $i = 1, \ldots, n$.

The midpoint rule is,

$$\int_a^b f(x)dx = h \sum_{i=1}^n f\left(\frac{x_i + x_{i-1}}{2}\right) + h^2 \frac{(b-a)}{24} f''(\xi). \tag{14.4}$$

The trapezoid rule is

$$\int_a^b f(x)dx = \frac{h}{2}[f(x_0) + 2f(x_1) + 2f(x_2) + \cdots + 2f(x_{n-1}) + f(x_n)]$$
$$-h^2 \frac{(b-a)}{12} f''(\xi). \tag{14.5}$$

Simpson's rule (n must be even) is

$$\int_a^b f(x)dx = \frac{h}{3}[f(x_0) + 4f(x_1) + 2f(x_2) + \cdots + 4f(x_{n-1}) + f(x_n)]$$
$$-h^4 \frac{(b-a)}{180} f^{iv}(\xi). \tag{14.6}$$

In each case, we have given an expression for the error in terms of a derivative of the function and the length h of the subintervals.

How can these methods be extended to higher dimensions? We will limit our discussion to two-dimensional methods over a rectangle $R = \{a \leq x \leq b, \ c \leq y \leq d\}$. Introduce mesh coordinates $x_i = a + ih$, $i = 0, \ldots, m$ with $h = (b - a)/m$ as before, and $y_j = c + jk$, $j = 0, \ldots, n$ where $k = (d - c)/n$. The mesh defines subrectangles $R_{i,j} = \{x_{i-1} \leq x \leq x_i, \ y_{j-1} \leq y \leq y_j\}$. The two-dimensional versions of the midpoint rule, the trapezoid rule, and Simpson's rule can be generated by polynomial interpolation for the simple rule, and then extended to a compound rule in the manner of Section 7.4. Since polynomial interpolation in several variables would take us too far afield, we will generate these rules by considering iterated integrals. The resulting compound rules have obvious geometric interpretations.

The compound midpoint rule is easily generalized to two dimensions by taking the midpoint of each subrectangle,

$$\mathbf{p}_{i,j} = \left(\frac{x_{i-1} + x_i}{2}, \frac{y_{j-1} + y_j}{2} \right),$$

and then forming the double sum

$$\sum_{i,j} f(\mathbf{p}_{i,j}) \text{area}(R_{i,j}) = hk \sum_{i,j} f(\mathbf{p}_{i,j}). \tag{14.7}$$

We can get an idea of the error in two dimensions by looking at functions $f(x, y)$ in the form of a product, $f(x, y) = r(x)s(y)$. Let us abbreviate the one-dimensional midpoint rule (14.4) as

$$\int_a^b f(x)dx = M_1(f) + E_{M_1}(f),$$

where the error term $E_{M_1}(f) = (b-a)(h^2/24)f''(\xi)$. Now suppose that $f(x, y) = r(x)s(y)$. Then by the Fubini theorem, we have

$$\iint_R f(x, y)dxdy = \int_a^b \int_c^d r(x)s(y)\, dydx$$

$$= \int_a^b r(x)dx \int_c^d s(y)dy$$

$$= \big(M_1(r) + E_{M_1}(r)\big)\big(M_1(s) + E_{M_1}(s)\big)$$

$$= M_1(r)M_1(s) + M_1(r)E_{M_1}(s) + M_1(s)E_{M_1}(r)$$

$$+ E_{M_1}(r)E_{M_1}(s).$$

The product $M_1(r)M_1(s)$ is just the midpoint rule M_2 for f on the two-dimensional rectangle. In fact,

$$M_1(r)M_1(s) = hk \sum_i r\left(\frac{x_i + x_{i-1}}{2}\right) \sum_j s\left(\frac{y_j + y_{j-1}}{2}\right)$$

$$= hk \sum_{i,j} f\left(\frac{x_i + x_{i-1}}{2}, \frac{y_j + y_{j-1}}{2}\right)$$

$$= hk \sum_{i,j} f(\mathbf{p}_{i,j}) = M_2(f),$$

where $\mathbf{p}_{i,j}$ are the midpoints of the subrectangles $R_{i,j}$. The leading terms in the error are

$$M_1(s)E_{M_1}(r) + M_1(r)E_{M_1}(s) = M_1(s)\frac{h^2}{24}(b - a)r''(\xi) + M_1(r)\frac{k^2}{24}(d - c)s''(\eta).$$
$$(14.8)$$

We have dropped the term $E_{M_1}(r)E_{M_1}(s)$ in the error because it is on the order of $h^2 k^2$ and hence much smaller than the other terms. We conclude that the error of the midpoint rule in two dimensions is $O(h^2 + k^2)$ when f is a product. This behavior of the error can easily be shown to be true for sums of products, $f(x, y) = \sum_j r_j(x)s_j(y)$. It can also be shown to be true for general functions $f(x, y)$ that have a sufficient number of continuous derivatives.

Now we turn to the trapezoid rule. The single panel trapezoid rule in one dimension is

$$T_1(f) = (b - a)\left(\frac{f(a) + f(b)}{2}\right).$$

We take the average value of f at the two endpoints of the interval and multiply by the length of the interval. In the compound trapezoid rule (14.5) we compute this average over each subinterval $[x_{i-1}, x_i]$ and then add over $i = 1, \dots, n$.

The averaging process just described suggests how to generalize the trapezoid rule to two dimensions. Let the subrectangle be

$$R_{i,j} = \{x_{i-1} \le x \le x_i, \; y_{j-1} \le y \le y_j\}.$$

Form the average of the values of f at the corners of this subrectangle, and multiply by the area hk of $R_{i,j}$:

$$hk\left(\frac{f(x_{i-1}, y_{j-1}) + f(x_i, y_{j-1}) + f(x_{i-1}, y_j) + f(x_i, y_j)}{4}\right).$$

Now we sum these averages over all of the subrectangles $R_{i,j}$. Thus our candidate for a two-dimensional trapezoid rule is

$$T_2(f) = \left(\frac{hk}{4}\right) \sum_{i=1, j=1}^{m,n} f(x_{i-1}, y_{j-1}) + f(x_i, y_{j-1}) + f(x_{i-1}, y_j) + f(x_i, y_j). \quad (14.9)$$

It is convenient for computation to write the one-dimensional trapezoid rule in the form of a scalar product. Let the *trapezoid vector* be the vector with $m+1$ components

$$\mathbf{t} = [1, 2, 2, \cdots, 2, 1],$$

and let $\mathbf{f}$ denote the $m+1$ vector of values

$$\mathbf{f} = [f(x_0), f(x_1), f(x_2), \dots, f(x_m)].$$

Then the one-dimensional trapezoid rule can be written

$$T_1(f) = \left(\frac{h}{2}\right)\langle \mathbf{t}, \mathbf{f}\rangle.$$

The two-dimensional trapezoid rule can conveniently be written as a summation over elements of a matrix. By analogy with an iterated integral, we can apply the one-dimensional trapezoid rule in an iterated fashion. Let $\mathbf{t}^x = [1, 2, \dots, 2, 1]$ be an $m+1$ vector, and let $\mathbf{t}^y = [1, 2, \dots, 2, 1]$ be an $n+1$ vector. First apply the trapezoid rule in the y variable:

$$T_1^y(f) = \left(\frac{k}{2}\right)\sum_{j=0}^{n} t_j^y f(x, y_j).$$

Then apply T_1 again to each term in this sum:

$$T_1^x(T_1^y(f)) = \left(\frac{hk}{4}\right) \sum_{i=0}^{m} t_i^x \sum_{j=0}^{n} t_j^y f(x_i, y_j)$$

$$\hspace{4.5cm} = \left(\frac{hk}{4}\right) \sum_{i,j=0}^{m,n} t_i^x t_j^y f(x_i, y_j).$$

(14.10)

The sum in (14.9) can be rewritten to agree with (14.10). Now let $\mathbf{f}$ denote the $(m + 1) \times (n + 1)$ matrix of function values

$$\mathbf{f} = [f(x_i, y_j)].$$

Let the *trapezoid matrix* be the $(m + 1) \times (n + 1)$ matrix

$$\mathbf{t}_2 = [t_i^x t_j^y] = \begin{bmatrix} 1 & 2 & 2 & \cdot & \cdot & 2 & 1 \\ 2 & 4 & 4 & \cdot & \cdot & 4 & 2 \\ 2 & 4 & 4 & \cdot & \cdot & 4 & 2 \\ \cdot & \cdot & \cdot & \cdot & \cdot & \cdot & \cdot \\ \cdot & \cdot & \cdot & \cdot & \cdot & \cdot & \cdot \\ 2 & 4 & 4 & \cdot & \cdot & 4 & 2 \\ 1 & 2 & 2 & \cdot & \cdot & 2 & 1 \end{bmatrix}.$$

If we think of both $\mathbf{t}^x$ and $\mathbf{t}^y$ as column vectors, the rank 1 matrix $\mathbf{t}_2$ can be written as the outer product

$$\mathbf{t}_2 = \mathbf{t}^x(\mathbf{t}^y)^T.$$

For two matrices $\mathbf{A}$ and $\mathbf{B}$ of the same dimensions, we use the notation $\mathbf{A} : \mathbf{B}$ to denote the sum

$$\mathbf{A} : \mathbf{B} = \sum_{i,j} a_{ij} b_{ij}.$$

If $\mathbf{A}$ and $\mathbf{B}$ are considered as vectors in $\mathbb{R}^{mn}$, then $\mathbf{A} : \mathbf{B}$ is simply the scalar product in $\mathbb{R}^{mn}$.

Now with this notation, we can write the two-dimensional trapezoid method as

$$T_2(f) = \left(\frac{hk}{4}\right)\mathbf{t}_2 : \mathbf{f} \tag{14.11}$$

where the right-hand side is the sum in (14.10).

By applying T_2 to a product $f(x, y) = r(x)s(y)$, we can deduce that the leading terms of the error expression for the two-dimensional trapezoid method are

$$E_{T_2}(f) = -\frac{h^2}{12}r''(\xi)T_1(s) - \frac{k^2}{12}s''(\eta)T_1(r)\cdots .$$

More generally, it can be shown that the error $E_{T_2}(f) = O(h^2 + k^2)$, just as in the case of the midpoint rule.

The trapezoid rule can be easily extended to any number of dimensions. A two- and three-dimensional Simpson's rule can be derived in a similar fashion and will be done in the exercises.

We can bound the error in each of the methods by using the expression for the error. For example, in the case of the one-dimensional midpoint rule, we have

$$|E_{M_1}(f)| \le (b - a)\frac{h^2}{24}\max_{[a,b]}|f''|.$$

If it is easy to compute the second derivative of f, this may give a useful estimate of the error. In the case of Simpson's rule, we would have to find a bound for the fourth derivative of f over the interval $[a, b]$, and this may not be practical.

In practice, we would like to be able to estimate the error by observing the computed values produced by the method, without having to resort to higher derivatives of the function. Here is how we can make an *estimate* of the error in the case of the midpoint or trapezoid rule. Let $T_1(f, n)$ be the result of using the trapezoid rule with n subintervals. Then

$$\int_a^b f(x)dx = T_1(f, n) + E_{T_1}(f, n),$$

where

$$E_{T_1}(f, n) = -(b - a)\frac{h^2}{12}f''(\xi).$$

Now if we double the number of subintervals, we replace h by $h/2$ in the error expression. Assume that f changes slowly, so that f'' is nearly constant. Then

$$E_{T_1}(f, 2n) = -(b - a)\frac{h^2}{4 \cdot 12} f''(\eta)$$

$$\approx (1/4)E_{T_1}(f, n).$$

Hence

$$\int_a^b f(x)dx = T_1(f, 2n) + E_{T_1}(f, 2n)$$

$$\approx T_1(f, 2n) + (1/4)E_{T_1}(f, n).$$

Comparing the two expressions for $\int_a^b f dx$, we have

$$T_1(f, n) + E_{T_1}(f, n) \approx T_1(f, 2n) + (1/4)E_{T_1}(f, n)$$

whence

$$E_{T_1}(f, n) \approx (4/3)[T_1(f, 2n) - T_1(f, n)]. \tag{14.12}$$

Thus we use (14.12) as an estimate of the error with n subintervals, and we take $T_1(f, 2n)$ as the estimate of the integral. This is not an error bound but rather an *error estimate*. The same argument applies to the midpoint rule.

The error expressions in the midpoint and trapezoid rules in two dimensions behave like $h^2 + k^2$. Hence replacing h by $h/2$ and k by $k/2$ approximately reduces the error by a factor of $1/4$. We can use the same error estimate (14.12) in two dimensions.

In the case of Simpson's rule, because the error goes as h^4, we can estimate the error in the same way as for the trapezoid rule, but without the factor of $4/3$. This will be seen in the exercises.

EXAMPLE 14.3: Let $f(x, y) = \sin(x + y)$. We will use the two-dimensional trapezoid rule to estimate the integral of f over the rectangle $R = \{0 \leq x \leq 2, 0 \leq y \leq 4\}$. With $n = 10, m = 20$, the result is $t_1 = 0.42903$. With $n = 20, m = 40$ the result is $t_2 = 0.43119$. The exact result is

$$\int \int_R \sin(x + y) \, dx dy = -\sin(6) + \sin(4) + \sin(2) = 0.43191.$$

The actual error is

$$t_1 - 0.43191 = -0.00287,$$

and the estimate of the error is

$$(4/3)(t_1 - t_2) = (4/3)(-0.00287) = -0.00383.$$

The methods described here for two- and three-dimensional integrals are not practical for higher dimensions, because the number of function evaluations becomes prohibitively large. Indeed, when $n = 4$, and we choose a mesh with 100 points on a side, the midpoint rule will require 10^8 function evaluations. Monte Carlo methods, that take a probabilistic approach to the value of the integral, are favored in this situation. See the book by W. H. Press and colleagues [P].

Exercises for 14.3

1. Let R be the rectangle $\{a \leq x \leq b, c \leq y \leq d\}$. Let $f(x, y)$ be continuous on R. Let $Q(x, y)$ be the bilinear function $q(x, y) = Axy + Bx + Cy + D$.

   a) Find the coefficients A, B, C, D so that Q agrees with (interpolates) f at the four corners of R.

   b) Show that the single-panel two-dimensional trapezoid rule arises by integrating Q over the rectangle R:

   $$\text{area}(R)\frac{f(a, c) + f(b, c) + f(a, d) + f(c, d)}{4} = \int \int_R Q(x, y)\, dx dy.$$

2. Let R be the rectangle $\{0 \leq x \leq h, 0 \leq y \leq k\}$, and let $f \in C^2(R)$. The midpoint of R is $\mathbf{p}_0 = (h/2, k/2)$. In two variables, Taylor's theorem with $n = 1$ (see equation (12.7)), can be written, with $\mathbf{p} = (x, y)$,

   $$\begin{aligned}
   f(\mathbf{p}) = {} & f(\mathbf{p}_0) + f_x(\mathbf{p}_0)(x - h/2) + f_y(\mathbf{p}_0)(y - k/2) \\
   & + (1/2)[f_{xx}(\mathbf{p}')(x - h/2)^2 + 2f_{xy}(\mathbf{p}')(x - h/2)(y - k/2) \\
   & + f_{yy}(\mathbf{p}')(y - k/2)^2],
   \end{aligned}$$

   where $\mathbf{p}'$ lies on the line between $\mathbf{p}_0$ and $\mathbf{p} = (x, y)$. Integrate this expansion over the rectangle R. The first term in the integrated expansion is $f(\mathbf{p}_0)hk$, which is the midpoint rule for R with $m = n = 1$.

a) Show that the next two terms involving $f_x(\mathbf{p}_0)$ and $f_y(\mathbf{p}_0)$ integrate to zero.

b) Show that the error, $\int \int_R f - hkf(\mathbf{p}_0)$, can be bounded in absolute value by

$$hk \left(\frac{h^2 + k^2}{24} + \frac{hk}{16} \right) \max_R \{|f_{xx}|, |f_{xy}|, |f_{yy}|\}.$$

c) Now let R be the general rectangle $R = \{a \leq x \leq b, c \leq y \leq d\}$. Partition R into equal-sized h by k subrectangles R_{ij}, $1 \leq i \leq m$, $1 \leq j \leq n$, with $h = (b - a)/n$, $k = (d - c)/m$. Use the error bound of part b) on each subrectangle, and then sum over i, j. Show that the error of the midpoint rule over R may be bounded

$$|E_{n,m}| \leq (b - a)(d - c) \left(\frac{h^2 + k^2}{24} + \frac{hk}{16} \right) \max_R \{|f_{xx}|, |f_{xy}|, |f_{yy}|\}.$$

Notice that the leading terms of this error bound are consistent with equation (14.8).

3. Use the approach of iterated integration to derive the two-dimensional form of Simpson's rule. The Simpson matrix s_2 is the $m+1 \times n+1$ matrix such that the two-dimensional rule can be expressed in the form

$$S_2(f) = \left(\frac{hk}{9} \right) s_2 : \mathbf{f}.$$

Show that $s_2 = (\mathbf{s}^x)' \cdot \mathbf{s}^y$ where $\mathbf{s}^x$ is the $m + 1$ Simpson vector in the x variable and $\mathbf{s}^y$ is the $n + 1$ Simpson vector in the y variable.

4. Write down formulas for the three-dimensional versions of the trapezoid and the midpoint rules.

5. a) Write a code to implement the two-dimensional trapezoid rule.

b) Apply your code to estimate the value of the following integrals with an error estimate of 10^{-4}.

$$(i) \ \int_0^2 \int_1^3 e^{\cos(xy)} \, dx dy \qquad (ii) \ \int_0^{\pi/2} \int_0^{\pi/2} \sqrt{\sin(x + y)} \, dx dy.$$

6. Let $S_1(f, n)$ (n even) be Simpson's rule with $h = (b-a)/n$. An expression for the error

$$E_{S_1(f,n)} = \int_a^b f(x)dx - S_1(f, n)$$

is given in (14.6). Show that we can estimate

$$E_{S_1(f,n)} \approx \frac{16}{15}(S_1(f, 2n) - S_1(f, n)).$$

14.4 Change of Variable in Multiple Integrals

For many reasons, it is important to be able to have a change of variable formula for multiple integrals. In particular, if we are integrating over a nonrectangular region, and we wish to use the trapezoid rule or Simpson's rule over a rectangle, we must be able to map the region onto a rectangle. The change of variable formulas also plays an important role in continuum mechanics and in statistics.

We recall the change of variable formula (7.15) in one dimension. Let $\varphi : [\alpha, \beta] \to \mathbb{R}$ be a C^1 function. Let $f : [c, d] \to \mathbb{R}$ be continuous, and suppose that $\varphi([\alpha, \beta]) \subset [c, d]$. Then

$$\int_{\varphi(\alpha)}^{\varphi(\beta)} f(x)dx = \int_\alpha^\beta f(\varphi(t))\varphi'(t)dt.$$

The function φ determines the limits of integration $\varphi(\alpha)$ and $\varphi(\beta)$. It could be that $\varphi(\alpha) < \varphi(\beta)$ or $\varphi(\alpha) > \varphi(\beta)$. However, in many situations, we start with an integral $[a, b]$, with $a < b$. We choose $\varphi : [\alpha, \beta] \to [a, b]$ with $\varphi(\alpha) = a$ and $\varphi(\beta) = b$ and assume that $\varphi'(t) \geq 0$. In this case we write

$$\int_a^b f(x)dx = \int_\alpha^\beta f(\varphi(t))\varphi'(t)dt.$$

On the other hand, if $\varphi(\alpha) = b$ and $\varphi(\beta) = a$, and we assume that $\varphi' \leq 0$, we have

$$\int_a^b f(x)dx = \int_\beta^\alpha f(\varphi(t))\varphi'(t)dt$$

$$= -\int_\alpha^\beta f(\varphi(t))\varphi'(t)dt.$$

We can combine both cases by writing the formula as

$$\int_a^b f(x)dx = \int_\alpha^\beta f(\varphi(t))|\varphi'(t)|dt. \tag{14.13}$$

This is the form in which we will seek the change of variable formula in higher dimensions.

Before we can begin to talk about a change of variable formula in higher dimensions, we must be sure that a change of variable will keep us within the class of Jordan domains.

At times it will be more convenient to use a generalized rectangle with all edges having the same length. We will call such a generalized rectangle a *cube*. In terms of the $\| \ \|_\infty$ norm, the cube centered at $\mathbf{a}$ with edges of length 2δ, is

$$K = \{\mathbf{x} : \|\mathbf{x} - \mathbf{a}\|_\infty \leq \delta\}.$$

Theorem 14.11 Let G be an open set in $\mathbb{R}^n$, and let $\mathbf{g} : G \to \mathbb{R}^n$ be a C^1 function that is one-to-one on G, with $\mathbf{Dg}(\mathbf{x})$ invertible for all $\mathbf{x} \in G$. If $D \subset G$ is a compact Jordan domain, then the image $\mathbf{g}(D)$ is also a compact Jordan domain.

Proof: Because $\mathbf{g}$ is continuous, the image set $\mathbf{g}(D)$ is compact. Hence $\mathbf{g}(D)$ is bounded. To show that $\mathbf{g}(D)$ is a Jordan domain, we must show that $\partial \mathbf{g}(D)$ is a set of zero Jordan content. Since D is a Jordan domain, ∂D is a set of zero Jordan content. Hence for a given $\varepsilon > 0$, there is a finite collection of generalized rectangles R_j such that $D \subset \cup_j R_j$ and $\sum \text{vol}(R_j) < \varepsilon$. Each of these generalized rectangles is contained in a finite union of cubes K_{jk} such that $\sum_k \text{vol}(K_{jk}) \leq 2\text{vol}(R_j)$. Thus we may assume that there is a finite collection of cubes K_l such that $\partial D \subset \cup_l K_l$ and $\sum_l \text{vol}(K_l) < \varepsilon$. Furthermore, by making the cubes smaller if necessary, we can assume that $\cup_l K_l \subset G$. Let $\mathbf{x}_l$ be the center of K_l so that

$$K_l = \{\mathbf{x} : \|\mathbf{x} - \mathbf{x}_l\|_\infty \leq \delta_l\}.$$

The volume of K_l is $(2\delta_l)^n$. The union of the cubes, $\cup_l K_l$, is a compact set, and $\mathbf{Dg}$ is continuous. Hence by the mean value theorem (Theorem 10.12), there is an $M \geq 0$ such that

$$\|\mathbf{g}(\mathbf{x}) - \mathbf{g}(\mathbf{y})\|_\infty \leq M\|\mathbf{x} - \mathbf{y}\|_\infty$$

for all $\mathbf{x}, \mathbf{y} \in \cup_l K_l$. Let $\mathbf{y}_l = \mathbf{g}(\mathbf{x}_l)$. We see that $\mathbf{g}(K_l)$ is contained in the cube

$$\tilde{K}_l = \{\mathbf{y} : \|\mathbf{y} - \mathbf{y}_l\|_\infty \leq M\delta_l\}.$$

Now we make the key observation that the boundary of $\mathbf{g}(D)$ is contained in the image $\mathbf{g}(\partial D)$ (exercise 1 of this section). Thus

$$\partial \mathbf{g}(D) \subset \mathbf{g}(\partial D) \subset \cup_l \mathbf{g}(K_l) \subset \cup_l \tilde{K}_l.$$

It follows that

$$\sum_l \mathrm{vol}(\tilde{K}_l) \leq \sum_l (2M\delta_l)^n = M^n \sum_l \mathrm{vol}(K_l) < M^n \varepsilon.$$

We conclude that $\mathbf{g}(\partial D)$ has zero Jordan content.

We shall begin our derivation of the change of variable formula by concentrating on the case of linear mappings. To get some idea of what to expect, recall what happens to area and volume under a linear transformation. Suppose that Q is a rectangle, $Q = \{0 \leq s \leq \alpha, 0 \leq t \leq \beta\}$, and that a linear tranformation $\mathbf{T}$ has the 2×2 matrix $\mathbf{A} = [a_{ij}]$ in the standard basis. Then the image $R = \mathbf{T}(Q)$ is the parallelogram with sides given by the vectors

$$\mathbf{p} = \alpha[a_{11}, a_{21}] \quad \text{and} \quad \mathbf{q} = \beta[a_{12}, a_{22}].$$

The area of R is

$$\begin{aligned}
\mathrm{area}(R) &= \|\mathbf{p} \times \mathbf{q}\| \\
&= \alpha\beta|a_{11}a_{22} - a_{12}a_{21}| \\
&= \mathrm{area}(Q)|\det(\mathbf{A})|.
\end{aligned}$$

In three dimensions, we consider a box $Q = \{0 \leq s \leq \alpha, 0 \leq t \leq \beta, 0 \leq u \leq \gamma\}$ and a linear transformation $\mathbf{T}$ with 3×3 matrix $\mathbf{A} = [a_{ij}], i, j = 1, 2, 3$. Now the image $R = \mathbf{T}(Q)$ is the parallelepiped with edges given by the vectors

$$\mathbf{p} = \alpha[a_{11}, a_{21}, a_{31}], \quad \mathbf{q} = \beta[a_{12}, a_{22}, a_{23}], \quad \mathbf{r} = \gamma[a_{13}, a_{23}, a_{33}].$$

The area of the face spanned by the vectors $\mathbf{p}$ and $\mathbf{q}$ is $\|\mathbf{t}\|$ where $\mathbf{t} = \mathbf{p} \times \mathbf{q}$. Next we compute the magnitude of the projection of $\mathbf{r}$ onto $\mathbf{t}$, given by

$$\frac{|\langle \mathbf{r}, \mathbf{t} \rangle|}{\|\mathbf{t}\|}.$$

Finally, the volume of R is computed by multiplying this magnitude by the area of the face spanned by $\mathbf{p}$ and $\mathbf{q}$.

$$\begin{aligned}
\text{volume}(R) &= \|\mathbf{t}\| \frac{|\langle \mathbf{t}, \mathbf{r} \rangle|}{\|\mathbf{t}\|} \\
&= |\langle \mathbf{r}, (\mathbf{p} \times \mathbf{q}) \rangle| \\
&= \alpha\beta\gamma |\det(\mathbf{A})| \\
&= |\det(\mathbf{A})| \text{volume}(Q).
\end{aligned}$$

The reader may already be convinced by these examples that a nonsingular linear mapping in any dimension changes the volume of a generalized rectangle by multiplying by the absolute value of the determinant. Because a Jordan domain can be approximated arbitrarily well by a union of nonoverlapping rectangles, it is not a great leap to believe that a linear mapping changes the volume of a Jordan domain by the same factor. Those readers who wish to go to the heart of the matter and who are impatient with details, should assume Theorem 14.15 and go directly to Lemma 14.16.

The road to Theorem 14.15 will be traveled in several steps. We will use the coordinates of the standard basis $\mathbf{e}_1 = [1, 0, \ldots, 1], \ldots, \mathbf{e}_n = [0, 0, \ldots, 0, 1]$.

We first consider the case of a linear transformation that changes only one coordinate.

Lemma 14.12 Let $\mathbf{T}$ be a nonsingular linear transformation that changes only one coordinate, and let $R \subset \mathbb{R}^n$ be a generalized rectangle. Let $\mathbf{A} = [a_{ij}]$, $i, j = 1, \ldots, n$ be the matrix of $\mathbf{T}$ in the standard basis. Then

$$\text{volume}(\mathbf{T}(R)) = |\det(\mathbf{A})| \, \text{volume}(R). \tag{14.14}$$

Proof: Without loss of generality, we assume that $\mathbf{T}$ changes only the first coordinate. Then the matrix of $\mathbf{T}$ is of the form

$$\mathbf{A} = \begin{bmatrix} a_{11} & a_{12} & \cdots & a_{1n} \\ 0 & 1 & \cdots & 0 \\ \cdot & \cdot & \cdots & \cdot \\ \cdot & \cdot & \cdots & \cdot \\ 0 & 0 & \cdots & 1 \end{bmatrix}.$$

We note that $\det \mathbf{A} = a_{11}$. Now write the generalized rectangle $R = J \times K$ where $J = [\alpha, \beta]$ and K is a generalized rectangle in $\mathbb{R}^{n-1}$. We split a vector $\mathbf{x} \in \mathbb{R}^n$ in the same way: $\mathbf{x} = (x_1, \hat{\mathbf{x}})$ where $\hat{\mathbf{x}} = (x_2, \dots, x_n)$. We also write the first row of $\mathbf{A}$ as $\mathbf{a} = (a_{11}, \hat{\mathbf{a}})$ where $\hat{\mathbf{a}} = (a_{12}, \cdots a_{1n})$. Thus $\mathbf{y} = \mathbf{Tx}$ means that $y_1 = a_{11}x_1 + \langle \hat{\mathbf{a}}, \hat{\mathbf{x}} \rangle$, and $\hat{\mathbf{y}} = \hat{\mathbf{x}}$.

For the moment, we assume $a_{11} > 0$. In most cases, $\mathbf{T}(R)$ is not a generalized rectangle. For each $\hat{\mathbf{y}} \in K$, the first coordinate y_1 runs between $a(\hat{\mathbf{y}}) = a_{11}\alpha + \langle \hat{\mathbf{a}}, \hat{\mathbf{y}} \rangle$ and $b(\hat{\mathbf{y}}) = a_{11}\beta + \langle \hat{\mathbf{a}}, \hat{\mathbf{y}} \rangle$. Hence $\mathbf{T}(R)$ is contained in the generalized rectangle $L \times K$ where

$$L = [\ \min_{\hat{\mathbf{y}} \in K} a(\hat{\mathbf{y}}), \quad \max_{\hat{\mathbf{y}} \in K} a(\hat{\mathbf{y}})\ ].$$

Let $\chi(\mathbf{y})$ be the characteristic function of $\mathbf{T}(R)$. In other words, $\chi(\mathbf{y}) = 1$ when $\mathbf{y} \in \mathbf{T}(R)$ and $\chi(\mathbf{y}) = 0$ when $\mathbf{y} \notin \mathbf{T}(R)$. Since a linear transformation is C^1, we know that $\mathbf{T}(R)$ is a Jordan domain that has volume, and the volume is given by

$$\text{vol}(\mathbf{T}(R)) = \int_{L \times K} \chi(\mathbf{y}) d\mathbf{y}.$$

Using the Fubini theorem, we can write this integral as an iterated integral,

$$\begin{aligned} \text{vol}(\mathbf{T}(R)) &= \int_K \int_L \chi(y_1, \hat{\mathbf{y}}) dy_1 d\hat{\mathbf{y}} \\ &= \int_K \int_{a(\hat{\mathbf{y}})}^{b(\hat{\mathbf{y}})} dy_1\ d\hat{\mathbf{y}} \\ &= a_{11}(\beta - \alpha) \int_K d\hat{\mathbf{y}} \\ &= a_{11}(\beta - \alpha)\text{vol}(K) \\ &= \det \mathbf{A}\ \text{vol}(R). \end{aligned}$$

If $a_{11} = \det \mathbf{A} < 0$, the same calculation yields $\text{vol}(\mathbf{T}(R)) = -\det \mathbf{A}\,\text{vol}(R)$. This finishes the proof of the lemma.

Lemma 14.13 Let $\mathbf{T}$ be a nonsingular linear transformation of $\mathbb{R}^n$ that changes only one coordinate. Let $\mathbf{A}$ be the matrix of $\mathbf{T}$. If $D \subset \mathbb{R}^n$ is a Jordan domain, then $\text{vol}(\mathbf{T}(D)) = |\det \mathbf{A}|\,\text{vol}(D)$.

Proof: According to Theorem 14.9, for each $\varepsilon > 0$, we may approximate D by a finite collection of nonoverlapping generalized rectangles R_j so that $D \subset \cup R_j$ and

$$\sum \text{vol}(R_j) = \text{vol}(\cup R_j) \leq \text{vol}(D) + \varepsilon/|\det \mathbf{A}|.$$

Now by Theorem 14.11, $\mathbf{T}(D)$ is a Jordan domain. Furthermore the R_j are nonoverlapping, so the sets $\mathbf{T}(R_j)$ intersect in sets of zero Jordan content. Hence by parts (ii) and (iii) of Theorem 14.9,

$$\text{vol}(\mathbf{T}(D)) \leq \text{vol}(\mathbf{T}(\cup R_j))$$

$$= \sum \text{vol}(\mathbf{T}(R_j))$$

$$= |\det \mathbf{A}| \sum \text{vol}(R_j).$$

The last equality is a consequence of Lemma 14.12. Therefore

$$\text{vol}(\mathbf{T}(D)) \leq |\det \mathbf{A}|\,\text{vol}(D) + \varepsilon. \tag{14.15}$$

Because $\mathbf{T}(D)$ is also a Jordan domain, we may approximate it by a finite union of nonoverlapping rectangles Q_i so that $T(D) \subset \cup Q_i$ and

$$\sum \text{vol}(Q_i) = \text{vol}(\cup Q_i) \leq \text{vol}(\mathbf{T}(D)) + \varepsilon.$$

Then $D = \mathbf{T}^{-1}(\mathbf{T}(D))$ so that

$$\text{vol}(D) \leq \text{vol}(\cup \mathbf{T}^{-1}(Q_i))$$

$$= \sum_i \frac{\text{vol}(Q_i)}{|\det \mathbf{A}|}$$

$$\leq \frac{1}{|\det \mathbf{A}|}(\text{vol}(\mathbf{T}(D)) + \varepsilon),$$

whence

$$|\det \mathbf{A}| \, \mathrm{vol}(D) \le \mathrm{vol}(\mathbf{T}(D)) + \varepsilon. \qquad (14.16)$$

Combining (14.15) and (14.16), we see that

$$|\det \mathbf{A}| \, \mathrm{vol}(D) - \varepsilon \le \mathrm{vol}(T(D)) \le |\det \mathbf{A}| \, \mathrm{vol}(D) + \varepsilon.$$

Because ε is arbitrary, Lemma 14.15 is proved.

The following lemma is proved in Appendix II.

Lemma 14.14 Let $\mathbf{A}$ be a nonsingular $n \times n$ matrix. Then $\mathbf{A}$ can be factored

$$\mathbf{A} = \mathbf{B}_n \mathbf{E}_n \mathbf{B}_{n-1} \mathbf{E}_{n-1} \cdots \mathbf{B}_1 \mathbf{E}_1 \qquad (14.17)$$

where $\mathbf{B}_j$ changes only the jth coordinate and $\mathbf{E}_j$ is an elementary matrix that interchanges two rows.

Theorem 14.15 Let $\mathbf{T}$ be a linear transformation of $\mathbb{R}^n$ with $n \times n$ nonsingular matrix $\mathbf{A}$. If D is a Jordan domain, then

$$\mathrm{vol}(\mathbf{T}(D)) = |\det \mathbf{A}| \, \mathrm{vol}(D).$$

Proof: Let the matrix $\mathbf{A}$ be factored as in equation (14.17). The linear transformation $\mathbf{S}_j$ with matrix $\mathbf{E}_j$ interchanges two coordinates, and hence preserves volume. Let $\mathbf{T}_j$ be the linear transformation with matrix $\mathbf{B}_j$. Then $\mathbf{T}$ can be factored

$$\mathbf{T} = \mathbf{T}_n \mathbf{S}_n \mathbf{T}_{n-1} \mathbf{S}_{n-1} \cdots \mathbf{T}_1 \mathbf{S}_1.$$

We will use Lemma 14.13 repeatedly. For a Jordan domain D,

$$\begin{aligned}
\mathrm{vol}(\mathbf{T}(D)) &= |\det \mathbf{B}_n| \mathrm{vol}(\mathbf{T}_{n-1} \mathbf{S}_{n-1} \cdots \mathbf{T}_1 \mathbf{S}_1(D)) \\
&= |\det \mathbf{B}_n| |\det \mathbf{B}_{n-1}| \mathrm{vol}(\mathbf{T}_{n-2} \mathbf{S}_{n-2} \cdots \mathbf{T}_1 \mathbf{S}_1(D)) \\
&= |\det \mathbf{B}_n| |\det \mathbf{B}_{n-1}| \cdots |\det \mathbf{B}_1| \, \mathrm{vol}(D) \\
&= |\det \mathbf{A}| \, \mathrm{vol}(D).
\end{aligned}$$

The proof of Theorem 14.15 is complete.

Our next goal is to extend the result of Theorem 14.15 to a nonlinear change of coordinates. First we must make an estimate that compares the effect of a nonlinear mapping $\mathbf{g}$ on a small volume with the effect of the linear approximation $\mathbf{Dg}$ on the same small volume.

Lemma 14.16 Let G be an open set in $\mathbb{R}^n$. Let $\mathbf{g} : G \to \mathbb{R}^n$ be C^1 and one-to-one, with $\mathbf{Dg}(\mathbf{x})$ invertible for all $\mathbf{x} \in G$. Let $\mathbf{x}_0 \in G$ be given, and let $\mathbf{A} = \mathbf{Dg}(\mathbf{x}_0)$. Then for each ε, there is a $\delta > 0$ such that, if $K \subset G$ is a cube with center $\mathbf{x}_0 \in K$ and edge of length 2δ, we have

$$|\det \mathbf{A}|(1 - \varepsilon)^n \mathrm{vol}(K) \le \mathrm{vol}(\mathbf{g}(K)) \le |\det \mathbf{A}|(1 + \varepsilon)^n \mathrm{vol}(K). \qquad (14.18)$$

Proof: First we simplify the problem using translations. Set $\tilde{\mathbf{g}}(\mathbf{x}) = g(\mathbf{x} + \mathbf{x}_0) - g(\mathbf{x}_0)$. We see that $\tilde{\mathbf{g}}(0) = 0$ and that, for any cube K centered at $\mathbf{x}_0$, $\mathrm{vol}(g(K)) = \mathrm{vol}(\tilde{\mathbf{g}}(K))$. We further simplify the problem as we did in the proof of the inverse function theorem. We set $\mathbf{h}(\mathbf{x}) = \mathbf{A}^{-1}\tilde{\mathbf{g}}(\mathbf{x})$. Since $\mathbf{D}\tilde{\mathbf{g}}(0) = \mathbf{Dg}(\mathbf{x}_0) = \mathbf{A}$, we see that $\mathbf{Dh}(0) = \mathbf{I}$.

The main idea is to use the linear approximation of $\mathbf{h}$ at $\mathbf{x} = 0$:

$$\mathbf{h}(\mathbf{x}) = \mathbf{x} + \mathbf{R}(\mathbf{x})$$

where $\mathbf{R}(\mathbf{x}) = o(\|\mathbf{x}\|)$. We have

$$\|\mathbf{x}\|_\infty - \|\mathbf{R}(\mathbf{x})\|_\infty \le \|\mathbf{h}(\mathbf{x})\|_\infty \le \|\mathbf{x}\|_\infty + \|\mathbf{R}(\mathbf{x})\|_\infty.$$

Now for $\varepsilon > 0$, there is a $\delta > 0$ such that $\|\mathbf{R}(\mathbf{x})\|_\infty \le \varepsilon\|\mathbf{x}\|_\infty$ when $\|\mathbf{x}\|_\infty \le \delta$. Hence for $\|\mathbf{x}\|_\infty \le \delta$,

$$(1 - \varepsilon)\|\mathbf{x}\|_\infty \le \|\mathbf{h}(\mathbf{x})\|_\infty \le (1 + \varepsilon)\|\mathbf{x}\|_\infty. \qquad (14.19)$$

The inequalities (14.19) imply that if K is a cube centered at 0 with $K \subset K_\delta$, we have

$$(1 - \varepsilon)K \subset \mathbf{h}(K) \subset (1 + \varepsilon)K. \qquad (14.20)$$

Now recall that $\tilde{\mathbf{g}}(\mathbf{x}) = \mathbf{T}(\mathbf{h}(\mathbf{x}))$ where $\mathbf{T}(\mathbf{x}) = \mathbf{Ax}$ is the linear transformation with matrix $\mathbf{A}$. Hence the inclusions (14.20) yield

$$(1 - \varepsilon)\mathbf{T}(K) \subset \tilde{\mathbf{g}}(K) \subset (1 + \varepsilon)\mathbf{T}(K)$$

so that

$$(1 - \varepsilon)^n \mathrm{vol}(\mathbf{T}(K)) \le \mathrm{vol}(\tilde{\mathbf{g}}(K)) \le (1 + \varepsilon)^n \mathrm{vol}(\mathbf{T}(K)) \qquad (14.21)$$

when $K \subset K_\delta$. However, by Theorem 14.15, we have $\text{vol}(\mathbf{T}(K)) = |\det \mathbf{A}|\text{vol}(K)$. If we substitute into the inequalities (14.21), we obtain

$$(1 - \varepsilon)^n |\det \mathbf{A}|\text{vol}(K) \leq \text{vol}(\tilde{\mathbf{g}}(K)) \leq (1 + \varepsilon)^n |\det \mathbf{A}|\text{vol}(K).$$

The lemma is proved, because $\text{vol}(\mathbf{g}(K)) = \text{vol}(\tilde{\mathbf{g}}(K))$.

Finally we get to the general result.

Theorem 14.17 (Change of variable) Let $\mathbf{g}(\mathbf{x})$ be a C^1 mapping of an open set $G \subset \mathbb{R}^n$ that is one-to-one with $\mathbf{Dg}(\mathbf{x})$ invertible for each $\mathbf{x} \in G$. Let $D \subset G$ be a compact Jordan domain.

Then for any continuous function $f : \mathbf{g}(D) \to \mathbb{R}$,

$$\int_{\mathbf{g}(D)} f(\mathbf{x}) \, d\mathbf{x} = \int_D f(\mathbf{g}(\mathbf{y}))|\det \mathbf{Dg}(\mathbf{y})| \, d\mathbf{y}. \tag{14.22}$$

Proof: We first show that the formula is valid when D is a cube K. Because K is compact and f is continuous, f is bounded below on K. By adding on a suitable constant, we can assume that $f \geq 0$. Now subdivide K into small nonoverlapping cubes K_j, $j = 1, \ldots N$. Then because $\mathbf{g}$ is one-to-one,

$$\int_{\mathbf{g}(K)} f(\mathbf{x}) \, d\mathbf{x} = \sum_{j=1}^N \int_{\mathbf{g}(K_j)} f(\mathbf{x}) \, d\mathbf{x}.$$

By the intermediate value theorem for integrals (Theorem 14.10) there is a point $\mathbf{x}_j \in \mathbf{g}(K_j)$ such that

$$\int_{\mathbf{g}(K_j)} f(\mathbf{x}) \, d\mathbf{x} = f(\mathbf{x}_j)\text{vol}(\mathbf{g}(K_j)).$$

Furthermore, because $\mathbf{g}$ is one-to-one, there is a unique point $\mathbf{y}_j \in K_j$ with $\mathbf{g}(\mathbf{y}_j) = \mathbf{x}_j$. Hence

$$\int_{\mathbf{g}(K_j)} f(\mathbf{x}) \, d\mathbf{x} = f(\mathbf{g}(\mathbf{y}_j))\text{vol}(\mathbf{g}(K_j)).$$

Now we want to use Lemma 14.16 to estimate the right-hand side of this last equation. Because $\mathbf{g} \in C^1$ and K is compact, $\mathbf{x} \to \mathbf{g(x)}$ and $\mathbf{x} \to \mathbf{Dg(x)}$ are both uniformly continuous on K (Theorem 9.4). This implies that for a given $\varepsilon > 0$, the $\delta > 0$ of Lemma 14.16 may be chosen independently of the point $\mathbf{x}_0$. We assume that all of the cubes K_j are chosen with the length of the longest edge less than 2δ. Let $\mathbf{y}_j^*$ be the center of K_j. Using the fact that $f \geq 0$, we have

$$(1 - \varepsilon)^n S_j \leq \int_{\mathbf{g}(K_j)} f(\mathbf{x}) \, dx \leq (1 + \varepsilon)^n S_j$$

where

$$S_j = f(\mathbf{g}(\mathbf{y}_j))| \det \mathbf{Dg}(\mathbf{y}_j^*)|\mathrm{vol}(K_j).$$

If we sum these inequalities over j, we obtain

$$(1 - \varepsilon)^n S_\delta \leq \int_{\mathbf{g}(K)} f(x) d\mathbf{x} \leq (1 + \varepsilon)^n S_\delta,$$

where

$$S_\delta = \sum_{j=1}^{N} f(\mathbf{g}(\mathbf{y}_j))| \det \mathbf{Dg}(\mathbf{y}_j^*)|\mathrm{vol}(K_j).$$

Because it may happen that $\mathbf{y}_j \neq \mathbf{y}_j^*$, this is not a usual Riemann sum for $h(\mathbf{y}) \equiv f(\mathbf{g(y)})| \det \mathbf{Dg(y)}|$ on K. However, h is continuous on K and hence uniformly continuous on K. This implies that

$$\lim_{\delta \to 0} S_\delta = S \equiv \int_K f(\mathbf{g(y)})| \det \mathbf{Dg(y)}| \, dy.$$

We conclude that, for each $\varepsilon > 0$,

$$(1 - \varepsilon)^n S \leq \int_{\mathbf{g}(K)} f(\mathbf{x}) \, dx \leq (1 + \varepsilon)^n S.$$

This proves the theorem for the case of a cube when $f \geq 0$, and in particular, for the case of a nonnegative constant function. For a general function f, we can write $f = \tilde{f} - C$ where $\tilde{f} \geq 0$ and C is a nonnegative constant. The result follows using the linearity of the integral.

The general case of a Jordan domain is handled by approximating D by a union of nonoverlapping cubes and then applying the previous result to each cube.

14.5 Applications of the Change of Variable Theorem

One of the important uses of the change of variable theorem is to map a domain into a rectangle or box so that numerical methods can be used.

EXAMPLE 14.4:   Here is an example in three dimensions where a change of variable is needed to make a numerical estimate of the value of an integral.   Suppose that the set $D = \{(x, y, z) : 0 \leq x \leq z, 0 \leq y \leq 1/z, 1 \leq z \leq 2\}$. This chunk of material is cut out of a larger piece with density function $\delta(x, y, z) = \exp(-(x^2 + y^2))$. The mass of the material contained in D is given by the integral

$$\int_D \delta(x, y, z)\, dx dy dz,$$

and the average density in this piece is given by

$$\frac{\int_D \delta(x, y, z)\, dx dy dz}{\mathrm{vol}(D)}.$$

The volume of the region can be calculated by hand, but we will need a numerical estimate of the mass integral.

First we observe that D is the image of the cube

$$K = \{0 \leq u, v \leq 1, 1 \leq w \leq 2\},$$

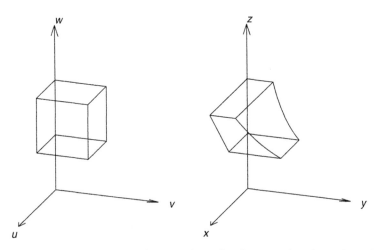

Figure 14.2 Cube K on the left, and image D under the mapping $(x, y, z) = g(u, v, w) = (uv, v/w, w)$ on the right

under the mapping $\mathbf{g} = (g_1, g_2, g_3)$, where

$$
\begin{aligned}
x &= g_1(u, v, w) = uw \\
y &= g_2(u, v, w) = v/w \\
z &= g_3(u, v, w) = w.
\end{aligned}
$$

See Figure 14.2. The Jacobian matrix of $\mathbf{g}$ is

$$
\mathbf{Dg}(u, v, w) = \begin{bmatrix} w & 0 & u \\ 0 & 1/w & -v/w^2 \\ 0 & 0 & 1 \end{bmatrix},
$$

with determinant

$$
\det \mathbf{Dg}(u, v, w) = 1 + v/w.
$$

Hence the volume of D can be computed,

$$\mathrm{vol}(D) = \int_D dxdydz = \int_K |\det \mathbf{Dg}(u, v, w)| \, dudvdw$$

$$= \int_1^2 \int_0^1 \int_0^1 (1 + v/w) dudvdw$$

$$= \int_1^2 \left(1 + \frac{1}{2w}\right) dw$$

$$= 1 + (1/2) \ln 2.$$

Making the same change of variable, the mass of D can be written

$$\mathrm{mass}(D) = \int_D \delta(x, y, z) = \int_D e^{-x^2 - y^2} dxdydz$$

$$= \int_K e^{-(uw)^2 - (v/w)^2} \left(1 + \frac{v}{w}\right) dudvdw. \tag{14.23}$$

We use a three-dimensional version of Simpson's rule to estimate this integral. First we use 20 subdivisions in each direction, with a result $s_1 = 0.651029491890$. We double the number of subdivisions in each direction and get the result $s_2 = 0.651029655067$. The error in s_1 is approximately $s_1 - s_2 = -1.6317 \times 10^{-7}$. Using s_2 as an estimate for the mass, we find that the average density is approximately 0.48347127833729.

If we are lucky, the domain R in xy space may be described in terms of the level curves of two functions $u = \varphi(x, y)$ and $v = \psi(x, y)$. For example

$$\tilde{D} = \{(x, y) : a \le \varphi(x, y) \le b\} \cap \{(x, y) : c \le \psi(x, y) \le d\}.$$

In this case, the level curves $\varphi(x, y) = $ constant and $\psi(x, y) = $ constant provide a new set of coordinate curves on R. It is natural to take the set

$$D = \{(u, v) : a \le u \le b, \ c \le v \le d\}.$$

The mapping $(x, y) \to \mathbf{h}(x, y)$ is the inverse of the mapping $\mathbf{g} : D \to \tilde{D}$ that we need to use Theorem 14.17. If it is difficult to solve for (x, y) in terms of (u, v), the following approach may work. Recall from Theorem 11.9 that

$$\mathbf{Dg}(\varphi(x, y), \psi(x, y)) = [\mathbf{Dh}(x, y)]^{-1}$$

whence

$$\det \mathbf{Dg}(\varphi(x, y), \psi(x, y)) = \frac{1}{\det \mathbf{Dh}(x, y)}.$$

If $\varphi(x, y)$ and $\psi(x, y)$ are given explicitly, it may be easy to calculate $\det \mathbf{Dh}(x, y)$. Then we only need to express $\det \mathbf{Dh}(x, y)$ in terms of u and v.

EXAMPLE 14.5: Let $\tilde{D} \subset \mathbb{R}^2$ be the region described by

$$\tilde{D} = \{(x, y) : 1 \le xy \le 3\} \cap \{1 \le x^2 - y^2 \le 4\}.$$

How do we find the area of $\tilde{D}$? The inverse mapping $\mathbf{h}(x, y) = (xy, x^2 - y^2)$ maps $\tilde{D}$ onto the rectangle $D = \{1 \le u \le 3, \ 1 \le v \le 4\}$. We have

$$\mathbf{Dh}(x, y) = \begin{bmatrix} y & x \\ 2x & -2y \end{bmatrix}$$

with

$$\det \mathbf{Dh}(x, y) = -2(x^2 + y^2).$$

To solve for x in terms of (u, v), we can try $y = u/x$ and substitute in the equation for ψ to find $x^4 - vx^2 - u^2 = 0$ whence

$$x^2 = \frac{v + \sqrt{v^2 + 4u^2}}{2}.$$

Similarly,

$$y^2 = \frac{-v + \sqrt{v^2 + 4u^2}}{2}.$$

The expressions for x and y are more complicated, but we see that

$$x^2 + y^2 = \sqrt{v^2 + 4v^2}.$$

Hence

$$\int\int_{\tilde{D}} dA(x,y) = \int_1^3 \int_1^4 \frac{dudv}{2\sqrt{v^2 + 4u^2}}.$$

This integral can be estimated easily using a two-dimensional version of Simpson's rule.

Polar and Spherical Coordinates

Polar coordinates and spherical coordinates can be understood as changes of variable. A set $F \subset \mathbb{R}^2$, described in terms of Cartesian coordinates x and y, can be thought of as the image of a set D, $f = g(D)$, where D is another set in $\mathbb{R}^2$, this time described in terms of polar coordinates. The mapping is

$$(x,y) = g(r,\theta) = (r\cos\theta, r\sin\theta), \quad r \geq 0, \ 0 \leq \theta < 2\pi.$$

The Jacobian matrix of g is

$$\mathbf{Dg}(r,\theta) = \begin{bmatrix} \cos\theta & -r\sin\theta \\ \sin\theta & r\cos\theta \end{bmatrix},$$

and $|\det \mathbf{Dg}(r,\theta)| = r\cos^2\theta + r\sin^2\theta = r$. Note that the mapping g maps the whole line $r = 0$ in the $r\theta$ plane into the point $(0,0)$ in the xy plane. The mapping g is not globally one-to-one, and $\det \mathbf{Dg} = 0$ on the line $r = 0$. However, for any bounded set D, the intersection of this line with D is a set of zero Jordan content. Theorem 4.17 may be extended to cover this case (see exercise 3 of this section). The formula (14.22) becomes the usual one for integration in polar coordinates,

$$\int\int_F f(x,y)\, dA(x,y) = \int\int_D f(r\cos\theta, r\sin\theta)r\, dr d\theta.$$

EXAMPLE 14.6: We consider an integral that is a slight variation of one usually done in multivariable calculus. Let S be the solid region bounded by the sphere of radius 2 on top and bottom and on the sides by the cylinder $\{(x-1)^2 + y^2 \leq 0.25\}$. The volume of this region is given by the integral

$$2 \int \int_A \int_0^{\sqrt{4-x^2-y^2}} dz \, dx dy = 2 \int \int_A \sqrt{4 - x^2 - y^2} \, dx dy,$$

where $A = \{(x, y) : (x-1)^2 + y^2 \leq 0.25\}$. To attempt to evaluate this integral, we first make the change of variable $x = 1 + \xi$ for which $dx dy = d\xi dy$. The integral is now taken over the disc centered at the origin, $\{\xi^2 + y^2 \leq 0.25\}$. Finally we put in polar coordinates, $\xi = r \cos \theta$, $y = r \sin \theta$, with $d\xi dy = r dr d\theta$. The original integral has now become

$$2 \int_0^{2\pi} \int_0^{0.5} \sqrt{3 - 2r \cos \theta - r^2} \, r dr d\theta.$$

This integral cannot be done analytically, so we turn to a two-dimensional Simpson's rule over the rectangle $[0, 0.5] \times [0, 2\pi]$. First we use Simpson's rule with $n = 10$ subintervals in the r variable and $m = 120$ subdivisions in the θ variable. This yields the result $s_1 = 5.35097715268604$. Now we double the number of subdivisions in each direction, taking $n = 20$ and $m = 240$. The result is $s_2 = 5.35097738068712$. The difference $s_1 - s_2 = -2.280010829025514 \times 10^{-7}$. Thus the first result s_1 approximates the exact result with an error on the order of 2×10^{-7}. We take s_2 as our estimate of the integral.

In the case of spherical coordinates, the mapping $(x, y, z) = \mathbf{g}(r, \phi, \theta)$ is

$$x = \rho \sin \phi \cos \theta$$

$$y = \rho \sin \phi \sin \theta$$

$$z = \rho \cos \phi,$$

$$\rho \geq 0, \quad 0 \leq \phi \leq \pi, \quad 0 \leq \theta \leq 2\pi.$$

We have

$$\mathbf{Dg}(\rho, \phi, \theta) = \begin{bmatrix} \sin\phi\cos\theta & \rho\cos\phi\cos\theta & -\rho\sin\phi\sin\theta \\ \sin\phi\sin\theta & \rho\cos\phi\sin\theta & \rho\sin\phi\cos\theta \\ \cos\phi & -\rho\sin\phi & 0 \end{bmatrix}.$$

The determinant is

$$\det \mathbf{Dg}(\rho, \phi, \theta) = \rho^2(\cos^2\theta + \sin^2\theta)(\sin^2\phi + \cos^2\phi)\sin\phi$$
$$= \rho^2\sin\phi.$$

Because the range of ϕ is $0 \le \phi \le \pi$, $\det \mathbf{Dg} \ge 0$. The formula (14.22) becomes

$$\int\int\int_F f(x, y, z)\, dV(x, y, z)$$

$$= \int\int\int_D f(\rho\sin\phi\cos\theta, \rho\sin\phi\sin\theta, \rho\cos\phi)\rho^2\sin\phi\, d\rho d\phi d\theta.$$

For dimensions $n > 3$, we will give a restricted version of the analogue of polar coordinates. We will indicate how to integrate functions that depend on the radial variable. For the remainder of this section, $\|x\|$ will denote the Euclidean norm $\|x\|_2$ in $\mathbb{R}^n$. We will denote the closed Euclidean ball of radius a by

$$B_n(a) = \{\mathbf{x} \in \mathbb{R}^n : \|\mathbf{x}\| \le a\}.$$

The linear transformation $\mathbf{x} \to a\mathbf{x}$ maps $B_n(1)$ onto $B_n(a)$. The matrix for this transformation is $a\mathbf{I}$, and its determinant is simply a^n. Consequently by Theorem 14.17,

$$\mathrm{vol}(B_n(a)) = a^n\mathrm{vol}(B_n(1)). \tag{14.24}$$

The volume between between two spherical shells of radius a and of radius $a + \Delta a$ is the difference of the volumes,

$$\mathrm{vol}(B_n(a + \Delta)) - \mathrm{vol}(B_n(a)) = [(a + \Delta a)^n - a^n]\mathrm{vol}(B_n(1)).$$

Using our intuitive notion of area for a two-dimensional surface in $\mathbb{R}^3$, we expect that the area of the sphere of radius a in $\mathbb{R}^n$ is

$$\Omega_n = \lim_{\Delta a \to 0} \frac{\text{vol}(B_n(a + \Delta a)) - \text{vol}(B_n(a))}{\Delta a} = \lim_{\Delta a \to 0} \frac{(a + \Delta a)^n - a^n}{\Delta a} \text{vol}(B_n(1))$$

$$= na^{n-1}\text{vol}(B_n(1)).$$

This expression turns out to be consistent with a more systematic definition for an $(n-1)$-dimensional surface in $\mathbb{R}^n$. Formulas for $\text{vol}(B_n(1))$ and Ω_n are given in terms of the gamma function. Recall from Chapter 7 that the function $\Gamma(x)$ is defined for $x > 0$ by the integral

$$\int_0^\infty u^{x-1}e^{-u}\, du.$$

In the exercises for this section, we will see that

$$\text{vol}(B_n(1)) = \frac{\pi^{n/2}}{\Gamma((n+2)/2)}$$

and

$$\Omega_n = \frac{2\pi^{n/2}}{\Gamma(n/2)}.$$

With this preparation, we begin by considering functions f that depend only on the Euclidean norm $\|\mathbf{x}\|$. The following lemma gives a simple way to evaluate the integral of this kind of function over the region between two concentric spheres.

Lemma 14.18 Let $f : [a, b] \to \mathbb{R}$ be continuous. Then with $\rho = \|\mathbf{x}\|$,

$$\int_{a \leq \rho \leq b} f(\rho)\, d\mathbf{x} = \Omega_n \int_a^b f(\rho)\rho^{n-1}\, d\rho.$$

Proof: Let $\Delta\rho = (b-a)/N$, and let $\rho_j = a + j\Delta\rho$, $j = 0, 1, \ldots, N$ be a partition of $[a, b]$. Then

$$C_j = \{\mathbf{x} : \rho_{j-1} \leq \|\mathbf{x}\| \leq \rho_j\}$$

is the part of $B_n(\rho_j)$ that is not contained in $B_n(\rho_{j-1})$. We may write

$$\int_{a \leq \rho \leq b} f(\rho) \, d\mathbf{x} = \sum_{j=1}^{N} \int_{C_j} f(\rho) \, d\mathbf{x}.$$

By the mean value theorem for integrals (Theorem 14.10), there is a $\sigma_j \in [\rho_{j-1}, \rho_j]$ such that

$$\int_{C_j} f(\rho) \, d\mathbf{x} = f(\sigma_j)\text{vol}(C_j)$$

$$= f(\sigma_j)[\text{vol}(B_n(\rho_j)) - \text{vol}(B_n(\rho_{j-1}))].$$

Using equation (14.24) and the mean value theorem for functions of one variable, there is a $\tau_j \in [\rho_{j-1}, \rho_j]$ such that

$$\text{vol}(B_n(\rho_j)) - \text{vol}(B_n(\rho_{j-1})) = (\rho_j^n - \rho_{j-1}^n)\text{vol}(B_n(1))$$

$$= n\text{vol}(B_n(1))\tau_j^{n-1}\Delta\rho$$

$$= \Omega_n \tau_j \Delta\rho.$$

Hence

$$\int_{B_n(a)} f(\rho) \, d\mathbf{x} = \Omega_n \sum_{j=1}^{N} f(\sigma_j)\tau_j^{n-1}\Delta\rho.$$

Strictly speaking, this is not a Riemann sum because it may be that $\sigma_j \neq \tau_j$. Nevertheless, because f is assumed to be continuous on $[a, b]$, it is uniformly continuous on $[a, b]$. This is enough to show that the sums in the previous equation converge to $\int_a^b f(\rho)\rho^{n-1} \, d\rho$. The lemma is proved.

Notice that in two dimensions, the lemma simply says that

$$\int \int_{a \leq r \leq b} f(r) \, dA = \int_0^{2\pi} \int_a^b f(r)r \, dr \, d\theta = 2\pi \int_a^b f(r)r \, dr.$$

In three dimensions,

$$\iiint_{a \le \rho \le b} f(\rho)dV = \int_0^{2\pi} \int_0^\pi \int_a^b f(\rho)\rho^2 \sin\phi \, d\rho d\phi d\theta = 4\pi \int_a^b f(\rho)\rho^2 d\rho.$$

Exercises for 14.4 and 14.5

1. This result was used in the proof of Theorem 14.11. Assuming the hypotheses of Theorem 14.11, show that $g(\partial D) = \partial g(D)$.

2. Let D be the square $\{-1 \le x, y \le 1\}$, and let $g(x, y) = (x, x^2 y)$. Verify that $\det(Dg) = 0$ on the line $x = 0$. By making a direct calculation of area$(g(D))$, show that

$$\text{area}(g(D)) = \int_D |\det(Dg)| dx dy.$$

3. The calculation of exercise 2 suggests that the hypotheses of Theorem 14.17 may be relaxed somewhat. Show that Theorem 14.17 still holds if $|\det(Dg(x))| = 0$ on a set of zero Jordan content.

4. The finite element method is an important procedure for generating numerical solutions of partial differential equations. In the process of using the FEM one must evaluate the integrals of functions over many triangles.

   a) Suppose a triangle T has vertices $p_0 = (x_0, y_0)$, $p_1 = (x_1, y_1)$ and $p_2 = (x_2, y_2)$. The triangle S with vertices $(0, 0)$, $(1, 0)$ and $(0, 1)$ is called the *reference* triangle. Find an affine change of variable $(x, y) = g(\xi, \eta)$ that maps S onto T, with $(0, 0)$ going to p_0, $(1, 0)$ going to p_1 and $(0, 1)$ going to p_2.

   b) Using the change of variable theorem (Theorem 14.17), write $\int_T f(x, y) dx dy$ as an integral over S.

   c) Find the function $P(\xi, \eta) = A\xi + B\eta + C$ that interpolates a function $h(\xi, \eta)$ at the vertices of the reference triangle S. Derive a quadrature rule that approximates $\int_S h(\xi, \eta) d\xi d\eta$ with $\int_S P(\xi, \eta) d\xi d\eta$.

   d) Using the change of variable formula of part b), derive a quadrature rule for $\int_T f(x, y) dx dy$ in terms of the values $f(p_0)$, $f(p_1)$ and $f(p_2)$.

5. Let $R = \{(x, y) : \varphi(x) \le y \le \psi(x), a \le x \le b\}$. Let $S = \{(u, v) : a \le u \le v, 0 \le v \le 1\}$. Find a mapping $g : S \to R$, and use the change of variable theorem to write $\int_R f(x, y) dx dy$ as an integral over S.

6. Find the area of the set R bounded by the lines $x + y = 2$ and $x + y = 4$, and the curves $y^2 - x^2 = 1$ and $y^2 - x^2 = 4$. Use the approach of Example 14.5.

7. Let D be the square $1 \le u, v \le 2$, and let $\mathbf{g}(u, v, t) = (t(u^2 - v^2) + (1 - t)u, \ 2tuv + (1 - t)v)$ be a family of mappings that depend on the parameter t. Let $\tilde{D}(t)$ be the image of D under the mapping $(u, v) \to \mathbf{g}(u, v, t)$.

   a) Sketch the set $\tilde{D}(t)$ for several values of t, $0 \le t \le 1$.

   b) Find an expression for the area of $\tilde{D}(t)$ as an integral over D with t as a parameter.

   c) Find an expression for $(d/dt)(\text{area}(\tilde{D}(t)))$.

8. Find the volume of the portion of the ball $x^2 + y^2 + z^2 \le 4$ that lies above the surface $z = e^x$.

   a) Let $F = \{x^2 + \exp(2x) + y^2 \le 4\}$. Find out what the boundary of F looks like by graphing $y = g(x) = \sqrt{4 - x^2 - \exp(2x)}$ on the interval $-2 \le x \le 1$. Find the points a and b where $g(x) = 0$ using a software rootfinder or your own Newton code.

   b) Write the required volume as a double integral over F

   $$\int_a^b \int_{-g(x)}^{g(x)} f(x, y) dy \, dx = 2 \int_a^b \int_0^{g(x)} f(x, y) dy \, dx.$$

   c) Make the change of variable $x = u$, $y = v\sqrt{4 - u^2 - \exp(2u)}$ to bring the integral into the form:

   $$2 \int_a^b \int_0^1 q(u, v) dv \, du.$$

   d) Use a two-dimensional Simpson code with $m = n = 50$ and then with $m = n = 100$. Estimate the error in your first result.

9. We determine the volume of the unit ball (in the Euclidean norm) in $\mathbb{R}^n$. For $\mathbf{x} \in \mathbb{R}^n$, we write $\mathbf{x} = (\hat{\mathbf{x}}, x_n)$ where $\hat{\mathbf{x}} \in \mathbb{R}^{n-1}$. Then

$$B_n(1) = \{(\hat{\mathbf{x}}, x_n) : \hat{\mathbf{x}} \in B_{n-1}(1) \text{ and } x_n^2 \le 1 - \|\hat{\mathbf{x}}\|^2\}.$$

a) Use the Fubini theorem and Lemma 14.18 to show that

$$\text{vol}(B_n(1)) = 2 \int_{B_{n-1}(1)} \int_0^{\sqrt{1-\|\hat{\mathbf{x}}\|^2}} dx_n d\hat{\mathbf{x}}$$

$$= 2(n-1)\text{vol}(B_{n-1}(1)) \int_0^1 \rho^{n-2}\sqrt{1-\rho^2}\, d\rho.$$

b) Make the substitution $\rho = \sin u, 0 \leq u \leq \pi/2$, and integrate by parts once to find

$$\int_0^1 \rho^{n-2}\sqrt{1-\rho^2}\, d\rho = \frac{1}{n-1} \int_0^{\pi/2} \sin^n u\, du.$$

Consequently we have the recursion relation

$$\text{vol}(B_n(1)) = (2 \int_0^{\pi/2} \sin^n u\, du)\, \text{vol}(B_{n-1}(1)). \qquad (14.25)$$

c) Use the reduction formula

$$\int_0^{\pi/2} \sin^n u\, du = \frac{n-1}{n} \int_0^{\pi/2} \sin^{n-2} u\, du.$$

Show that for n even,

$$\int_0^{\pi/2} \sin^n u\, du = \frac{(n-1)(n-3)\cdots 3\cdot 1}{n(n-2)\cdots 4\cdot 2}(\frac{\pi}{2}),$$

and that for n odd,

$$\int_0^{\pi/2} \sin^n u\, du = \frac{(n-1)(n-3)\cdots 4\cdot 2}{n(n-2)\cdots 5\cdot 3}.$$

d) Use these expressions in the recursion relation (14.25) to show that for n even, $n = 2k$,

$$\text{vol}(B_n(1)) = \frac{\pi^k}{k!},$$

and that for n odd, $n = 2k - 1$,

$$\text{vol}(B_n(1)) = \pi^{k-1} \frac{4^k k!}{(2k)!}.$$

The two expressions for n even and n odd may be combined into one formula using the gamma function $\Gamma(x)$. It has the important property that $\Gamma(x + 1) = x\Gamma(x)$ for $x > 0$. In particular, $\Gamma(1) = 1$, and $\Gamma(1/2) = \sqrt{\pi}$. It follows that $\Gamma(n + 1) = n!$. Using these properties of $\Gamma(x)$, the formula for the volume of $B_n(1)$ can be expressed

$$\text{vol}(B_n(1)) = \frac{\pi^{n/2}}{\Gamma((n + 2)/2)}.$$

The area of the unit sphere in $\mathbb{R}^n$ is

$$\Omega_n = n\text{vol}(B_n(1)) = \frac{2\pi^{n/2}}{\Gamma(n/2)}.$$

14.6 Improper Integrals in Several Variables

Improper integrals arise when either the function is unbounded on a bounded set, or when the function is bounded but the set of integration is unbounded. Before considering functions on $\mathbb{R}^n$, we review some of the basic results for improper integrals in one dimension that were seen in Chapter 7.

In one dimension, we said (Definition 7.8),

Definition Let $f(x)$ be integrable on the interval $[r, b]$ for all $r > a$. We define

$$\int_a^b f(x)dx = \lim_{r \downarrow a} \int_r^b f(x)dx$$

when this limit exists.

If f is integrable on the interval $[a, r]$ for all $r > a$, we define

$$\int_a^\infty f(x)dx = \lim_{r \to \infty} \int_a^r f(x)dx$$

when this limit exists. In both cases we say that the improper integral *converges*. In either case, if the improper integral $\int |f(x)|$ converges, we say that the integral $\int f$ converges *absolutely*.

The two kinds of convergence are related in Theorem 7.21 which we restate here.

Theorem If an improper integral is absolutely convergent, it is convergent.

The following examples were discussed in Chapter 7.

(i) $\int_0^1 x^{-p} \, dx$ converges for $p < 1$ and diverges for $p \geq 1$, and

(ii) $\int_1^\infty x^{-p} \, dx$ converges for $p > 1$ and diverges for $p \leq 1$.

The basic comparison test was stated in Theorem 7.20.

Theorem If $0 \leq f(x) \leq g(x)$ for $x \geq a$ and $\int_a^\infty g(x)dx$ converges, then the improper integral $\int_a^\infty f(x)dx$ also converges.

A similar comparison test holds for unbounded functions on a finite interval.

Our goal in this section is to find analogous results for higher dimensions. For integrals taken over an exterior region, we can frame our definition for convergence of improper integrals in n dimensions as follows. We continue to use $\|\mathbf{x}\|$ to denote the Euclidean norm $\|\mathbf{x}\|_2$.

Definition 14.5 Let $f(\mathbf{x})$ be continuous on the set $\{\|\mathbf{x}\| \geq a\}$. We say the improper integral

$$\int_{\|\mathbf{x}\| \geq a} f(\mathbf{x})\, d\mathbf{x}$$

converges (absolutely) if

$$\lim_{r \to \infty} \int_{a \leq \|\mathbf{x}\| \leq r} |f(\mathbf{x})|\, d\mathbf{x}$$

exists. In this case we set

$$\int_{a \leq \|\mathbf{x}\|} f(\mathbf{x})\, d\mathbf{x} = \lim_{r \to \infty} \int_{a \leq \|\mathbf{x}\| \leq r} f(\mathbf{x})\, d\mathbf{x}.$$

We need to make this stronger definition of convergence in terms of $|f(\mathbf{x})|$ because cancellation in the integrals can occur in so many different ways in higher dimensions.

When we deal with improper integrals over all of $\mathbb{R}^n$, we may also use cubes. Recall that $K_a = \{\mathbf{x} \in \mathbb{R}^n : \|\mathbf{x}\|_\infty \leq a\}$. Of course, K_a is the "ball" of radius a in the infinity norm. Then because $B_n(a) \subset K_a \subset B_n(2a)$, we have

$$\int_{B_n(a)} |f(\mathbf{x})|\, d\mathbf{x} \leq \int_{K_a} |f(\mathbf{x})|\, d\mathbf{x} \leq \int_{B_n(2a)} |f(\mathbf{x})|\, d\mathbf{x}.$$

Hence the improper integral $\int_{\mathbb{R}^n} f(\mathbf{x})\, d\mathbf{x}$ converges in the sense of Definition 14.5 if and only if

$$\lim_{a \to \infty} \int_{K_a} |f(\mathbf{x})|\, d\mathbf{x}$$

exists and in this case has the same limit as the integrals over $B_n(a)$.

In higher dimensions, functions can become unbounded (blow up) at finite points in many different ways. For example, a function on $\mathbb{R}^3$ can become unbounded on a two-dimensional surface, on a curve or at a point. We limit ourselves to improper integrals in which the function becomes unbounded at a single isolated point x_0. In the next definition we let

$$D_r = \{x : r \leq \|x - x_0\| \leq a\}$$

and

$$D_0 = \cup_{r>0} D_r.$$

Definition 14.6 Let $f(x)$ be continuous on the set D_0. We say that the improper integral

$$\int_{D_0} f(x)\, dx$$

converges (absolutely) if

$$\lim_{r \downarrow 0} \int_{D_r} |f(x)|\, dx$$

exists and we set

$$\int_{D_0} f(x)\, dx = \lim_{r \downarrow 0} \int_{D_r} f(x)\, dx.$$

Now we take $f(\|x\|) = \|x\|^{-p}$ in Lemma 14.18 and compare with our known results from one dimension. We immediately deduce

Theorem 14.19

(i) The improper integral

$$\int_{\|x\| \geq 1} \frac{dx}{\|x\|^p} = \Omega_n \int_1^\infty \rho^{-p} \rho^{n-1}\, d\rho$$

converges for $p > n$ and diverges for $p \leq n$.

(ii) The improper integral

$$\int_{\|\mathbf{x}\|\leq 1} \frac{d\mathbf{x}}{\|\mathbf{x}\|^p} = \Omega_n \int_0^1 \rho^{-p}\rho^{n-1}\, d\rho$$

converges for $p < n$ and diverges for $p \geq n$.

A useful test for convergence over an exterior region is given next.

Theorem 14.20 Let $f(\mathbf{x})$ be continuous on the exterior region $\|\mathbf{x}\| \geq a$. Suppose that there is a function $g(\rho) \geq 0$, continuous for $\rho \geq a$, with

(i) $|f(\mathbf{x})| \leq g(\|\mathbf{x}\|)$ for $\|\mathbf{x}\| \geq a$, and such that
(ii) $\int_{\|\mathbf{x}\|\geq a} g(\|\mathbf{x}\|)\, d\mathbf{x}$ converges.

Then the improper integral $\int_{\|\mathbf{x}\|\geq a} f(\mathbf{x})\, d\mathbf{x}$ converges (absolutely).

A similar convergence test is useful for functions that become unbounded at a single isolated point $\mathbf{x}_0$.

Theorem 14.21 Let $f(\mathbf{x})$ be continuous on $D_r = \{r \leq \|\mathbf{x} - \mathbf{x}_0\| \leq a\}$ for each $r > 0$. Let $D_0 = \cup_{r>0} D_r$. Suppose that there is a function $g(\rho) \geq 0$, continuous on $0 < \rho \leq a$, with

(i) $f(\mathbf{x}) \leq g(\|\mathbf{x} - \mathbf{x}_0\|)$ on D_0 and such that
(ii) $\int_{D_0} g(\|\mathbf{x} - \mathbf{x}_0\|)\, d\mathbf{x}$ converges.

Then the improper integral

$$\int_{D_0} f(\mathbf{x})\, d\mathbf{x}$$

converges (absolutely).

EXAMPLE 14.7:   Consider the function

$$f(x,y) = \frac{xy + \sin(x^2)}{(x^2 + y^2)^{3/2}},$$

which is continuous on $x^2 + y^2 > 0$. Letting $r = \sqrt{x^2 + y^2}$ be the Euclidean distance from the origin in $\mathbb{R}^2$, we can bound f by

$$|f(x,y)| \le \frac{|xy| + x^2}{r^3} \le \frac{2r^2}{r^3} = \frac{2}{r}.$$

In view of the comparison test (Theorem 14.21), we conclude that for any $a > 0$, the improper integral

$$\int_{0 < r \le a} f(x,y) \, dA(x,y)$$

converges.

EXAMPLE 14.8: An important multidimensional integral for statistics and for the theory of diffusion is

$$\int_{\mathbb{R}^n} e^{-\|\mathbf{x}\|^2} \, d\mathbf{x}. \tag{14.26}$$

We will show that this integral converges, and then we will use an elegant argument to evaluate it.

First, it is easy to show that this integral converges because, by Lemma 14.18

$$\int_{\|\mathbf{x}\| \le a} e^{-\|\mathbf{x}\|^2} \, d\mathbf{x} = \Omega_n \int_0^a \rho^{n-1} e^{-\rho^2} d\rho.$$

Using l'Hôpital's rule we find that, for each $k > 0$, there is a constant C_k such that

$$e^{-\rho^2} \le C_k \rho^{-k} \quad \text{for } \rho \ge 1.$$

Then choosing $k = n + 1$, we have

$$\rho^{n-1}e^{-\rho^2} \le C_{n+1}\rho^{-2}.$$

for $\rho \ge 1$. Therefore

$$\int_0^a \rho^{n-1}e^{-\rho^2} d\rho \le \int_0^1 \rho^{n-1}e^{-\rho^2} d\rho + C_{n+1}\int_1^a \rho^{-2}d\rho,$$

which implies that

$$\lim_{a\to\infty}\int_0^a \rho^{n-1}e^{-\rho^2}\,d\rho$$

exists. This means that the integral (14.26) converges.

When $n = 2$, (14.26) can be evaluated using polar coordinates:

$$\int_{\mathbb{R}^2} e^{-x^2-y^2}\,dxdy = \int_0^{2\pi}\int_0^\infty e^{-r^2}r\,drd\theta$$

$$= 2\pi(-1/2)e^{-r^2}\Big|_0^\infty \qquad (14.27)$$

$$= \pi.$$

To evaluate the integral (14.26) for general n we will use the fact that the integral (14.26) is also the limit of the integrals over cubes:

$$\int_{\mathbb{R}^n} e^{-\|\mathbf{x}\|^2}\,d\mathbf{x} = \lim_{a\to\infty}\int_{K_a} e^{-\|\mathbf{x}\|^2}\,d\mathbf{x}.$$

Using the Fubini theorem and the fact that

$$e^{-\|\mathbf{x}\|^2} = e^{-x_1^2}\cdots e^{-x_n^2},$$

we find that

$$\int_{K_a} e^{-\|\mathbf{x}\|^2}\, d\mathbf{x} = \int_{-a}^{a} e^{-x_1^2}\, dx_1 \int_{-a}^{a} e^{-x_2^2}\, dx_2 \cdots \int_{-a}^{a} e^{-x_n^2}\, dx_n$$

$$= \left[\int_{-a}^{a} e^{-x^2}\, dx \right]^n.$$

Now we combine this last equation with (14.27) to deduce that

$$\int_{\mathbb{R}^n} e^{-\|\mathbf{x}\|^2}\, d\mathbf{x} = \pi^{n/2}. \tag{14.28}$$

Exercises for 14.6

1. Determine which of the following improper integrals converges.

a)

$$\int \int_{R^2} \frac{\cos(x + y^2)}{(1 + x^2 + y^2)^{3/2}}\, dx dy$$

b)

$$\int \int \int_{x^2+y^2+z^2<1} \frac{|x|^{1/2}}{(x^2 + y^2 + z^2)^{3/2}}\, dx dy dz$$

c)

$$\int \int \int_{x^2+y^2+z^2>1} (x^2 + y)e^{-x^2-y^2-z^2}\, dx dy dz$$

2. In this exercise, $\|\mathbf{x}\| = \|\mathbf{x}\|_2$. For $\mathbf{x} \in \mathbb{R}^3$, let $\rho(\mathbf{x})$ be a charge density that is continuous and such that $\rho(\mathbf{x}) = 0$ for $\|\mathbf{x}\| \geq 1$.

a) Show that the electrostatic potential, given by

$$\varphi(\mathbf{x}) = \frac{1}{4\pi} \int \int \int_{\mathbb{R}^3} \frac{\rho(\mathbf{y})}{\|\mathbf{x} - \mathbf{y}\|} \, d\mathbf{y},$$

is a convergent integral for each $\mathbf{x} \in \mathbb{R}^3$.

b) Show that there is a constant $C > 0$ such that

$$|\varphi(\mathbf{x})| \leq \frac{C}{\|\mathbf{x}\|}$$

for $\|\mathbf{x}\| \geq 2$.

3. Let $f(x, y) = r^{-\alpha}$ where $r = \sqrt{x^2 + y^2}$. Let $D = \{(x, y) : r \leq 1\}$. Show that there are values of α such that

a) f is unbounded at $(0, 0)$, but

b) the integrals

$$\int_D f(x, y)\,dx\,dy, \quad \int_D |f_x(x, y)|\,dx\,dy, \quad \text{and} \quad \int_D |f_y(x, y)|\,dx\,dy$$

are all convergent.

4. Let $f(x, y) = y^{-3/2}e^{x-y}$, and let $D = \{(x, y) : 0 \leq x \leq y \leq 1\}$. The function f is singular on the whole x axis, so the convergence tests we have developed do not apply to $\int \int_D f$. Nevertheless, we can take a direct approach. Let $D_\varepsilon = D \cap \{y \geq \varepsilon\}$.

a) Does

$$\int_D f\,dx\,dy = \lim_{\varepsilon \to 0} \int \int_{D_\varepsilon} f$$

exist?

b) Let $E = \{(x, y) : 0 \leq y \leq x \leq 1\}$ and $E_\varepsilon = E \cap \{y \geq \varepsilon\}$. Does

$$\int \int_E f = \lim_{\varepsilon \to 0} \int \int_{E_\varepsilon} f$$

exist?

14.7 Applications in Probability

The change of variable theorem (Theorem 14.17) and improper integrals are important tools in the study of random variables.

A continuous *random variable* X is a real-valued function defined on subsets of a probability space. The *probability density function* (p.d.f.) of the random variable X is a nonegative function f defined on $\mathbb{R}$, such that f is integrable on each bounded interval, and

$$\int_{\mathbb{R}} f(x)dx = \lim_{a \to \infty} \int_{-a}^{a} f(x)dx = 1.$$

The probability that X lies in the interval $[a, b]$ is given by the integral

$$Pr[a \leq X \leq b] = \int_{a}^{b} f(x)dx.$$

The *distribution function* of X is

$$G(x) = Pr[X \leq x] = \int_{-\infty}^{x} f(\xi)d\xi.$$

An example of a continuous random variable is the height of individuals in a large population.

EXAMPLE 14.9:

Let $\sigma > 0$ and $\mu \in \mathbb{R}$, and set

$$f_{\sigma,\mu}(x) = \frac{1}{\sigma\sqrt{2\pi}} e^{-(x-\mu)^2/2\sigma^2}. \qquad (14.29)$$

We claim that $\int_{\mathbb{R}} f_{\sigma,\mu}(x)dx = 1$ for any choice of $\sigma > 0$ and $\mu \in \mathbb{R}$.

In fact, if we make the change of variable $y = (x - \mu)/\sigma\sqrt{2}$, and use equation (14.28), we find that

$$\int_R f_{\sigma,\mu}(x)dx = \frac{1}{\sigma\sqrt{2\pi}} \int_{\mathbb{R}} e^{-(x-\mu)^2/2\sigma^2}$$

$$= \frac{1}{\sqrt{\pi}} \int_{\mathbb{R}} e^{-y^2} dy$$

$$= 1.$$

Furthermore, for each $\sigma > 0$ and μ, and for each n,

$$|x|^n f_{\sigma,\mu}(x) \to 0 \quad \text{as} \quad |x| \to \infty.$$

Thus the following improper integral converges absolutely. It may be evaluated, again using the change of variable $y = (x - \mu)/\sigma\sqrt{2}$:

$$\int_{\mathbb{R}} x f_{\sigma,\mu}(x)dx = \frac{1}{\sigma\sqrt{2\pi}} \int_{\mathbb{R}} x e^{-(x-\mu)^2/2\sigma^2} dx$$

$$= \frac{1}{\sqrt{\pi}} \int_{\mathbb{R}} (\mu + \sigma\sqrt{2}y)e^{-y^2} dy$$

$$= \mu.$$

We have used the fact that ye^{-y^2} is an odd function which implies that $\int_{\mathbb{R}} ye^{-y^2} dy = 0$. The random variable with probability density (14.29) is said to be normally distributed with mean μ and standard deviation σ.

There are many situations that involve several random variables. For example, X may be the random variable that is the height of individuals and Y may be the weight of individuals in a large population. A *joint* p.d.f. for two random variables X and Y is a function $f(x, y) \geq 0$, that is integrable on each square $K_a = \{|x|, |y| \leq a\}$, and such that $\int_{\mathbb{R}^2} f(x, y) dx dy = 1$. The probability that $a \leq X \leq b$ and $c \leq Y \leq d$ is given by

$$Pr[a \leq X \leq b,\ c \leq Y \leq d] = \int_a^b \int_c^d f(x, y) dx dy.$$

The joint distribution function of X and Y is

$$G(x, y) = \int_{-\infty}^x \int_{-\infty}^y f(\xi, \eta)\, d\xi d\eta.$$

We shall assume that, for each x, $y \to f(x, y)$ is integrable on $\mathbb{R}$, and for each y, $x \to f(x, y)$ is integrable on $\mathbb{R}$. Then we define the *marginal* p.d.f. of X to be

$$f_1(x) = \int_{\mathbb{R}} f(x, y) dy$$

and the *marginal* p.d.f. of Y to be

$$f_2(y) = \int_{\mathbb{R}} f(x, y) dx.$$

We often form functions of the random variables, and then we must find the p.d.f. of the new random variable. For example, if X_1 is the height of the husband in a married couple, and X_2 is the height of the wife in the couple, a new random variable is $Y = (X_1 + X_2)/2$, which is the average of these heights. If we know the joint p.d.f. of X_1 and X_2, how do we find the p.d.f. of Y? We outline how this is done in the next paragraph.

Suppose that $f(x_1, x_2)$ is the joint p.d.f. of two continuous random variables X_1 and X_2. Let F be the set of points where $f > 0$. Let $Y_1 = u(X_1, X_2)$ be a new random variable. We suppose that there is another function $v(x_1, x_2)$ such that

$(x_1, x_2) \rightarrow (y_1, y_2) = (u(x_1, x_2), v(x_1, x_2))$ is a one-to-one, C^1 map on F. Denote this map by $\mathbf{y} = \mathbf{h}(\mathbf{x})$, and let $E = \mathbf{h}(F)$. If $A \subset F$ and $B = \mathbf{h}(A) \subset E$, then the event $(X_1, X_2) \in A$ is equivalent to the event $(Y_1, Y_2) \in B$. Hence

$$Pr[(Y_1, Y_2) \in B] = Pr[(X_1, X_2) \in A] = \int_A f(x_1, x_2) dx_1 dx_2.$$

However, we want to express the probability for Y_1, Y_2 in terms of an integral over B. Let $\mathbf{x} = \mathbf{g}(\mathbf{y})$ be the inverse mapping to $\mathbf{h}$: $\mathbf{h} \circ \mathbf{g}(\mathbf{y}) = \mathbf{y}$ for all $\mathbf{y} \in E$, and $\mathbf{g} \circ \mathbf{h}(\mathbf{x}) = \mathbf{x}$ for all $\mathbf{x} \in F$. We assume that $\mathbf{Dg}(\mathbf{y})$ is invertible for $\mathbf{y} \in E$. Then, using the change of variable theorem (Theorem 14.17) we see that for every set $B \subset E$,

$$Pr[(Y_1, Y_2) \in B] = \int_A f(\mathbf{x}) d\mathbf{x}$$

$$= \int_B f(\mathbf{g}(\mathbf{y})) |\det \mathbf{Dg}(\mathbf{y})| d\mathbf{y}.$$

It follows that the joint p.d.f. for (Y_1, Y_2) is

$$\hat{f}(\mathbf{y}) = \begin{cases} f(\mathbf{g}(\mathbf{y})) |\det \mathbf{Dg}(\mathbf{y})|, & \mathbf{y} \in E \\ 0, & \mathbf{y} \notin E \end{cases}.$$

The marginal p.d.f. of Y_1, which is what we seek, is

$$\hat{f}_1(y_1) = \int_{\mathbb{R}} \hat{f}(y_1, y_2) dy_2$$

$$= \int_{\mathbb{R}} f(\mathbf{g}(y_1, y_2)) |\det \mathbf{Dg}(y_1, y_2)| dy_2.$$

EXAMPLE 14.10: Let the random variables X_1 and X_2 have the joint p.d.f.

$$f(x_1, x_2) = (1/\pi) \exp(-(x_1 - x_2)^2 - x_2^2).$$

Equation (14.28) tells us that

$$\int\int_{\mathbb{R}^2} e^{-x^2-y^2}\, dx dy = \pi.$$

It is easy to show that $\int\int_{\mathbb{R}^2} f(x_1, x_2)\, dx_1 dx_2 = 1$ using the change of variable $y_1 = x_1 - x_2$, $y_2 = x_2$. We wish to find the p.d.f. of the random variable $Y_1 = (X_1 + X_1)/2$. As the mapping $\mathbf{h}$, we take $y_1 = (x_1 + x_1)/2$, $y_2 = x_2$. It follows that the inverse of $\mathbf{h}$ is $(x_1, x_2) = \mathbf{g}(y_1, y_1) = (2y_1 - y_2, y_2)$. We have $\det \mathbf{Dg}(\mathbf{y}) \equiv 2$, so the joint p.d.f. of the random variables $Y_1 = (X_1 + X_2)/2$ and $Y_2 = X_2$ is

$$\hat{f}(y_1, y_2) = 2f(2y_1 - y_2, y_2)$$
$$= (2/\pi)e^{-4(y_1-y_2)^2-y_2^2}.$$

The marginal p.d.f. of Y_1 is

$$\hat{f}_1(y_1) = \int_{\mathbb{R}} \hat{f}(y_1, y_2) dy_2$$

$$= (2/\pi)\int_{\mathbb{R}} e^{-4(y_1-y_2)^2-y_2^2}\, dy_2.$$

We say that the random variables X_1 and X_2 are (stochastically) *independent* if the joint p.d.f. $f(x_1, x_2)$ is a product,

$$f(x_1, x_2) = g(x_1)h(x_2).$$

In this case, using the Fubini theorem,

$$Pr[a \leq X_1 \leq b,\ c \leq X_2 \leq d] = \int_a^b \int_c^d f(x_1, x_2)dx_1 dx_2$$

$$= \int_a^b g(x_1)dx_1 \int_c^d h(x_2)dx_2$$

$$= Pr[a \leq X_1 \leq b]\, Pr[c \leq X_2 \leq d].$$

For example, the normal p.d.f. in two variables,

$$f(x_1, x_2) = (1/\pi)e^{-x_1^2 - x_2^2} = (1/\sqrt{\pi})e^{-x_1^2}(1/\sqrt{\pi})e^{-x_2^2}$$

is the product of two one-dimensional normal density functions. Independent random variables can arise by repeating an experiment. The random variable X_1 can represent the outcome on the first trial, and X_2, the outcome on the second trial. Of course, this idea can be extended to any number of trials.

EXAMPLE 14.11:   Let X_1 and X_2 be the independent random variables that arise by taking two random samples from a distribution that has the p.d.f.

$$h(x) = \frac{1}{\sqrt{2\pi}}e^{-x^2/2}.$$

The joint p.d.f. of X_1 and X_2 is

$$f(x_1, x_2) = \frac{1}{2\pi}e^{-(x_1^2 + x_2^2)/2}.$$

Let $Y_1 = (X_1 + X_2)/2$ and let $Y_2 = (X_1 - X_2)^2$. The new random variable Y_1 is the mean of the sample and Y_2 is twice the variance of the sample. The associated mapping is $\mathbf{y} = \mathbf{h}(\mathbf{x})$ given by

$$y_1 = \frac{x_1 + x_2}{2}, \qquad y_2 = \frac{(x_1 - x_2)^2}{2}.$$

Let $F = \mathbb{R}^2$ and let $E = \{(y_1, y_2) \in \mathbb{R}^2 : y_2 \geq 0\}$. The function $\mathbf{h}$ maps F onto E, but it is not one-to-one. To be able to use the techniques we have developed so far, we will redefine f to be zero on the line $x_1 = x_2$. This will not affect the values of any integrals of f. The set of points F where $f > 0$ can be expressed as a disjoint union $F = F_1 \cup F_2$ where

$$F_1 = \{(x_1, x_2) : x_2 > x_1\} \quad \text{and} \quad F_2 = \{(x_1, x_2) : x_2 < x_1\}.$$

Now $\mathbf{h}$ maps F onto a modified $E = \{(y_1, y_2) : y_2 > 0\}$. If $A \subset F$ is an event, and $B = \mathbf{h}(A) \subset E$, we can break A down into two mutually exclusive events,

$$A = A_1 \cup A_2$$

where $A_1 = A \cap F_1$ and $A_2 = A \cap F_2$. Then

$$
\begin{aligned}
Pr[(Y_1, Y_2) \in B] &= Pr[(X_1, X_2) \in A] \\
&= Pr[(X_1, X_2) \in A_1 \text{ or } (X_1, X_2) \in A_2] \\
&= Pr[(X_1, X_2) \in A_1] + Pr[(X_1, X_2) \in A_2].
\end{aligned}
$$

The mapping $\mathbf{h}$ is not one-to-one, but it has one inverse $\mathbf{g}_1 : E \to F_1$, given by

$$x_1 = y_1 - \sqrt{\frac{y_2}{2}}, \quad x_2 = y_1 + \sqrt{\frac{y_2}{2}},$$

and another inverse $\mathbf{g}_2 : E \to F_2$, given by

$$x_1 = y_1 + \sqrt{\frac{y_2}{2}}, \quad x_2 = y_1 - \sqrt{\frac{y_2}{2}}.$$

Hence for $B \subset E$,

$$
Pr[(Y_1, Y_2) \in B] = \int_{A_1} f(\mathbf{x})dx + \int_{A_2} f(\mathbf{x})dx
$$

$$
= \int_B f(\mathbf{g}_1(\mathbf{y})) |\det \mathbf{Dg}_1(\mathbf{y})| \, d\mathbf{y} + \int_B f(\mathbf{g}_2(\mathbf{y})) |\det \mathbf{Dg}_2(\mathbf{y})| \, d\mathbf{y}.
$$

Now it is easy to compute

$$|\det \mathbf{Dg}_1| = |\det \mathbf{Dg}_2| = \frac{1}{\sqrt{2y_2}},$$

so

$$Pr[(Y_1, Y_2) \in B] = \int_B [f(\mathbf{g}_1(\mathbf{y})) + f(\mathbf{g}_2(\mathbf{y}))] \frac{d\mathbf{y}}{\sqrt{2y_2}}.$$

Because $f(\mathbf{g}_1(\mathbf{y})) = f(\mathbf{g}_2(\mathbf{y}))$, we conclude that the joint p.d.f. for Y_1 and Y_2 is

$$\hat{f}(y_1, y_2) = \frac{1}{\pi\sqrt{2y_2}} \exp\left[-\frac{(y_1 - \sqrt{y_2/2})^2}{2} - \frac{(y_1 + \sqrt{y_2/2})^2}{2} \right]$$

$$= \frac{1}{\pi\sqrt{2y_2}} \exp(-y_1^2 - y_2/2)$$

$$= \frac{1}{\pi\sqrt{2y_2}} e^{-y_2/2} e^{-y_1^2}.$$

In this example, we have the interesting result that the sample mean Y_1 and the sample variance Y_2 are again independent random variables.

Exercises for 14.7

1. Let X_1 and X_2 be independent random variables with the joint p.d.f. $f(x_1, x_2) = f_1(x_1)f_2(x_2)$. Let $Y_1 = u(X_1)$ and $Y_2 = v(X_2)$ where u and v are C^1 functions. Show that Y_1 and Y_2 are independent by computing the joint p.d.f. of Y_1 and Y_2.

2. Let X_1 and X_2 be random variables, each with the uniform p.d.f. $h(x) = 1/9$, $1 < x < 10$, and $h = 0$ elsewhere. We assume that X_1 and X_2 are independent, so the joint p.d.f. is $f(x_1, x_2) = h(x_1)h(x_2)$. Let $Y_1 = X_1 X_2$.

   a) Find the p.d.f. of Y_1.
   b) Let $A = \{1 < y_1 < 2\} \cup \{10 < y_1 < 20\}$. Show that $Pr[Y_1 \in A] > 1/9$. This problem is of interest for analyzing the random error in floating point computations.

3. Let X_1 and X_2 be two random variables with the same p.d.f. $h(x) = \exp(-x^2/4)/\sqrt{4\pi}$. We assume that X_1 and X_2 are independent. Let $Y_1 = X_1^2 + X_2^2$ and $Y_2 = X_2$.

a) Find the joint p.d.f. of Y_1 and Y_2.
b) Find the marginal p.d.f. of Y_1.

4. Let X_1, X_2, X_3 be a random sample from a distribution that has p.d.f. $h(x) = \exp(-x)$ for $x > 0$, and zero elsewhere. Let

$$Y_1 = \frac{X_1}{X_1 + X_2}, \quad Y_2 = \frac{X_1 + X_2}{X_1 + X_2 + X_3}, \quad Y_3 = X_1 + X_2 + X_3.$$

Show that the joint p.d.f. of Y_1, Y_2, Y_3 has the form

$$\hat{f}(y_1, y_2, y_3) = \hat{f}_1(y_1)\hat{f}_2(y_2)\hat{f}_3(y_3),$$

thereby making Y_1, Y_2, Y_3 mutually stochastically independent.

Chapter 15

Applications of Integration to Differential Equations

Chapter Overview

In this chapter we develop some aspects of integration that have important applications to the theory of differential equations. Section 1 discusses results regarding integrals that depend on a parameter. In particular, we find criteria that allow one to differentiate under the integral. Section 2 uses the results of Section 1 and the idea of convolution to make smooth approximations to a function. In particular, we derive the Weierstrass theorem about uniform polynomial approximation of continuous function. In Section 3 we derive the diffusion equation and show how solutions may be constructed using convolution with the fundamental solution. In Section 4 we derive the Euler equations for incompressible, inviscid flow.

15.1 Interchanging Limits and Integrals

In many important cases, the solutions of linear partial differential equations are given in terms of integrals involving parameters. To verify that these formulas indeed yield solutions, and to derive additional properties of the solutions, we need to differentiate with respect to the parameters. In this section we prove results dealing with this question, both for integrals over bounded sets, and for improper integrals.

By way of motivation, suppose that $t \to f(\mathbf{x}, t)$ is a family of functions on a generalized rectangle $R \subset \mathbb{R}^n$ that depends on the parameter t in either a continuous or a C^1 fashion. If P is a partition of R with subrectangles R_j and points $\mathbf{x}_j \in R_j$, then the Riemann sum

$$S(f, P, t) = \sum_j f(\mathbf{x}_j, t)\mathrm{vol}(R_j)$$

is a continuous function or a C^1 function depending on the regularity of f. It seems plausible that $\int_R f(\mathbf{x}, t) \, d\mathbf{x}$, which is the limit of the Riemann sums, should be a continuous or C^1 function of t.

Theorem 15.1 Let $D \subset \mathbb{R}^n$ be a compact Jordan domain and let $I = [a, b]$ be a closed interval with $a < b$. Let $f(\mathbf{x}, t) : D \times I \to \mathbb{R}$ be continuous. Let $g : D \to \mathbb{R}$ be bounded and continuous except possibly on a set of Jordan content zero. Then

$$t \to \int_D f(\mathbf{x}, t)g(\mathbf{x}) \, d\mathbf{x}$$

is continuous on I.

Proof: For each $t \in I$, $f(\mathbf{x}, t)g(\mathbf{x})$ is continuous on D except possibly on a set of Jordan content zero and is bounded. Hence for each $t \in I$, $f(\mathbf{x}, t)g(\mathbf{x})$ is Riemann integrable on D. In addition, f is continuous on the compact set $D \times I$, and hence by Theorem 9.14, it is uniformly continuous on $D \times I$. Let $t_0 \in I$ and $\varepsilon > 0$ be given. Then there is $\delta > 0$ such that

$$|f(\mathbf{x}, t) - f(\mathbf{x}, t_0)| \leq \varepsilon$$

for all $\mathbf{x} \in D$ and all t with $|t - t_0| \leq \delta$. Consequently,

$$\left| \int_D f(\mathbf{x}, t)g(\mathbf{x}) \, d\mathbf{x} - \int_D f(\mathbf{x}, t_0)g(\mathbf{x}) \, d\mathbf{x} \right| \leq \int_D |f(\mathbf{x}, t) - f(\mathbf{x}, t_0)||g(\mathbf{x})| \, d\mathbf{x}$$

$$\leq \varepsilon M \mathrm{vol}(D)$$

where $M = \sup_D |g|$. The proof of Theorem 15.1 is complete.

Theorem 15.2 Let $D \subset \mathbb{R}^n$ be a compact Jordan domain and $I = [a, b]$ an interval with $a < b$. Now suppose that $f : D \times I \to \mathbb{R}$ and $f_t : D \times I \to \mathbb{R}$ are both continuous. Let $g : D \to \mathbb{R}$ be bounded and continuous except possibly on a set of Jordan content zero. Then the function $t \to \int_D f(\mathbf{x}, t) g(\mathbf{x}) \, d\mathbf{x}$ is C^1 and we can differentiate under the integral sign:

$$\frac{\partial}{\partial t} \int_D f(\mathbf{x}, t) g(\mathbf{x}) \, d\mathbf{x} = \int_D f_t(\mathbf{x}, t) g(\mathbf{x}) \, d\mathbf{x}.$$

Proof: Let $I_t = \{a \le s \le t\}$ for $a \le t \le b$. We can use the Fubini Theorem 14.6 to compute the double integral, integrating first in s:

$$\int_{D \times I_t} f_t(\mathbf{x}, s) g(\mathbf{x}) \, ds d\mathbf{x} = \int_D \int_a^t f_t(\mathbf{x}, s) g(\mathbf{x}) \, ds d\mathbf{x}$$

$$= \int_D [f(\mathbf{x}, t) - f(\mathbf{x}, a)] g(\mathbf{x}) \, d\mathbf{x}.$$

On the other hand, if we integrate first in $\mathbf{x}$ we have

$$\int_{D \times I_t} f_t(\mathbf{x}, s) g(\mathbf{x}) \, d\mathbf{x} ds = \int_a^t \int_D f_t(\mathbf{x}, s) g(\mathbf{x}) \, d\mathbf{x} ds.$$

When we equate these two expressions, we have

$$\int_a^t \int_D f_t(\mathbf{x}, s) g(\mathbf{x}) \, d\mathbf{x} ds = \int_D [f(\mathbf{x}, t) - f(\mathbf{x}, a)] g(\mathbf{x}) \, d\mathbf{x}.$$

Now f_t is continuous on $D \times I$ so by Theorem 15.1, the function $s \to \int_D f_t(\mathbf{x}, s) g(\mathbf{x}) \, d\mathbf{x}$ is continuous. Therefore by the fundamental theorem of calculus in one variable we conclude

$$\int_D f_t(\mathbf{x}, t) g(\mathbf{x}) \, d\mathbf{x} = \frac{\partial}{\partial t} \int_a^t \int_D f(\mathbf{x}, s) g(\mathbf{x}) \, d\mathbf{x} ds$$

$$= \frac{\partial}{\partial t} \int_D f(\mathbf{x}, t) g(\mathbf{x}) \, d\mathbf{x}.$$

The following corollary in one dimension anticipates Theorem 15.7 which will be derived using the change of variable formula for multiple integrals.

Corollary Let $f(x, t)$ and $f_t(x, t)$ be defined and continuous for $(x, t) \in [\alpha_0, \beta_0] \times [a, b]$. Let $\alpha(t)$ and $\beta(t)$ be C^1 on $[a, b]$ with $\alpha_0 \leq \alpha(t) < \beta(t) \leq \beta_0$. Then

$$\frac{d}{dt} \int_{\alpha(t)}^{\beta(t)} f(x, t)\, dx = f(\beta(t), t)\beta'(t) - f(\alpha(t), t)\alpha'(t) + \int_{\alpha(t)}^{\beta(t)} f_t(x, t)\, dx.$$

Proof: For $\alpha_0 \leq u < v \leq \beta_0$ and $a \leq t \leq b$, set

$$G(u, v, t) = \int_u^v f(x, t)\, dx.$$

Then using the chain rule we have

$$\begin{aligned}
\frac{d}{dt} \int_{\alpha(t)}^{\beta(t)} f(x, t)\, dx &= \frac{d}{dt} G(\alpha(t), \beta(t), t) \\
&= G_u(\alpha(t), \beta(t), t)\alpha'(t) + G_v(\alpha(t), \beta(t), t)\beta'(t) \\
&\quad + G_t(\alpha(t), \beta(t), t).
\end{aligned}$$

From the fundamental theorem of calculus we see that

$$G_v(u, v, t) = f(v, t) \quad \text{and} \quad G_u(u, v, t) = -f(u, t),$$

and from Theorem 15.2 we see that

$$G_t(u, v, t) = \int_u^v f_t(x, t)\, dx.$$

The corollary is proved.

Theorem 15.2 and its corollary are often used to show that certain integrals with a parameter are solutions of a differential equation.

EXAMPLE 15.1: The principle of Duhamel.

Duhamel's principle is used to construct the solution of an inhomogeneous differential equation as an integral of solutions of a homogeneous equation.

Consider the inhomogeneous initial value problem

$$y''(t) + \omega^2 y(t) = q(t) \tag{15.1}$$

$$y(0) = 0 \tag{15.2}$$

$$y'(0) = 0. \tag{15.3}$$

For each $s \leq t$, let $z(t, s)$ be the solution of the homogeneous initial value problem

$$z_{tt} + \omega^2 z = 0, \quad z(s, s) = 0, \quad z_t(s, s) = q(s).$$

We shall show that

$$w(t) = \int_0^t z(t, s)\, ds$$

is the solution of the initial value problem (15.1)–(15.3). First we note that $w(0) = 0$. We use the corollary to differentiate w once,

$$\frac{d}{dt} w = z(t, t) + \int_0^t z_t(t, s)\, ds$$

$$= \int_0^t z_t(t, s)\, ds,$$

from which it follows that $w'(0) = 0$ as well. Now we differentiate a second time:

$$\frac{d^2}{dt^2} w = z_t(t, t) + \int_0^t z_{tt}(t, s)\, ds$$

$$= q(t) + \int_0^t (-\omega^2 z(t, s))\, ds$$

$$= q(t) - \omega^2 w(t),$$

which shows that w satisfies (15.1)–(15.3). In this case of constant coefficients, we can easily find that

$$z(t, s) = (1/w) \sin(w(t - s))q(s)$$

which yields the explicit representation of the solution of (15.1)–(15.3),

$$y(t) = \int_0^t \frac{1}{w} \sin(w(t - s))q(s) \, ds.$$

Integral representations of solutions of differential equations are often more useful for numerical evaluation then power series representations.

EXAMPLE 15.2:   The solutions of Bessel's equation are important in the study of vibrations of a circular membrane. Bessel's equation of order zero is

$$r^2 y''(r) + r y'(r) + r^2 y(r) = 0, \quad r \geq 0. \tag{15.4}$$

The unique solution of this equation with $y(0) = 1$ is called the Bessel function of the first kind of order zero, denoted $J_0(r)$. We shall use Theorem 15.2 to verify that J_0 has the integral representation

$$J_0(r) = \frac{1}{\pi} \int_0^\pi \cos(r \sin \theta) \, d\theta.$$

For the moment we let $u(r)$ denote the right side of this equation. First we note that because of the normalizing factor of $(1/\pi)$, $u(0) = 1$. Next we use Theorem 15.2 to calculate

$$u'(r) = -\frac{1}{\pi} \int_0^\pi \sin(r \sin \theta) \sin \theta \, d\theta$$

and

$$u''(r) = -\frac{1}{\pi} \int_0^\pi \cos(r \sin \theta) \sin^2 \theta \, d\theta.$$

Before we can add up the terms in the equation, we must integrate by parts once (with respect to θ) in the expression for u':

$$u'(r) = -\frac{1}{\pi}\left[-\cos\theta\sin(r\sin\theta)\Big|_0^\pi + \int_0^\pi r\cos(r\sin\theta)\cos^2\theta\,d\theta\right]$$

$$= -\frac{r}{\pi}\int_0^\pi \cos(r\sin\theta)\cos^2\theta\,d\theta.$$

Then we have

$$r^2u''(r) + ru'(r) + r^2u(r) = \frac{r^2}{\pi}\int_0^\pi \cos(r\sin\theta)[-\sin^2\theta - \cos^2\theta + 1]\,d\theta$$

$$= 0.$$

Because there is only one solution of (15.4) with $y(0) = 1$, we conclude that $J_0(r) = u(r)$.

Bessel functions are used in many situations in the solution of partial differential equations. Some examples are given in the book of Cooper, [Coo].

Improper Integrals with a Parameter

The results for improper integrals depending on a parameter are similar to those for infinite series of functions. The finite integrals $\int_c^r f(x,t)dx$ correspond to the partial sums $\sum_{k=1}^n f_k(x,t)$. As was the case with series, the key idea is uniform convergence. For brevity, we shall state results where $x \in \mathbb{R}$, and t is a single parameter. They are easily extended to the case of several parameters $t_1, \ldots, t_p$ and $\mathbf{x} \in \mathbb{R}^n$.

Definition 15.1 Let $I \subset \mathbb{R}$ be an interval. For each $t \in I$, let $x \rightarrow f(x,t)$ be a bounded integrable function on $[c,r]$ for each $r \leq c$. Assume that for each $t \in I$, the improper integral $\int_c^\infty f(x,t)dx$ converges to $F(t)$. We shall say that the improper integral $\int_0^\infty f(x,t)dx$ converges *uniformly* on I to $F(t)$ if, for each $\varepsilon > 0$, there is an $R \geq c$ such that

$$\left|F(t) - \int_0^r f(x,t)dx\right| < \varepsilon$$

for all $r \geq R$ and all $t \in I$.

The analogue of the Weierstrass test (Theorem 8.16) is a comparison test.

Theorem 15.3 Let $I \subset \mathbb{R}$ be an interval, and for each $t \in I$, let $x \to f(x, t)$ be bounded and integrable on $[c, r]$ for each $r \geq c$. If there is a function $h(x)$ such that, for all $t \in I$, $|f(x, t)| \leq h(x)$ and $\int_c^\infty h(x)dx$ converges, then the integral $\int_c^\infty f(x, t)dx$ converges uniformly on I to $F(t)$.

Proof: In view of Theorem 7.20, we see that $\int_c^\infty f(x, t)dx$ converges absolutely to $F(t)$ each $t \in I$. Now

$$\left| F(t) - \int_c^r f(x, t)dx \right| \leq \int_r^\infty |f(x, t)|dx \leq \int_r^\infty |h(x)|dx.$$

Given $\varepsilon > 0$, we may choose $R > c$ such that $\int_r^\infty h(x)dx < \varepsilon$ for $r \geq R$. It follows that

$$\left| F(t) - \int_c^r f(x, t)dx \right| < \varepsilon$$

for $r \geq R$ and all $t \in I$.

Theorem 15.1 takes the following form for improper integrals.

Theorem 15.4 Let $I \subset \mathbb{R}$ be an interval and suppose that $f : [c, \infty) \times I$ is continuous and that $g : [c, \infty) \to \mathbb{R}$ is continuous on each interval $[c, r]$, $r \geq c$, except at possibly a finite number of points. If the improper integral $\int_c^\infty f(x, t)g(x)dx$ converges uniformly on I to $F(t)$, then $F(t)$ is continuous on I.

Proof: By Theorem 15.1, the function

$$F_r(t) \equiv \int_c^r f(x, t)g(x)dx$$

is continuous on I for each $r \geq c$.   According to Definition 15.1, $F(t) = \lim_{r \to \infty} F_r(t)$, uniformly for $t \in I$. Theorem 8.10 can be extended easily to cover the case of a continuous index. We conclude that $F(t)$ is continuous on I.

In a similar manner we have the analogue for improper integrals of Theorem 15.2.

Theorem 15.5 Let $I \subset \mathbb{R}$ be an interval and suppose that $f : I \times [c, \infty) \to \mathbb{R}$ and $f_t : I \times [c, \infty) \to \mathbb{R}$ are both continuous. Let $g : [c, \infty) \to \mathbb{R}$ be continuous on each interval $[c, r]$, $r \geq c$, except possibly at a finite number of points. If the improper integral $\int_c^\infty f(x, t)g(x)dx$ converges uniformly to $F(t)$ and $\int_c^\infty f_t(x, t)g(x)dx$ converges uniformly for $t \in I$, then F is C^1 on I and

$$F'(t) = \int_c^\infty f(x, t)g(x)dx.$$

Proof: Let F_r be defined as in the proof of Theorem 15.4. Then by Theorem 15.2, F_r is in $C^1(I)$ for each $r \geq c$, and

$$F_r'(t) = \int_c^r f_t(x, t)g(x)dx.$$

Since F_r converges uniformly to F on I and $F_r'(t)$ converges uniformly on I to $\int_c^\infty f_t(x, t)g(x)dx$, we conclude from Theorem 8.12 that F is in $C^1(I)$ and that the assertion of the theorem holds.

EXAMPLE 15.3: The Laplace transform is a very useful tool in the study of differential equations. It has the form

$$F(t) = \int_0^\infty e^{-xt} f(x)dx.$$

A function $f(x)$ is said to be of *exponential type* if there exist constants M and α such that

$$|f(x)| \leq Me^{\alpha x} \quad \text{for } x \geq 0.$$

Then for $t > \alpha$ the integral for $F(t)$ converges and it converges uniformly on $[\alpha + \delta, \infty)$ for each $\delta > 0$. Using Theorem 15.5, it is easy to show that F is C^∞ on $t > \alpha$ with

$$F^{(n)}(t) = \int_0^\infty (-x)^n e^{-xt} f(x)dx.$$

In particular let us look at the Laplace transform of the square wave function. Let

$$f_0(x) = \begin{cases} 1 & \text{for } 1 \leq x < 2 \\ 0 & \text{elsewhere} \end{cases}$$

and set

$$f(x) = \sum_{k=0}^{\infty} f_0(x - 2k).$$

Then our results tell us that the Laplace transform of this function is C^{∞} for $t > 0$.

Since solutions of the diffusion equation are written in the form

$$u(x, t) = \int_{\mathbb{R}} G(x - y, t) f(y) dy,$$

we shall see many examples of this kind of improper integral in Section 15.3.

Exercises for 15.1

1. Use the corollary to Theorem 15.2 to calculate the derivative of the following expressions.

a) $f(t) = \int_{t^2}^{t} \sin^2(t - s) ds$
b) $g(x) = \int_{1}^{e^x} \log(xt) dt$.

2. Let $k(x, t)$ be a C^1 function on $R = \{0 \le t \le 1, \ x \in \mathbb{R}\}$ such that $k(x, t) > 0$ and $k(x, t) \le 1/|x|$ for $|x| \ge 1$. Let $f(x) = \int_{0}^{1} k(x, t) dt$.

a) Show that $f(x)$ attains a maximum value in at least one point x_0.
b) Show that $\int_{0}^{1} k_x(x_0, t) dt = 0$.

3. Bessel's equation of order n is

$$r^2 y'' + r y' + (r^2 - n^2) y = 0.$$

Using Example 15.2 as a guide, show that

$$J_n(r) = \frac{1}{\pi} \int_{0}^{\pi} \cos(r \sin \theta - n\theta) \, d\theta$$

is a solution of Bessel's equation of order n with $J_n(0) = 1$. This solution is called the Bessel function of the first kind of order n.

4. The trapezoid rule is extremely accurate when applied to the integral of a C^∞ periodic function over its period. As we saw in Chapter 7, the error tends to zero faster than h^n for any n. Evaluate the integral (15.4) which represents $J_0(r)$ using the trapezoid rule with $h = \pi/100$ for various values of r. Compare the values you compute with those produced by a mathematical software package.

5. The principle of Duhamel (Example 15.1) is valid for quite general differential equations. Let $a(t)$, $b(t)$, and $c(t)$ be continuous on $t \geq 0$ with $a(t) > 0$. Let $z(t, s)$ be the unique C^2 solution of the initial value problem

$$a(t)z_{tt} + b(t)z_t + c(t)z = 0 \quad \text{for } t \geq s$$

$$z(s, s) = 0, \quad z_t(s, s) = \frac{1}{a(s)}.$$

Show that if $q(t)$ is continuous on $t \geq 0$, then

$$y(t) = \int_0^t z(t, s)q(s)ds$$

satisfies the inhomogeneous differential equation and initial conditions

$$a(t)y'' + b(t)y' + c(t)y(t) = q(t)$$

$$y(0) = 0 \quad y'(0) = 0.$$

6. Let $g(x)$ be a C^1 function on $\mathbb{R}$. Show that

$$u(x, t) = \frac{1}{2} \int_{x-t}^{x+t} g(y)dy$$

solves the initial value problem for the wave equation

$$u_{tt} - u_{xx} = 0, \quad u(x, 0) = 0, \quad u_t(x, 0) = g(x).$$

7. The principle of Duhamel can also be applied to solutions of inhomogeneous partial differential equations. Let $w(x, t, s)$ be the solution of
the initial value problem

$$w_{tt} - w_{xx} = 0, \quad \text{for } t \geq s$$

$$w(x, s, s) = 0, \quad w_t(x, s, s) = q(x, s).$$

Assume that $q(x, t)$ is continuous on $\mathbb{R}^2$ with $q(x, t) = 0$ for $|x| \geq a > 0$.
According to the result of exercise 6,

$$w(x, t, s) = \frac{1}{2} \int_{x-(t-s)}^{x+(t-s)} q(y, s) dy.$$

Consider the initial value problem for the wave equation

$$u_{tt} - u_{xx} = q(x, t)$$

$$u(x, 0) = 0, \quad u_t(x, 0) = 0.$$

a) Show that the solution of this initial value problem is given by the
integral

$$u(x, t) = \int_0^t w(x, t, s) ds.$$

b) Show that the integral for u can be written as double integral of
$q(y, s)/2$ over the backward triangle

$$D(x, t) = \{(y, s) : |y - x| \leq (t - s), \ 0 \leq s \leq t\}.$$

8. For $t > 0$, let

$$F(t) = \int_0^\infty e^{-xt} dx.$$

F is the Laplace transform of the function $f(x) \equiv 1$.

a) Verify the formula $F(t) = 1/t$ for $t > 0$.
b) Justify the formula

$$\frac{n!}{t^{n+1}} = \int_0^\infty x^n e^{-xt} dx.$$

9. Let $f(x)$ be bounded and continuous on $[0, \infty)$. Let

$$F(t) = \int_0^\infty \frac{t f(x) dx}{t^2 + x^2}.$$

a) Show that for each $\delta > 0$, the integral converges uniformly on $[\delta, \infty)$.
b) Show that $\lim_{t \downarrow 0} F(t) = (\pi/2) f(0)$.
c) If $f(0) \neq 0$, does the integral converge uniformly on $[0, \infty)$?

15.2 Approximation by Smooth Functions

In many situations we can establish some formula for functions which have a certain number of derivatives, and we would like to show that this formula also holds for functions which are only continuous. Sometimes this can be accomplished by making smooth approximations to the function in question. We shall use the idea of the convolution to accomplish this.

Definition 15.2 Let $f, g : \mathbb{R}^n \to \mathbb{R}$. The *convolution*, denoted $f * g$, is the integral

$$(f * g)(\mathbf{x}) = \int_{\mathbb{R}^n} f(\mathbf{y}) g(\mathbf{x} - \mathbf{y}) d\mathbf{y} \tag{15.5}$$

when this integral converges.

The convolution is defined for various classes of functions. For our purposes, we shall require that (i) f is bounded, f is continuous except possibly on a set of Jordan content zero, and (ii) g is continuous and $\int_{\mathbb{R}^n} |g(\mathbf{x})| d\mathbf{x}$ converges.

In this case the integral will converge for all $\mathbf{x}$. An important feature of the convolution is that it is commutative. Make the change of variable $\mathbf{z} = \mathbf{x} - \mathbf{y}$ in (15.5). The mapping $\mathbf{y} \rightarrow \mathbf{x} - \mathbf{z}$ has Jacobian matrix $-\mathbf{I}$ and $|\det(-\mathbf{I})| = |(-1)^n| = 1$. Hence by the change of variable theorem, we see that

$$(f * g)(\mathbf{x}) = \int_{\mathbb{R}^n} f(\mathbf{y})g(\mathbf{x} - \mathbf{y}) \, d\mathbf{y} = \int_{\mathbb{R}^n} f(\mathbf{x} - \mathbf{z})g(\mathbf{z}) \, d\mathbf{z} = (g * f)(\mathbf{x}).$$

If $g \geq 0$ and $\int_{\mathbb{R}^n} g(\mathbf{x}) \, d\mathbf{x} = 1$, then $(f * g)(\mathbf{x}) = \int f(\mathbf{y})g(\mathbf{x} - \mathbf{y}) \, d\mathbf{y}$ is a weighted average of the values of $f(\mathbf{y})$. If g_k is a sequence of functions with $\int g_k \, d\mathbf{x} = 1$ and $g_k = 0$ for $\|\mathbf{x}\| \geq 1/k$, then the average $f * g_k$ is taken over a smaller and smaller set of points $\|\mathbf{x} - \mathbf{y}\| \leq 1/k$ as $k \rightarrow \infty$. This suggests that the averages $f * g_k$ should converge to $f(\mathbf{x})$. We can also think of $f * g$ as the sum of translates of g. If g is a function in C^k, then it is plausible that $f * g$ should also be in class C^k.

To make these speculations concrete, we make a definition.

Definition 15.3 A sequence of continuous functions $g_k : \mathbb{R}^n \rightarrow \mathbb{R}$ is an *approximate identity* if

(i) $g_k(\mathbf{x}) \geq 0$.
(ii) g_k is continuous and integrable on $\mathbb{R}^n$ for each k with $\int_{\mathbb{R}^n} g_k(\mathbf{x}) d\mathbf{x} = 1$.
(iii) For each $\delta > 0$,

$$\lim_{k \to \infty} \int_{\|\mathbf{x}\| \geq \delta} g_k(\mathbf{x}) \, d\mathbf{x} = 0.$$

We shall give some examples of approximate identities after the next theorem.

Theorem 15.6 Let $f : \mathbb{R}^n \rightarrow \mathbb{R}$ be bounded, and continuous except perhaps on set of Jordan content zero. Let the sequence of functions g_k be an approximate identity and set

$$f_k(\mathbf{x}) = \int_{\mathbb{R}^n} f(\mathbf{y})g_k(\mathbf{x} - \mathbf{y}) \, d\mathbf{y}.$$

Then

(i) f_k is bounded and continuous on $\mathbb{R}^n$.

(ii) If f is continuous at the point $\mathbf{x}$, $f_k(\mathbf{x})$ converges to $f(\mathbf{x})$.

(iii) If f is continuous for all $\mathbf{x}$, and $f(\mathbf{x}) = 0$ outside some bounded set, then the sequence f_k converges uniformly on $\mathbb{R}^n$ to f.

Proof: Let $M = \sup_{\mathbb{R}^n} |f(\mathbf{x})|$. Then since $\int_{\mathbb{R}^n} g_k(\mathbf{x})d\mathbf{x} = 1$, we see that

$$|f_k(\mathbf{x})| \leq \int_{\mathbb{R}^n} |f(\mathbf{x} - \mathbf{y})|g_k(\mathbf{y})\, d\mathbf{y}$$

$$\leq M \int_{\mathbb{R}^n} g(\mathbf{y})d\mathbf{y} = M.$$

Next we fix $\mathbf{x}_0 \in R^n$ and fix the index k. Assume that $\|\mathbf{x} - \mathbf{x}_0\| \leq 1$.

$$|f_k(\mathbf{x}) - f_k(\mathbf{x}_0)| = \left| \int_{\mathbb{R}^n} f(\mathbf{y})[g_k(\mathbf{x} - \mathbf{y}) - g_k(\mathbf{x}_0 - \mathbf{y})]\, d\mathbf{y} \right|$$

$$\leq M \int_{\mathbb{R}^n} |g_k(\mathbf{x} - \mathbf{y}) - g_k(\mathbf{x}_0 - \mathbf{y})|\, d\mathbf{y}.$$

Let $\varepsilon > 0$ be given. Because the integral $\int_{\mathbb{R}^n} g_k$ converges, there is an $R > 0$ such that $\int_{\|\mathbf{y}\| \geq R} g_k < \varepsilon$. Let $R_1 = R + \|\mathbf{x}_0\| + 1$. We note that $\|\mathbf{z} - \mathbf{x}\| \geq R_1$ implies $\|\mathbf{z}\| \geq R$. Hence if we make the change of variable $\mathbf{z} = \mathbf{x} - \mathbf{y}$ we have

$$\int_{\|\mathbf{y}\| \geq R_1} g_k(\mathbf{x} - \mathbf{y})\, d\mathbf{y} = \int_{\|\mathbf{z} - \mathbf{x}\| \geq R_1} g_k(\mathbf{z})\, d\mathbf{z}$$

$$\leq \int_{\|\mathbf{z}\| \geq R} g_k(\mathbf{z})\, d\mathbf{z} < \varepsilon.$$

In the same fashion we also have

$$\int_{\|\mathbf{y}\| \geq R_1} g_k(\mathbf{x}_0 - \mathbf{y})\, d\mathbf{y} < \varepsilon.$$

Thus we see that

$$|f_k(\mathbf{x}) - f_k(\mathbf{x}_0)| \leq M \int_{B_1} |g_k(\mathbf{x} - \mathbf{y}) - g_k(\mathbf{x}_0 - \mathbf{y})| \, d\mathbf{y} + 2M\varepsilon$$

where $B_1 = \{\|\mathbf{y}\| \leq R_1\}$. The function $(\mathbf{x}, \mathbf{y}) \rightarrow g_k(\mathbf{x} - \mathbf{y})$ is continuous on $\mathbb{R}^n \times \mathbb{R}^n$ and hence is uniformly continuous on the compact subset $\{\|\mathbf{x} - \mathbf{x}_0\| \leq 1\} \times B_1$. Thus there is a $\delta > 0$, $\delta \leq 1$, such that

$$|g_k(\mathbf{x} - \mathbf{y}) - g_k(\mathbf{x}_0 - \mathbf{y})| \leq \varepsilon$$

for $\|\mathbf{x} - \mathbf{x}_0\| \leq \delta$ and all $\mathbf{y} \in B_1$. We conclude that for $\|\mathbf{x} - \mathbf{x}_0\| \leq \delta$,

$$|f_k(\mathbf{x}) - f_k(\mathbf{x}_0)| \leq \varepsilon M \mathrm{vol}(B_1) + 2M\varepsilon.$$

This establishes (i).

Now we turn to (ii). Because the sequence g_k is an approximate identity,

$$\int_{\mathbb{R}^n} g_k(\mathbf{x} - \mathbf{y}) \, d\mathbf{y} = \int_{\mathbb{R}^n} g_k(\mathbf{y}) d\mathbf{y} = 1.$$

Hence

$$f(\mathbf{x}) - f_k(\mathbf{x}) = \int_{\mathbb{R}_n} [f(\mathbf{x}) - f(\mathbf{y})] g_k(\mathbf{x} - \mathbf{y}) \, d\mathbf{y}.$$

Let $\varepsilon > 0$ be given. Since $\mathbf{x}$ is a point where f is continuous, there is a $\delta > 0$ such that $|f(\mathbf{x}) - f(\mathbf{y})| < \varepsilon$ for $\|\mathbf{x} - \mathbf{y}\| < \delta$. We write

$$f(\mathbf{x}) - f_k(\mathbf{x}) = I_1 + I_2$$

where

$$I_1 = \int_{\|\mathbf{x}-\mathbf{y}\| \leq \delta} [f(\mathbf{x}) - f(\mathbf{y})] g_k(\mathbf{x} - \mathbf{y}) \, d\mathbf{y}$$

and

$$I_2 = \int_{\|\mathbf{x}-\mathbf{y}\| \geq \delta} [f(\mathbf{x}) - f(\mathbf{y})] g_k(\mathbf{x} - \mathbf{y}) \, d\mathbf{y}.$$

Because $|f(\mathbf{x}) - f(\mathbf{y})| < \varepsilon$ for $\|\mathbf{x} - \mathbf{y}\| \leq \delta$, we see that

$$|I_1| \leq \varepsilon \int_{\mathbb{R}^n} g_k(\mathbf{x} - \mathbf{y}) \, d\mathbf{y} = \varepsilon.$$

With δ now fixed, and with $M = \sup |f|$, we have

$$|I_2| \leq \int_{\|\mathbf{x}-\mathbf{y}\| \geq \delta} (|f(\mathbf{x})| + |f(\mathbf{y})|) g_k(\mathbf{x} - \mathbf{y}) \, d\mathbf{y}$$

$$\leq 2M \int_{\|\mathbf{x}-\mathbf{y}\| \geq \delta} g_k(\mathbf{x} - \mathbf{y}) \, d\mathbf{y} = 2M \int_{\|\mathbf{x}\| \geq \delta} g_k(\mathbf{x}) \, d\mathbf{x}.$$

Thus

$$|f(\mathbf{x}) - f_k(\mathbf{x})| \leq |I_1| + |I_2| \leq \varepsilon + 2M \int_{\|\mathbf{x}\| \geq \delta} g_k(\mathbf{x}) \, d\mathbf{x}.$$

Now choose K so that for $k \geq K$, $\int_{\|\mathbf{x}\| \geq \delta} g_k(\mathbf{x}) \, d\mathbf{x} \leq \varepsilon/(2M)$. Then for $k \geq K$,

$$|f(\mathbf{x}) - f_k(\mathbf{x})| \leq 2\varepsilon.$$

The assertion (ii) is proved.

Finally we note that if f is continuous, and $f = 0$ outside a bounded set, then f is uniformly continuous. Hence the choice of δ in the previous paragraph is independent of $\mathbf{x}$. The choice of K, which depends on δ is then also independent of $\mathbf{x}$. The theorem is proved.

EXAMPLE 15.4: As a first example of an approximate identity, we take the functions

$$g_k(\mathbf{x}) = \begin{cases} A_k(1 - k\|\mathbf{x}\|) & \text{for } \|\mathbf{x}\| \leq 1/k \\ 0 & \text{for } \|\mathbf{x}\| \geq 1/k \end{cases}.$$

Here $\|\mathbf{x}\| = \|\mathbf{x}\|_2$ and A_k is chosen so that $\int_{\mathbb{R}^n} g_k = 1$. For $k \geq 1/\delta$, $\int_{\|\mathbf{x}\| \geq \delta} g_k \, d\mathbf{x} = 0$ so that the sequence of functions g_k is an approximate identity. In an exercise for this section we will see how convolution with this function produces C^1 approximations, even though g_k is not C^1.

EXAMPLE 15.5:   Let

$$\varphi(\mathbf{x}) = \frac{1}{\pi^{n/2}} e^{-\|\mathbf{x}\|^2}$$

with $\|\mathbf{x}\| = \|\mathbf{x}\|_2$. Then we take the functions

$$g_k(\mathbf{x}) = k^n \varphi(k\mathbf{x}).$$

Using the change of variable $\mathbf{y} = k\mathbf{x}$ and Theorem 14.28 we see that $\int_{\mathbb{R}^n} g_k(\mathbf{x})\, d\mathbf{x} = 1$.
Next we see that for $\delta > 0$,

$$\int_{\|\mathbf{x}\|\geq\delta} g_k(\mathbf{x})\, d\mathbf{x} = k^n \int_{\|\mathbf{x}\|\geq\delta} \varphi(k\mathbf{x})\, d\mathbf{x}$$

$$= \frac{\Omega_n}{\pi^{n/2}} \int_{k\delta}^{\infty} e^{-\rho^2} \rho^{n-1}\, d\rho.$$

The last integral clearly tends to zero as $k \to \infty$. Thus the functions g_k constitute an approximate identity. The functions g_k are C^∞. If we assume that f satisfies the hypotheses of Theorem 15.3, we may use Theorem 15.5 to differentiate under the integral sign to any order:

$$\frac{\partial^m}{\partial x_j^m} f_k(\mathbf{x}) = \int_{\mathbb{R}^n} \frac{\partial^m}{\partial x_j^m} g_k(\mathbf{x} - \mathbf{y}) f(\mathbf{y})\, \mathbf{y}.$$

In this case the approximations f_k are C^∞.
   To get a clearer picture of these smooth approximations, we consider an example in one dimension. Let $f(x) = 1$ for $0 < x < 1$ and $f = 0$ elsewhere. The smooth approximations are

$$f_k(x) = \int_0^1 g_k(x - y)\, dy = \frac{k}{\sqrt{\pi}} \int_0^1 e^{-k^2(x-y)^2}\, dy.$$

We make the change of variable $z = k(x - y)$ and find that

$$f_k(x) = \frac{1}{\sqrt{\pi}} \int_{k(x-1)}^{kx} e^{-z^2}\, dz. \tag{15.6}$$

These integrals cannot be computed in terms of elementary functions. Instead we shall rewrite the integral in terms of a well known function from statistics which is computed in tables and is available in most software packages.

The error function erf(x) is defined for $x \geq 0$ by the improper integral

$$\mathrm{erf}(x) = \frac{2}{\sqrt{\pi}} \int_0^x e^{-s^2} ds. \qquad (15.7)$$

The constant $2/\sqrt{\pi}$ is chosen so that $0 \leq \mathrm{erf}(x) < 1$ with $\lim_{x \to \infty} \mathrm{erf}(x) = 1$.

Now for $x \geq 1$, the integral (15.6) can be rewritten as the difference

$$f_k(x) = \frac{1}{\sqrt{\pi}} \int_0^{kx} e^{-z^2} dz - \frac{1}{\sqrt{\pi}} \int_0^{k(x-1)} e^{-z^2} dz$$

$$= (1/2)[\mathrm{erf}(kx) - \mathrm{erf}(k(x-1))].$$

We can make similar manipulations for other values of x to arrive at the formulas

$$f_k(x) = \begin{cases} (1/2)[\mathrm{erf}(k(1-x)) - \mathrm{erf}(-kx)], & x \leq 0 \\ (1/2)[\mathrm{erf}(kx) + \mathrm{erf}(k(1-x))], & 0 < x < 1. \\ (1/2)[\mathrm{erf}(kx) - \mathrm{erf}(k(x-1))], & x \geq 1 \end{cases}$$

From these formulas we see that

$$\lim_{k \to \infty} f_k(x) = \begin{cases} 0 & \text{for } x < 0 \text{ and } x > 1 \\ 1/2 & \text{for } x = 0 \text{ and } x = 1. \\ 1 & \text{for } 0 < x < 1 \end{cases}$$

The graph of f_k is shown in Figure 15.1 for several values of k.

The Weierstrass Approximation Theorem

The Weierstrass approximation theorem states that a continuous function $f(\mathbf{x})$ on a compact set can be approximated uniformly by polynomials. We shall use convolution and an approximate identity to prove this result.

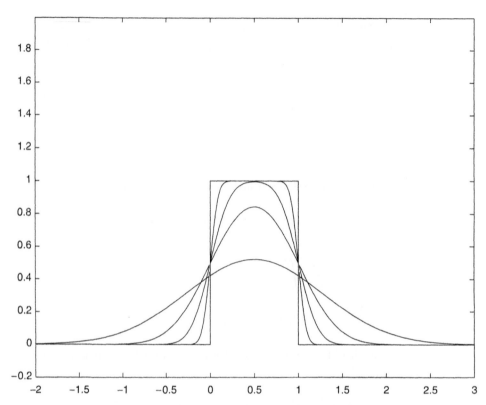

Figure 15.1 Smooth approximations f_k to the function $f(x) = 1$ for $0 < x < 1$ and $f(x) = 0$ elsewhere. Graphs of f_k are shown for $k = 1, 2, 4, 10$.

To simplify the calculations we shall restrict our attention to functions of one variable. Note that, in the convolution, if we take $g(x) = x^p$ for some power p, we have

$$(f * g)(x) = \int_{\mathbb{R}} f(y)(x - y)^p dy$$

$$= \int_{\mathbb{R}} f(y)[x^p + px^{p-1}(-y) + \cdots + px(-y)^{p-1} + (-y)^p] dy$$

$$= \sum_{j=0}^{p} c_j x^{p-j}$$

where

$$c_j = \begin{pmatrix} p \\ j \end{pmatrix} \int_{\mathbb{R}} f(y)(-y)^j \, dy.$$

Therefore, if g is a polynomial, the convolution $f * g$ will also be a polynomial. It is clear, however, that we cannot we use polynomials for the functions g_k in an approximate identity. On the other hand, because we are only interested in approximating f on a finite interval $[a, b]$, we do not need for $f * g$ to be a polynomial on all of $\mathbb{R}$. If f is continuous on $[a, b]$ and $f(x) = 0$ for $x < a$ or $x > b$, the convolution integral,

$$(f * g)(x) = \int_a^b f(y)g(x - y) \, dy,$$

uses only the values of g on the interval $[x - b, x - a]$. Furthermore, the values of $(f * g)(x)$ for $a \le x \le b$ are determined by the values of g in the symmetric interval $\{|x| \le b - a\}$. Hence if g equals a polynomial on $\{|x| \le b - a\}$, then $(f * g)(x)$ will be a polynomial for $a \le x \le b$. These observations lead to a proof of

Theorem 15.7 (Weierstrass polynomial approximation)
Let $f(x)$ be continuous on the interval $[a, b]$. Then there is a sequence of polynomials $f_k(x)$ such that f_k converges uniformly to f on $[a, b]$.

Proof: First, a small technical detail. The function f is given to us on the interval $[a, b]$ and we must extend it to be a continuous function $\tilde{f}$ on $\mathbb{R}$ with $\tilde{f}(x) = 0$ outside a larger interval. One simple way to do this is to add on a linear piece at either end to take the function to zero. For example,

$$\tilde{f}(x) = \begin{cases} f(a)(x + 1 - a) & \text{for } a - 1 \le x \le a \\ f(x) & \text{for } a \le x \le b \\ f(b)(1 + b - x) & \text{for } b \le x \le b + 1 \\ 0 & \text{for } x \le a - 1 \text{ and } x \ge b + 1. \end{cases}$$

If we approximate $\tilde{f}$ uniformly by polynomials on $[a - 1, b + 1]$, then, of course, we have uniform approximation of f on $[a, b]$. Thus we can assume from now on that f is continuous on $\mathbb{R}$ and $f = 0$ outside the interval $[a, b]$.

Now we must construct an approximate identity that is customized to be equal to a polynomial on the interval $\{|x| \leq b - a\}$. It will suffice to do this for the interval $[-1, 1]$. Let

$$h_k(x) = \begin{cases} (1 - x^2)^k / c_k & \text{for } |x| \leq 1 \\ 0 & \text{for } |x| > 1 \end{cases}.$$

The constant $c_k = \int_{-1}^{1}(1-x^2)^k dx$. The functions h_k are continuous, nonnegative, and c_k is chosen so that $\int_{\mathbb{R}} h_k(x)dx = 1$. It remains to show that for each $\delta, 0 < \delta < 1$, $\int_{|x| \geq \delta} h_k(x)dx$ tends to zero as $k \to \infty$. Now because $(1 - x^2)^k$ is monotone decreasing on $[0, 1]$, we can estimate

$$\int_{|x| \geq \delta} h_k(x)dx = \frac{2}{c_k}\int_{\delta}^{1}(1 - x^2)^k dx \leq \frac{2}{c_k}(1 - \delta)(1 - \delta^2)^k.$$

We must show that the quotient $(1 - \delta^2)^k / c_k$ tends to zero as $k \to \infty$ and for this we need a lower bound for c_k. Recall the Bernoulli inequality, $(1 + x)^k \geq 1 + kx$ for $x \geq -1$, which implies that $(1 - x^2)^k \geq 1 - kx^2$ for $|x| \leq 1$. This estimate is only useful for $|x| \leq k^{-1/2}$, but that is enough because

$$c_k = \int_{-1}^{1}(1 - x^2)^k dx \geq 2\int_{0}^{k^{-1/2}}(1 - x^2)^k dx$$

$$\geq 2\int_{0}^{k^{-1/2}}(1 - kx^2)dx = \frac{2}{3}k^{-1/2}.$$

Therefore

$$\frac{(1 - \delta^2)^k}{c_k} \leq (3/2)k^{1/2}e^{k \log(1-\delta^2)}.$$

The constant $\log(1 - \delta^2) < 0$. We have seen from l'Hôpital's rule that $k^p e^{-Ck}$ tends to zero as $k \to \infty$ for any power p whenever $C > 0$. We conclude that for $\delta > 0$,

$$\lim_{k \to \infty} \int_{|x| \geq \delta} h_k(x)dx = 0.$$

For the general interval $[a, b]$, the functions

$$g_k(x) = (b - a)h_k((b - a)x)$$

constitute an approximate identity with the property that $f_k * g_k$ is a polynomial on $[a, b]$. By Theorem 15.3, f_k converge uniformly to f. The proof is complete.

Exercises for 15.2

1. Let $g_k(x) = k(1 - k|x|)$ for $|x| \le 1/k$ and $g_k(x) = 0$ otherwise. We have verified that g_k is an approximate identity. Let $f(x) = 1$ for $0 \le x \le 1$ and $f = 0$ otherwise. Compute the convolutions $f_k = f * g_k$ and show directly that f_k converges to f at points were f is continuous. Show further that f_k is actually C^1.

2. Suppose that f is continuous except at $x = 0$ and that $f_r = \lim_{x \downarrow 0} f(x)$ and $f_l = \lim_{x \uparrow 0} f(x)$ both exist. Let g_k be an approximate identity with $g_k(-y) = g_k(y)$. Let $f_k = f * g_k$. Show that

$$\lim_{k \to \infty} f_k(0) = \frac{f_r + f_l}{2}.$$

3. Let

$$f(x) = \begin{cases} f_l & \text{for } x < 0 \\ f_r & \text{for } x > 0 \end{cases}.$$

Let g_k be an approximate identity. Show that $f_k = f * g_k$ is differentiable at $x = 0$ with $f_k'(0) = (f_r - f_l)g_k(0)$.

4. Let

$$f(x) = \begin{cases} 0 & \text{for } x < 0 \\ 2x & \text{for } 0 \le x < 1/2 \\ 2(1 - x) & \text{for } 1/2 \le x < 1 \\ 0 & \text{for } x > 1 \end{cases}.$$

Take $h_k(x) = (1/c_k)(1 - x^2)^k$ for $|x| \le 1$ and $h_k = 0$ elsewhere. Compute the approximations of Theorem 15.7, $f_k = f * h_k$, for $k = 1$. Use a symbolic software package to compute f_k for higher values of k. Plot the functions f_k on $-0.5 \le x \le 1.5$. Notice that the convergence is very slow.

5. Suppose that f is continuous on $[a, b]$ and that $\int_a^b x^k f(x)dx = 0$ for $k = 1, 2, \dots$. Use the Weierstrass approximation theorem to show that $f \equiv 0$.

15.3 Diffusion

The derivation of many of the partial differential equations of mathematical physics starts with an integral conservation law that states that some macroscopic quantity such as mass, momentum, or energy is conserved. There are two basic ideas that link the macroscopic description with the partial differential equation which is the microscopic description. The first is given in the

Theorem 15.8 (Null Theorem) Let $f(\mathbf{x})$ be continuous on an open set $G \subset \mathbb{R}^n$. Suppose that $\int_D f(\mathbf{x})dx = 0$ for all subsets $D \subset G$. Then $f(\mathbf{x}) \equiv 0$ in G.

Proof: Let $\mathbf{x}_0 \in G$. Since G is open, there is an $r > 0$ such that the ball $B(\mathbf{x}_0, \delta) = \{\mathbf{x} : \|\mathbf{x} - \mathbf{x}_0\| < \delta\} \subset G$ for $\delta \leq r$. The norm can be any one of the equivalent norms. By the corollary to Theorem 14.10,

$$f(\mathbf{x}_0) = \lim_{\delta \to 0} \frac{\int_{B(\mathbf{x}_0, \delta)} f(\mathbf{x})\, dx}{\mathrm{vol}(B(\mathbf{x}_0, \delta))} = 0.$$

The second basic idea is the divergence theorem which we state here.

Divergence Theorem

Let $G \subset \mathbb{R}^n$ be an open set and let $\mathbf{f} : G \to \mathbb{R}^n$ be in $C^1(G)$. Let $D \subset G$ be a compact set with smooth boundary surface S. Let $\mathbf{n}$ be the exterior unit normal at each point of S. Then

$$\int_D \mathrm{div}(\mathbf{f})\, dx = \int_S \mathbf{f} \cdot \mathbf{n}\, dS.$$

Here we have used the more customary calculus notation $\mathbf{f} \cdot \mathbf{n}$ for $\langle \mathbf{f}, \mathbf{n} \rangle$.

We shall give two examples of partial differential equations derived in this manner. The first is the partial differential equation that governs diffusion. The second, discussed in Section 15.4, is the Euler equations which describe fluid flow.

Now we shall assume that $\mathbf{x} = (x, y, z) \in \mathbb{R}^3$. Let $u(\mathbf{x}, t)$ be the *concentration* at the point $\mathbf{x}$ and at time t of some chemical that is diffusing through a motionless fluid. The concentration has the units of mass/volume. Let $D \subset \mathbb{R}^3$ be a compact set with a smooth boundary. The mass of the chemical contained in D at time t is

$$\int_D u(\mathbf{x}, t)\, d\mathbf{x}$$

and

$$\frac{d}{dt} \int_D u(\mathbf{x}, t)\, d\mathbf{x}$$

is the time rate of change of the mass of chemical contained in D. The *flux* $\mathbf{f} = (f_1, f_1, f_3)$ is a vector field that describes the rate at which the chemical is diffusing through a surface in the fluid. The flux has the units of mass/area/time. Let S be a piece of two-dimensional surface in $\mathbb{R}^3$ and suppose that $\mathbf{n}(\mathbf{x})$ is a unit normal to the surface at the point $\mathbf{x} \in S$. Then $\mathbf{f}(\mathbf{x}) \cdot \mathbf{n}(\mathbf{x})$ is the rate at which the chemical is flowing through the surface in the direction specified by $\mathbf{n}(\mathbf{x})$. The rate at which the chemical is flowing through the piece of surface S is

$$\int_S \mathbf{f} \cdot \mathbf{n}\, dS.$$

Now suppose that D is a compact set with smooth bounding surface S and let $\mathbf{n}$ be the exterior unit normal to S. In the absence of any sources or sinks of the chemical, the time rate of change of the mass of chemical in D equals the negative of the rate at which the chemical is flowing outward through the boundary:

$$\frac{d}{dt} \int_D u(\mathbf{x}, t)\, d\mathbf{x} = - \int_{\partial D} \mathbf{f} \cdot \mathbf{n}\, dS. \tag{15.8}$$

The minus sign in (15.8) is necessary because if $\mathbf{f} \cdot \mathbf{n} > 0$, the chemical is leaving the set D, which implies that the mass of chemical in D is decreasing.

Now we make the important additional assumptions on the concentration u and the flux $\mathbf{f}$ that both of these functions are C^1 throughout the fluid (this may be all of $\mathbb{R}^3$). This assumption allows us to apply Theorem 15.2 to differentiate under the integral sign on the left of equation (15.8). It also allows us to use the divergence theorem on the right side of (15.8). Hence with these assumptions (15.8) can be rewritten

$$\int_D [u_t(\mathbf{x}, t) + \mathrm{div}(\mathbf{f})]\, d\mathbf{x} = 0.$$

Since D is an arbitrary compact set, we can use Theorem 15.5 to conclude that u and $\mathbf{f}$ satisfy the partial differential equation

$$u_t(\mathbf{x}, t) + \operatorname{div}(\mathbf{f}) = 0. \tag{15.9}$$

At this point, we need to make an assumption about the nature of diffusion. It seems evident that the chemical should diffuse from a region where the concentration is higher to where it is lower. Fick's law of diffusion quantifies this effect. It says that $\mathbf{f}$ is proportional to the negative of ∇u:

$$\mathbf{f} = -\gamma \nabla u.$$

The constant of proportionality γ is called the *diffusion constant* for this chemical in this ambient fluid. If we assume that u is C^2, and use the fact that $\operatorname{div}(\nabla u) = \Delta u$, we see that (15.9) becomes

$$u_t - \gamma \Delta u = 0. \tag{15.10}$$

Equation (15.10) is the well known linear diffusion equation. It also describes the flow of heat in a body and is often called the heat equation. Although we derived the diffusion equation for $\mathbf{x} \in \mathbb{R}^3$, is it often considered for functions $u(\mathbf{x}, t)$ where $\mathbf{x} \in \mathbb{R}^n$ and $t \geq 0$. In this case

$$\Delta u = \sum_{j=1}^{n} \frac{\partial^2 u}{\partial x_j^2}.$$

The function

$$G(\mathbf{x}, t) = \frac{1}{(4\pi\gamma t)^{n/2}} e^{-\|\mathbf{x}\|^2/(4\gamma t)} \tag{15.11}$$

is called the *fundamental solution* of the diffusion equation in $\mathbb{R}^n$. Here $\|\mathbf{x}\| = \|\mathbf{x}\|_2$ is the Euclidean norm. We leave it as an exercise to show that G satisfies (15.10) for any $t > 0$ and any $\mathbf{x} \in \mathbb{R}^n$. Because of the normalizing factor $(4\pi\gamma t)^{-n/2}$, $\int_{\mathbb{R}^n} G(\mathbf{x}, t) \, d\mathbf{x} = 1$ for all $t > 0$.

Diffusion can also be described in probabilistic terms, starting with the random motion of the individual particles of the diffusing chemical. In one space dimension G is

$$G(x, t) = \frac{1}{\sqrt{4\pi\gamma t}} e^{-x^2/(4\gamma t)}.$$

We recall from probability theory that G is the normal probability density function with mean $\mu = 0$ and standard deviation $\sigma = \sqrt{2\gamma t}$.

Note also that

$$\lim_{t \downarrow 0} G(\mathbf{x}, t) = 0 \ \text{ for } \mathbf{x} \neq 0,$$

but that $G(\mathbf{0}, t) = (4\pi\gamma t)^{-n/2}$ blows up as t tends to 0. The fundamental solution G describes diffusion in the idealized situation of a unit amount of the chemical concentrated at the single point $\mathbf{x} = 0$.

We can construct bounded solutions of the diffusion equation using integrals of G. Let $f(\mathbf{y})$ be a bounded function which is continuous except possibly on a set of Jordan content zero. Then we use the convolution in the spatial variable to define

$$u(\mathbf{x}, t) = \int_{\mathbb{R}^n} G(\mathbf{x} - \mathbf{y}, t) f(\mathbf{y}) \, d\mathbf{y}. \tag{15.12}$$

The function $(\mathbf{x}, t) \to G(\mathbf{x} - \mathbf{y}, t)$ is a copy of the fundamental solution translated to the point $\mathbf{y}$. The function u of (15.12) is just a superposition of these copies with weights $f(\mathbf{y})$. We shall use our results about differentiation of improper integrals to verify that (15.12) does indeed provide a solution of the diffusion equation.

Theorem 15.9 Suppose that $f(x)$ is continuous on $\mathbb{R}^n$ except on a set of Jordan content zero and is bounded. Then we have

(i) The function u given by (15.12) is in $C^\infty(\mathbb{R}^n \times \{t > 0\})$ and is a solution of the diffusion equation for $t > 0$ and all $\mathbf{x} \in \mathbb{R}^n$.

(ii) If $\mathbf{x}_0$ is point of continuity of f, then u is continuous at $(\mathbf{x}_0, 0)$. In particular,

$$\lim_{t \downarrow 0} u(\mathbf{x}_0, t) = f(\mathbf{x}_0).$$

Proof: We must establish the uniform convergence of the improper integral in (15.12) using Theorem 15.3. For $R > 0$ and $\delta > 0$, let

$$K(R, \delta) = \{(\mathbf{x}, t) : \|\mathbf{x}\| \leq R \text{ and } t \geq \delta\}.$$

Now for $\|\mathbf{x}\| \leq R \leq \|\mathbf{y}\|$, $\|\mathbf{y} - \mathbf{x}\| \geq \|\mathbf{y}\| - R$. Hence for $t \geq \delta$,

$$e^{-\frac{\|\mathbf{x} - \mathbf{y}\|^2}{4\gamma t}} \leq e^{-\frac{(\|\mathbf{y}\| - R)^2}{4\gamma \delta}}.$$

It follows that for $(\mathbf{x}, t) \in K(R, \delta)$, there is a constant A, depending on δ, such that for $\|\mathbf{y}\| \geq R$,

$$G(\mathbf{x} - \mathbf{y}, t) \leq A e^{-\frac{(\|\mathbf{y}\| - R)^2}{4\gamma\delta}}.$$

Since the function f is assumed bounded on $\mathbb{R}^n$, the integral (15.12) converges uniformly for $(\mathbf{x}, t) \in K(R, \delta)$. The function $G(\mathbf{x} - \mathbf{y}, t)$ is continuous in $\mathbf{x}, \mathbf{y}$ and t for $t > 0$. We conclude by Theorem 15.4 that $u(\mathbf{x}, t)$ is continuous on $K(R, \delta)$ for each $R > 0$ and $\delta > 0$. Consequently, u is continuous on all of $\{t > 0\} \times \mathbb{R}^n$.

The derivatives $\partial^k / \partial x_j^k G(\mathbf{x} - \mathbf{y}, t)$ and $\partial^l / \partial t^l G(\mathbf{x} - \mathbf{y}, t)$ can all be estimated in a similar manner. Thus by Theorem 15.5, u is in $C^\infty(\mathbb{R}^n \times \{t > 0\})$. We differentiate under the integral sign to deduce

$$u_t - \gamma \Delta u = \int_{\mathbb{R}^n} [G_t(\mathbf{x} - \mathbf{y}, t) - \gamma \Delta_x G(\mathbf{x} - \mathbf{y}, t)] f(\mathbf{y}) \, d\mathbf{y} = 0.$$

We have proved (i).

Now we turn to (ii). Since f is continuous at $\mathbf{x}_0$, for each $\varepsilon > 0$, there is a $\delta > 0$ such that

$$|f(\mathbf{y}) - f(\mathbf{x}_0)| < \varepsilon \quad \text{for } \|\mathbf{y} - \mathbf{x}_0\| < \delta.$$

We use the fact that $\int_{\mathbb{R}^n} G(\mathbf{x} - \mathbf{y}, t) d\mathbf{y} = 1$ for all $\mathbf{x}$ and all $t > 0$. Hence

$$u(\mathbf{x}, t) - f(\mathbf{x}_0) = \int_{\mathbb{R}^n} [f(\mathbf{y}) - f(\mathbf{x}_0)] G(\mathbf{x} - \mathbf{y}, t) d\mathbf{y}.$$

Now we break the integral up into two parts:

$$|u(\mathbf{x}, t) - f(\mathbf{x}_0)| \leq \int_{\|\mathbf{y} - \mathbf{x}\| < \delta/2} |f(\mathbf{y}) - f(\mathbf{x}_0| G(\mathbf{x} - \mathbf{y}, t) d\mathbf{y}$$

$$+ \int_{\|\mathbf{y} - \mathbf{x}\| \geq \delta/2} |f(\mathbf{y}) - f(\mathbf{x}_0| G(\mathbf{x} - \mathbf{y}, t) d\mathbf{y}.$$

Note that

$$\|\mathbf{y} - \mathbf{x}_0\| \leq \|\mathbf{y} - \mathbf{x}\| + \|\mathbf{x} - \mathbf{x}_0\|.$$

Hence if we choose $\|\mathbf{x} - \mathbf{x}_0\| < \delta/2$, and $\|\mathbf{y} - \mathbf{x}\| < \delta/2$, then $\|\mathbf{y} - \mathbf{x}_0\| < \delta$. This means that in the first integral, $|f(\mathbf{y}) - f(\mathbf{x}_0)| < \varepsilon$. Consequently for $\|\mathbf{x} - \mathbf{x}_0\| < \delta/2$, we have

$$|u(\mathbf{x}, t) - f(\mathbf{x_0})| \le \varepsilon + 2M \int_{\|\mathbf{y}-\mathbf{x}\| \ge \delta/2} G(\mathbf{x} - \mathbf{y}, t) d\mathbf{y}$$

where M is the bound for $|f(y)|$. This inequality holds for all $t > 0$. Finally we must show that there is a $\tau > 0$ such that $0 < t < \tau$ implies that the remaining integral is smaller than ε. We make a change of variable in the second integral, setting $\mathbf{z} = (\mathbf{y} - \mathbf{x})/\sqrt{4\gamma t}$ whence $d\mathbf{y} = (4\gamma t)^{n/2} d\mathbf{z}$. Since $G(\mathbf{x} - \mathbf{y}, t) = G(\mathbf{y} - \mathbf{x}, t)$, the integral becomes

$$\frac{2M}{(4\pi)^{n/2}} \int_{\|\mathbf{z}\| \ge \delta/(\sqrt{4\gamma t})} e^{-\|\mathbf{z}\|^2} d\mathbf{z} = \frac{2M\Omega_n}{(4\pi)^{n/2}} \int_{\rho \ge \delta/(\sqrt{4\gamma t})} e^{-\rho^2} \rho^{n-1} d\rho.$$

Now as t decreases to zero, the lower limit of the integral increases to ∞. It follows that there is a $\tau > 0$ so that the second integral is less than $\varepsilon/2$ for $t < \tau$.

To summarize, we have shown that there is a $\delta > 0$ and a $\tau > 0$ such that for $\|\mathbf{x} - \mathbf{x_0}\| < \delta/2$ and $0 < t < \tau$,

$$|u(\mathbf{x}, t) - f(\mathbf{x_0})| < \varepsilon + \varepsilon = 2\varepsilon.$$

The proof of the theorem is complete.

Remark Using the maximum principle for solutions of the diffusion equation, it can be shown that the solution given by (15.12) is the unique bounded solution of the initial value problem.

It is an important property of diffusion that even if the initial data f have discontinuities, the solution $u(\mathbf{x}, t)$ will have derivatives of all orders for $t > 0$. Mathematically, this is a consequence of formula (15.12) which shows that the solution is given as the convolution of f with a smooth function, in the same manner as in Example 15.5.

EXAMPLE 15.6: Solutions of the heat equation given by the formula (15.12) are not, as a rule, easy to compute. However, we can use Example 15.5 to construct a solution for the initial data $f(x) = 1$ for $0 < x < 1$ and $f = 0$ elsewhere. The solution of (15.10) for this f is just the family of smooth approximations f_k with the index k replaced by the continuous parameter $1/\sqrt{4\gamma t}$. We are interested in the behavior, both as t tends to 0 and as t increases. From Example 15.5, we see that

$$u(x, t) = \begin{cases} (1/2)[\text{erf}((1-x)/\sqrt{4\gamma t}) - \text{erf}(-x/\sqrt{4\gamma t})], & x \le 0 \\ (1/2)[\text{erf}(x/\sqrt{4\gamma t}) + \text{erf}((1-x)/\sqrt{4\gamma t})], & 0 < x < 1 \\ (1/2)[\text{erf}(x/\sqrt{4\gamma t}) - \text{erf}((x-1)/\sqrt{4\gamma t})], & x \ge 1. \end{cases}$$

We see that $\int_{\mathbb{R}} u(x, t)dx = \int_{\mathbb{R}} f(x)dx = 1$ for all $t > 0$ and for each x, $\lim_{t \to \infty} u(x, t) = 0$. We shall see in the exercises that similar properties hold for a wide class of solutions of the diffusion equation.

For a fuller discussion of the diffusion equation, the reader can consult the book of Cooper [Coo] or that of McOwen [M].

Exercises for 15.3

1. Let $f(\mathbf{x})$ be a continuous function on $\mathbb{R}^n$ with $f = 0$ outside some bounded set. Let $u(\mathbf{x}, t) = \int_{\mathbb{R}^n} G(\mathbf{x} - \mathbf{y}, t) f(\mathbf{y}) \, d\mathbf{y}$ where G is the fundamental solution of the heat equation.

   a) Show that there is a constant $M > 0$ such that $|u(\mathbf{x}, t)| \leq M$ for all $\mathbf{x} \in \mathbb{R}^n$ and $t \geq 0$.

   b) Show that $\lim_{t \to \infty} u(\mathbf{x}, t) = 0$, uniformly in $\mathbf{x} \in \mathbb{R}^n$.

2. For $k = 1, 2, \ldots$, let

$$f_k(x) = \begin{cases} k/2 & \text{for } |x| \leq 1/k \\ 0 & \text{for } |x| > 1/k \end{cases}.$$

This sequence of initial data represents a unit amount of chemical concentrated in a shorter and shorter interval $|x| \leq 1/k$. Let $u_k(x, t) = \int G(x - y, t) f_k(y) dy$ be solutions of the diffusion equation. Show that for each $\tau > 0$, $u_k(x, t)$ converges uniformly on $\{(x, t) : t \geq \tau > 0\}$ to $G(x, t)$. This result justifies the interpretation given for $G(x, t)$.

3. We did not need to require the initial data f to be nonnegative for our discussion of the formula (15.12). However when speaking of a chemical concentration it is natural to assume that $f(x) \geq 0$ and we shall do so for this exercise.

   Let $f : \mathbb{R} \to \mathbb{R}$ be continuous, $f \geq 0$, with $f = 0$ for $|x| \geq a$. Let $u(x, t)$ be given by formula (15.12).

   a) Use (15.12) to show that for each $\delta > 0$ there is a positive constant $A(\delta)$ such that for $|x| > a$ and $t \geq \delta$,

$$|u(x, t)|, |u_t(x, t)|, |u_x(x, t)|, |u_{xx}(x, t)| \leq A(\delta)e^{-(|x|-a)}.$$

   These estimates show that the improper integrals

$$\int_{\mathbb{R}} u(x, t)dx, \quad \int_{\mathbb{R}} u_t(x, t)dx, \quad \int_{\mathbb{R}} u_x(x, t)dx, \quad \text{and} \quad \int_{\mathbb{R}} u_{xx}(x, t)dx$$

all converge absolutely and uniformly with respect to t for $t \geq \delta$. Theorem 15.5 then justifies the differentiation of the improper integral

$$\frac{d}{dt} \int_{\mathbb{R}} u(x,t)dx = \int_{\mathbb{R}} u_t(x,t)dx.$$

For the next parts of the exercise, use the result of part a) and the fact that u satisfies the heat equation $u_t - \gamma u_{xx} = 0$.

b) Let $Q(t) = \int_{\mathbb{R}} u(x,t)dx$ be the total amount of chemical. Show that $Q(t)$ is constant, $Q(t) \equiv \int_{\mathbb{R}} f(x)dx$, by showing $(d/dt)Q = 0$.

c) Assume that $Q > 0$. Let

$$m(t) = \frac{1}{Q} \int_{\mathbb{R}} xu(x,t)dx.$$

Show that m is also independent of t.

d) Let

$$p(t) = \frac{1}{Q} \int_{\mathbb{R}} x^2 u(x,t)dx.$$

Show that $p(t) = p(0) + 2\gamma t$.

4. Continue with the terms of exercise 3.

a) Use integration by parts to show that

$$\int_{\mathbb{R}} x^2 e^{-x^2} dx = \frac{1}{2} \int_{\mathbb{R}} e^{-x^2} dx.$$

b) Find m and $p(t)$ for the fundamental solution $G(x,t)$.

c) Evaluate the integral

$$\int_{-\sqrt{p(t)}}^{\sqrt{p(t)}} G(x,t)dx$$

using the error function. The amount of the chemical contained in the interval $[-\sqrt{p(t)}, \sqrt{p(t)}]$ is a constant which does not depend on γ.

5. The steady-state diffusion equation is $\Delta u = 0$. In two dimensions it is $u_{xx} + u_{yy} = 0$. Let $f(y)$ be a bounded function on $\mathbb{R}$ that is continuous except at a finite number of points. For $x > 0$, define

$$u(x, y) = \frac{x}{\pi} \int_{\mathbb{R}} \frac{f(\eta)d\eta}{x^2 + (y - \eta)^2}.$$

a) Show that the integral for u converges uniformly for $x \geq \delta > 0$ and for $|y| \leq R$.

b) Show that the integrals for the derivatives u_x, u_y, u_{xx} and u_{yy} all converge uniformly for $x \geq \delta > 0$ and $|y| \leq R$. Conclude that $u_{xx} + u_{yy} = 0$ for $x > 0$ and $y \in \mathbb{R}$.

c) Show that if f is continuous at a point y_0, then $u(x, y)$ is continuous at $(0, y_0)$. In particular, show that $\lim_{x \downarrow 0} u(x, y_0) = f(y_0)$. The proof is similar to that of part (ii) of Theorem 15.9.

6. Solutions of the steady-state heat equation, $\Delta u(\mathbf{x}) = 0$, exhibit an important property, known as the *mean value property*. We state it in two dimensions. Let C_r be the circle of radius r centered at (x_0, y_0). If $\Delta u = 0$, then

$$\frac{1}{2\pi r} \int_{C_r} u\,ds = u(x_0, y_0).$$

That is, the value of u at any point is always the average of its values around any circle centered at that point. The integral around the circle C_r is expressed

$$\int_{C_r} u\,ds = \int_0^{2\pi} u(x_0 + r\cos\theta, y_0 + r\sin\theta)r\,d\theta.$$

a) Show that if u is C^2 and $\Delta u = 0$, then

$$\frac{d}{dr}\left(\frac{1}{2\pi r}\int_{C_r} u\,ds\right) = 0.$$

b) Show that

$$\lim_{r \to 0} \frac{1}{2\pi r}\int_{C_r} u\,ds = u(x_0, y_0).$$

c) Combine a) and b) to deduce the mean value property.

15.4 Fluid Flow

The Kinematics of Fluid Flow

By a fluid we mean either a gas or a liquid. We shall first investigate how certain quantities move with the fluid for a given velocity field. In the next section we shall apply the laws of physics to determine which velocity fields can represent a fluid flow.

In the Eulerian description of a fluid, the observer views the fluid at fixed points $\mathbf{x} = (x, y, z)$ in space. Her observation point does not move with the fluid. At each point (x, y, z), and at each time t, we assume given the velocity of the fluid,

$$\mathbf{q}(\mathbf{x}, t) = \mathbf{q}(x, y, z, t) = (u(x, y, z, t), v(x, y, z, t), w(x, y, z, t)).$$

The x component of velocity is u, the y component is v, and the z component is w. Now from our observer's point of view, if we wish to follow the path of a fluid particle, we need functions $x(t)$, $y(t)$, and $z(t)$ to locate the position of the particle as a function of t. The curve $t \to (x(t), y(t), z(t))$ parameterizes the path of the particle and the velocity of the fluid particle is $(x'(t), y'(t), z'(t))$. This expression must be the same as the given velocity vector at the location $(x(t), y(t), z(t))$ and time t. Hence $x(t), y(t), z(t)$ must satisfy the system of ordinary differential equations

$$\frac{dx}{dt} = u(x(t), y(t), z(t), t)$$

$$\frac{dy}{dt} = v(x(t), y(t), z(t), t) \qquad (15.13)$$

$$\frac{dz}{dt} = w(x(t), y(t), z(t), t).$$

Let G be an open set of $\mathbb{R}^3$. We can think of the points in G as fluid particles. Let a velocity field $\mathbf{q}(\mathbf{x}, t)$ be given on $Q = G \times [0, \infty)$. Now for each point $\mathbf{y} = (\xi, \eta, \zeta) \in G$, we want to follow the path of the fluid particle that starts at $\mathbf{y}$ at time $t = 0$. We assume that there exists a unique solution of the equations (15.13) with the initial conditions $x(0) = \xi$, $y(0) = \eta$, $z(0) = \zeta$. We denote these solutions

$$\mathbf{x}(\mathbf{y}, t) = (x(\xi, \eta, \zeta, t), \ y(\xi, \eta, \zeta, t), z(\xi, \eta, \zeta, t)).$$

Assuming the solution exists for a sufficiently long time and does not leave G, the mapping $\mathbf{y} \to \mathbf{x}(\mathbf{y}, t)$ generates a *flow*, which is a mapping $\varphi_t : G \to G$. For each $t \geq 0$, φ_t is a one-to-one map of G into G which has an inverse on its range. We assume that the solutions $x(\xi, \eta, \zeta, t)$, $y(\xi, \eta, \zeta, t)$, $z(\xi, \eta, \zeta, t)$ of the system (15.13) are C^2 functions of their arguments and that the Jacobian matrix $\mathbf{D}\varphi_t(\mathbf{y})$ is invertible for each $(\mathbf{y}, t) \in Q$.

Now let $D(0)$ be an open subset of G such that $\bar{D}(0)$ is compact. Assume the boundary $\partial D(0)$ is smooth. Let $D(t)$ be the image of $D(0)$ under the flow φ_t. $D(t)$ consists of the fluid particles at time t that occupied the set $D(0)$ at time $t = 0$. We are interested in how the flow deforms the set $D(0)$ and what quantities are preserved by the flow. In particular, we want to know how the volume of $D(0)$ is changed by the flow. We shall prove a general result that will allow us to answer this question, and to derive the Euler equations of fluid flow.

Theorem 15.10 Let $f(\mathbf{x}, t) = f(x, y, z, t)$ be a C^1 function on $Q = G \times [0, \infty)$. Then

$$\frac{d}{dt} \int_{D(t)} f(\mathbf{x}, t) \, d\mathbf{x} = \int_{D(t)} [f_t + \text{div}(f\mathbf{q})] \, d\mathbf{x}$$

$$= \int_{D(t)} [f_t + \nabla f \cdot \mathbf{q} + f \text{div}(\mathbf{q})] \, d\mathbf{x}. \qquad (15.14)$$

Before we enter into the proof of Theorem 15.10, we need to do a brief calculation with determinants.

Lemma 15.11 Let $\mathbf{A}(t)$ be an $n \times n$ matrix whose elements are C^1 functions of t,

$$\mathbf{A}(t) = [a_{ij}(t)].$$

Denote the jth column of $\mathbf{A}(t)$ by $\mathbf{a}_j(t) = (a_{1j}(t), \ldots, a_{nj}(t))$. We shall write $\mathbf{A}(t) = [\mathbf{a}_1(t), \mathbf{a}_2(t), \ldots, \mathbf{a}_n(t)]$. Let $J(t) = \det \mathbf{A}(t)$. Then

$$\frac{d}{dt} J(t) = \det[\mathbf{a}_1'(t), \mathbf{a}_2(t), \ldots \mathbf{a}_n(t)]$$

$$+ \det[\mathbf{a}_1(t), \mathbf{a}_2'(t), \ldots, \mathbf{a}_n(t)]$$

$$+ \cdots \cdots \qquad (15.15)$$

$$+ \cdots \cdots$$

$$+ \det[\mathbf{a}_1(t), \mathbf{a}_2(t), \ldots, \mathbf{a}_n'(t)].$$

The same result holds if $\mathbf{a}_i$ are the rows of $\mathbf{A}$.

Proof: We must start with the difference quotients $(J(t+h)-J(t))/h$ and rewrite them in the manner in which we derive the Leibniz rule for differentiating a product.

$$J(t+h) \ - \ J(t)$$
$$= \ \det[\mathbf{a}_1(t+h), \ldots, \mathbf{a}_n(t+h)] - \det[\mathbf{a}_1(t), \ldots, \mathbf{a}_n(t)]$$
$$= \ \det[\mathbf{a}_1(t+h), \ldots, \mathbf{a}_n(t+h)] - \det[\mathbf{a}_1(t), \mathbf{a}_2(t+h), \ldots, \mathbf{a}_n(t+h)]$$
$$+ \det[\mathbf{a}_1(t), \mathbf{a}_2(t+h), \ldots, \mathbf{a}_n(t+h)] - \det[\mathbf{a}_1(t), \mathbf{a}_2(t), \ldots, \mathbf{a}_n(t+h)]$$
$$+ \ \cdots \cdots$$
$$+ \ \cdots \cdots$$
$$+ \det[\mathbf{a}_1(t), \mathbf{a}_2(t), \ldots, \mathbf{a}_n(t+h)] - \det[\mathbf{a}_1(t), \mathbf{a}_2(t), \ldots, \mathbf{a}_n(t)].$$

Remember that the determinant is linear in each column when the others are fixed. Hence we can write the difference quotient as

$$\frac{J(t+h) - J(t)}{h} = \det\left[\frac{\mathbf{a}_1(t+h) - \mathbf{a}_1(t)}{h}, \mathbf{a}_2(t+h), \ldots, \mathbf{a}_n(t+h)\right]$$

$$+ \det\left[\mathbf{a}_1(t), \frac{\mathbf{a}_2(t+h) - \mathbf{a}_2(t)}{h}, \ldots, \mathbf{a}_n(t+h)\right]$$

$$+ \ \cdots \cdots$$
$$+ \ \cdots \cdots$$

$$+ \det\left[\mathbf{a}_1(t), \mathbf{a}_2(t), \ldots, \frac{\mathbf{a}_n(t+h) - \mathbf{a}_n(t)}{h}\right].$$

Taking the limit as $h \to 0$, we see that (15.15) results.

Proof of Theorem 15.10: The difficulty in computing the derivative on the left side of (15.14) comes from the fact that the boundary of $D(t)$ is changing with t. To compute this derivative using Theorem 15.2, we must express the integral over $D(t)$ as an integral over a fixed set. The flow map is the function that takes $D(0)$ onto $D(t)$:

$$\varphi_t : D(0) \to D(t).$$

We use this mapping and the change of variable theorem to pull the integral over $D(t)$ back to one over $D(0)$. Let $J(\mathbf{y}, t) = \det \mathbf{D}\varphi_t(\mathbf{y})$. We have assumed that $\mathbf{D}\varphi_t(\mathbf{y})$ is invertible for all $(\mathbf{y}, t) \in Q$ and that $\varphi_t(\mathbf{y}) = \varphi(\mathbf{y}, t)$ is C^2 in all variables. Now $J(\mathbf{y}, 0) = 1$ and $J(\mathbf{y}, t)$ never changes sign. Hence $J(\mathbf{y}, t) > 0$ for all t. Then we can write

$$\int_{D(t)} f(\mathbf{x}, t) \, d\mathbf{x} = \int_{D(0)} f(\varphi_t(\mathbf{y}), t) J(\mathbf{y}, t) \, d\mathbf{y}.$$

Since $\varphi_t(\mathbf{y}) = \mathbf{x}(\mathbf{y}, t)$,

$$J(\mathbf{y}, t) = \det \left[\frac{\partial \mathbf{x}}{\partial \mathbf{y}} \right] = \det \begin{bmatrix} x_\xi(\xi, \eta, \zeta, t) & x_\eta(\xi, \eta, \zeta, t) & x_\zeta(\xi, \eta, \zeta, t) \\ y_\xi(\xi, \eta, \zeta, t) & y_\eta(\xi, \eta, \zeta, t) & y_\zeta(\xi, \eta, \zeta, t) \\ z_\xi(\xi, \eta, \zeta, t) & z_\eta(\xi, \eta, \zeta, t) & z_\zeta(\xi, \eta, \zeta, t) \end{bmatrix}.$$

We shall apply Lemma 15.11 to compute $\partial J/\partial t$, using the rows rather than the columns. Let the rows of J be denoted $\mathbf{r}_1, \mathbf{r}_2, \mathbf{r}_3$. When we differentiate the first row of J with respect to t, we obtain

$$\begin{aligned} (\partial/\partial t)\mathbf{r}_1 &= ((x_\xi)_t, (x_\eta)_t, (x_\zeta)_t) \\ &= ((x_t)_\xi, (x_t)_\eta, (x_t)_\zeta) \\ &= (u_\xi, u_\eta, u_\zeta). \end{aligned} \tag{15.16}$$

Here we have used the fact that the second derivatives of x are continuous. This allows us to interchange the order of differentiation. We also used the fact that x is a solution of the system of differential equations (15.13). Now finally, we must compute the derivatives u_ξ, u_η, and u_ζ. Remember that $u = u(x, y, z, t)$ and $x = x(\xi, \eta, \zeta, t)$, $y = y(\xi, \eta, \zeta, t)$, and $z = z(\xi, \eta, \zeta, t)$. Hence by the chain rule,

$$\begin{aligned} u_\xi &= u_x x_\xi + u_y y_\xi + u_z z_\xi \\ u_\eta &= u_x x_\eta + u_y y_\eta + u_z z_\eta \\ u_\zeta &= u_x x_\zeta + u_y y_\zeta + u_z z_\zeta. \end{aligned}$$

We substitute these expressions into (15.16) to yield

$$(\partial/\partial t)\mathbf{r}_1 = u_x \mathbf{r}_1 + u_y \mathbf{r}_2 + + u_z \mathbf{r}_3.$$

Similarly,

$$(\partial/\partial t)\mathbf{r}_2 = v_x\mathbf{r}_1 + v_y\mathbf{r}_2 + v_z\mathbf{r}_3$$

and

$$(\partial/\partial t)\mathbf{r}_3 = w_x\mathbf{r}_1 + w_y\mathbf{r}_2 + w_z\mathbf{r}_3.$$

Now using Lemma 15.11,

$$(\partial/\partial t)J(t) = \det \begin{bmatrix} u_x\mathbf{r}_1 + u_y\mathbf{r}_2 + u_z\mathbf{r}_3 \\ \mathbf{r}_2 \\ \mathbf{r}_3 \end{bmatrix}$$

$$+ \det \begin{bmatrix} \mathbf{r}_1 \\ v_x\mathbf{r}_1 + v_y\mathbf{r}_2 + v_z\mathbf{r}_3 \\ \mathbf{r}_3 \end{bmatrix}$$

$$+ \det \begin{bmatrix} \mathbf{r}_1 \\ \mathbf{r}_2 \\ w_x\mathbf{r}_1 + w_y\mathbf{r}_2 + w_z\mathbf{r}_3 \end{bmatrix}.$$

Because a determinant with two rows equal is zero, the first determinant equals $u_x J$, the second equals $v_y J$ and the third equals $w_z J$. We conclude that

$$(\partial/\partial t)J(\mathbf{y}, t) = (u_x + v_y + w_z)J(\mathbf{y}, t) = \mathrm{div}(\mathbf{q})J. \qquad (15.17)$$

The second ingredient we shall need to establish (15.14) comes from another application of the chain rule:

$$\begin{aligned} (\partial/\partial t)f(\mathbf{x}(\mathbf{y}, t), t) &= (\partial/\partial t)f(x(\xi, \eta, \zeta, t), y(\xi, \eta, \zeta, t), z(\xi, \eta, \zeta, t), t) \\ &= f_t + f_x x_t + f_y y_t + f_z z_t \\ &= f_t + u f_x + v f_y + w f_z = f_t + \mathbf{q} \cdot \nabla f. \qquad (15.18) \end{aligned}$$

The derivative we have computed is the rate of change of the quantity $f(\mathbf{x}, t)$ along the particle path that starts at $\mathbf{y}$. It is referred to as the *material derivative*.

Finally we are ready to establish (15.14). Using Theorem 15.2,

$$\frac{d}{dt} \int_{D(t)} f(\mathbf{x}, t)\, d\mathbf{x} = \frac{d}{dt} \int_{D(0)} f(\mathbf{x}(\mathbf{y}, t), t) J(\mathbf{y}, t)\, d\mathbf{y}$$

$$= \int_{D(0)} \frac{\partial}{\partial t} [f(\mathbf{x}(\mathbf{y}, t), t) J(\mathbf{y}, t)]\, d\mathbf{y}.$$

Now continuing with the aid of (15.17) and (15.18), we find

$$\frac{d}{dt} \int_{D(t)} f(\mathbf{x}, t)\, d\mathbf{x} = \int_{D(0)} [(f_t + \mathbf{q} \cdot \nabla f) J(\mathbf{y}, t) + f(\partial/\partial t) J(\mathbf{y}, t)]\, d\mathbf{y}$$

$$= \int_{D(0)} [f_t + \mathbf{q} \cdot \nabla f + f \,\mathrm{div}(\mathbf{q})] J(\mathbf{y}, t)\, d\mathbf{y}$$

$$= \int_{D(0)} [f_t + \mathrm{div}(\mathbf{q} f)] J(\mathbf{y}, t)\, d\mathbf{y}.$$

Finally we use the change of variable theorem one more time to transform the last integral to $D(t)$. Thus

$$\frac{d}{dt} \int_{D(t)} f(\mathbf{x}, t)\, d\mathbf{x} = \int_{D(t)} [f_t + \mathrm{div}(f\mathbf{q})]\, d\mathbf{x}.$$

The proof of Theorem 15.10 is complete.

The divergence part of the integral on the right of (15.14) can be expressed as an integral over the boundary of $D(t)$ with an application of the divergence theorem,

$$\frac{d}{dt} \int_{D(t)} f(\mathbf{x}, t)\, d\mathbf{x} = \int_{D(t)} f_t\, d\mathbf{x} + \int_{\partial D(t)} f(\mathbf{q} \cdot \mathbf{n})\, dS. \qquad (15.19)$$

The exterior unit normal to $\partial D(t)$ is $\mathbf{n}$ and dS is the element of surface area on $\partial D(t)$. The boundary integral is the flux of the quantity $f\mathbf{q}$ through the boundary surface $\partial D(t)$. Equation (15.19) is the higher dimensional analogue of the corollary to Theorem 15.2. It is called the *Reynolds transport theorem*.

Now we deduce a first consequence of (15.14) by taking $f \equiv 1$. Then

$$\frac{d}{dt} \mathrm{vol}(D(t)) = \int_{D(t)} \mathrm{div}(\mathbf{q})\, d\mathbf{x} = \int_{\partial D(t)} \mathbf{q} \cdot \mathbf{n}\, dS. \qquad (15.20)$$

Equation (15.20) tells us that if $\operatorname{div}(\mathbf{q}) > 0$, then the volume of $D(t)$ is increasing, and if $\operatorname{div}(\mathbf{q}) < 0$, the volume is decreasing. If $\operatorname{div}(\mathbf{q}) = 0$, the volume is conserved, even though the shape may be changing. This provides a good geometrical interpretation of the divergence.

EXAMPLE 15.7: We consider an example in two dimensions. Let the velocity field be given by

$$\mathbf{q}(x, y) = \left(\frac{1}{1 + x^2}, 0 \right).$$

The solutions of the system (15.13) are given implicitly as

$$x + x^3/3 = t + \xi + \xi^3/3, \quad y \equiv \eta.$$

We have $\operatorname{div}(\mathbf{q}) = -2x/(1 + x^2)$. The geometric interpretation of (15.20) may be seen as follows. Let $D(0)$ be a square of fluid particles contained in the half-plane $x < 0$. We note that $\operatorname{div}(\mathbf{q}) > 0$ for $x < 0$. This means that as t increases from zero, the square of particles will be moved to the right and lengthened in the x direction. After the square of particles moves to the right of $x = 0$, it will be compressed in the x direction. The sets $D(t)$ are plotted in Figure 15.2 for several values of t.

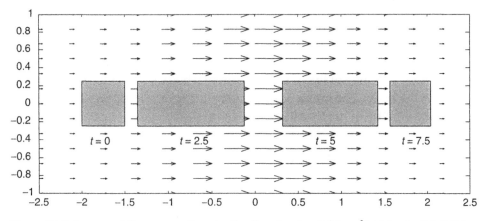

Figure 15.2 Images of the square D under the flow $(u, v) = (1/(1 + x^2), 0)$ for $t = 2.5, 5, 7.5$

The Equations of Fluid Flow

The fluid will be described in terms of five quantities,

$$\rho(\mathbf{x}, t) \quad \text{the density,}$$
$$p(\mathbf{x}, t) \quad \text{the presssure,}$$
$$\mathbf{q} = (u, v, w) \quad \text{the three components of velocity.}$$

We shall assume that all these functions are C^1 functions of x, y, z, t. The mass of fluid contained in $D(t)$ is

$$\int_{D(t)} \rho(\mathbf{x}, t) \, d\mathbf{x}.$$

The first equation of fluid flow that we derive is a consequence of the physical principle of *conservation of mass* and (15.14). If the mass of fluid contained in $D(t)$ is constant, (15.14) with $f = \rho$ yields

$$0 = \frac{d}{dt} \int_{D(t)} \rho(\mathbf{x}, t) \, d\mathbf{x} = \int_{D(t)} [\rho_t + \text{div}(\rho \mathbf{q})] \, d\mathbf{x}.$$

Since $D(t)$ is an arbitrary volume, the Null Theorem implies that ρ and $\mathbf{q}$ satisfy the partial differential equation

$$\rho_t + \text{div}(\rho \mathbf{q}) = 0. \tag{15.21}$$

This is the differential form of the principle of conservation of mass. Equation (15.21) is often called the *continuity equation*.

The next physical principle we invoke is Newton's second law which states that the rate of change of linear momentum of a body equals the sum of forces acting on the body. The momentum density is a vector field, $\rho \mathbf{q}$. The momentum of the volume of fluid contained in $D(t)$ is

$$\mathbf{m}(t) \equiv \int_{D(t)} \rho(\mathbf{x}, t) \mathbf{q}(\mathbf{x}, t) \, d\mathbf{x}$$

$$= \left(\int_{D(t)} \rho u \, d\mathbf{x}, \int_{D(t)} \rho v \, d\mathbf{x}, \int_{D(t)} \rho w \, d\mathbf{x} \right).$$

We use (15.14) on each component of $\mathbf{m}$, taking first $f = \rho u$, then $f = \rho v$, and then $f = \rho w$, to deduce

$$\frac{d}{dt}\mathbf{m}_1(t) = \int_{D(t)} [(\rho u)_t + \operatorname{div}(\rho u \mathbf{q})]\, d\mathbf{x},$$

$$\frac{d}{dt}\mathbf{m}_2(t) = \int_{D(t)} [(\rho v)_t + \operatorname{div}(\rho v \mathbf{q})]\, d\mathbf{x},$$ (15.22)

$$\frac{d}{dt}\mathbf{m}_3(t) = \int_{D(t)} [(\rho w)_t + \operatorname{div}(\rho w \mathbf{q})]\, d\mathbf{x}.$$

This expression must be equated to the sum of forces acting on the volume of fluid contained in $D(t)$. These forces are classified as *body forces* and *surface forces*. Body forces, such as gravity, act on every point in $D(t)$. We shall lump all body forces into a body force density $\mathbf{f}$, which has the units of force per unit mass. The total body force acting on $D(t)$ is

$$\int_{D(t)} \rho \mathbf{f}\, d\mathbf{x}.$$ (15.23)

The surface forces depend on the type of fluid. We shall assume that the fluid has no viscosity. In this case the only force acting on the surface of a volume V is that exerted by the fluid exterior to V. This surface force is exerted normal to the surface of V. It therefore has the form $-p\mathbf{n}$ where $\mathbf{n}$ is the exterior unit normal to ∂V, and p is a scalar function, called the pressure. p has the units of force per unit area. The total surface force on the volume of fluid $D(t)$ is therefore

$$-\int_{\partial D(t)} p\mathbf{n}\, dS.$$ (15.24)

This boundary integral can be transformed into a volume integral by the use of the divergence theorem. Let the exterior unit normal vector to $\partial D(t)$ be $\mathbf{n} = (n_1, n_2, n_3)$. The components of (15.24) can be written

$$\int_{\partial D(t)} p n_1\, dS = \int_{D(t)} p_x\, d\mathbf{x}$$

$$\int_{\partial D(t)} p n_2\, dS = \int_{D(t)} p_y\, d\mathbf{x}$$

$$\int_{\partial D(t)} pn_3 dS = \int_{D(t)} p_y\, d\mathbf{x}$$

or more concisely in vector notation,

$$-\int_{\partial D(t)} p\mathbf{n}\, dS = -\int_{D(t)} \nabla p\, d\mathbf{x}. \tag{15.25}$$

Then combining the body force (15.23) and the surface force (15.25), Newton's second law for a nonviscous fluid becomes

$$\frac{d}{dt}\mathbf{m}(t) = \int_{D(t)} [\rho \mathbf{f} - \nabla p]\, d\mathbf{x}. \tag{15.26}$$

Now using (15.22) to substitute in the left side of (15.26) and employing the Null Theorem on each component, we arrive at three more PDEs

$$\begin{aligned}
(\rho u)_t + \operatorname{div}(\rho u \mathbf{q}) + p_x &= \rho f_1 \\
(\rho v)_t + \operatorname{div}(\rho v \mathbf{q}) + p_y &= \rho f_2 \\
(\rho w)_t + \operatorname{div}(\rho w \mathbf{q}) + p_z &= \rho f_3.
\end{aligned} \tag{15.27}$$

These equations may be simplified by taking into account the equation of continuity (15.21). Let us expand the first equation of (15.27). It becomes

$$\rho_t u + \rho u_t + u\nabla\rho \cdot \mathbf{q} + \rho\nabla u \cdot \mathbf{q} + \rho u\, \operatorname{div}(\mathbf{q}) + p_x = \rho f_1. \tag{15.28}$$

But multiplying (15.21) by u we see that

$$\rho_t u + u\nabla\rho \cdot \mathbf{q} + \rho u\, \operatorname{div}(\mathbf{q}) = 0. \tag{15.29}$$

Thus using (15.29) we see that (15.28) reduces to

$$\rho u_t + \rho\nabla u \cdot \mathbf{q} + p_x = \rho f_1.$$

In the same way the second and third equations of (15.27) can be reduced to

$$\begin{aligned}
\rho v_t + \rho\nabla v \cdot \mathbf{q} + p_y &= \rho f_2 \\
\rho w_t + \rho\nabla w \cdot \mathbf{q} + p_z &= \rho f_3.
\end{aligned}$$

The last three equations can be expressed more succinctly in vector notation. Together with the equation of continuity we have

$$\rho_t + \operatorname{div}(\rho\mathbf{q}) = 0 \tag{15.30}$$

$$\mathbf{q}_t + \nabla\mathbf{q} \cdot \mathbf{q} + \nabla p/\rho = \mathbf{f}. \tag{15.31}$$

These are the *Euler* equations for a *nonviscous* fluid. They are a system of four PDE's in the five unknowns $\rho, p, \mathbf{q} = (u, v, w)$. This system is not solvable unless we reduce the number of unknowns by making other physical assumptions, or by introducing another equation. For example, the density of water is practically constant, so that for many purposes in hydrodynamics, we can assume that ρ is constant. This assumption yields a system of four PDE's in four unknowns:

$$\operatorname{div}(\mathbf{q}) = 0 \tag{15.32}$$

$$\mathbf{q}_t + \nabla\mathbf{q} \cdot \mathbf{q} + \nabla p/\rho = \mathbf{f}. \tag{15.33}$$

Equations (15.32) and (15.33) are the Euler equations for *nonviscous, incompressible* flow. We say incompressible because the continuity equation (15.21) has become (15.32) and by (15.20) the volume of a body of fluid $D(t)$ will remain constant. This system has been studied extensively in many contexts. It has also been found that the assumption of constant density (incompressible flow) is valid for subsonic flow of air around an airplane wing. The system (15.32)-(15.33) is still a challenge to mathematicians and scientists trying to understand the properties of their solutions. Numerical studies have been very important for design of ship hulls, automobiles bodies and airplane wings, and in the study of turbulence. Turbulence is a very active field of study today.

A classical introduction to fluid dynamics is provided in the book of Batechlor, [B]. A briefer, more concise mathematical treatment is given in the book of Chorin and Marsden, [Ch].

Exercises for 15.4

In the following exercises we shall consider steady-state, incompressible flow with no body forces, i.e, $\mathbf{f} = 0$. Thus ρ is constant and $p = p(x, y, z)$ while

$q = (u(x, y, z), v(x, y, z), w(x, y, z))$. The Euler equations become

$$\text{div}(q) = 0$$
$$\rho \nabla q \cdot q + \nabla p = 0.$$

A *streamline* is a particle path, which is a solution of the autonomous system

$$\frac{dx}{dt} = u(x, y, z)$$

$$\frac{dy}{dt} = v(x, y, z)$$

$$\frac{dz}{dt} = w(x, y, z).$$

Define

$$e(x, y, z) = \left(\frac{\rho}{2}\right) \|q\|^2 + p(x, y, z).$$

1. Show that e is constant along a streamline. This is Bernoulli's law. It says that when the fluid speeds up, the pressure decreases.
2. Rotating flow in two dimensions. Let $q(x, y) = (-y, x)/r^\alpha$ where $r = \sqrt{x^2 + y^2}$.

   a) Show that $\text{div}(q) = 0$.
   b) Calculate $\nabla q \cdot q$ and find the pressure p to solve Euler's equations. Your answer will have different forms depending on whether $\alpha = 1$ or $\alpha \neq 1$.

3. Radial flow in two dimensions. Let $q(x, y) = (x, y)/r^\alpha$.

   a) Show that $\text{div}(q) = 0$ only for $\alpha = 2$. Find the pressure in that case.
   b) In three dimensions, let $q(x, y, z) = (x, y, z)/r^\alpha$ where $r = \sqrt{x^2 + y^2 + z^2}$. For what value of α is $\text{div}(q) = 0$? What is the pressure in that case?

4. The *vorticity* is the vector field $w = \nabla \times q$. For steady-state two-dimensional flow, show that w is constant along streamlines. This result is not true in three dimensions, making the analysis and computation of three-dimensional flows more difficult than for two-dimensional flows.

Appendix II

A Matrix Factorization

Here we present a factorization of a nonsingular matrix into simpler factors. We use the notation e_j, $j = 1, \ldots, n$ to denote the standard basis in $\mathbb{R}^n$. We will refer to coordinates in the standard basis.

Theorem Let $\mathbf{A}$ be an $n \times n$ nonsingular matrix. Then $\mathbf{A}$ can be factored

$$\mathbf{A} = \mathbf{B}_n \mathbf{E}_n \mathbf{B}_{n-1} \mathbf{E}_{n-1} \cdots \mathbf{B}_1 \mathbf{E}_1$$

where $\mathbf{B}_j$ changes only the jth coordinate and $\mathbf{E}_j$ is an elementary matrix that results when two rows of the identity are interchanged.

Proof: Before giving the formal proof, we illustrate the procedure for the case of the first coordinate. Let $\mathbf{B}_1$ be the matrix that results from replacing the first row of the identity with the first row of $\mathbf{A}$. For any $\mathbf{x} \in \mathbb{R}^n$, $(\mathbf{B}\mathbf{x})_1 = (\mathbf{A}\mathbf{x})_1$, and $\mathbf{B}_1$ fixes the remaining coordinates. For the moment, we assume that $\mathbf{B}_1$ is invertible, and we set $\mathbf{A}_1 = \mathbf{A}\mathbf{B}_1^{-1}$. The matrix $\mathbf{A}_1$ is invertible because it is the product of two invertible matrices. To show that $\mathbf{A}_1$ fixes the first coordinate, let $\mathbf{y} \in \mathbb{R}^n$. There is a unique $\mathbf{x}$ such that $\mathbf{B}_1\mathbf{x} = \mathbf{y}$. It follows that

$$(\mathbf{A}_1\mathbf{y})_1 = (\mathbf{A}\mathbf{x})_1 = (\mathbf{B}_1\mathbf{x})_1 = \mathbf{y}_1.$$

We have factored $\mathbf{A} = \mathbf{A}_1\mathbf{B}_1$ where $\mathbf{B}_1$ changes only the first coordinate, and $\mathbf{A}_1$ is invertible and fixes the first coordinate, provided that $\mathbf{B}_1$ is invertible. If

$\mathbf{B}_1$ is not invertible, it is because $a_{11} = 0$. If $a_{11} = 0$, we know that there is some $j > 1$ such that $a_{1j} \neq 0$ because we have assumed that $\mathbf{A}$ is nonsingular. Let $\mathbf{E}_1$ be the elementary matrix that results by interchanging rows 1 and j of the identity. Then $\mathbf{AE}_1$ is the same as $\mathbf{A}$, but with columns 1 and j interchanged. Now we let $\mathbf{B}_1$ be the matrix that results from replacing the first row of the identity with the first row of $\mathbf{AE}_1$, and we set $\mathbf{A}_1 = \mathbf{AE}_1\mathbf{B}_1^{-1}$.

The proof now proceeds by induction. We assume that for some m, $2 \leq m \leq n$, we have the achieved the factorization

$$\mathbf{A} = \mathbf{A}_{m-1}\mathbf{B}_{m-1}\mathbf{E}_{m-1} \cdots \mathbf{B}_1\mathbf{E}_1$$

where each $\mathbf{B}_j$ changes only the jth coordinate and $\mathbf{A}_{m-1}$ is an invertible matrix that fixes the first $m - 1$ coordinates. Thus $\mathbf{A}_{m-1}$ looks like the identity in rows $1 \leq i \leq m-1$. Hence $\det \mathbf{A}_{m-1} = \det \tilde{\mathbf{A}}_{m-1}$ where $\tilde{\mathbf{A}}_{m-1}$ is the $(n-m+1)\times(n-m+1)$ matrix in the lower right corner. Since $\mathbf{A}_{m-1}$ is assumed nonsingular, there must be some element $\tilde{a}_{jm}$, $j \geq m$, in the top row of $\tilde{\mathbf{A}}_{m-1}$ that is nonzero. Let $\mathbf{E}_m$ be that elementary matrix that exchanges row j and row m. Then $\mathbf{A}_{m-1}\mathbf{E}_m$ is the same as $\mathbf{A}_{m-1}$ except that columns m and j are interchanged.

Let $\mathbf{Q}_m$ be the orthogonal projection onto the one-dimensional subspace spanned by $\mathbf{e}_m$, $\mathbf{Q}_m\mathbf{x} = \langle \mathbf{e}_m, \mathbf{x}\rangle\mathbf{e}_m$, and let $\mathbf{P}_m = \mathbf{I} - \mathbf{Q}_m$. Finally we set

$$\mathbf{B}_m = \mathbf{P}_m + \mathbf{Q}_m\mathbf{A}_{m-1}\mathbf{E}_m.$$

Now $\mathbf{B}_m$ changes only the mth coordinate, or equivalently, it fixes all but the mth coordinate. To see this, multiply this last equation on the left by $\mathbf{P}_m$, and use the fact that $\mathbf{P}_m^2 = \mathbf{I}$ and $\mathbf{P}_m\mathbf{Q}_m = 0$. This implies that

$$\begin{aligned}\mathbf{P}_m\mathbf{B}_m &= \mathbf{P}_m^2 + \mathbf{P}_m\mathbf{Q}_m\mathbf{A}_{m-1}\mathbf{E}_m \\ &= \mathbf{P}_m.\end{aligned}$$

The matrix $\mathbf{B}_m$ is also nonsingular. We have just seen that $\mathbf{B}_m$ changes only the mth coordinate. Hence if $\mathbf{x} = (x_1, \dots, x_{m-1}, 0, x_{m+1}, \dots, x_n)$, $\mathbf{B}_m\mathbf{x} = \mathbf{x}$. Suppose now that $\mathbf{x} = \lambda\mathbf{e}_m$. Then $\mathbf{E}_m\mathbf{e}_m = \mathbf{e}_j$, so that

$$\begin{aligned}\mathbf{B}_m\mathbf{x} &= \lambda\mathbf{Q}_m\mathbf{A}_{m-1}\mathbf{e}_j \\ &= \lambda\mathbf{Q}_m[\,j^{th}\text{column of } \mathbf{A}_{m-1}] \\ &= \lambda\tilde{a}_{jm}.\end{aligned}$$

By assumption, $\tilde{a}_{jm} \neq 0$, which implies that $\mathbf{B}_m\mathbf{x} = 0$ if and only if $\lambda = 0$. Finally we set

$$\mathbf{A}_m = \mathbf{A}_{m-1}\mathbf{E}_m\mathbf{B}_m^{-1}.$$

This yields the factorization $\mathbf{A}_{m-1} = \mathbf{A}_m \mathbf{B}_m \mathbf{E}_m$ because $\mathbf{E}_m^2 = \mathbf{I}$. The matrix $\mathbf{A}_m$ is invertible because it is the product of invertible matrices.

It remains for us to show that $\mathbf{A}_m$ fixes the first m coordinates. Let $\mathbf{y} \in \mathbb{R}^n$. Because $\mathbf{B}_m$ is nonsingular, there is a unique $\mathbf{x}$ such that $\mathbf{y} = \mathbf{B}_m \mathbf{x}$. Now remember that $\mathbf{B}_m$, $\mathbf{E}_m$ and $\mathbf{A}_{m-1}$ fix the first $m-1$ coordinates. The ith coordinate of $\mathbf{A}_m \mathbf{y}$ is $\langle \mathbf{e}_i, \mathbf{A}_m \mathbf{y} \rangle$. For $i < m$, we have

$$
\begin{aligned}
\langle \mathbf{e}_i, \mathbf{A}_m \mathbf{y} \rangle &= \langle \mathbf{e}_i, \mathbf{A}_{m-1} \mathbf{E}_m \mathbf{x} \rangle = \langle \mathbf{e}_i, \mathbf{E}_m \mathbf{x} \rangle \\
&= \langle \mathbf{e}_i, \mathbf{x} \rangle = \langle \mathbf{e}_i, \mathbf{B}_m \mathbf{x} \rangle \\
&= \langle \mathbf{e}_i, \mathbf{y} \rangle.
\end{aligned}
$$

The defining equation for $\mathbf{B}_m$ implies that $\mathbf{A}_m$ also fixes the mth coordinate. In fact,

$$
\begin{aligned}
\langle \mathbf{e}_m, \mathbf{A}_m \mathbf{y} \rangle &= \langle \mathbf{e}_m, \mathbf{A}_{m-1} \mathbf{E}_m \mathbf{x} \rangle = \langle \mathbf{e}_m, \mathbf{Q}_m \mathbf{A}_{m-1} \mathbf{E}_m \mathbf{x} \rangle \\
&= \langle \mathbf{e}_m, (\mathbf{B}_m - \mathbf{P}_m) \mathbf{x} \rangle = \langle \mathbf{e}_m, \mathbf{B}_m \mathbf{x} \rangle \\
&= \langle \mathbf{e}_m, \mathbf{y} \rangle.
\end{aligned}
$$

The proof is complete.

EXAMPLES: Let $\mathbf{A}$ be the 2×2 matrix

$$
\mathbf{A} = \begin{bmatrix} a & b \\ c & d \end{bmatrix}.
$$

Assume $a \neq 0$ and $ad - bc \neq 0$. We set

$$
\mathbf{B}_1 = \begin{bmatrix} a & b \\ 0 & 1 \end{bmatrix}.
$$

Then

$$
\mathbf{A}_1 = \mathbf{A} \mathbf{B}_1^{-1} = \begin{bmatrix} 1 & 0 \\ c/a & (ad - bd)/a \end{bmatrix}.
$$

After setting $\mathbf{B}_2 = \mathbf{A}_1$, we have $\mathbf{A} = \mathbf{B}_2 \mathbf{B}_1$ where $\mathbf{B}_1$ changes only the first coordinate, and $\mathbf{B}_2$ changes only the second coordinate.

For a 3×3 example, we take

$$A = \begin{bmatrix} 1 & 2 & 0 \\ -1 & 0 & 3 \\ 1 & 4 & 7 \end{bmatrix}.$$

The factorization of **A** is

$$\mathbf{A} = \mathbf{B}_3\mathbf{B}_2\mathbf{B}_1 = \begin{bmatrix} 1 & 0 & 0 \\ 0 & 1 & 0 \\ 2 & 1 & 4 \end{bmatrix} \begin{bmatrix} 1 & 0 & 0 \\ -1 & 2 & 3 \\ 0 & 0 & 1 \end{bmatrix} \begin{bmatrix} 1 & 2 & 0 \\ 0 & 1 & 0 \\ 0 & 0 & 1 \end{bmatrix}.$$

Solutions to Selected Exercises

Chapter I

Section 1.1

1. To prove (v), first show that, if $ab > 0$, then either both $a > 0$ and $b > 0$, or both $a < 0$ and $b < 0$. Then use the fact that $a(1/a) = 1 > 0$.
2. To prove (vi), use (v) and (iii) to deduce that $1 < b/a$. Then use (iii) and (v) again with $1/b$.
3. Use (ii) of Theorem 1.2 to deduce $2a = a + a < a + b < 2b$. Then use (iii).
4. Use exercise 4 to deduce $0 \leq (a - b)^2$. Then expand $(a - b)^2$.

Section 1.2

2. Let $S = \{x_1, \ldots, x_n\}$. Then $|x_1| + \cdots + |x_n|$ is an upper bound for S. Hence S has a least upper bound b.
3. Assume that the least upper bound property holds. Let T be the set of lower bounds of S. Show that T is bounded above. Then show that $a = \sup T$ is the greatest lower bound of S.
4. First use the corollary to Theorem 1.4 to show that, if $a \geq 0$ and $a \leq 1/n$ for all $n \in \mathbb{N}$, then $a = 0$.
6. For all $x \in A \cup B$, either $x \leq \sup A$ or $x \leq \sup B$.

Section 1.3

1. a) Use the fact that for $0 \le x \le 1$, we have $x \le \sqrt{x}$; and that for $x \ge 1$, we have $\sqrt{x} \le x$. b) A lower estimate is

$$f(x) \ge \begin{cases} x & \text{for } 0 \le x \le 1 \\ x^{-3/2} & \text{for } x \ge 1 \end{cases}.$$

2.

$$x^2 \cos x - 4 \cos 2 = x^2(\cos x - \cos 2) + (x - 2)(x + 2) \cos 2.$$

3.

$$\frac{1}{1 + x^2} - \frac{1}{y^2 + 1} = \frac{(y - x)(y + x)}{(x^2 + 1)(y^2 + 1)}.$$

4. Use the fact that for $x \ge 1$, $\sqrt{x^4 + 4} \le \sqrt{x^4 + 4x^4} = x^2\sqrt{5}$.
5. Use the reverse triangle inequality to find that for $2 \le |x| \le 3$, $|x - 1| \ge |\,|x| - 1| \ge 1$.
6. The reverse triangle inequality implies that for $|x|$ sufficiently large,

$$|p_n(x)| \ge |x|^n(a_n - |a_{n-1}|/|x| - \cdots - |a_0|/|x|^n).$$

Section 1.4

4. Use a) to prove part b).

Section 1.5

1. If J satisfies the condition, use the least upper bound property and the greatest lower bound property to show that $(a, b) \subset J$ where $a = \inf J$ and $b = \sup J$.
2. For each $x \in A$, $c + f(x) \le c + \sup_A f$. To show equality, use the fact that, for each $\varepsilon > 0$, there is an $x \in A$ with $f(x) > \sup_A f - \varepsilon$.
4. No. Consider $f(x) = x$ on $[-1, 1]$.
5. Start with $h(x, y) \le \sup_Y h(x, y)$, and then take $\inf_X$ on both sides.

Chapter 2

Section 2.1

1. d) Use the fact that $\sqrt{a} - \sqrt{b} = (a - b)/(\sqrt{a} + \sqrt{b})$.
2. No. Check the values of $\sin(2n\pi/3)$.
4. $||a_n| - |L|| \le |a_n - L|$.
6. For n sufficiently large, $a_{n+1} \le r'a_n$ where $r' = (1 + r)/2$.
7. If $A > B$, show that for n sufficiently large, $a_n > b_n$.
9. For $\varepsilon > 0$ given, there is an $N > 0$ such that for $n > N$, $|a_n - L| \le \varepsilon$. Show that for $n > N$,

$$|\sigma_n - L| \le \frac{|a_1 - L| + \cdots + |a_N - L|}{n} + \varepsilon.$$

10. Show that $|a_n| \le (1 - |a|)/2$ for n sufficiently large.
11. Note that for $n > c$, the factor $c/n < 1$.
13. The possible limits of the sequence are $L = 0$ and $L = 1/2$. Assume that $|a_0 - 1/2| \le c/2$ where $c < 1$.

Section 2.2

2. Use induction to show $\{a_n\}$ is monotone increasing with limit $L = 4$.
3. To show the sequence is bounded, use the formula for the geometric series (with $r = 1/2$)

$$1 + r + r^2 + \cdots + r^n = \frac{1 - r^{n+1}}{1 - r}.$$

5. Find the possible limit(s) of the sequence. Show that $\{y_n\}$ is monotone increasing using an induction argument.
6. b) Use induction to show that for n even,

$$a_1 < a_3 < \cdots < a_{n-1} < a_n < a_{n-2} < \cdots < a_2.$$

c) Using methods associated with difference equations, we find that

$$a_n = (1/3)a_1(1 - 4(-1/2)^n) + (2/3)a_2(1 + 2(-1/2)^n).$$

The sequence converges to $(a_1 + 2a_2)/3 = 7/3$ when $a_1 = 1$ and $a_2 = 3$.

7. First show that $0 < x_1 < 2/A$ implies $x_2 > 0$. Then use $x_{n+1} = 1/A - A(x_n - 1/A)^2$ to show that $x_n \leq 1/A$ for $n \geq 2$.

Section 2.3

2. Use the result of Example 2.3.

Chapter 3

Section 3.1

1. Use the reverse triangle inequality.
3. Use the standard manipulation for square roots:

$$\sqrt{a} - \sqrt{b} = \frac{a - b}{\sqrt{a} + \sqrt{b}}.$$

6. Show that if f is monotone increasing, $\lim_{x \downarrow a} f(x) = \inf_{(a,b)} f$.
10. Show first that for x sufficiently large, $x^2|f(x)| \leq |L| + 1$.
11. Use the fact that the rationals are dense in $\mathbb{R}$.

Section 3.2

2. First find a $\delta_0 > 0$ such that $|x| \geq |a|/2$ for $|x - a| \leq \delta_0$.
4. Use the fact that the rationals are dense in the reals.
6. If $f(x_0) > g(x_0)$, use exercise 5 to show that there is a $\delta > 0$ such that $h(x) = f(x)$ for $|x - x_0| < \delta$.
7. b) Show that if $x_0 \in I$, there is an $\varepsilon > 0$ such that $\{|y - f(x_0)| < \varepsilon\} \subset J$. Then use the definition of continuity to show there is a $\delta > 0$ such that $\{|x - x_0| < \delta\} \subset I$.

Section 3.3

1. Because f is continuous on $[a, b]$, it has a minimizer $x_* \in [a, b]$.
3. Find a pair of sequences x_n, y_n with $|x_n - y_n| \leq 1/n$ and $|x_n^2 - y_n^2| \geq 1$.

4.

$$|f(x) - f(y)| = \frac{|x - y||1 - xy|}{(1 + x^2)(1 + y^2)}.$$

Find a bound, independent of x and y, for the quantity

$$\frac{|1 - xy|}{(1 + x^2)(1 + y^2)}.$$

Sections 3.4 and 3.5

2. After five iterations, the minimizer $x_* \in [a_5, b_5]$ where $a_5 = 0.8639$ and $b_5 = 0.8820$. Ten interations are needed to make the interval have length less than 0.005. In this case $a_{10} = 0.8541$ and $b_{10} = 0.8582$.
3. The minimum occurs at $x_* = 0.6799$.
5. Check $f(.1), f(1)$ and a third point.
6. Verify that $f(x) - x < 0$ for x sufficiently negative, and $f(x) - x > 0$ for x sufficiently positive.
7. Verify that $\min_I f \le (f(x_1) + \cdots f(x_n))/n \le \max_I f$.
8. Unique real root at 2.2056.

Chapter 4

Section 4.1

2. $\lim_{x \to 0}(g(x)/x) = 1$. The linear approximation is $l(x) = x$ and $|R(x)| \le |x|^{4/3}$.
4. Instead of trying to solve the equation $f(x) = y$, we solve the linear equation $f(x_0) + f'(x_0)(x - x_0) = y$.
5. For $p > 0$, $f(0) = A$ and the difference quotients

$$\frac{f(x) - f(0)}{x} = B + |x|^{p-1} g(x).$$

6. The function $|f(x)|$ will be differentiable at x if $f(x) \neq 0$. But there are cases where $|f(x)|$ is differentiable when $f(x) = 0$.
7. They are the averages of the slopes of the secant lines from left of a and from right of a.
9. In the case that $f'(0) > 0$, there is a $\delta > 0$ such that $f(x)/x > f'(0)/2 > 0$ for $|x| \leq \delta$. Show that the range of f contains the interval $[-\varepsilon, \varepsilon]$ where $\varepsilon = \delta f'(0)/2$.

Section 4.2

1. b) Use the intermediate value theorem to show that $g(x) = 1 + x^2 - (2/3)x^3$ has a zero in the interval $(1, 2)$.
2. a) Show that $f'(x) > 0$ for $x \geq 2$ and that $f'(x) < 0$ for $0 < x \leq 1$. Then use the mean value theorem.
4. Use the mean value theorem to show that $f(x) < x$ for x sufficiently large.
5. Use the mean value theorem on the interval from x to $\pi/4$. a) Make upper and lower estimates on the derivative $- \sin(\theta)$.
6. c) Show that for each y there are numbers x_1 and x_2 (depending on y) such that $f(x_1) < y < f(x_2)$. Then use the intermediate value theorem.
8. The mean value theorem implies that

$$\frac{f(x + h) - f(x)}{h} = f'(\theta_x)$$

for $x < \theta_x < x + h$.
12. Calculate $(d/dt)H(u(t), v(t))$ using the chain rule, and then use the differential equations.

Sections 4.3 and 4.4

2. Use the form of the Cauchy mean value theorem of part b).
3. Check that $\lim_{x \to 0} f'(x)$ does not exist.
6. If $y(x) \leq 0$ for some $x \in (0, 1)$, then $\min y$ is attained at some point $x_0 \in (0, 1)$.

Chapter 5

Section 5.1

1. Calculate the derivative at $x = 0$ using difference quotients.
5. About $x = 19$.
6. The approximation is $g(x) = 1 + x^2/2$.
10. Suppose $n\pi \leq s \leq n\pi + 1/n\pi$. By the mean value theorem, there is a θ, $n\pi < \theta < s$ such that $s \tan s = \sec^2(\theta)s(s - n\pi)$.
11. b) Use the Taylor polynomial P_1 to approximate $f(x) = \sin x - x \cos x$ at $x = 3\pi/2$ and the fact that $f''(x) \leq 0$ for $\pi \leq x \leq 3\pi/2$.
13. Show that $(AC)'(x) \leq 0$ for $x \geq 1$.

Section 5.2

2. a) The Euler method, $y_{n+1} = y_n + hf(x_n, y_n)$. c) The backward Euler method, $y_{n+1} = y_n + hf(x_{n+1}, y_{n+1})$.

Sections 5.3 and 5.4

1. The interpolating polynomial is

$$Q_2(x) = \left(\frac{1}{\pi/4 - 1}\right)\left[\frac{4}{\pi\sqrt{2}}(x^2 - x) - x^2 + (\pi/4)x\right].$$

The error, when $0 \leq x \leq 1$ is

$$|\sin x - Q_2(x)| \leq (1/6)|x(x - \pi/4)(x - 1)| \leq 0.0170.$$

When x is allowed outside the interval $[0, 1]$ the error gets much bigger. For example when $x = 2$, the error is 2.929.

2. Use equation (5.12).
3. $Q_2(x) = (1/2)[g(-1)(x^2 - x) - 2g(0)(x^2 - 1) + g(1)(x^2 + x)]$.
4. $A = (1/2)[2g(0) + g(1) - g(-1)]$.
9. Assuming that f satisfies the condition on the derivative, show that for $z < x < y$

$$\frac{f(y) - f(x)}{y - x} \geq \frac{f(x) - f(z)}{x - z}.$$

Then solve for $f(x)$ in this inequality to show that f is convex.

Chapter 6

Sections 6.1 and 6.2

1. a) Look at the maximum value of f. b) For $0 \leq c \leq 1$ the only fixed point is $x = 0$. For $1 \leq c \leq 4$, the fixed points are $x = 0$ and $x = (c-1)/c$.
2. a) The iterates converge to the fixed point $x_* = 1/2$. b) The iterates appear to oscillate between two points. They do not converge to the fixed point $x_* = (c-1)/c = 0.6875$. The derivative $|f'(x_*)| > 1$.
4. Verify that $|g'(x)| \leq 1/2$ for $0 \leq x \leq \pi^2/4$.
5. a) The root lies near 0.6. b) $g'(0.6) \approx -1.5$.
7. a) $(d/dx)(g(x, \lambda) - x) = \lambda \cos x - 1 < 0$ for $x \geq \pi/2$. b) $x_*(\lambda) = g(x_*(\lambda), \lambda) \leq \lambda$. d) For $\lambda \geq \pi/2$, there is a θ, $\pi/2 < \theta < \lambda$, such that

$$\cos \lambda = \pi/2 - \lambda - (\lambda - \pi/2)^2 (\cos \theta)/2.$$

   e) $\lambda(\pi/2 - \lambda) = -1$ for $\lambda = 2.057$.
9. Use the mean value theorem on the difference $x_{n+1} - x_* = g(x_n) - g(x_*)$.

Section 6.3

2. a) Newton's method with starting value $x_0 = .5$ converges to $x_* = 0.58853274398186$ in four iterations. This is full double precision. c) If $x_0 \geq 1.8$, the Newton iterates may converge to another root at approximately 15.707.
3. a) Check $f(0.1)$, $f(2)$ and $f(4)$. b) The first iterate x_1 is positive for $x_0 < 1/e$ and negative for $x_0 > 1/e$. c) The roots are 0.15859 and 3.14619.
4. For $\lambda = 1.6$, $x_* = 1.59934789053$, takes three iterations.
   For $\lambda = 1.8$, $x_* = 1.765862664164$, takes 4 iterations.
   For $\lambda = 2$, $x_* = 1.895494267033$, takes 5 iterations.
5. Use the mean value theorem to write $f'(x_n) = f''(\eta)(x_n - x_*)$ and substitute in (6.11).
7. Using the error expression for Taylor polynomials, we have

$$g(x_n) - g(x_*) = (1/2)g''(\xi)(x_n - x_*)^2$$

for some number ξ between x_n and x_*.

Chapter 7

Section 7.1

2. That f is integrable follows from Theorem 7.6. To compute the value of $\int_a^b f$, construct a partition $P_\varepsilon = \{a = x_0 < x_1 < \cdots < x_{2m+1} = b\}$ with $x_1 = y_1 - \varepsilon$, $x_2 = y_1 + \varepsilon$, $x_3 = y_2 - \varepsilon$, $x_4 = y_2 + \varepsilon$, etc. Determine $\inf_\varepsilon U(f, P_\varepsilon)$.

3. a) For $c > 0$ use exercise 3 of Section 1.5. When $c < 0$, you will need to prove the appropriate analogue. b) Use exercise 2 of section 1.5.

4. Show that for any partition P, $L(f, P) \leq L(g, P)$ whence

$$\int_a^b f = \sup L(f, P) \leq \sup L(g, P) = \int_a^b g.$$

5. Write the hypothesis as $f(x) \leq g(x) + \delta$ for all $x \in [a, b]$, and $g(x) \leq f(x) + \delta$ for all $x \in [a, b]$. Then use exercise 4 and exercise 3b.

6. First show that if f is bounded and integrable on $[a, b]$, and $\varepsilon > 0$ is given, then there is a partition P_n of equally spaced points $x_j = a + jh$, $j = 0, \ldots n$ such that $U(f, P_n) - L(f, P_n) < \varepsilon$.

7. Suppose $a < c < b$. Since f is continuous, and $f(c) > 0$, there is some interval $[c - \delta, c + \delta] \subset [a, b]$ on which $f(x) \geq f(c)/2$. Use this to show that there is a partition P_* such that $L(f, P_*) > 0$.

Section 7.2

3. a) If $f(c) < 0$ for some $c \in I$, use the technique of exercise 7 of Section 7.1 to show that there is a $\delta > 0$ such that $\int_{c-\delta}^{c+\delta} f < 0$.

4. Take $g = f$ and use exercise 7 of Section 7.1.

6. Write

$$\int_a^b f\,dx - \sum_{j-1}^n f(s_j)h = \sum_{j=1}^n \int_{x_{j-1}}^{x_j} [f(x) - f(s_j)]dx.$$

Use the mean value theorem to estimate each of the integrals over the interval $[x_{j-1}, x_j]$.

Section 7.3

2. The integral is the composite function $F(\varphi(x))$ where $F(x) = \int_a^x f(s)ds$.
3. $f(\beta(x))\beta'(x) - f(\alpha(x))\alpha'(x)$.
6. Write the integral as

$$\int_a^{a+p} f(x)dx = \int_a^p f(x)dx + \int_p^{a+p} f(x)dx.$$

Make a change of variable and use the periodicity to show that

$$\int_p^{p+a} f(x)dx = \int_0^a f(x)dx.$$

7. Let $v(t) = A + c\int_0^t |u(s)|\,ds$ and show that v satisfies the differential inequality $v' \leq cv$.
8. Clearly $f(0) = 0$. If $f(t) \neq 0$ for some $t > 0$, $f'(t) = 1$. Let $t_0 = \sup\{s : f(t) = 0 \text{ on } [0, s]\}$.
11. Because f is continuous and $f(0) = 0$, for $\varepsilon > 0$ given, there is a $\delta > 0$ such that $|f(x)| \leq \varepsilon$ for $0 \leq x \leq \delta$. Write the integral as

$$\int_0^1 f(x^n)dx = \int_0^{\delta^{1/n}} f(x^n)dx + \int_{\delta^{1/n}}^1 f(x^n)dx.$$

Section 7.4

2. With $x_1 = a$, $x_2 = a + (b - a)/3$, $x_3 = a + 2(b - a)/3$ and $x_4 = b$, the quadrature rule is

$$M(f) = \frac{b-a}{8}[f(x_1) + 3f(x_2) + 3f(x_3) + f(x_4)].$$

3. Since the constant K is independent of f and of $[a, b]$, make a special choice of f and integrate over $[0, 1]$.

7. b) Note that $f(0) = 1/2$ and use the intermediate value theorem. Also for $c \geq 4$,

$$\int_0^1 \frac{dx}{1 + e^{cx}} \leq \int_0^1 \frac{dx}{1 + e^{4x}} \leq \int_0^1 e^{-4x} dx.$$

c) One can use Simpson's rule to approximate the integral, choosing the number of panels large enough so that the error is less than 10^{-6}. Then use bisection to locate the root in an interval of length less than 10^{-4}. The value $c_* = 2.437511$.

Section 7.5

1. The integrals a), d), and f) converge. To show that e) diverges, use the Taylor expansion for $\sin x$ at $x = \pi/2$.
2. Write $x^p e^{-kx} = (x^p e^{-kx/2})e^{-kx/2}$ and use l'Hôpital's rule to show $x^p e^{-kx/2}$ is bounded on $[0, \infty)$.
3. b) converges for $\alpha + \beta > -1$.
6. $p = 6$ makes the integral $\int_{10}^{\infty} |g(t)|dt \leq 10^{-6}$, then one must deal with a highly oscillatory integrand $\cos(t^6)$. Another approach is to integrate by parts, putting the derivatives on $1/(1 + x^2)$.

Chapter 8

Section 8.1

2. If $|b_k| \leq M$ for all k, then the partial sums $\sum_{k=1}^{n} |a_k b_k| \leq M \sum_{k=1}^{n} |a_k|$ are bounded.
3. a) Show that for k sufficiently large, $a_k^2 \leq |a_k|$. b) The alternating series $\sum_k a_k$ with $a_k = (-1)^k/\sqrt{k}$ converges, but $\sum_k a_k^2$ does not converge.
5. a), c), d), and f) converge. Use the root test on d).
7. Make comparison with appropriate integrals as in the proof of Theorem 8.3.

Section 8.2

3. For $\varepsilon > 0$, $|g_n(x)| \le \varepsilon$ for $1 - \varepsilon \le x \le 1$ and all n.
4. Use the usual device to estimate $\sqrt{x^2 + 1/n} - |x| = \sqrt{x^2 + 1/n} - \sqrt{x^2}$.
5. b) No. There is a sequence $x_n \to \infty$ such that $f_n(x_n) = 1$ for all n.
7. Show first that for each $x_0 \in [a, b]$, there is a $\delta > 0$ such that f_n converges uniformly to f on $|x - x_0| \le \delta$. Then cover $[a, b]$ with a finite number of such intervals.
9. For given $\varepsilon > 0$, choose $\delta = \varepsilon/M$. Then choose N so that

$$\int_\delta^1 |f_n(x) - f(x)| dx \le \varepsilon \text{ for } n \ge N.$$

11. a) Use the formula for the sum of a geometric series.

Section 8.3

1. Use the ratio test, Theorem 8.6.
3. Series is $\sum_{k=0}^\infty (-1)^k x^{2k}$. Radius of convergence $R = 1$.
4. Series is $(2/\sqrt{\pi})[x - x^3/3 + x^4/(5 \cdot 2!) - x^7/(7 \cdot 3!) + \cdots]$.
5. Use the ratio test.
6. d)

$$\frac{1}{1 + 2x} = \left(\frac{1}{3}\right)\frac{1}{1 + (2/3)(x - 1)} = \frac{1}{3}\sum_{k=0}^\infty (-1)^k \left(\frac{2(x - 1)}{3}\right)^k$$

converges for $|x - 1| < 3/2$.
7. Use ratio test for a), b), c). a) $R = 3$; b) $R = 1$; c) $R = e^{-1}$. Use root test for d); $R = 1/2$.
8. $g(0) = 1/f(0)$, $g'(0) = -f'(0)/f(0)^2$, etc.
9. The recursion relation for the coefficients is $a_{k+2} = -a_k/(k+2)$ for $k \ge 0$. Thus with a_0 and a_1 given

$$a_2 = -\frac{a_0}{2}, \quad a_4 = \frac{a_0}{2 \cdot 4}, \quad a_6 = -\frac{a_0}{2 \cdot 4 \cdot 6}$$

$$a_3 = -\frac{a_3}{3}, \quad a_5 = \frac{a_1}{3 \cdot 5}, \quad a_7 = -\frac{a_1}{3 \cdot 5 \cdot 7}.$$

Chapter 9

Section 9.1

2. b) The norm is $\|\mathbf{x}\| = \sqrt{a_1 x_1^2 + \cdots a_n x_n^2}$.
3. $\mathbf{y} = (1, 1, \ldots, 1)$.
4. Since $|x_j| \leq \sqrt{x_1^2 + \cdots x_n^2}$ for each $j = 1, \ldots n$, it follows that $\|\mathbf{x}\|_\infty \leq \|\mathbf{x}\|_2$. $\|\mathbf{x}\|_2 \leq \|\mathbf{x}\|_1$ because $x_1^2 + \cdots x_n^2 \leq (|x_1| + \cdots |x_2|)^2$. To prove $\|\mathbf{x}\|_1 \leq \sqrt{n}\|\mathbf{x}\|_2$, use the Schwartz inequality as in exercise 3.

Section 9.2

1. a) neither, b) open, c) closed, d) closed, e) neither.
4. (i) The intersection of the closed sets F and K is closed. (ii) K compact implies K is bounded, whence $F \cap K$ is also bounded. If each set $K_1, \ldots, K_p$ is bounded, then so is $\cup_{k=1}^p K_k$.
5. Let $\mathbf{x}_k \in \bar{A}$ and suppose $\mathbf{x}_k$ converges to $\mathbf{a}$. For each k, there is a $\mathbf{y}_k \in A$ such that $\|\mathbf{x}_k - \mathbf{y}_k\| \leq 1/k$. Show that $\mathbf{y}_k$ also converges to $\mathbf{a}$.
6. a) Suppose that $d = 0$. Then for each k, there are points $\mathbf{x}_k \in K$ and $\mathbf{y}_k \in F$ with $\|\mathbf{x}_k - \mathbf{y}_k\| \leq 1/k$. Apply Bolzano-Weierstrass to the points $\mathbf{x}_k$.
8. If $\mathbf{a}, \mathbf{b} \in A$, there are sequences $\mathbf{a}_k, \mathbf{b}_k \in A$ with $\mathbf{a}_k$ converging to $\mathbf{a}$, and $\mathbf{b}_k$ converging to $\mathbf{b}$. The points on the line segment from $\mathbf{a}_k$ to $\mathbf{b}_k$ are contained in A and converge to the points on the line segment from $\mathbf{a}$ to $\mathbf{b}$.

Section 9.3

1. If A is not closed, there is a point $\mathbf{a} \in \partial A$ with $\mathbf{a} \notin A$. Construct a function that blows up at $\mathbf{x} = \mathbf{a}$ but is continuous on A.
2. $p + q > 1$.
3. a) Use the (ε, δ) formulation of continuity. b) Use the sequential definition of continuity.
5. Both sets S and T are unbounded, but $S \cap T$ is compact.

Chapter 10

Section 10.1

1. a) $f_y = |x|$ exists for all (x, y). b) $f_x = y$ for $x > 0$ and $f_y = -y$ for $x < 0$.
   $f_x(0, y)$ does not exist for $y \neq 0$. c) f does have a linear approximation
   at $(0, 0)$.
3. a) $l(x, y) = -1 + (1/2)(x + y)$.
   b) $l(x, y) = (1 - \log 2)x - y/2$.
   c) $l(x, y, z) = e^{13/4}(x + y + 3z - 9/2)$.
   d) $l(\mathbf{x}) = -5 + 2\langle \mathbf{v}, \mathbf{x} \rangle$.
4. a) $2.5/\sqrt{13}$. b) $(1.5 - \log(2))/\sqrt{2}$. c) $3 \exp(13/4)$. d) $-2\|\mathbf{v}\|_2$.
6. If $\mathbf{v} = (1, v_2)$, with $1/2 \leq v_2 \leq 2$, the difference quotients $(f(h\mathbf{v}) -$
   $f(0, 0))/h \geq 0$ for $h \geq 0$.
7. Use Theorem 10.7.

Section 10.2

1. If $S = \max_{1 \leq i \leq n} \sum_j |a_{ij}|$ it is easy to show that $\|\mathbf{A}\|_\infty \leq S$. To show
   equality, let k be a row index such that $\sum_j |a_{kj}| = S$. Let
   $$\mathbf{z} = (z_1, z_2, \dots, z_n)$$
   such that $z_j = 1$ if $a_{kj} \geq 0$ and $z_j = -1$ if $a_{kj} < 0$. Use $\mathbf{z}$ to show
   $\|\mathbf{A}\|_\infty = S$.
3. First show the sup quantity is less than or equal to $\|\mathbf{A}\|_2$. To show
   equality, take $\mathbf{y} = \mathbf{A}\mathbf{x}$.

Section 10.3

1. a) $\mathbf{Df}(t) = [2, t(1 + t^2)^{-1/2}, e^t]^T$. The linear approximation provides the
   tangent line to the curve.
   b) $Df(x, y) = \nabla f(x, y) = (2 \sin(x + y^2) + 2x \cos(x + y^2), 4xy \cos(x + y^2))$.
   The linear approximation provides the tangent plane approximation
   to the graph of f.
   c)

   $$\mathbf{Df}(x, y) = \begin{bmatrix} 1 & 2y \\ -\sin(x + y) & -\sin(x + y) \\ 2x/y & -x^2/y^2 \end{bmatrix}.$$

The linear approximation provides a tangent plane approximation to the image set which is a two-dimensional surface in $\mathbb{R}^3$.

2. b) The linear approximation at $(0,0)$ is $\mathbf{l}(x,y) = (x+y, y)$.

c) $\|\mathbf{f}(h,h) - \mathbf{l}(h,h)\|_1 = (3/2)h^2$.

3. For $(x,y) \in Q_r$, $\|\mathbf{Df}(x,y)\|_\infty \leq 2(x+y) \leq 4r$ and $\|\mathbf{Df}(x,y)\|_1 \leq \max\{2x+y, x+2y\} \leq 3r$.

5.

$$[F_s, F_t] = [f_x, f_y, f_t] \begin{bmatrix} x_s & x_t \\ y_s & y_1 \\ 0 & 1 \end{bmatrix}.$$

7.

$$\begin{bmatrix} u' \\ v' \end{bmatrix} = - \begin{bmatrix} f_x & f_y \\ g_x & g_y \end{bmatrix}^{-1} \begin{bmatrix} f_t \\ g_t \end{bmatrix}.$$

Chapter 11

Sections 11.1 and 11.2

3. $\mathbf{x}_*(t) - \mathbf{x}_*(s) = \mathbf{g}(\mathbf{x}_*(t), t) - \mathbf{g}(\mathbf{x}_*(s), s)$. Add and subtract the term $\mathbf{g}(\mathbf{x}_*(t), s)$ to deduce

$$\|\mathbf{x}_*(t) - \mathbf{x}_*(s)\| \leq \|\mathbf{g}(\mathbf{x}_*(t), t) - \mathbf{g}(\mathbf{x}_*(t), s)\| + c\|\mathbf{x}_*(t) - \mathbf{x}_*(s)\|.$$

4. First show that $\|\mathbf{Dg}\|_F^2 \leq (1/2)\cos^2(x+y) + (1/4)(x+y)^2$. Then show that

$$\max_{0 \leq s \leq 2} (1/2)\cos^2 s + s^2/4 < 1.$$

5. a) $\|\mathbf{Dg}(x,y)\|_1 = 0.6\exp(-(x+y)) + (1/3)|\cos(y+1)| \leq 0.6 + 1/3$. On the other hand, $\|\mathbf{Dg}(0,0)\|_\infty = 1.2$. b) After 10 iterations, the fixed point is $x_* = 0.28247$ and $y_* = 0.47089$. The rate of convergence in either the one-norm or the infinity-norm is about 0.57.

Section 11.3

3. The root $\mathbf{p}* = (0.906376, 1.226594)$.
4. Use Theorems 11.6 and 11.7.
5. a)

$$\|\mathbf{Df}(0,0)^{-1}\|_\infty \leq \frac{1}{|\Delta(\lambda)|} \max\{A(\lambda), B(\lambda)\}$$

where

$$A(\lambda) = |\lambda + \beta_y(0,0)| + |1 + \beta_x(0,0)|$$

and

$$B(\lambda) = |1 + \alpha_y(0,0)| + |\lambda + \alpha_x(0,0)|$$

and

$$\Delta(\lambda) = (\lambda + \alpha_x(0,0))(\lambda + \beta_y(0,0)) - (1 + \beta_x(0,0))(1 + \alpha_y(0,0)).$$

b) $L \leq 8M$.   c) Show that for $|\lambda|$ sufficiently large, the quantity r of Theorem 11.6 satisfies $r \leq 1/3$.
7. Starting at $(1,0,0,0)$ the Newton iterates converge to $A = 0.319, B = 0.783, C = 0.535$, and $D = 0.491$.

Sections 11.4 and 11.5

1. $\mathbf{f}(x, y) = \mathbf{f}(x, y + 2\pi)$.
2. a)

$$\langle \mathbf{Df}(\mathbf{x})\mathbf{v},\ \mathbf{v} \rangle \geq \lambda_0 \|\mathbf{v}\|_2^2 - |\langle \mathbf{DN}(\mathbf{x})\mathbf{v},\ \mathbf{v} \rangle|.$$

b) Here $\mathbf{A}$ has eigenvalues $\lambda = 4$ and $\lambda = 6$.
3. $\mathbf{Dg}(\mathbf{y}) = \mathbf{Df}(\mathbf{g}(\mathbf{y}))^{-1}$. Since $\mathbf{x} \to \mathbf{Df}(\mathbf{x})$ is C^1 and $\mathbf{y} \to \mathbf{g}(\mathbf{y})$ is C^1, the left side is C^1, which makes $\mathbf{g}$ C^2.
5. a) The matrix $\mathbf{Df}_{x,y,z}(0,0,-1)$ has no invertible 2×2 submatrix.   b) Yes, $x = \sqrt{3 - y^2}$ and $z \equiv 2$.
6. a) $f_y(1, 1) \neq 0$ so a function $y = g(x)$ exists on a interval containing $x = 1$.

7. b) When $\lambda = 1.5$, the roots are $(x_1, y_1) = (0.8306, 2.3985)$ and $(x_1, y_2) = (1.1637, 0.8773)$. c) The critical value of λ is $\lambda_* = 0.7828$ and in this case the single root is $(0.8024, 1.3445)$.

9. Another solution curve $\theta(\omega)$ bifurcates from the constant solution $\theta(\omega) \equiv 0$ when $\omega = \sqrt{g/R}$.

Section 11.6

1. a) The condition $l < 1$ means that at least one spring is stretched, i.e., $d > l$ or $d_2 > l$. Use the first equation of (11.35) to show that both must be stretched because $-1 < x < 1$. b) $x_0 = (k_2 - k_1)(1 - l)/(k_1 + k_2)$.

2. a) Because $p_y = q_x = 0$ when $y = 0$, the equations that determine the linear approximation to the solution when the mass is small are

$$p_x(x_0, 0)(x - x_0) = 0 \quad \text{and} \quad q_y(x_0, y)y = -mg.$$

where

$$p_x(x_0, 0) = k_1 + k_2, \quad q_y(x_0, 0) = k_1 + k_2 - l\left(\frac{k_1(1 - x_0) + k_2(1 + x_0)}{1 - x_0^2}\right).$$

b) Use a second-order approximation. You will need to compute the second derivatives of p and q and evaluate at $(x_0, 0)$.

3. a) $x_\infty = (k_2 - k_1)/(k_2 + k_1)$. c) $y = -mg/(k_2 + k_2) - l$.

4.

m	x	y
.25	.2858	-1.2189
.5	.3190	-2.0913
.75	.3275	-2.9266
1.00	.3305	-3.7519
3.00	.3332	-10.2972

5. The solution is approximately $(x, y) = (1.25, 0.15)$; it does not lie in H. The right spring is much stiffer than the left spring. The right spring is compressed, whereas the left spring is stretched.

7. b) The equilibrium points when $m_1 = m_2$ are $(x_1, y_1) = (-1/3, 0)$ and $(x_2, y_2) = (1/3, 0)$. In this equilibrium position $d_1 = d_2 = d_3 = 2/3$.

e) The matrix $\mathbf{Df}(\mathbf{x}_0, \mathbf{y}_0)$, with $\mathbf{x}_0 = (-1/3, 1/3)$ and $\mathbf{y}_0 = (0, 0)$, is

$$\mathbf{Df}(\mathbf{x}_0, \mathbf{y}_0) = k \begin{bmatrix} 2 & 0 & -3l/2 & 0 \\ 0 & 2 - 3l & 0 & -(1 - 3l/2) \\ -3l/2 & 0 & 2 & 0 \\ 0 & -(1 - 3l/2) & 0 & 2 - 3l \end{bmatrix}.$$

The solution of the linear equations is $x_1 = -1/3$, $x_2 = 1/3$, and

$$y_1 = -\frac{(2m_1 + m_2)g}{3k(1 - 3l/2)} \quad \text{and} \quad y_2 = -\frac{(m_1 + 2m_2)g}{3k(1 - 3l/2)}.$$

Chapter 12

Section 12.1

1. The linear approximation $l(x, y) = \sqrt{3}(1 + x/3 + y/6)$. The quadratic approximation $q(x, y) = l(x, y) + (x^2 - xy/2 - y^2/8)/(3^{3/2})$. For E_1 the power is $p = 2$ and for E_2, the power is $p = 3$.
2. b) Critical point at $(10/9, -23/9)$ is a saddle point. c) Critical points at $\pm(2^{-1/8}, 2^{-5/8})$. Both are minima. d) There are 9 critical points: $(0, 0)$, a maximum; $(0, \pm 1)$, $(\pm 1/\sqrt{2}, 0)$, saddle points; $(\pm 1/\sqrt{2}, \pm 1)$, minima.
3. a) $(3, -3, 3)$ is a maximum, $(-3, 3, -3)$ is a minimum. b) Critical points are $\pm(2^{-3/2}, 2^{-1/2}, -2^{-1/2})$, both are saddle points.
5. a) Quadratic approximation at $(0, 0)$ is $q(x, y) = \sigma x^2 + y^2$. d) When $\sigma = 0$, $f(x, 2x^2) = -x^4$.
6. $W(a, b) = b(a - b)/3 + 11/12$.

Section 12.2

4. b) If $f(\mathbf{x}) < 0$, $\mathbf{x} \in F^o$ because f is continuous. If $\mathbf{x} \in F^o$, there is a $\delta > 0$ such that $\mathbf{x} \pm \delta \mathbf{u} \in F$ for any unit vector $\mathbf{u}$. Then use the fact that f is strictly convex on F.

Section 12.3

1. The critical points of the potential $V(x)$ are $x = -1/2, 0, 1$. The critical points $x = -1/2$ and $x = 1$ are linearly stable, and the critical point at $x = 0$ is unstable.
2. The potential is

$$V(x_1, x_2) = (1/2)x_1^2 + (3/2)x_2^2 + \frac{2}{1 + (x_1 - x_2)^2}.$$

Critical points are $(0, 0)$, $(-0.8582, 0.2861)$, $(0.8582, -0.2862)$. The first is a saddle point of V, and hence unstable. The other two are minima of V and hence linearly stable.

Section 12.4

1. $\mathbf{p}_1 = (x_1, y_1) = (-0.2018, 1.0214)$, $f(x_1, y_1) = 0.1261$. $\mathbf{p}_2 = (x_2, y_2) = (0.1746, 0.1746)$, $f(x_2, y_2) = 0.0110$.
5. Starting at $\mathbf{p}_0 = (5, 4)$, the steepest descent method takes 20 iterations to arrive at $(1.8742, 0.9347)$ where the function value is 0.1437.

Section 12.5

1. If $\mathbf{Ax} = \lambda \mathbf{x}$ and $\mathbf{Ay} = \mu \mathbf{x}$, with $\lambda \neq \mu$, then $\lambda \langle \mathbf{x}, \mathbf{y} \rangle = \langle \mathbf{Ax}, \mathbf{y} \rangle = \langle \mathbf{x}, \mathbf{Ay} \rangle = \mu \langle \mathbf{x}, \mathbf{y} \rangle$ implies that $\langle \mathbf{x}, \mathbf{y} \rangle = 0$.
2. $|\langle \mathbf{Ax}, \mathbf{y} \rangle| \leq \sqrt{\langle \mathbf{Ax}, \mathbf{x} \rangle} \sqrt{\langle \mathbf{Ay}, \mathbf{y} \rangle}$.

Section 12.6

2. In this case $\mathbf{g}(c) = (c, Ac^2 + 1)^T$, and equation (5.33) becomes

$$c_k = c_{k-1} - \frac{(2A + 1)c_{k-1} + 2A^2 c_{k-1}^3}{1 + 4A^2 c_{k-1}^2}.$$

3. Minimum squared error is attained at $c_* = 7.784$, $r_* = 0.2029$. $E^2(c_*, r_*) = 0.0017$. c) $t = 24.6235$.

4. Minimum squared error is attained at $\beta_* = 208.93, \gamma_* = 0.0562. \ E^2(\beta_*, \gamma_*) = 631.0073$.

Chapter 13

Section 13.1

1. a) The maximum value of $f|S$ is attained at four points, $(\sqrt{2}, \pm 3/\sqrt{2})$ and $(-\sqrt{2}, \pm 3/\sqrt{2})$. The minimum value of $f|S$ is zero and it is attained at four points, $(\pm 2, 0)$ and at $(0, \pm 3)$. b) At the points where the minimum is attained, the level curve $f = 0$ is not tangent to the constraint curve $g = 0$. The Lagrange multiplier at these points is $\lambda^* = 0$.
2. There are six extreme points.

λ	x	y	$f(x, y)$
10	6	0	252
-8	-6	0	-180
3	0	6	72
-3	0	-6	-72
2.9302	1.2868	5.8604	70.8766
-2.7302	-2.4868	-5.4604	-63.4633

3. The maximum of $f|S$ occurs at $(0, 1, 0)$ with $\lambda = 1/2$ and $\mu = 1$. The minimum of $f|S$ occurs at $(0, -1, 0)$ with $\lambda = -1/2$ and $\mu = 1$.
4. The four critical points are

x_1	y_1	x_2	y_2	d
2.3056	1.7349	1.6013	0.7990	1.3719
2.3056	1.7349	0.3987	-0.7990	10.0571
-1.7900	-2.2347	1.7805	0.6252	20.9269
-1.7900	-2.2347	0.2195	-0.6252	6.6285

Section 13.2

2. The four critical points of $f|S$ are

λ	μ	x	y	z	f
0.0000	−1.0000	−0.3956	1.8956	−0.5000	−4.2500
0.0000	−1.0000	1.8956	−0.3956	−0.5000	−4.2500
−0.6943	−0.6887	1.1151	1.1151	−1.2301	−3.7169
0.9165	0.0220	−0.4484	−0.4484	1.8968	1.4947

3. There are eight real critical points.

λ	μ	x_1	y_1	x_2	y_2	d
0.1563	2.5000	−2.0000	2.0000	0.5000	−0.5000	12.5000
0.0938	−1.5000	2.0000	−2.0000	0.5000	−0.5000	4.5000
0.0313	−0.5000	−2.0000	2.0000	−1.5000	1.5000	0.5000
0.2188	3.5000	2.0000	−2.0000	−1.5000	1.5000	24.5000
0.6237	5.0415	−3.8194	−4.1528	0.3213	1.6513	50.8325
0.6237	5.0415	4.1528	3.8194	−1.6513	−0.3213	50.8325
0.3763	−3.0415	4.1528	3.8194	0.6513	1.3213	18.5009
0.3763	−3.0415	−3.8194	−4.1528	−1.3213	−0.6513	18.5009

b) The global minimum occurs at $(x_1, y_1, x_2, y_2) = (-2, 2, -3/2, 3/2)$.
A local minimum occurs at $(2, -2, 1/2, -1/2)$. The other two solutions
of the Lagrange equations that lie on the line $x+y = 0$ are saddle points.
c) The Hessian matrix of L, with respect to the variables x_1, y_1, x_2, y_2,
taken in that order, is

$$
\mathbf{D}^2 L = \begin{bmatrix}
2 - 10\lambda & 6\lambda & -2 & 0 \\
6\lambda & 2 - 10\lambda & 0 & -2 \\
-2 & 0 & 2 - 2\mu & 0 \\
0 & -2 & 0 & 2 - 2\mu
\end{bmatrix}.
$$

The eigenvalues of $\mathbf{D}^2 L$ at the global minimum are all positive so that
$\mathbf{D}^2 L$ is actually positive definite on all of $\mathbb{R}^4$. d) The tangent space to
S at points (x_1, y_1, x_2, y_2) with $x_1 + y_1 = 0$ and $x_2 + y_2 = 0$ consists

of the vectors $\mathbf{v} = (a, a, b, b)$. The Hessian $\mathbf{D}^2 L$ evaluated at the local minimum $(2, -2, 1/2, -1/2)$ is positive definite on T.

4. b) $L(\mathbf{r}(t), \lambda^*) = f(\mathbf{r}(t))$ for $|t| < 1$. c) Use the fact that $\nabla_x L(\mathbf{a}, \lambda^*) = 0$.

5. According to Lemma 13.5 and (13.22), $\mathbf{H} \equiv \mathbf{D}_{x,\lambda} L(\mathbf{a}, \lambda^*)$ is nonsingular. Then evaluate $\langle \mathbf{Hv}, \mathbf{v} \rangle$ for $\mathbf{v} = (\nabla g(\mathbf{a}), y)$. Show that there are choices of y such that $\langle \mathbf{Hv}, \mathbf{v} \rangle > 0$ and $\langle \mathbf{Hv}, \mathbf{v} \rangle < 0$.

Section 13.3

1. The maximizer is $\mathbf{p}^* = (0, 0)$. The KKT conditions are not satisfied at $\mathbf{p}^*$. The hypothesis $\nabla g(\mathbf{p}^*) \neq 0$ is not satisfied.

2. The minimizer is found at $(0, \sqrt{2})$.

3. The maximizer is found at $((1 - \sqrt{3})/2, 0, \sqrt{3}/2)$. The constraint $h_1 \leq 0$ is active and the constraint $h_2 \leq 0$ is inactive.

Section 13.4

1. The maximum profit lies on the line $y = 2x$. Hence $z(x) = 2^{3/4}x$. Substituting in the equation $\alpha z R'(z) = \pi_x = rx$ yields $(1 + z^2(x))^{3/2} = 2^{3/4}(\alpha/r)$. Finally $x_0 = [2^{-3/2}(2^{1/2}(25/4)^{2/3} - 1)]^{1/2} = 1.1589$. $y_0 = 2x_0 = 2.3177$, and $z_0 = 1.9490$. Since $zR'' + R' = (1 - 2z^2)/(1 + z^2)^{5/2}$, we see that (13.74) is satisfied.

4. Consider two cases: (x, y) is a multiple of $\mathbf{v}$, and (x, y) is not a multiple of $\mathbf{v}$.

6. Let $h(a, y) = m - \pi(y, a) = m - R(y, a) + C(y) + a$. The problem is to find max R subject to the constraints $h \leq 0$, and $-a \leq 0$. The Lagrange function is $L = R(a, y) - \lambda(-a) - \mu h(a, y)$. The KKT equations are

$$L_y = 0$$
$$L_a = 0$$
$$\lambda, \mu \geq 0$$
$$\lambda(-a) = 0$$
$$\mu h = 0$$
$$(-a) \leq 0$$
$$h \leq 0.$$

7. a) The minimium cost is attained at the point $(x_*, y_*, z_*, \lambda_*) = (5.7307, 17.7745, 19.8725, -394.9153)$.

Chapter 14

Section 14.1

1. For any partition P of the rectangle R, $U(f,P) \geq 0$ because there are points with rational coordinates in every subrectangle R_{ij}.
2. Use the fact that f is uniformly continuous on R.
3. Let $\varepsilon > 0$ be given. For a sufficiently fine partition of R, the diagonal $y = x$ of the square is contained in a collection of subrectangles R_{ij} with total area less than ε.
6. Use the result of exercise 5.

Section 14.2

3. Let D be contained in generalized rectangles R_1 and R_2. Then $D \subset R = R_1 \cap R_2$. Show that R is a generalized rectangle and that

$$\int_{R_1} \chi_D = \int_R \chi_D = \int_{R_2} \chi_D.$$

5. Break the analysis into cases. First suppose that $\mathbf{x} \in \partial(A \cup B)$ and that $\mathbf{x} \in A \cup B$. Show that there is a sequence of points $\mathbf{x}_k \in A' \cap B'$ that converges to $\mathbf{x}$. If $\mathbf{x} \notin A \cup B$, there is a sequence $\mathbf{x}_k \in A \cup B$ converging to $\mathbf{x}$. At least one of A or B must contain an infinite number of these $\mathbf{x}_k$.

Section 14.3

1. Let $\alpha = b - a$ and $\beta = d - c$. Then doing the interpolation on the rectangle $[0, \alpha] \times [0, \beta]$ yields

$$A = \frac{f(\alpha, \beta) + f(0,0) - f(\alpha, 0) - f(0, \beta)}{\alpha \beta}, \quad D = f(0,0)$$

$$B = \frac{f(\alpha, 0) - f(0,0)}{\alpha}, \quad C = \frac{f(0, \beta) - f(0,0)}{\beta}.$$

3. For $m = 4$, $n = 6$, the Simpson matrix is the 5×7 matrix

$$
S_2 = \begin{bmatrix}
1 & 4 & 2 & 4 & 2 & 4 & 1 \\
4 & 16 & 8 & 16 & 8 & 16 & 4 \\
2 & 8 & 4 & 8 & 4 & 8 & 2 \\
4 & 16 & 8 & 16 & 8 & 16 & 4 \\
1 & 4 & 2 & 4 & 2 & 4 & 1
\end{bmatrix}.
$$

5. (i) With $m = n = 200$, trapezoid yields 4.827276, and with $m = n = 400$ yields, 4.827241, the difference being 3.5×10^{-5}.
   (ii) With $m = n = 100$, trapezoid yields 2.194767 and with $m = n = 200$ it yields 2.194831, the difference being 6.4×10^{-5}.

Sections 14.4 and 14.5

1. To show $\partial g(D) \subset g(\partial D)$, let $\mathbf{y}_0 \in \partial g(D)$. Then there is a sequence $\mathbf{w}_k \in g(D)$ with $\mathbf{w}_k$ converging to $\mathbf{y}_0$. Let $x_k = g^{-1}(\mathbf{y}_k)$. The inverse function theorem implies that $\mathbf{x}_k$ converges to $\mathbf{x}_0 = g^{-1}(\mathbf{y}_0)$.

4. a)

$$
g(\xi, \eta) = \begin{bmatrix} x_0 \\ y_0 \end{bmatrix} - \begin{bmatrix} x_1 - x_0 & x_2 - x_0 \\ y_1 - y_0 & y_2 - y_0 \end{bmatrix} \begin{bmatrix} \xi \\ \eta \end{bmatrix}.
$$

b)

$$
\int\int_T f(x, y)\, dxdy = \|(\mathbf{p}_1 - \mathbf{p}_0) \times (\mathbf{p}_2 - \mathbf{p}_0)\| \int\int_S f(g(\xi, \eta))\, d\xi d\eta.
$$

c)

$$
\int\int_S P(\xi, \eta)\, d\xi d\eta = \frac{h(0,0) + h(1,0) + h(0,1)}{6}.
$$

d)

$$
\text{area}(T)\frac{f(\mathbf{p}_0) + f(\mathbf{p}_1) + f(\mathbf{p}_2)}{6}.
$$

5. $g(u, v) = (u, (1 - v)\varphi(u) + v\psi(u))$.
6. $h(x, y) = (x + y, y^2 - x^2)$.

$$\text{area}(R) = \int \int_D \frac{dudv}{2(x + y)} = \int \int_D \frac{dudv}{2u}$$

where $D = \{(u, v) : 2 \le u \le 4, 1 \le v \le 4\}$.

8. a) $a = -0.19954$ and $b = 0.6393$. b) The volume is given by

$$2 \int_a^b \int_0^{g(x)} \left[\sqrt{4 - x^2 - y^2} - e^x\right] dydx.$$

c) After changing variable to map into a rectangle, Simpson's rule with $m = n = 50$ yields 6.1213. With $m = n = 100$, it yields 6.1211. An estimate for the error is thus 2×10^{-4}.

Section 14.6

1. a), b), and c) all converge
2. a) Because ρ is continuous and $\rho(\mathbf{x}) = 0$ for $\|\mathbf{x}\| \ge 1$, ρ is bounded. b) For $\|\mathbf{x}\| \ge 2$ and $\|\mathbf{y}\| \le 1$, $\|\mathbf{x} - \mathbf{y}\| \ge \|\mathbf{x}\|/2$.
4. a)

$$\int \int_{D_\varepsilon} f = \int_\varepsilon^1 y^{-3/2}(1 - e^{-y})dy$$

and the limit as ε tends to zero exists. b) The integral does not converge.

Section 14.7

2. Use the change of variable $(y_1, y_2) = (x_1 x_2, x_2)$. The set $F = \{(x_1, x_2) : 1 \le x_1, x_2 \le 10\}$ is mapped in a one-to-one fashion onto the set $E = E_1 \cup E_2$ where $E_1 = \{(y_1, y_2) : 1 \le y_2 \le y_1 \le 10\}$ and $E_2 = \{(y_1, y_2) : 1 \le y_1/10 \le y_2 \le 10\}$. The joint p.d.f. is $\hat{f}(y_1, y_2) = 1/(81y_2)$

for $(y_1, y_2) \in E$. The marginal p.d.f. is

$$\hat{f}_1(y_1) = \begin{cases} (1/81)\log(y_1), & 1 \le y_1 \le 10 \\ (1/81)\log(100/y_1), & 10 \le y_1 \le 100. \end{cases}$$

3. The mapping $\mathbf{h}(\mathbf{x}) = (x_1^2 + x_2^2, x_2)$ is not one-to-one on $\mathbb{R}^2$. There are two inverse mappings, $\mathbf{g}_\pm(\mathbf{y}) = (\pm\sqrt{y_1 - y_2^2}, y_2)$. The joint p.d.f. for Y_1 and Y_2 is

$$\hat{f}(\mathbf{y}) = [f(\mathbf{g}_+(\mathbf{y})) + f(\mathbf{g}_-(\mathbf{y}))] = \frac{1}{4\pi}(y_1 - y_2^2)^{-1/2} e^{-y_1/4}$$

defined for $|y_2| \le \sqrt{y_1}$. The marginal p.d.f. of Y_1 is $\hat{f}_1(y_1) = \exp(-y_1/4)/4$.
4. The inverse mapping is $\mathbf{g}(\mathbf{y}) = (y_1y_2y_3, (1 - y_1)y_2y_3, (1 - y_2)y_3)$. The joint p.d.f. for (Y_1, Y_2, Y_3) is $\hat{f}(\mathbf{y}) = y_2y_3^2 e^{-y_3}$.

Chapter 15

Section 15.1

1. a) $f'(t) = \int_{t^2}^t 2\sin(t - s)\cos(t - s)ds - 2t\sin^2(t - t^2)$.
   b) $g'(x) = e^x(\log x + x) + (e^x - 1)/x$.
2. f is continuous on $\mathbb{R}$, $f(x) > 0$ and $f(x)$ tends to zero as $|x| \to \infty$. Hence f attains a maximum at some point x_0. f' may be calculated by differentiating under the integral.
4. The trapezoid rule with $h = \pi/100$ yields 0.01998585030422 for the value of $J_0(100)$ which agrees to all digits with the value from MATLAB.
6. $u_t(x, t) = (g(x + t) + g(x - t))/2$ and $u_x(x, t) = (g(x + t) - g(x - t))/2$.
8. b) The integral $\int_0^\infty x^n e^{-xt}dx$ converges uniformly on $t \ge \delta > 0$ for each n. Use integration by parts to justify the formula for $n = 1$. Then use integration by parts and induction to justify the formula for general n.

9. b) First verify that the statement is true when f is a nonzero constant function, using the fact that for $t > 0$, $\int_0^\infty t\,dx/(t^2 + x^2) = \pi/2$. Then for $\varepsilon > 0$, choose $\delta > 0$ so that $|f(x) - f(0)| < \varepsilon$ when $0 \le x \le \delta$. Write the integral as

$$F(t) - (\pi/2)f(0) = \int_0^\delta \frac{t(f(x) - f(0))dx}{x^2 + t^2} + \int_\delta^\infty \frac{t(f(x) - f(0))dx}{x^2 + t^2}.$$

Section 15.2

1. $f_k(x) = \int_{x-1}^x g_k(y)dy.$

$$f_k(x) = \begin{cases} 0 & x < -1/k \\ 1/2 + k(x + kx^2/2) & -1/k \le x \le 0 \\ 1/2 + k(x - kx^2/2) & 0 \le x \le 1/k \\ 1 & 1/k < x < 1 - 1/k \\ 1/2 - k(x - 1 + (k/2)(x-1)^2) & 1 - 1/k \le x \le 1 \\ 1/2 - k(x - 1 - (k/2)(x-1)^2) & 1 \le x \le x + 1/k \\ 0 & x > 1 + 1/k \end{cases}.$$

2. Since g_k is even, $\int_0^\infty g_k(x)dx = 1/2$.

$$f_k(0) - (f_r + f_l)/2 = \int_{-\infty}^0 (f(y) - f_l)g_k(y)dy + \int_0^\infty (f(y) - f_r)g_k(y)dy.$$

4. When $k = 1$, the normalizing constant $c_1 = 4/3$ and the approximate polynomial is $-3x^2/8 + 3x/8 + 17/64$.
5. In this case all the approximating polynomials $f_k = 0$.

Section 15.3

1. b) If $M = \sup|f|$, then

$$|u(\mathbf{x}, t)| \le \frac{Ma^n\Omega_n}{(4\pi\gamma t)^{n/2}}.$$

3. b) For $t > 0$,

$$\frac{dQ}{dt} = \int_{\mathbb{R}} u_t(x, t)dx = \int_{\mathbb{R}} u_{xx}(x, t)dx = u_x(x, t)\Big|_{-\infty}^{\infty} = 0.$$

c) Differentiate as in part b) and then integrate by parts.

4. b) $m = 0$ and $p(t) = 4\gamma t$.

5. a) For $x \geq \delta > 0$ and $|y| \leq R \leq |\eta|$,

$$\frac{|f(\eta)|}{x^2 + (y - \eta)^2} \leq \frac{M}{\delta^2 + (|\eta| - R)^2}.$$

Section 15.4

1. Note that $(d/dt)u^2(x(t), y(t), z(t)) = 2u\mathbf{q} \cdot \nabla u$.

2. b) When $\alpha \neq 1$, $p = r^{2(1-\alpha)}/2(1 - \alpha) + C$. When $\alpha = 1$, $p = \log r + C$.

3. b) Only for $\alpha = 3$. $p = -(1/2)r^{-4} + C$.

4. Show $d\omega/dt = uv_{xx} + vv_{xy} - uu_{xy} - vu_{yy}$. Differentiate the first equation of $\mathbf{q} \cdot \nabla\mathbf{q} + \nabla p = 0$ with respect to y and the second with respect to x, and subtract.

References

[A] H. Averch and L. Johnson, "Behavior of the firm under regulatory constraint," *American Economic Review* **52**(1962), p 1052–1069.

[B] G. K. Batchelor, *Introduction to Fluid Dynamics*, Cambridge University Press, London, 1967.

[Ba] D. M. Bates and D. B. Watts, *Nonlinear Regression Analysis and its Applications*, Wiley, New York, 1988.

[Bk] G. Birkhoff and S. MacLane, *A Survey of Modern Algebra*, Fifth Edition, A. K. Peters, Wellesley, MA, 1997.

[Bl] L. Blume and C. P. Simon, *Mathematics for Economists*, W. W. Norton, New York, 1994.

[Ch] A. J. Chorin and J. E. Marsden, *A Mathematical Introduction to Fluid Mechanics*, Springer-Verlag, New York, 1990.

[Col] J. S. Coleman, *Introduction to Mathematical Sociology*, The Free Press of Glencoe, Collier-Macmillan Limited, London, 1964.

[Coo] J. Cooper, *Introduction to Partial Differential Equations with MATLAB*, Birkhauser, Boston, 1998.

[Da] K. R. Davidson and A. P. Donsig, *Real Analysis and Real Applications*, Prentice-Hall, Upper Saddle River, NJ, 2002.

[De] J. E. Dennis, Jr. and R. B. Schnabel, *Numerical Optimization and Nonlinear Equations*, SIAM Classics in Applied Mathematics, SIAM, Philadelphia, 1996.

[Ha] P. Halmos, *Naive Set Theory*, D. Van Nostrand, New York, 1960.

[He] I. N. Herstein, *Topics in Algebra*, Second Edition, Wiley, New York, 1975.

[Ki] D. Kincaid and W. Cheney, *Numerical Analysis*, Brooks/Cole, Pacific Grove, CA, 1991.

[Kr] S. G. Krantz, *Real Analysis and Foundations*, Studies in Advanced Mathematics, CRC Press, Boca Raton, FL, 1991.

[Lu] D. G. Luenberger, *Introduction to Linear and Nonlinear Programming*, Addison Wesley, Reading, MA, 1973.

[M] R. McOwen, *Partial Differential Equations, Methods and Applications*, Prentice-Hall, Upper Saddle River, NJ, 1996.

[N] S. Nash and A. Sofer, *Linear and Nonlinear Programming*, McGraw-Hill, New York, 1996.

[O] J. M. Ortega and W. C. Rheinboldt, *Iterative Solution of Nonlinear Equations in Several Variables*, SIAM Classics in Applied Mathematics, SIAM, Philadelphia, 2000.

[P] W. H. Press, S. A. Teukolsky, W.T. Vetterling, and B. P. Flannery, *Numerical Recipes in Fortran, The Art of Scientific Computing*, Cambridge University Press, Cambridge, UK, 1986.

[R] T. Rockafellar, *Convex Analysis*, Princeton University Press, Princeton, NJ, 1970.

[St] G. Strang, *Linear Algebra and its Applications*, Third Edition, Harcourt Brace Jovanovich, Orlando, FL, 1988.

[Str] R. Strichartz, *The Way of Analysis*, Jones and Bartlett, Sudbury, MA, 2000.

Index